Discrete Mathematics

Discrete Mathematics

N Ch S N Iyengar
Professor
Department of Computer Applications
Vellore Institute of Technology
(Deemed University), Vellore

K A Venkatesh
Head
Department of Computer Applications
Alliance Business Academy, Bangalore

V M Chandrasekaran
Assistant Professor
Department of Mathematics
Vellore Institute of Technology
(Deemed University), Vellore

P S Arunachalam
Senior Lecturer
Department of Mathematics
SRM Engineering College, Chennai

VIKAS® PUBLISHING HOUSE PVT LTD

VIKAS® PUBLISHING HOUSE PRIVATE LIMITED
E-28, Sector-8, Noida-201301 (UP), India
Phone: +91-120-4078900 • Fax: +91-120-4078999
Registered Office: A-27, 2nd Floor, Mohan Co-operative Industrial Estate, New Delhi-110044
E-mail: helpline@vikaspublishing.com • Website: www.vikaspublishing.com

SALES AND MARKETING OFFICES OF VIKAS® AND S CHAND AND COMPANY LIMITED

Chennai	: Ph: 044-2363 2120, chennai@schandpublishing.com
Guwahati	: Ph: 0361-2738 811, 2735 640, guwahati@schandpublishing.com
Hyderabad	: Ph: 040-4018 6018, hyderabad@schandpublishing.com
Jalandhar	: Ph: 0181-4645 630, jalandhar@schandpublishing.com
Kolkata	: Ph: 033-2335 7458, 2335 3914, kolkata@schandpublishing.com
Lucknow	: Ph: 0522-4003 633, lucknow@schandpublishing.com
Mumbai	: Ph: 022-2500 0297, mumbai@schandpublishing.com
Patna	: Ph: 0612-2260 011, patna@schandpublishing.com

Discrete Mathematics
ISBN: 978-81259-1362-7

Product Code: V3DIM50MATH10ENAA03O

First Published 2003
Reprinted several times between 2003 and 2020, 2022, 2023
Reprint 2025

PRINTED IN INDIA
By Vikas Publishing House Private Limited, Plot 20/4, Site-IV, Industrial Area Sahibabad, Ghaziabad-201010 and Published by Vikas Publishing House Private Limited, E-28, Sector-8, Noida-201301

Preface

Discrete Mathematics is taught as a core subject in almost all undergraduate courses in applied sciences, especially mathematics and computer science. This book is a suitable textbook for students of B.Sc., M.Sc., B.E. (Computer Science) and MCA. Topics such as Mathematical Logic, Set Theory, Algebraic Systems, Boolean Algebra and Graph Theory that are essential to the study of Computer Science are discussed in detail. This book comprehensively covers all the topics in the subject and is rich in pedagogical features such as solved problems, exercises and objective questions.

We hope that the book will prove useful to all the students and teachers. We welcome feedback on the book.

Authors

Contents

2
Matrix Algebra

3
Mathematical Logic

4
Induction, Recursion and Recurrence Relations

5
Algebraic Systems

6
Lattices and Boolean Algebra

7
Graph Theory

8
Formal Languages and Automata

CHAPTER

1

Set Theory

INTRODUCTION

The concept of set plays a fundamental role in the development of discrete structures. The word "set" is synonymous with collection. A set is a collection of objects, but not mathematically. Consider a set of some even numbers, i.e., $A = \{x : x \text{ is some even number}\}$.

Is 4 a member of the set A? We are not in a position to say so. Hence, a set may be defined as a collection of well -defined objects. The basic idea is to identify the objects in the environment based on the distinctions of these objects. In general, objects or their concepts are associated with properties or attributes. Sets are always denoted by capital letters, and elements of sets by lower case letters. The symbolic notation for denoting that a is a member of the set A is $a \in A$.

The symbol '$\in$' denotes set membership and is read 'belongs to.' The symbol '$\notin$' means does not belong to.

Set It is defined as a collection of well-defined objects.

Example

i. $A = \{x : x \text{ is a flower}\}$

ii. $B = \{x : x \text{ is a vowel in the English alphabet}\}$

iii. $N = \{1, 2, 3, 4....\}$, set of natural numbers.

Cardinality The number of distinct elements in a set is called the cardinality of the set.

Example i. $A = \{a, b, c\}$

$\#A = |A| = n(A) = 3$

ii. $B = \{x : 2 \leq x \leq 6\}$

$n(B) = \#B = |B| = 6$

(#/11/$n \rightarrow$ denotes cardinality)

Equivalent and equal sets Let A and B be two sets. The set A and B are called as

i. equivalent sets, if $\#A = \#B$.

ii. equal sets, if the sets have the same elements.

Example $A = \{a, b, c\}$ $B = \{1, 2, 3\}$ $C = \{a, b, c\}$. A & B are equivalent whereas A & C are equal sets.

Subset and Superset Let A & B be two sets. We call A is a subset of B if every element of A is an element of B and is denoted by $A \subseteq B$. Here B is called superset of A.

Example

i. $A = \{1, 2, 3, 4\}, B = \{1, 2, 3, 4, 5, 6\}$. Clearly $A \subseteq B$

ii. $A = \{x : x \in N\}; B = \{x : x \in Z\}$. Clearly $A \subseteq B$

iii. $A = \{x : x$ is a vowel in the English alphabet$\}$ and $B = \{x : x$ is an English alphabet$\}$ clearly $A \subseteq B$

iv. $A = \{\alpha, \beta, \gamma\}$, $B = \{\alpha, \beta, \gamma\}$ clearly $A \subseteq B$

In the above examples, (i), (ii), (iii) are proper subsets, whereas (iv) is improper.

Empty set or Null set A set is called empty or null set if its cardinality is 0. It is denoted by ϕ or $\{\}$.

Example $A = \{x : x \in N, x < 2$ and $x < 4\}$

$$\Rightarrow A = \{\}$$

Note: i. Empty set is a subset of every set, i.e., $\phi \subseteq A$, for any set A.

ii. For any set $A, A \subseteq A$.

Universal Set A set μ is called an universal set if μ is the superset of all the sets which are under our discussion.

Example $\mu = \{1, 2, 3, 4, 5, 6, 7, 8, 9, 10\}$

$A = \{1, 3, 5, 7, 9\}; B = \{2, 4, 6, 8, 10\}$

$C = \{5, 10\}; D = \{3, 6, 9\}; E = \{4, 8\}$

Clearly μ is the superset of all the sets A, B, C, D, E and hence μ is the universal set.

OPERATIONS ON SETS

Union Let A and B be two sets. The union of A and B is defined as the set of all elements of A and the set of all elements of B and the common elements being taken once, and is given by

$$A \cup B = \{x : x \in A \text{ or } x \in B\}$$

Example $A = \{2, 4, 6, 8, 10\}$ and $B = \{4, 8, 12\}$

$$A \cup B = \{2, 4, 6, 8, 10, 12\}$$

Intersection Let A and B be sets. The intersection of A and B is defined as the set of all elements common to both sets A and B, and is given by

$$A \cap B = \{x : x \in A \text{ and } x \in B\}$$

Example

(i) $A = \{x : x$ is an English alphabet$\}$ and $B = \{x : x$ is a vowel in English alphabet$\}$

$A \cap B = \{a, e, i, o, u\}$

(ii) $A = \{1, 2, 3, 4\}$, $B = \{2, 4\}$

$A \cap B = \{2, 4\}$

(iii) $A = \{a, b, c\}; B = \{1, 2, 3, 4\}$

$A \cap B = \{\}$

Disjoint set Two sets are said to be disjoint if $A \cap B = \{\}$.

Example As in (iii) above

Difference set Let A & B be two sets. The difference of A with respect to B is the set of all elements that are in A and not in B.

i.e., $A - B = \{x/x \in A \text{ and } x \notin B\}$

Example $A = \{a, b, c, d, e, f, g\}$ and $B = \{a, b, c, d, h, i, j\}$

$A - B = \{e, f, g\}$ and $B - A = \{h, i, j\}$ [Note that $A - B \neq B - A$]

Complement Let A be a set. The complement of A is denoted by A^c (with respect to the universal set) and is given by $A^c = \mu\backslash A$.

$A^c = \{x / x \in \cup \text{ and } x \notin A\}$

Example $\mu = \{1, 2, 3, 4, 5, 6, 7, 8, 9, 10\}$

$A = \{1, 3, 5, 7, 9\}, B = \{2, 4, 6, 8, 10\}, C = \{3, 6, 9\}$

$A^c = \mu\backslash A = \{2, 4, 6, 8, 10\}$

$B^c = \mu\backslash B = \{1, 3, 4, 7, 9\}$

$C^c = \mu\backslash C = \{1, 2, 4, 5, 7, 8, 10\}$

Venn Diagram

A diagrammatic representation of set operation is called venn diagram. Now we shall represent every operation of sets as a Venn diagram. (Square or rectangle represents an universal set and circle or ellipse represent sets)

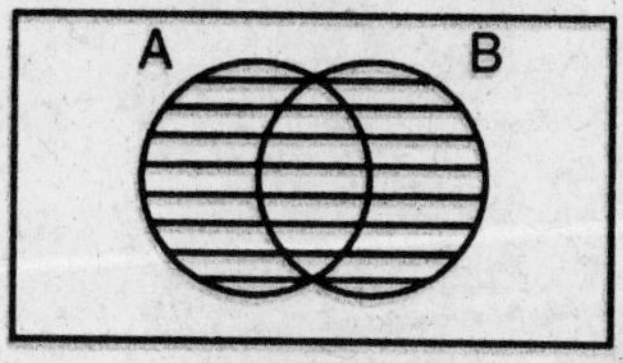

AUB

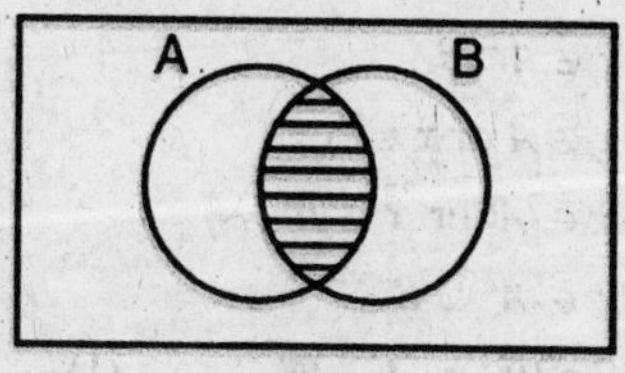

A∩B

μ

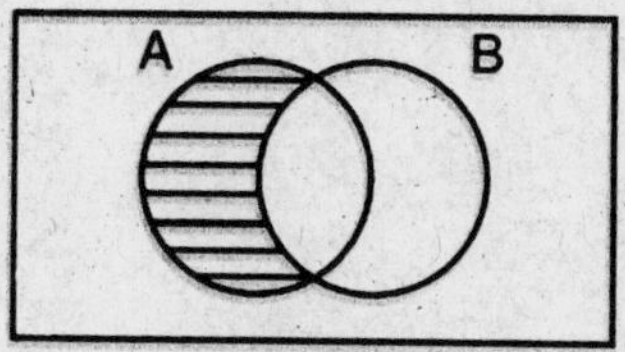

A - B

A^C

EXERCISE I

1. Give an example of a set whose elements are sets.
2. If $A = \{1, 2\}$, determine if the following statements are true or false.

   a. $\{b\} \in A$ b. $\{a\} \subseteq A$

   c. $\phi \in A$ d. $a \subseteq A$

3. Determine which of the following sets is an empty set.
   a. $\{a / a \text{ is an odd integer and } a^2 = 4\}$
   b. $\{a / a \text{ is an integer and } a < 1\}$
   c. $\{a / a \text{ is an integer and } a + 9 = 9\}$
4. Give an example for which $A \cup B = A \cup C$ and $B \neq C$.
5. Let A be a non-empty set. $P(A)$ denotes the set of all subsets of A. $P(A)$ is called as power set of A. Find the power set of the following sets.
   i. $A = \{a, b\}$ ii. $B = \{1, 2, 3\}$ iii. $\{\}$

Union of sets

Let A and B be two sets. The cardinality of $A \cup B$ is given as

$$\#(A \cup B) = \#A + \#B - \#(A \cap B)$$

Note: The above identity can be extended to n sets as

$$\#(A_1 \cup A_2 \cup \ldots \cup A_n) = \Sigma_i \# A_i - \sum_{\substack{i<j \\ i \neq j}} \#(A_i \cap A_j) + \ldots + (-1)^{n-1} \#(A_1 \cup A_2 \cup \ldots \cup A_n)$$

De Morgan's Law Let A and B be two sets. Then

i. $(A \cap B)^c = A^c \cup B^c$

ii. $(A \cap B)^c = A^c \cap B^c$

Proof Let $x \in (A \cap B)^c$

$\Rightarrow x \notin A \cap B$

$\Rightarrow x \notin A$ or $x \notin B$

$\Rightarrow x \in A^c$ or $x \notin B^c$

$\Rightarrow x \in A^c \cup B^c$

$\therefore (A \cap B)^c \subseteq A^c \cup B^c$ (1)

Now let $y \in A^c \cup B^c$

$\Rightarrow y \in A^c$ or $y \in B^c$

$\Rightarrow y \notin A$ or $y \notin B$

$\Rightarrow y \notin A \cap B$

$\Rightarrow y \in (A \cap B)^c$

$\Rightarrow A^c \cup B^c \subseteq (A \cap B)^c$ (2)

From (1) and (2), $(A \cap B)^c = A^c \cup B^c$.

Similarly we can prove that the other identity $(A \cup B)^c = A^c \cap B^c$.

Set Identities

Table 1 Set Identities

Identity	Law
$A \cup \phi = A$, where ϕ = null set $A \cap \mu = A$, where μ = universal set	Identity laws
$A \cup \mu = \mu$ $A \cap \phi = \phi$	Domination laws
$A \cup A = A$ $A \cap A = A$	Idempotent laws
$(A^C)^C = A$	Complementation law
$A \cup B = B \cup A$ $A \cap B = B \cap A$	Commutative laws
$A \cap (B \cup C) = (A \cap B) \cup (A \cap C)$ $A \cup (B \cap C) = (A \cup B) \cap (A \cup B)$	Distributive laws
$(A \cap B)^C = A^C \cup B^C$ $(A \cup B)^C = A^C \cap B^C$	Demorgan's laws

Representation of a Set in a Computer

We wish to represent a set A in a computer operator.

i. The set A is of finite order
   i.e., # A is finite

ii. An arbitrary order should be specified in the set A.
   For example $A = \{1, 2, 3, 4, 5, ..., n\}$

In computer it is represented as 11111...1, a bit string of length n, all the bits are 1.

Example 1.1 Let the universal set be $\mu = \{1, 2, ..., 10\}$. Express each of the following sets with bit strings where the *i*th bit in the string is 1 if i is in the set and 0 otherwise.

i. $\{3, 4, 5\}$ ii. $\{2, 3, 4, 7, 8, 9\}$

Solution

i. The bit string that represents the set of integers $\{3, 4, 5\}$ in μ has one bit in the third, fourth and fifth positions, and zero elsewhere.
   $\therefore$ It is 0011100000

ii. Similarly $\{2, 3, 4, 7, 8, 9\}$ is 0111001110.

Symmetric Difference of Sets

Let A and B be two sets. Their symmetric difference is defined as

$$A \Delta B = (A \backslash B) \cup (B \backslash A) \text{ and it is denoted by } A \oplus B$$

Min set Let P and Q be subsets of the universal set μ. Construct all possible sets which is an intersection of P and its complement or Q and its complement (but not a set of P and P^c or Q and Q^c).

$P \cap Q, P \cap Q^c, P^c \cap Q$ and $P^c \cap Q^c$.

The above four sets are called as min sets with respect to P and Q.

Max set Let P and Q be subsets of the universal set μ. Construct all possible sets which are union of P and its complement or Q and its complement, (but not a set of P and P^c or Q and Q^c)

i.e., $P \cup Q; P \cup Q^c; P^c \cup Q; P^c \cup Q^c$

The above four sets are called as max sets.

Min term Let $\{P_i\}_{1 \le i \le n}$ a family of subsets be a set X. A set is of the form $\sum_{i=1}^{n} S_i$, where each S_i is P_i or P_i^c is called as **min set** or **min term,** generated by $P_1, P_2, \ldots, P_n$

Example 1.2 Let $\mu = \{1, 2, \ldots, 10\}$, $A = \{2, 4, 6\}$, $B = \{3, 5, 7, 9\}$. Find the min term and max term with respect to A and B.

Solution $A^c = \{1, 3, 5, 7, 8, 9, 10\}$ and $B^c = \{1, 2, 4, 6, 8, 10\}$

The min terms are

a) $A \cap B = \{\}$

b) $A^c \cap B = \{3, 5, 7, 9\}$

c) $A \cap B^c = \{2, 4, 6\}$

d) $A^c \cap B^c = \{8, 10\}$

The max terms are

a) $A \cup B = \{2, 3, 4, 5, 6, 7, 9\}$

b) $A^c \cup B = \{1, 3, 5, 7, 8, 9, 10\}$

c) $A \cup B^c = \{1, 2, 4, 6, 8, 10\}$

d) $A^c \cup B^c = \{1, 2, 3, 4, 5, 6, 7, 8, 9, 10\}$

Note: If n subsets of a set are given, then the number of min terms (or) max terms is 2^n.

Partition of a set Let S be a non-empty set. The family of a sets $\{A_1, A_2, \ldots, A_n\}$ is a partition of the set S if

i. $S = \bigcup_{i=1}^{n} A_i$

ii. $A_i \cap A_j = \emptyset$ if $i \neq j$

Example $S = \{1, 2, 3, 4, 5, 6, 7\}$, $A_1 = \{1, 3, 5, 7\}$, $A_2 = \{2, 4, 6\}$

Clearly $\{A_1, A_2\}$ is a partition of S.

Note: Every singleton subset constitutes a partition.

Dual (duality principle) Let S be any identity involving sets and the set operations $\cup$, $\cap$ and C. If T is obtained from S by applying the substitutions $\cup \to \cap$; $\cap \to \cup$, $\emptyset \to \mu$; $\mu \to \emptyset$, then T is also true and T is called as dual of S.

Example The dual of $A \cup (B \cap A) = A$ is

$$A \cap (B \cup A) = A$$

ii. The relation array can be viewed graphically as elements of sets represented by models, and an ordered pair is represented by an edge between the vertices that correspond to the pair elements, with an arrow pointing to the second element of the pair.

Example 1.5 Let $A = \{1, 2, 3, 4\}$, $B = \{1, 4, 9, 16\}$ and the relation $R = \{(1, 1), (2, 4), (3, 9), (4, 16)\}$. Draw the relation graph.

Solution First we shall write the relation matrix M_R.

$$M_R = \begin{bmatrix} 1 & 0 & 0 & 0 \\ 0 & 1 & 0 & 0 \\ 0 & 0 & 1 & 0 \\ 0 & 0 & 0 & 1 \end{bmatrix}$$

Now we shall draw the relation graph G_R.

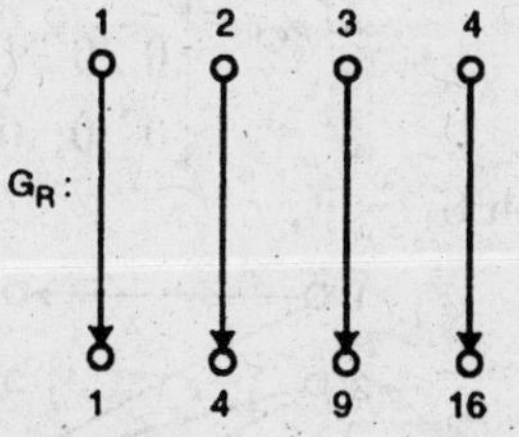

Example 1.6 Let $A = \{a, b, c, d\}$, $B = \{a, e, f, d\}$ and let $R = \{(a, e), (a, f), (b, e), (c, f), (b, d), (d, d), (d, a)\}$. Draw the relation graph.

Solution First we shall write the relation matrix M_R.

$$M_R = \begin{bmatrix} 0 & 1 & 1 & 0 \\ 0 & 1 & 0 & 1 \\ 0 & 0 & 1 & 0 \\ 1 & 0 & 0 & 1 \end{bmatrix}$$

The relation graph G_R is given as

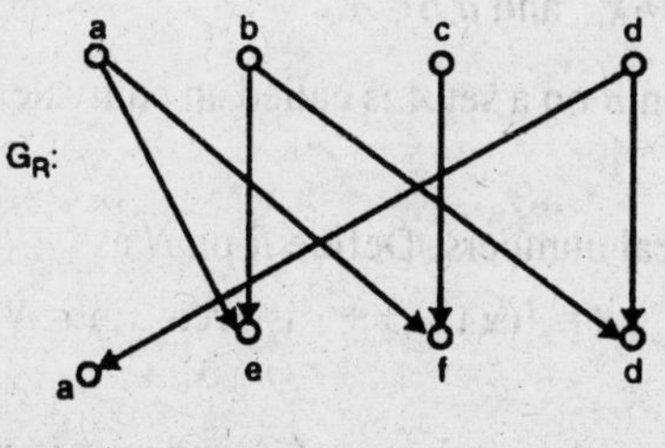

Composition of Two Relations

Let R be a binary relation from the set A to the set B and S be a binary relation from the set B to the set C, then the ordered pair (R, S) is said to be composable. If (R, S) is a composable pair of binary relations, the composite $R \circ S$ and R and S, is a binary relation from the set A to the set C, such that, for $a \in A$ and $c \in C$ a $(R \circ S)_C$ if for some $b \in B$, both aRb and bSc.

Example 1.7 $A = \{1, 2, 3, 4\}$, $B = \{1, 3, 9, 10\}$ $C = \{5, 6, 7, 8\}$, $R = \{(1, 1), (1, 3), (2, 9), (2, 10), (3, 3), (4, 10)\}$ $S = \{(1, 5), (3, 7), (9, 7), (10, 8)\}$. Find $R \circ S$ and its relation graph.

Solution

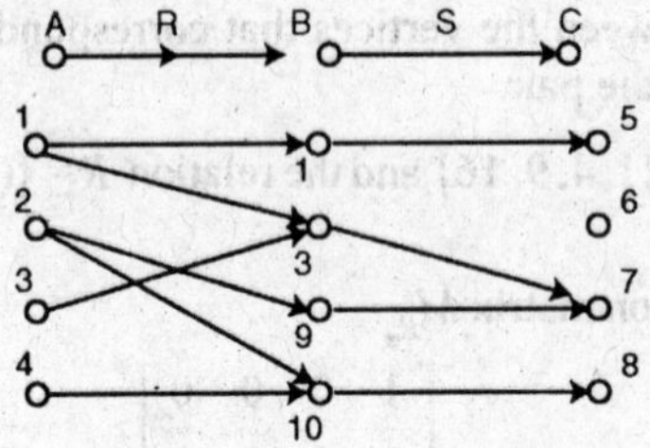

$R \circ S = \{(1,5), (1,7), (2,7), (2,8), (3,7), (4,8)\}$

The corresponding matrix is

$$M_{R\circ S} = \begin{bmatrix} 1 & 0 & 1 & 0 \\ 0 & 0 & 1 & 1 \\ 0 & 0 & 1 & 0 \\ 0 & 0 & 0 & 1 \end{bmatrix}$$

and the corresponding relation graph $G_{R\circ S}$ is,

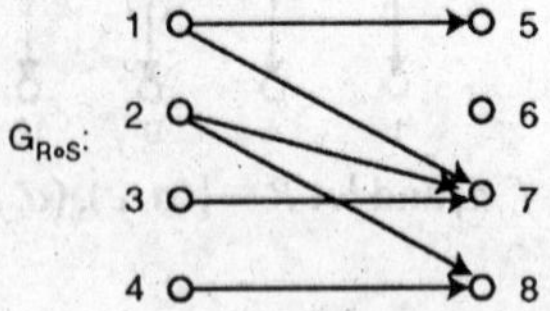

Properties of a Binary Relation

Let R be a relation on a set A (i.e., $R \subseteq A \times A$). R is called

i. reflexive if $aRa, \forall\, a \in A$

ii. symmetric if aRb then $bRa, \forall\, a, b \in A$.

iii. transitive if aRb and bRc then $aRc, \forall\, a, b, c \in A$.

iv. irreflexive if $a \not R a$, and $a \in$ A.

v. antisymmetric if aRb then $b \not R a$, and $a, b \in A$.

Equivalence relation A relation R on a set A is called an equivalence relation if R is reflexive, symmetric and transitive.

Example Let N be the set of natural numbers. Define R on N as

$$R = \{(x,y) / x + y \text{ is even}, x, y \in N\}$$

Proof Let $x \in N$. Now $x + x = 2x$.

Clearly $2x$ is even. Therefore R is reflexive. Let $x, y \in N$ and $x + y$ is even. Clearly $y + x$ is also even and hence R is symmetric.

Now if $x + y$ is even and $y + z$ is even then we have to prove that $x + z$ is even.
Since x + y and y + z are even, both $(x + y)$ and $(y + z)$ are divisible by 2.
$\therefore$ $(x + y) + (y + z)$ is also divisible by 2 i.e., $x + (y + y) + z$ is divisible by 2.
$\therefore$ $(x + z)$ is divisible by 2.
Hence R is transitive, and an equivalence relation.

Note: From the relation graph or relation matrix, the kind of relation can be identified.

SOLVED PROBLEMS

1. The relation R on a set is represented by

$$M_R = \begin{bmatrix} 1 & 1 & 0 \\ 1 & 1 & 1 \\ 0 & 1 & 1 \end{bmatrix}$$

Is R reflexive, symmetric or antisymmetric?

Solution In the matrix M_R, the diagonal elements are 1. Therefore R is reflexive. Since the matrix M_R is symmetric, the relation R is also symmetric.

2. The relation R and R_1 on a set is represented by

a. $M_R = \begin{bmatrix} 1 & 0 & 0 \\ 0 & 1 & 0 \\ 1 & 0 & 1 \end{bmatrix}$ b. $M_{R1} = \begin{bmatrix} 1 & 1 & 1 \\ 1 & 0 & 1 \\ 1 & 1 & 1 \end{bmatrix}$

Are the relations R and R_1 reflexive, symmetric, antisymmetric, and or transitive ?

Solution

a. Since the matrix M_R is symmetric and its diagonal entries are 1. The relation R is symmetric and reflexive. Since R is not antisymmetric R is transitive.

b. The relation R_1 is not reflexive.

R_1 is symmetric [$\because$ M_{R1} is symmetric] and R_1 is transitive.

3. Draw the relation graph for the following relations.

a. $R = \{(1, 1), (1, 3), (2, 1), (2, 3), (2, 4), (3, 1), (3, 2), (4, 1)\}$ on the set $X = \{1, 2, 3, 4\}$.

b. $R_1 = \{(1, 1), (1, 2), (1, 3), (2, 2), (2, 3), (3, 3)\}$ on the set $Y = \{1, 2, 3\}$

Solution

a. The relation graph G_R of R is drawn as:

The vertices of G_R are 1, 2, 3, 4.

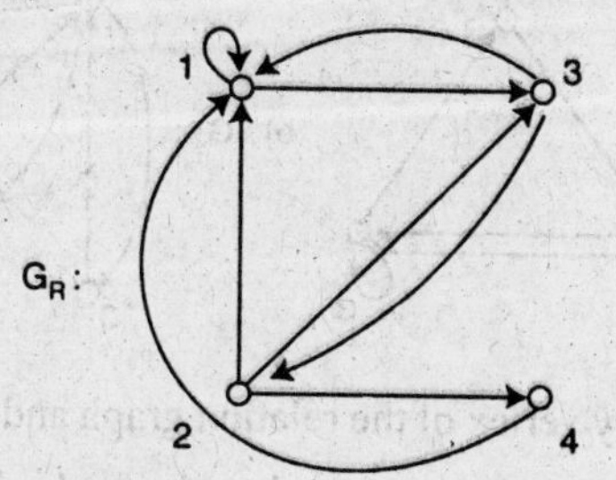

b. The relation graph G_{R1} of R_1 is drawn as.

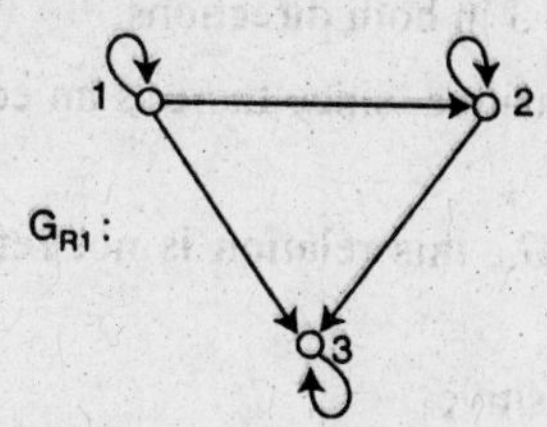

4. Let R be the relation represented by

$$M_R = \begin{bmatrix} 0 & 1 & 1 \\ 1 & 1 & 0 \\ 1 & 0 & 1 \end{bmatrix}$$

Find the relation matrices representing (a) R^{-1} (b) R^c (c) R^2

Solution

a. To get the inverse relation matrix (M^{R-1}) of a relation matrix (M_R) just write the transpose of M_R.

$$\therefore M_{R-1} = \begin{bmatrix} 0 & 1 & 1 \\ 1 & 1 & 0 \\ 1 & 0 & 1 \end{bmatrix}$$

b. To find the complement relation matrix, replace 0 by 1 and 1 by 0 in the given relation matrix.

$$\therefore M_{R^c} = \begin{bmatrix} 1 & 0 & 0 \\ 0 & 0 & 1 \\ 0 & 1 & 0 \end{bmatrix}$$

c. To find the relation matrix of R^2

Now $R^2 = R \circ R$.

If the relation matrix M_R is known, then $M_{R^2} = M_R . M_R$ (the matrix multiplication)

$$\therefore M_{R^2} = \begin{bmatrix} 0 & 1 & 1 \\ 1 & 1 & 0 \\ 1 & 0 & 1 \end{bmatrix} \begin{bmatrix} 0 & 1 & 1 \\ 1 & 1 & 0 \\ 1 & 0 & 1 \end{bmatrix} = \begin{bmatrix} 1 & 1 & 1 \\ 1 & 1 & 1 \\ 1 & 1 & 1 \end{bmatrix}$$

5. Find whether the relations for the directed graphs shown in the following figures are reflexive, symmetric, antisymmetric and \or transitive.

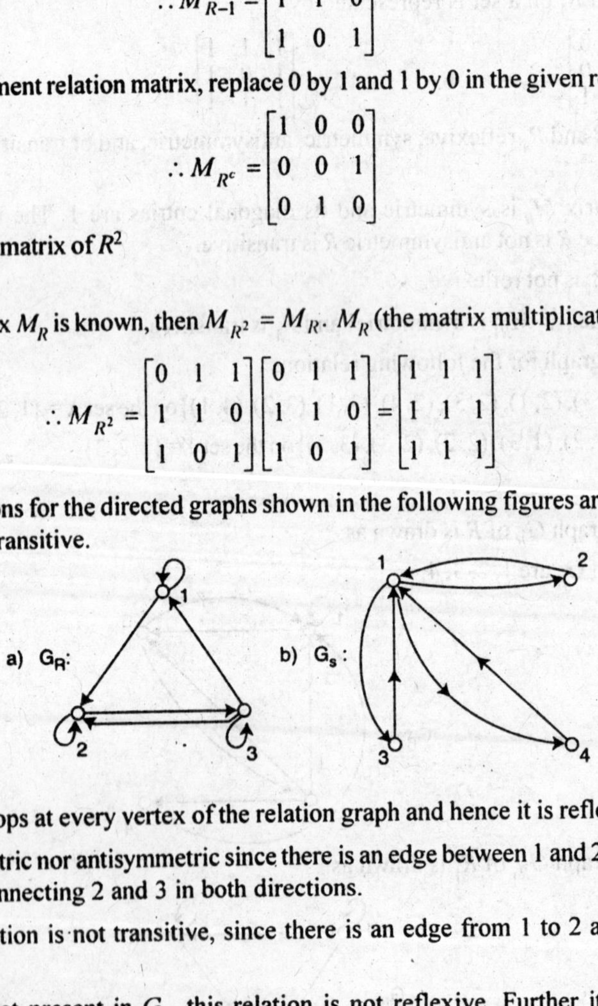

Solution

a. In G_R, there are loops at every vertex of the relation graph and hence it is reflexive.

It is neither symmetric nor antisymmetric since there is an edge between 1 and 2 but not from 2 to 1, but there are edges connecting 2 and 3 in both directions.

Moreover the relation is not transitive, since there is an edge from 1 to 2 and 2 to 3, but no edge from 1 to 3.

b. Since loops are not present in G_S, this relation is not reflexive. Further it is symmetric and not antisymmetric.

Moreover the relation is not transitive.

Equivalence class Let R be an equivalence relation on a set A. Let $x \in A$. The equivalence class of a is given by

$$[a]_R = \{x \in A \,/\, (a, x) \in R\}$$

Note: $[a]_R \neq \varnothing$, because $a \in [a]$.

Example 1.8 a) Prove that any two equivalence classes are identical or disjoint.

Solution First we shall prove that $(a,b) \in R$. This implies that $[a]_R = [b]_R$

Suppose $(a, b) \in R$

Claim $[a] = [b]$

Let $x \in [a] \Leftrightarrow (x,a) \in R$

$\Leftrightarrow (x,b) \in R$ [$\because$ $(x,a) \in R$ and $(a,b) \in R$ and R is transitive]

$\Leftrightarrow x \in [b]$

$\Leftrightarrow [a] \subseteq [b]$

$\therefore$ $[a] = [b]$

Now suppose $[a]$, $[b]$ are two equivalence classes.

Claim $[a] = [b]$ or $[a] \cap [b] = \varnothing$

If $[a] \cap [b] = \varnothing$ then nothing to prove.

Suppose $[a] \cap [b] \neq \varnothing$ then $x [a] \cap [b]$

$\Rightarrow x \in [a]$ and $x \in [b]$

$\Rightarrow (x,a) \in R$ and $(x,b) \in R$

$\Rightarrow [x] = [a]$ and $[x] = [b]$

$\Rightarrow [a] = [b]$

$\therefore$ $[a] \cap [b] = \varnothing$ or $[a] = [b]$

i.e., any two equivalance classes are identical or disjoint.

Example 1.8 b) Prove that an equivalance relation induces a partition and a partition induces an equivalance relation.

Solution Let $\{A_i / i \in z\}$ is a partition of a set A. Define a relation R on A by $(a,b) \in R$ if $a, b \in A_i$ for some i.

Claim R is an equivalence relation on A.

Let $a \in A$

$\Rightarrow a \in A_i$ for some i.

$\Rightarrow a, a \in A_i$ for some i

$\Rightarrow (a, a) \in R$.

$\therefore$ R is a reflexive relation on A.

Suppose $(a, b) \in R$. then by the definition of R,

$a, b \in A_i$ for some i.

$\therefore$ $b, a \in A_i$ for some i

$\Rightarrow (b, a) \in R$.

$\therefore$ R is a symmetric relation on A.

Suppose $(a, b) \in R$ and $(b, c) \in R$ then $a, b \in A_i$ and $b, c \in A_j$ for some i and j

Here $b \in A_i$ and $b \in A_j$

$\therefore A_i \cap A_j \neq \varnothing \Rightarrow A_i = A_j$, otherwise $\{A_i\}_{i \in I}$ is not a partition and hence $a, b, c \in A_i$

$\therefore (a, c) \in R$.

$\therefore R$ is a transitive relation on A.

R is also an equivalance relation on A.

Further we can also show that $A_i = [a]^{a \in A}$,

conversely, we can assume that R is an equivalance relation on set A.

Claim R induces a partition for A.

Let $x \in A$, $[x] = \{y \in A / (y, x) \in R\}$ and for any $x, y \in A$, we have

$$[x] \cap [y] = \varnothing \text{ or } [x] = [y]$$

$\therefore A = \cup_{x \in A} [x]$ i.e., $\{[x] / x \in A\}$ is a partition for A.

EXERCISE III

1. Write down all elements of R from $A = \{0, 1, 2, 3, 4, 5\}$ to $B = \{0, 1, 2, 3, 4\}$ where $(a, b) \in R$ if
   a. $a = b$ b. $a + b = 5$ c. $a < b$
   d. gcd of $aRb = 1$ e. lcm of $(a, b) = 2$ f. $a - b < 0$
2. Find whether the relation R on the set of people is reflexive, symmetric, antisymmetric and transitive, where $(a, b) \in R$ iff
   a. a is smaller than b b. a and b have a common grandparent
   c. a and b were born in the same hospital
3. Find whether the relation R on the set of all integers is reflexive, symmetric, antisymmetric and transitive, where $(x, y) \in R$ iff
   a. $x \neq y$ b. $xy \lneq 1$ c. $x \equiv y \pmod 5$
   d. $x = y^2$ e. x is a multiple of y
4. How many different relations are there from a set of n elements to a set of m elements.
5. Give an example of a relation which is symmetric and antisymmetric.
6. Let R and S be two relations which are reflexive, symmetric and transitive. Then show that $R \cap S$ is also reflexive, symmetric and transitive.
7. Let $R = \{(1, 2), (2, 3), (3, 4)\}$ and $S = \{(1, 1), (1, 2), (2, 1), (2, 2(, (2, 3), (3, 1), (3, 2), (3, 3), (3, 4)\}$ be relations from $\{1, 2, 3\}$ to $\{1, 2, 3, 4\}$.
   Find a) $R \cup S$ b) $R \cap S$ c) $R_1 - R_2$ d) $R_2 - R_1$
8. Let M_R and M_S be the relation matrices of the relations R and S
   $$M_R = \begin{bmatrix} 0 & 0 & 1 \\ 1 & 0 & 1 \\ 1 & 1 & 1 \end{bmatrix}; \ M_S = \begin{bmatrix} 1 & 1 & 0 \\ 1 & 1 & 0 \\ 0 & 1 & 1 \end{bmatrix}$$
   Find the relation matrices for the following relations.
   a. $R \cup S$ b. $R \cap S$ c. $R \circ S$ d. $S \circ R$
9. Draw the relation graphs for the following relations. $\{(a, a), (a, b), (b, c), (c, d), (d, a), (d, b)\}$
10. Identify the relation kinds from the graph obtained from problem no. 9.

11. Let $X = \{1, 2, 3, ..., 7\}$ and $R = \{(x, y) / x - y \text{ is divisible by } 3\}$

    Prove that R is an equivalance relation and draw the relation graph.

12. Let $x = \{1, 2, 3, 4, 5\}$ and let $P = \{(1, 2), (3), (4, 5)\}$. Prove that the partition P defines an equivalance relation on X.

13. A relation on a set that is reflexive and symmetric is called a compatible relation.
    a. Let A be a set of people and R be a binary relation on A such that $(a, b) \in R$ if a is a friend of b. Prove that R is a compatible relation
    b. Let A be a set of Latin words and R be a binary relation on A such that two words in A are related if they have one or more letters in common. Prove that R is a compatible relation.

Closures of a Relation

Let R be any relation on a set A. (R may be or may not be symmetric, reflexive and transitive). Let S be any other relation on A such that S contains R, and S is a subset of every relation containing R. Then S is called closure of R.

Reflexive closure Let R be a relation on a set X. The reflexive closure S of R is obtained by adding to R all pairs of the form (a,a) which are not in R, $a \in A$. Now S is reflexive, contains R and is in any reflexive relation containing R.

Example 1.9 $R = \{(a, a), (a, b), (b, a), (c, b)\}$ on the set $A = \{a, b, c\}$. Find the reflexive closure of R.

Solution By inspection, we can see that $(b, b), (c, c)$ are not in R.

$\therefore S = \{(a, a), (a, b), (b, a), (c, b), (b, b), (c, c)\}$

Then S is the reflexive closure of R.

Note: To obtain a reflexive closure of a relation R, just add the diagonal relation elements to R. i.e., Let $D = \{(a,c) / a \in A\}$, the diagonal relation S is the reflexive closure of R, then $S = R \cup D$

Symmetric closure Let R be a relation on set A. The symmetric closure S of R is obtained by adding to R all pairs of the form (b, a), if $(a, b) \in R$ and $(b, a) \notin R$

In other words, symmetric closure of R can be obtained as $S = R \cup R^{-1}$

Example 1.10 Let $R = \{(a, a), (a, b), (b, b), (c, c), (c, b)\}$ on $\{a, b, c\}$. Find the symmetric closure of R.

Solution Clearly R is not symmetric.

Now $S = R \cup R^{-1} = \{(a, a), (a, b), (b, b), (c, a), (c, b), (b, a), (a, c), (b, c)\}$

Clearly S is symmetric and it contains R and is in any symmetric relation containing R.

Paths in Directed Graphs

Now we shall study about paths in directed graphs and relationship between a relation on a set, before transitive closure.

Path A path from vertex a to vertex b in a directed graph G is a sequence of one or more edges (v_0, v_1), $(v_1, v_2), ..., (v_{n-1}, v_n)$ in G, with $v_0 = a$; $v_n = b$ i.e., a sequence of edges whose terminal vertex is same as the intial vertex of the next edge in this path. This path is denoted by $v_0, v_1, ..., v_n$ of length n.

Let R be a relation on a set $A = \{1, 2, .., n\}$ and G_R be the corresponding relation graph whose vertices are $a = 1$, $v_1 = 2, ..., b = n$. There is a path in G_R from a to b if there is a sequence of vertices a, $v_1, v_2, ..., v_{n-1}, b$ with $(a, v_1) \in R, (v_1, v_2) \in R, (v_2, v_3) \in R, ..., (v_{n-1}, b) \in R$

Result Let R be a relation on a set A. Then

i. $R^2 = R \circ R,\ R^n = R^{n-1} \circ R$

ii. $R^n \subset R$

iii. In G_R, the relational graph of R, there is a path of length n from a to b if $(a,b) \in R^n$.

Connectivity relation Let R be a relatiion on set A. The connectivity relation R^* consists of the pairs (a,b) such that there is a path between a and b in R.

$$\text{i.e., } R^* = \bigcup_{n=1}^{\infty} R^n$$

Result The transitive closure of a relation R equals the connectivity relation R^*.

Let R be a relation on a set A.

Claim R^* is the transitive closure of R. To prove that

i. R^* is transitive and

ii. S is a transitive relation on A with $R \subseteq S$. Then $R^* \subseteq S$.

By definition, $R^* = \bigcup_{i=1}^{\infty} R^i$

$\therefore$ R^* contains R.

i. If $(a,b) \in R^*$ and $(b,c) \in R^*$, then there are paths from a to b and from b to c in R. Thus we obtain a path from a to c by starting with the path a to b and following it with the path b to c.

$\therefore$ $(b,c) \in R^*$.

i.e., R^* is transitive.

ii. Let S be a transitive relation containing R.

Since S is transitive, S^n is transitive.

Further $S^n \subseteq S$. Since, $S^* = \bigcup_{i=1}^{\infty} S^i$, and $S^i \subseteq S$, $S^* \subseteq S$.

Since any path in R is also a path in S, $R^* \subseteq S^*$, if $R \subseteq S$.

Now we have, $R^* \subseteq S^*$ and $S^* \subseteq S$.

$$\Rightarrow R^* \subseteq S^*$$

i.e. any transitive relation that contains R must also contain R^*.

$\therefore$ R^* is the transitive closure of R.

Transitive closure Let M_R be the relation matrix of a relation R on the set A of n elements. Then the **transitive closure** matrix M_{R^*} is given by

$$M_{R^*} = M_R \vee M_{R^2} \vee M_{R^3} \vee \ldots \vee M_{R^n}$$

Example 1.11

i. Find the transitive closure of a relation R on the set $\{a,b,c\}$, whose relation matrix M_R is given as

$$M_R = \begin{bmatrix} 1 & 0 & 1 \\ 1 & 1 & 0 \\ 1 & 1 & 1 \end{bmatrix}$$

Solution Let R^* be the transitive closure of R. The relation matrix M_{R^*} of R^* is given as

$$M_{R^*} = M_R \vee M_{R2} \vee M_{R3}$$

Now

$$M_{R2} = \begin{bmatrix} 1 & 0 & 1 \\ 1 & 1 & 0 \\ 1 & 1 & 1 \end{bmatrix}; \quad M_{R3} = \begin{bmatrix} 1 & 1 & 1 \\ 1 & 1 & 1 \\ 1 & 1 & 1 \end{bmatrix}$$

$$M_{R*} = \begin{bmatrix} 1 & 0 & 1 \\ 1 & 1 & 0 \\ 1 & 1 & 1 \end{bmatrix} v \begin{bmatrix} 1 & 0 & 1 \\ 1 & 1 & 0 \\ 1 & 1 & 1 \end{bmatrix} v \begin{bmatrix} 1 & 1 & 1 \\ 1 & 1 & 1 \\ 1 & 1 & 1 \end{bmatrix} = \begin{bmatrix} 1 & 1 & 1 \\ 1 & 1 & 1 \\ 1 & 1 & 1 \end{bmatrix}$$

ii. **Find the transitive closure matrix of the relation R whose relation matrix is given as**

$$M_R = \begin{bmatrix} 1 & 0 & 1 \\ 0 & 1 & 0 \\ 1 & 1 & 0 \end{bmatrix}$$

Solution **Let R^* be the transitive closure of R and $M_R{}^*$ be the corresponding relation matrix.**

we have $M_{R*} = M_R \, v \, M_{R2} \, v \, M_{R3}$

$$\text{Now } M_{R^2} = \begin{bmatrix} 1 & 1 & 1 \\ 0 & 1 & 0 \\ 1 & 1 & 1 \end{bmatrix}; M_{R^3} = \begin{bmatrix} 1 & 1 & 1 \\ 0 & 1 & 0 \\ 1 & 1 & 1 \end{bmatrix}$$

$$M_{R^*} = \begin{bmatrix} 1 & 0 & 1 \\ 0 & 1 & 0 \\ 1 & 1 & 0 \end{bmatrix} v \begin{bmatrix} 1 & 1 & 1 \\ 0 & 1 & 0 \\ 1 & 1 & 1 \end{bmatrix} v \begin{bmatrix} 1 & 1 & 1 \\ 0 & 1 & 0 \\ 1 & 1 & 1 \end{bmatrix} = \begin{bmatrix} 1 & 1 & 1 \\ 0 & 1 & 0 \\ 1 & 1 & 1 \end{bmatrix}$$

Note: The transitive closure can be obtained by the following algorithm.

Transitive closure (M_R; $0-1$ $n \times n$ matrix)

$A \leftarrow M_R$

$B \leftarrow A$

for $i \leftarrow 2$ to n

begin

$A \leftarrow A \,.\, M_R$

$B \leftarrow B v A$

end (B is the required matrix R^*)

Partial Ordering

A relation R on a set A is said to be a partial order relation if R is (i) reflexive (ii) antisymmetric and (iii) transitive.

Example **Let S be non-empty. Define R as $\subseteq$ (contained in or equal to) $P(S)$. Clearly ($P(S)$, $\subseteq$, is a partial order set. If**

i. $A \in P(S), A \subseteq A$

ii. $A, B \in P(S), A \subseteq B$ and $B \subseteq A \Rightarrow A = B$

iii. $A, B, C, \in P(S)$,

whenever $A \subseteq B$ and $B \subseteq C \Rightarrow A \subseteq C$. $\quad \therefore \subseteq$ is a partial order relation on $P(S)$.

2. Let $\leq$ be a partial order on the set of integers and a, b, c be integers. Clearly
   i. $a \leq a$; where $a \in z$
   ii. If $a \leq b$ and $b \leq a$ then $a = b$, where $a, b \in z$.
   iii. If $a \leq b$ and $b \leq c$ then $a \leq c$, where $a, b, c \in z$.
   $\therefore$ $(z, \leq)$ is a partial order set (poset).

Comparable integers Two elements of a poset $(s, \leq)$ are said to be comparable if $a \leq b$ or $b \leq a$. Otherwise a and b are said to be incomparable.

Example In the poset $(z_+, 1)$, the integers 2 and 8 are comparable whereas 3 and 5 are incomparable.

[a / b is read as a divides b and $a / b \Rightarrow b = k\, a$ for some $k \in z$]

Chain A poset $(s, \leq)$ is called a chain if $\leq$ is total order relation.

(If every two elements of S are comparable then S is called a totally released set.)

Example The poset $(z, \leq)$ is a chain, whereas the poset $(z_+, 1)$ is not a chain.

Lexicographic Order (Dictionary Order)

The lexicographic ordering of $\leq$ on $A_1 \times A_2$ is defined by indicating that one pair is less than the other if the first entry of the first pair is less than the first entry of the second pair or if the first entries are equal, but the second entry of the first pair is less than the second entry of the second pair.

i.e., $(a,b) \leq (c,d)$ if $a < c$ or if $a = c$ and $b < d$.

Note: i. The above relation $\leq$ is a partial order relation on $A_1 \times A_2$

ii. Using the dictionary order, one can prove that the complex number system is an ordered field.

iii. Lexicographic ordering can be defined on the cartesian product of n posets $(A_1, \leq_1)$, $(A_2, \leq_2)$, ..., $(A_n, \leq_n)$.

We can define partial ordering $\leq$ on $A_1 \times A_2 \times .... \times A_n$ by $(a_1, a_2, ..., a_n) \leq (b_1 b_2 ... b_n)$ if $a_1 < b_1$ or if there exists an integer k such that

$$a_1 = b_1;\ a_2 = b_2, ..., a_k = b_k \text{ and } a_{k+1} <_{k+1} b_{k+1}\ (k \leq n)$$

Example 1.12 Find the lexicographic ordering of the following strings.

i. open, opener, opera, operand.

ii. zoo, zoom, zoology.

Solution

i. Clearly open $\leq$ opener $\leq$ opera $\leq$ operand. Because all the 4 strings have the same letters in the first 3 positions the first 2 strings have the same letters in the first 4 positions, but the second string has 2 more letters. The second and third string differs in the 4th position and so on.

ii. On similar conditions the lexicographic order is zoo $\leq$ zoology $\leq$ zoom.

Hasse Diagram

We can represent the poset as a directed graph. Every poset consisting of a set and a relation can be represented as a graph. We have to do minor modification in this relational graph. Since partial ordering relation is reflexive and transitive the relational graph will consist of self loops and edges corresponding to the transitive relations.

In the relational graph, the self loops and edges corresponding to the pairs (a, c) are removed whenever (a, b) and (b, c) are present. Finally, each edge is arranged so that its initial vertex is below its terminal vertex. All the arrows on the directed edges, are removed since all the edges point upward towards their terminal vertex.

Example 1.13 Draw the Hasse diagram representing the partial ordering $\{(a, b) / a/b\}$ on $\{2, 4, 5, 10, 12, 20, 25\}$.

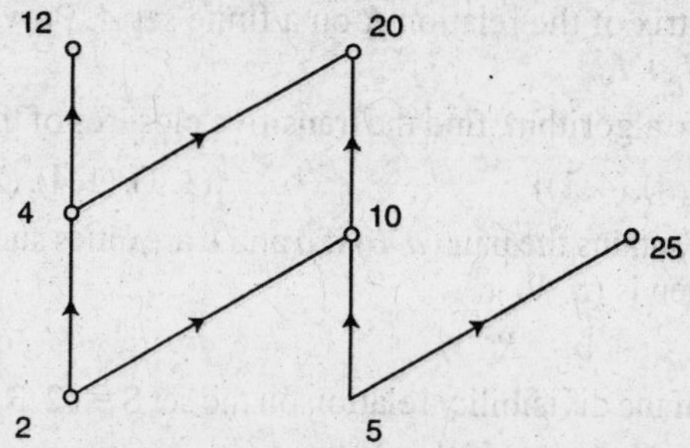

Solution Here the relation set = {(2, 4), (2, 12), (4, 12), (5, 10), (5, 20), (5, 25), (10, 20)}

Maximal and minimal elements An element a of a poset $(S, \leq)$ is called maximal element if there is no $b \in S$ such that $a < b$. Similarly an element a of a poset $(S, \leq)$ is called minimal if there is no $b \in S$ such that $b < a$.

Note: In the Hasse diagram, it is easy to spot the maximal and minimal element, because they are the 'top' and 'bottom' elements in that diagram.

Upper and lower bound An element υ of a poset $(s, \leq)$ is called an upper bound of S if $a \leq \upsilon \ \forall \ a \in S$. Similarly an element l of a poset $(S, \leq)$ is called an lower bound of S if $l \leq a \ \forall \ a \in S$.

The element x is called the least upper bound of the set S if (i) x is an upper bound of S and (ii) $x \leq z$ whenever z is an upper bound of S. Similarly, the element y is called the greatest lower bound of S if

i. y is a lower bound of S and
ii. $z \leq y$, whenever z is a lower bound of S.

Example 1.14 Find the greatest lower bound and the least upper bound of the sets {3, 9, 12} and {1, 2, 4, 5, 10} if they exist in the poset $(z_+, 1)$.

Solution An integer is a lower bound of {3, 9, 12} if 3, 9 and 12 are divisible by this integer. Such integers are 1 and 3 only. Clearly 3 is the greatest lower bound of {3, 9, 12}.

Similarly the greatest lower bound for {1, 2, 4, 5, 10} is 1.

An integer is an upper bound for {3, 9, 12} if it is divisible by 3, 9 and 12.

The integers with this properly are those divisible by the lcm of 3, 9 and12, which is 36. Hence 36 is the least upper bound for the set {3, 9, 12}.

Similarly 20 is the least upper bound for the set {1, 2, 4, 5, 10}.

Lattice A poset in which every pair of elements have both a least upper bound and a greatest lower bound is called a lattice.

Example

i. The poset ({1, 2, 4, 8},1) is a lattice.
ii. The poset $(P(S), \subseteq)$ is a lattice.

Here $A \cup B$ = least upper bound of A and B and $A \cap B$ = greatest lower bound of A and B, for any $A \subseteq S$, $B \subseteq S$.

EXERCISE IV

1. Let R be the relation on $\{a,b,c,d\}$ containing the ordered pairs (a, b), (b, b), (b, c), (c, a), (c, c) and (d, a). Find the
   a. reflexive closure of R b. symmetric closure of R
2. Let M_R be the relation matrix of the relation R on a finite set A. Prove that the matrix that represents the reflexive closure of R is $M_R \vee I_R$.
3. Using the transitive closure algorithm, find the transitive closures of the following relations on $\{1, 2, 3, 4\}$
   a. $\{(1,2),(2,1),(2,3),(3,4),(4,1)\}$ b. $\{(1,1),(1,4),(2,1),(2,3),(3,1),(3,2),(3,4),(4,2)\}$
4. Let R be the relation that contains the pair (a, b) if a and b are cities such that there is a point-to-point (non-stop) bus from a to b. When is (a, b) in
   a. R^2 b. R^3 c. R^*
5. Draw the Hasse diagram of the divisibility relation on the set $S = \{2, 3, 5, 9, 12, 15, 18\}$.
6. Give an example of a poset that has 3 maximal elements.
7. Give an example of a poset with 5 elements that is a lattice and an example of a poset with 5 elements that is not a lattice.

N-ary Relations and their Applications

A relation R is called an n-ary relation if $R \subseteq A_1 \times A_2 \times \ldots \times A_n$, where $A_1, A_2, \ldots, A_n$ are sets. These sets are called as domains of the relation and n is called its degree.

Note: If $n=2$ then R is a binary relation. If $n=3$ then R is a ternary relation. If $n=4$ then R is a quaternery relation.

Relations and Database

We shall study the relational data model based on the concept of a relation here.

A database is a collection of records, which are n-tuples made up of fields. The fields are the entries of the n-tuples. For example, a database of employees records is made up of fields containing name, empnumber, dept and salary. Here the employees records are represented as 4-tuples of the form (EMPNAME, EMPNUMBER, DEPT, SALARY)

Let us consider an example of 3 such records

(RAM, P001, PURCHASE, 10000)

(RAJA, M002, MARKETING, 12000)

(ASHA, G001, FRONTOFFICE, 7000)

Table An n-ary relation among the sets $A_1, A_2, \ldots, A_n$ is known as table on $A_1, A_2, \ldots, A_n$. The sets $A_1, A_2, \ldots, A_n$ are called the domain of the table.

A domain is a primarykey when no two n-tuples in the relation have the same value from this domain.

Combinations of domains also uniquely identify n-tuples in an n-ary relation. When the values of a set of domain determines an n-tuple in a relation, the cartesian product of these domains is called the composite key.

Table i

Empnumber	Empname	Dept	Salary (Rs)
P001	Ram	Purchase	10,000
M001	Raja	Marketing	12,000
M002	Ravi	Marketing	11,000
G001	Asha	Frontoffice	7,000

In empnumber, empname and salary are primary keys whereas dept is not, because marketing appears twice. (Here we assume that 4-tuples will be added in future.) Further the domains empname and salary constitute a composite key [with respect to table (i)].

Table ii

Empname	Salary (Rs)
Ram	10,000
Raja	12,000
Ravi	11,000
Asha	7,000

Table iii

Empname	Dept	Designation
Ram	Purchase	Purchase Officer
Raja	Marketing	Marketing Manager
Ravi	Marketing	Marketing Executive
Asha	Frontoffice	Receptionist

In general, there are many operations on *n*-ary relations that can be used to form new *n*-ary relations. Among these operations, we will discuss two important operations (i) Join and (ii) Project.

The Join operation is used to combine two tables when these two tables have the same domains (some identical fields).

The project operations are used to form new *n*-ary relations by deleting the same domains in every record of the relation.

Let *R* be a table of degree *m* and *S* be a table of degree *n*. The join operation of *R*:*S* is a table of degree $m + n - p$ (where $p \leq m$ and $p \leq n$) that consists of all $(m + n - p)$ tuples $(a_1, a_2, ..., a_{m-p}, c_1, ..., c_p, b_1, ..., b_{n-p})$, where *m* tuple$(a_1, a_2, ..., a_{m-p}, c_1, ..., c_p)$ belongs to *R* and *n*-tuple $c_1, ..., c_p, b_1, ..., b_{n-p})$ belongs to *S*.

The projection $P_{i_1, i_2, ..., i_m}$ deletes $n - m$ of the components of an *n*-tuple, leaving the i_1th, i_2th, ...i_mth components (where $m \leq n$ and n is degree of table on which the project operation is applied)

Example 1.15

1. What relation results when the project operation $P_{1,4}$ is applied to the relation in table *i*?

   Solution By definition of project operation, the second and third columns of table i are deleted and pairs representing empname and salary are obtained. Table ii is the result.

2. What is the table obtained when projection $P_{1,3}$ is applied to table iii?

   Solution By definition of projection, the second column (dept) is deleted and the remaining columns retained. The following table is the resulting table

Table iv

Empname	Designation
Ram	Purchase Officer
Raja	Marketing Manager
Ravi	Marketing Executive
Asha	Receptionist

3. What is the table obtained when join J_2 is applied to table i and table ii?

Solution The Join operator J_p produces a new relation from two relations (tables) by combining all the m-tuples of the first table with all n-tuples of the second table, where the last P components of m-tuples coincide with the first P components of n-tuples.

Table v

Train name	Train no	Destination
Chennai Exp	6024	Chennai
Coimbatore - Kurla	1014	Kurla
Lalbagh	2607	Chennai
Rajdhani	2430	New - Delhi
Mysore - Thanjavur	6232	Thanjavur

Table vi

Train no	Destination	Origin	Section
6024	Chennai	Bangalore	Jolarpet
1014	Kurla	Coimbatore	Jolarpet
2607	Chennai	Bangalore	Jolarpet
2430	New- Delhi	Trivandrum	Salem
6232	Thanjavur	Mysore	Salem

The last two coloumns of table v and the first two columns of table vi are identical. Hence the join operation is possible. The following is the resulting table.

Table vii

Train name	Train no	Destination	Origin	Section
Chennai Exp	6024	Chennai	Bangalore	Jolarpet
Coimbatore- Kurla	1014	Kurla	Coimbatore	Jolarpet
Lalbagh	2607	Chennai	Bangalore	Jolarpet
Rajdhani	2430	New Delhi	Trivandrum	Salem
Mysore - Thanjavur	6232	Thanjavur	Mysore	Salem

EXERCISE V

1. Assuming that no new n-tuples are added, find a composite key with two fields (domains) containing train no. field in table vii.
2. What do you obtain when you apply the projection $P_{1,3,4}$ to the 6-tuples
   a. (15, 001, PRO, Hospital, Bangalore, General)
   b. (01, 005, CMO, Nursing home, Madurai, Speciality)
3. Display the table produced by applying the projection $P_{2,4,5}$ to the table vii.
4. Which projection mapping is used to delete the second, fourth and fifth components of a 5-tuple?

5. How many domains (components / fields) are there in the n-tuples in the table obtained by applying the join operator J_3 to two tables with 6-tuples and 7-tuples respectively?
6. Construct the table obtained by applying the join operator J_2 to the following tables.

Faculty name	Dept	Course code
Iyengar	Comp.Sci	CS002
Chandresekar	Physics	Ph003
Venkatesh	Mathematics	MA001
SaiGanesh	Chemistry	Ch005

Dept	Course code	Room No
Comp.Sci	CS002	C - 200
Mathematics	MA001	C - 201
Chemistry	Ch005	M - 203
Physics	Ph003	P - 200

FUNCTIONS

A function or mapping from a set A to a set B is a 'method' which pairs elements of the set A with unique elements of the set B and we denote $f: A \rightarrow B$ to indicate that f is a function from the set A to the set B.

B is called the codomain of the function f and A is called its domain. Also, for each element a of A, f defines an element b of B. We write $a \xrightarrow{f} f(a)$ or $a \xrightarrow{f} b, a \in A, b \in B$

Example

1. The relation $f = \{(1, d), (2, c), (3, a)\}$ from $A = \{1, 2, 3\}$ to $B = \{a, c, d\}$ is a function from A to B. The domain of f is A and the codomain of f is B.
2. The relation $f = \{(a, b), (a, c), (b, d),\}$ from $A = \{a, b\}$ to $B = \{b, c, d\}$ is not a function.

Range of function Let $f: A \rightarrow B$ be a function. The range of the function

$R(f) = \{f(a) / a \in A\}$. (Note that $R(f) \leq B$).

Note:

1. From Example 1: $R(f)$ is $\{d, c, a\}$
2. Let $f: IR^+ \rightarrow IR$ be $f(x) = x^2$ (R^+, the set of + ve real numbers). Clearly f is a function, whose domain is the set of positive real numbers and the codomain is the real numbers.

$$R(f) = \{x^2 / x \in R^+\} = \{1, 4, 9, ...\}$$

Let $f: A \rightarrow B$ be a function. f is said to be

i. **One-to-One (1-1) function** if $x_1 \neq x_2$ then $f(x_1) \neq f(x_2)$, $\forall\, x_1, x_2 \in A$.

or

Whenever $f(x_1) = f(x_2)$ then $x_1 = x_2$. This function is also known as injective function.

ii. **Onto (Subjective) function** if for every element y in the codomain B, atleast one element x in the domain A such that $f(x) = Y$.

or

if $R(f)$ = codomain B.

iii. **Bijective Function** if f is both 1–1 and onto function

iv. **Constant Function** if every element of the domain is mapped to unique element of the domain. (or the domain consists of only one element)

v. **Into function** if atleast one element of the codomain is not mapped by any element of the domain.

vi. **Identify Function** if $f(x) = x$, $\forall\, x \in B$, in this case $A \leq B$.
(Some times it is defined as $f: A \to A$ & $f(x) = x$, $\forall\, x \in A$.

Example

i. Let $f: R^+ \to R$ be a function defined as $f(x) = 2(x+2)$ clearly f is 1–1 because if $2(x+2) = 2(y+2)$
$\Rightarrow 2x+4 = 2y+4$
$\Rightarrow 2x = 2y \Rightarrow x = y$
$\therefore f$ is 1–1

ii. Define $f: R \to R^+$ by $f(x) = e^x$, $\forall\, x \in R$. Clearly f is 1-1. Because if $f(x_1) = f(x_2)$
$\Rightarrow e^{x_1} = e^{x_2}$
$\Rightarrow e^{x_1 - x_2} = 1$
$\Rightarrow x_1 - x_2 = 0$
$\Rightarrow x_1 = x_2$
$\therefore f$ is 1–1

iii. Let A = {5, 6, 7} and $B = \{a, b\}$. Then the mapping $f: A \to B$ is defined as $f(5) = a$; $f(6) = b$; $f(7) = a$. Clearly f is not 1–1. But f is onto.

iv. Consider the example (ii). if $f: R \to R^+$, defined by $f(x) = e^x$ is onto. Let x be any element IR^+ then $\log y \in IR$ such that $f(\log y) = e^{\log y} = y$

v. Define $f: Z_+ \to Z_+$ as $f(n) = n^2$, $\forall\, n \in Z_+$. Clearly f is an into mapping (for example 3 is not mapped by any element of Z_+) and 1–1 mapping but f is not onto.

vi. Define $f: Z \to Z$ by $f(n) = n+1$, $\forall\, n \in z$. Clearly f is 1–1 and onto For if, (i) $f(n) = f(m) \Rightarrow n+1, = m+1$
$\Rightarrow n = m \therefore f$ is 1–1

ii. If n is any element of Z then $n-1 \in Z$ such that $f(n-1) = n-1+1 = n$. Hence f is onto.

Note: A one-one mapping of a set S onto itself is sometimes called a permutation of the set S.

Inverse Functions and Compositions of Function

Let f be a bijective function from the set A to the set B. The inverse function of f is the function that assigns to an element $b \in B$ the unique element a in A such that $f(a) = b$. The inverse function of f is denoted by f^{-1}. Hence $f^{-1}(b) = a$ when $f(a) = b$.

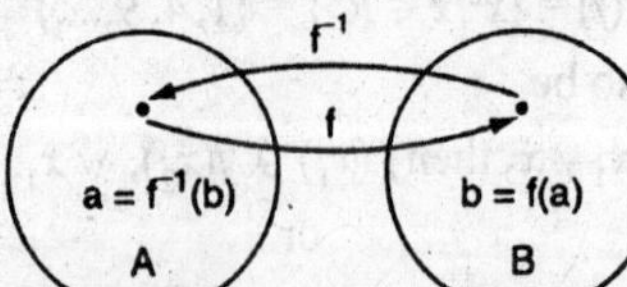

The function f^{-1} is the inverse function of f.

Note: A bijective function is called invertible since we can define an inverse of this function.

Example 1.16

i. Define $f: Z \to Z$ by $f(n) = n + 1$. Is f invertible, and if it is, what is its inverse?

Solution The function f has an inverse, since it is a bijective function refer example (iv)] Let y the image, of x, so that $y = x + 1$. Then $x = y - 1$ i.e., $y - 1$ is the unique element of Z that is sent to y by f. Hence $f^{-1} = y - 1$.

ii. Let $A = \{a, b, c\}$, & $B = \{5, 6, 7\}$. Define $F: A \to B$ as $f(a) = 5; f(b) = 6; f(c) = 7$. Is f invertible, and if it is, what is its inverse?

Solution Clearly the given function is bijective. The inverse function f^{-1} of f is given as $f^{-1}(5) = a$; $f^{-1}(6) = b; f^{-1}(7) = c$.

iii. Define $f: Z \to Z$ by $f(x) = x^2$. Is f invertible?

Solution Since $f(-2) = f(2) = 4$, f is not 1–1. If an inverse function were defined, it would have to assign two elements to 2. Hence f is not invertible.

Definition Let g be a function from the set A to the set B and let f be a function from the set B to the set C. The composition of the functions f and g denoted by $(f \circ g)$ is given by

$$(f \circ g)(x) = f(g(x)), \forall x \in A$$

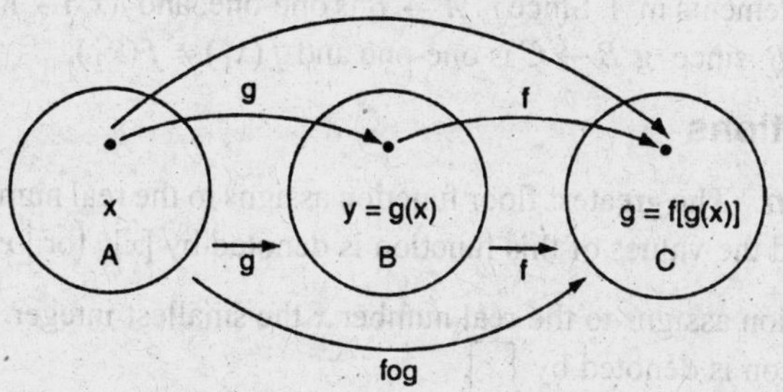

Example 1.17 Let $f: Z \to Z$ be a function defined by $f(x) = 2x + 3$: Let $g: Z \to Z$ be a function defined by $g(x) = 3x + 2$. Find (i) $f \circ g$ (ii) $g \circ f$.

Solution Both $f \circ g$ and $g \circ f$ are defined. Further,

i. $(f \circ g)(x) = f(g(x)) = f(3x+2)$
$= 2(3x+2) + 3 = 6x + 7$

ii. $(g \circ f)(x) = g(f(x)) = g(2x + 3)$
$= 3(2x + 3) + 2 = 6x + 11$

Eventhough $f \circ g$ and $g \circ f$ are defined, $f \circ g$ and $g \circ f$ need not be equal. i.e., the commutative law does not hold for the composition of functions.

Example 1.18 Let $A = \{1, 2, 3\}$, $B = \{x, y\}$, $C = \{a\}$ Let $f: A \to B$ be defined by $f(1) = x; f(2) = y; f(3) = x$. Let $g: B \to C$ be defined by $g(x) = a; g(y) = a$. Find (i) $f \circ g$ if possible (ii) $g \circ f$, if possible.

Solution

i. $(f \circ g)(x) = f(g(x))$ (by defenition) But f cannot be applied on C & hence $f \circ g$ is meaningless.

ii. $(g \circ f): A \to C$ is meaningful. Now $(g \circ f)(x) = g(f(x)), \forall x \in A$.

$\therefore (g \circ f)(1) = g(f(1)) = g(x) = a$
$(g \circ f)(2) = g(f(2)) = g(y) = a$
$(g \circ f)(3) = g(f(3)) = g(x) = a$

Result If $f: A \to B, g: B \to C$ and $h: C \to D$, then $(h \circ g) \text{ of} = h \circ (g \circ f)$.

Proof By the definition $[(h \circ g) \circ f](x) = (h \circ g)(f(x))$

$$= h(g(f(x))) \quad \to (1)$$

$$\text{and } [h \circ (g \circ f)](x) = h[(g \circ f)(x)]$$

$$= h[g(f(x))] \quad \to (2)$$

From (1) & (2), $(h \circ g) \text{ of} = h \circ (g \circ f)$

Result Let $f: A \to B$ and $g: B \to C$. Then

i. $g \circ f$ is onto if both f & g are onto.

ii. $g \circ f$ is 1–1 if both f & g are 1–1.

Proof

i. Let $Z \in C$. Since $g: B \to C$ is onto, F an element $y \in B$ such $g(y) = Z$. Since $f. A \to B$ is onto, for an element $x \in B$, such that $f(x) = y$

$\therefore (g \circ f)(x) = g(f(x)) = g(y) = Z$

$\therefore (g \circ f)$ is onto.

ii. Let $x_1 \neq x_2$ be two elements in A. Since $f: A \to B$ is one-one, and $f(x_1) \neq f(x_2)$, $g(f(x_1)) = g(f(x_2))$. Thus $g \circ f$ is one-one. In B. since $g:B \to C$ is one-one and $f(x_1) \neq f(x_2)$,

Some Important Functions

Greatest integer function The greatest floor function assigns to the real number x the largest integer that is less than or equal to x. and the values of this function is denoted by $[x]$. (or $\lfloor x \rfloor$)

Ceiling The ceiling function assigns to the real number x the smallest integer that is greater than or equal to x. The value of this function is denoted by $\lceil x \rceil$

Examples

i. $\lfloor x \rfloor = \lfloor 1/2 \rfloor = 0$; $\lceil x \rceil = \lceil 1/2 \rceil = 1$

ii. $\lfloor 5.6 \rfloor = 5$; $\lceil 5.6 \rceil = 6$

iii. $\lfloor 4.1 \rfloor = 4$; $\lceil 4.1 \rceil = 5$

iv. $\lfloor 3 \rfloor = 3$; $\lceil 3 \rceil = 3$

Example 1.19 Data stored on a computer disk or transmitted over a data network are represented as a string of bytes. How many bytes are required to encode 500 bits of data?

Solution To find the number of bytes needed, we determine the smallest integer that is atleast as large as the quotient when 500 is divided by 8, the number of bits in a byte.

$$\left\lceil \frac{500}{8} \right\rceil = \lceil 62.5 \rceil = 63 \text{ bytes is required}$$

Example 1.20 In ATM (asynchronous tranfer mode), data are organized into cells of 53 bytes. How many ATM cells can be transmitted in 2 minutes over a connection that transmits data at the rate of 200 KB per second.

Solution In 2 minutes, this connection can transmit $200000 \times 60 \times 2 = 2{,}40{,}00{,}000$ bits. Since each ATM cell is 53 bytes long $53 \times 8 = 424$ bits long.

No of ATM cells that can be transmitted in 2 minutes in the given connection is $\left\lceil \frac{2,40,00,000}{424} \right\rceil$

Modulus operator If x is a non-negative integer and y is a positive integer, we define x mod y to be the remainder when x is divided by y.

Example 11 mod 2 = 1; 5 mod 1 = 0; 365 mod 7 = 1.

Example 1.21

i. What day of the week will it be 365 days from friday?

Solution

i. 365 mod 7 = 1. Thus 365 days from friday, it will be saturday. [since 7 days after friday is friday. In general if K>o, $K \in Z$, after 7K days it is friday again]. Another important application of mod operator is ISBN (International standard Book Number). ISBN is a code of 10 characters separated by dashes such as 0-333-40733-7. It consists of four parts, a group code, a publisher code, a code that identifies the books among those books published by the particular publisher uniquely and a check character. This check character is used to validate an ISBN. [For example, 0 - 333 - 40736 - 7, the group code is 0, which identifics the book as one from an English speaking country. The publisher code 333 identifies the book as one published by Macmillan. The code 40736 uniquely identifies the book among those published by Macmillan. The check character is n mod 11, where n is the sum of the 1st digit plus two times the 2nd digit plus three times the 3rd digit,... nine times the ninth digit. If the value is 10, the check characters is x. In our example,

$n = 0 + 2.3 + 3.3 + 4.3 + 5.4 + 6.0 + 7.7 + 8.3 + 9.6 = 0 + 6 + 9 + 12 + 20 + 0 + 49 + 24 + 54 = 174$

$\therefore n$ mod 11 = 174 mod 11 = 7.

ii. ISBN is 0 - 07 - 003575 - x. Here 0-stands for the book as one from an English-speaking country. The publisher code 07 identifies the book as one published by McGraw Hill. The code 003575 uniquely identifies the book as one published by McGraw Hill. The check character is n mod 11, where

$n = 0 + 2.0 + 3.7 + 4.0 + 5.0 + 6.3 + 7.5 + 8.7 + 9.5 = 0 + 0 + 21 + 0 + 0 + 18 + 35 + 56 + 45$

$n = 175$

$\therefore n$ mod 11 = 175 mod 11 = 10.

$\therefore$ The check character is 10 i.e, it is represented as x.

Hashing Function

The central computer (server) at an university maintains records for each student. How many memory locations can be alloted so that the student records can be retrieved fastly? The solution to this problem is to use suitably chosen hasing function. Records are identified using a key, which uniquely identifies each student's record. For example, student records are identified using the Register number / Enrolment number of the student as the key. A hashing function h assigns memory location $h(p)$ to the record that has p as its key.

The most common hashing function is the function $h(p) = p$ mod q, where q is the number of available memory locations. [The hashing function should be *onto* so that all memory allocations are possible] [For example $q = 11$, the record of the student with register number 15 is assigned to memory location 4, since $h(15) = 4$ mod 11. Similarly, since $h(122) = 122$ mod 11 = 2, the record of the student with register number 121 is assigned to memory location 2.

Since a hashing function is not 1-1, more than one file may be assigned to a memory location. When this happens, we say that a collision occurs. To resolve a collision is to assign the first free location following the occupied memory location assigned by the hashing function. For instance, after making two earlier assignments we assign location 5 to the record of the student with register number 256.

CRYPTOLOGY

Cryptology is the study of secret messages. Here we will discuss Caesar cipher method. In this method,we can make meassages secret by shifting each letter (by K letters) forward in the alphabet (sending the last K letter of the alphabet to the first K letters).

For example, $K = 3$, then A is sent to D, B is sent to E, ..., X is sent to A. This is an example of encryption, that is the process of making a message secret. This can be represented mathematically as $f(p) = (p + k) \bmod 26$.

Note: The above function f that assigns to the non-negative integer p, $p \leq 25$, the integer $f(p)$ in the set $\{0, 1, ..., 25\}$ with $f(p) = (p + k) \bmod 26$.

Here we are replacing A by 0, K by 10, M by 12, Z by 25.

Example 1.22

1. What is the secret message produced from the message "NO ENTRY" using the Caesar cipher? (take $k=3$).

   Solution First replace the letters in the message with numbers. This yields 1341 413191724

   Now replace each of these numbers p by $f(p) = (p + k) \bmod 26$. This results in 1617 71622201

   Translating this back to letters produces the encrypted message "QR HQWUB".

2. What is the secret message produced from the message "BEST OF LUCK" using Caesar cipher method? (take $k = 5$)

   Solution First replace the letters in the message with numbers. This yields 141819 145 1120210

   Now replace each of these numbers p by $f(p) = (p + k) \bmod 26$. This results in 692324 1910 1625715

   Translating this back to letters produces the encrypted message. "GJXY TK QZHP"

Decryption

To recover the original message encrypted by Caesar cipher, the function f^{-1}, the inverse of f, is used. Note that the function f^{-1} sends an integer p from $\{0,1,...., 25\}$ + $f^{-1}(p) = (p - k) \bmod 26$. The process of finding the original message from the encrypted message is called decryption. Caesar's method and shift ciphers do not provide a high level of security. One approach that slightly enhances the security is to use a function of the form $f(p) = (kp + m) \bmod 26$ where k & m are integers, chosen such that 'f' is a bijection. Such a mapping is called an affine transformation.

Example 1.23

1. What letter replaces the letter M when the function $f(p) = (5p + 3) \bmod 26$ is used for encryption?

   Solution First, observe that 12 represents M. Then, using the encryption function specified, it follows that $f(12) = (5.12+3) \bmod 26 = 11$, since 11 represents J, M is replaced by J in the encrypted message.

2. What letter replaces the letter x when the function $f(p) = (2p + 5) \bmod 26$ is used for encryption?

   Solution First, observe that 23 represents x. Then, using the encryption function specified, it follows that $f(23) = (2.23 + 5) \bmod 26 = 25$. Since 25 represents Z, X is replaced by Z in the encrypted message.

EXERCISE VI

1. Find whether f is a function from the set of all integers Z to the set of all real members R if

   a. $f(n) = \pm n$; b. $f(n) = \sqrt{n^2 + 1}$ c. $f(n) = (n^2 - 4)^{-1}$

2. Find whether each of the following functions from $\{1, 2, 3, 4\}$ to itself is 1–1.
   a. $f(1)=2; f(2)=1; f(3)=3; f(4)=4$ b. $f(1)=2; f(2)=2; f(3)=4; f(4)=3$
   c. $f(1)=4; f(2)=2; f(3)=3; f(4)=4$.
3. Which functions in problem 2 are onto?
4. Find whether each of the following functions from Z to Z is 1–1.
   a. $F(x)=x-1$ b. $F(x)=x^2+1$ c. $F(x)=x^3$ d. $F(x)=\left\lceil \frac{x}{2} \right\rceil$
5. Which functions in problem 4 are onto?
6. Find whether each of the following functions is a bijection from R to R.
   a. $f(x)=-3+4$ b. $f(x)=-3x^2+7$ c. $f(x)=(x+1)/(x+2)$
   d. $f(x)=x^5+1$ e. $f(x)=2x+1$ f. $f(x)=x^2+1$ (g) $f(x)=x^3$
7. Let $A=\{-1, 0, 2, 4, 9\}$. Find $f(s)$ if
   a. $f(x)=1$ b. $f(x)=3x+1$ c. $f(x)=\left\lceil \frac{x}{2} \right\rceil$
8. If f and $f \circ g$ are 1–1, does it follow that g is 1–1? Justify your answer.
9. Find $f \circ g$ and $g \circ f$ where $f(x)=x^2+1$; $g(x)=x+2$ are functions from R to R.
10. Find (i) $f \circ g$ (ii) $g \circ f$ (iii) $f \circ f$ (iv) $g \circ g$ where $f(x)=x^2+3x+2$; $g(x)=2x-3$ are functions from R to R.
11. How many ATM cells can be transmitted in 15 seconds over a link operating at the following rates
    a. 1 megabit per second (1 Megabit = 1000000 bits) b. 128 KB per second
12. Find the inverse of the function $f(x)=x^3+1$.
13. If f is an invertible function from B to C and g is an invertible function from A to B. Prove that the inverse of the composition $(f \circ g)$ is given by $(f \circ g)^{-1} = g^{-1} o f^{-1}$.
14. The first nine digits of the ISBN of the 2nd edition of Discrete Mathematics by C.L. Liu are 0-07-100544. What is the check digit for that book.
15. The ISBN of Introduction to languages and the theory of computation (2nd ed.) is 0-07-46 Q 722-3, where Q is a digit. Find the value of Q.
16. Find whether the check character of the ISBN for the book "Systems - Programming" by Donovan was computed correctly by the publisher.
17. Which memory locations are assigned by the hashing function $h(k) = K \bmod 100$ to the records of students with the register numbers?
    a. 10457 b. 100321 c. 560076
18. A parking lot has 26 slots, numbered from 0 to 25, reserved for visitors. Visitors are assigned slots using the hashing function $h(k) = k \bmod 26$, where k is the number formed from the first two digits on a visitor's car number plate.
    i. What slots are assigned by the hashing function to cars that have the following first two digits on their number plates?
       01, 11, 21, 22, 99.
    ii. Describe a procedure that visitors should follow in a free parking space when collisions happen.
19. Encrypt the message "DO NOT PASS THROUGH" by translating the letters into numbers, applying the encryption function specified, and then translating the numbers back into letters
    a. $f(p)=(p+3) \bmod 26$ b. $f(p)=(p+9) \bmod 26$ c. $f(p)=(3p+11) \bmod 26$.
20. Decrypt the following messages encrypted using caeser cipher method $f^{-1}(p) = (p-3)$ mad 26.
    a. QR HQWUB b. PHHW BRX LQ WKH SDUN

MATHEMATICAL INDUCTION

Mathematical Induction is an important technique to prove assertions like '$n<2^n$, for all positive integers'. This technique is extensively used to prove results of discrete objects.

A proof by mathematical induction that $p(n)$ is true for every positive integer n consists of two steps

i. Basis step: The proposition $p(1)$ is shown to be true.

ii. Inductive step: The implication $p(n)\rightarrow p(n+1)$ is shown to be true for every positive integer n. Here, the statement $p(n)$ for a fixed positive integer n is called the inductive hypothesis.

Note: In a proof by mathematical induction it is not assumed that $p(n)$ is true for all positive integers. It is only shown that if it is assumed that $p(n)$ is true, then $p(n+1)$ is also true.

Example 1.24 Prove that $1+2+3+....+n=\frac{1}{2}n(n+1)$.

Solution Let $p(n)$ be the proposition that the sum of first n natural numbers is $\frac{1}{2}n(n+1)$. We must first complete the basis step; i.e. we have to show that $p(1)$ is true. Then we must show that $p(n+1)$ is true when $p(n)$ is assumed to be true.

Basis step: $p(1)$ states that the first natural number is 1. This is true.

Inductive step: To complete the inductive step we must show that the proposition $p(n)\rightarrow p(n+1)$ is true for every positive integer n.

Suppose $p(n)$ is true for a positive integer n;

$$\text{i.e. } 1+2+3+...+n=\frac{1}{2}n(n+1) \qquad \rightarrow (1)$$

Claim $p(n+1)$ is true, assuming that $p(n)$ is true. Adding $(n+1)$ both sides of the equation (1), we get

$$[1+2+3+...+n]=\frac{1}{2}n\ (n+1)+(n+1) \text{ i.e. } 1+2+...+n+(n+1)=\frac{1}{2}n\ (n+1)+(n+2)$$

$\therefore$ $p(n+1)$ is true, when $s(n)$ is true.

$\therefore$ $p(n)$ is true for all positive integers of n.

Example 1.25 Prove that the product of three consecutive integers is divisible by 3!

Solution Let $p(n)$ be the statement that $\frac{n(n+1)(n+2)}{3!}$ is an integer.

Basis $p(1)$ is the proposition that $\frac{1.2.3}{3!}$ is true.

Inductive step Assume that $p(n)$ is true. i.e. $\frac{n(n+1)(n+2)}{3!}$ is an integer.

Claim $\frac{(n+1)(n+2)(n+3)}{3!}$ is an integer, when $p(n)$ is true.

$$\text{Now } \frac{(n+1)(n+2)(n+3)}{3!}-\frac{n(n+1)(n+2)}{3!}=\frac{1}{6}(n+1)(n+2)(n+3-n)$$

$$=\frac{1}{2}(n+1)(n+2)$$

Since $(n+1)$ & $(n+2)$ are consecutive, one of them must be even and hence the final product is an integer.

$\therefore$ $p(n)$ is true for all positive integers.

Example 1.26 Prove that $1+2+2^2+...+2^n=2^{n+1}-1$ for all non-negative integers of n.

Solution Let $p(n)$ be the statement that this formula is correct for the integer n.

Basis $p(0)$ is true since $2^0 = 1 = 2^1 - 1$

Inductive step Assume that $p(n)$ is true i.e., $1 + 2 + 2^2 + ... + 2^n = 2^{n+1} - 1$ (1)

Claim $p(n + 1)$ is true.

Now $1 + 2 + 2^2 + ... + 2^n + 2^{n+1} = (1 + 2 + 2^2 + ... + 2^n) + 2^{n+1}$

$= (2^{n+1} - 1) + 2^{n+1}$ [by (1)]

$= 2.2^{n+1} - 1 = 2^{n+2} - 1$

$\therefore$ $p(n)$ is true for all $\forall\ n \geq$ o.

Example 1.27 Prove that $a + ar + ar^2 + ... + ar^n = \dfrac{ar^{n+1} - a}{r - 1}$, if $r \neq 1$.

Solution Let $p(n)$ be the statement that the sum of the first $(n + 1)$ terms of a glometric progression in this formula is correct.

Basis step: $p(0)$ is true, since $a = \dfrac{ar - a}{r - 1}$

Inductive step Assume that $p(n)$ is true.

i.e. $a + ar + ar^2 + ... + ar^n = \dfrac{ar^{n+1} - a}{r - 1}$ (1)

Claim $p(n)$ is true.

Add ar^{n+1} to both sides of equation (1), we get,

$$a + ar + ar^2 + ... + ar^n + ar^{n+1} = \frac{ar^{n+1} - a}{r - 1} + ar^{n+1} \quad (2)$$

Consider $\dfrac{ar^{n+1} - a}{r - 1} + ar^{n+1} = \dfrac{ar^{n+1} - a}{r - 1} + \dfrac{ar^{n+2} - ar^{n+1}}{r - 1} = \dfrac{ar^{n+2} - a}{r - 1}$ (3)

Now equation (2) becomes

$$a + ar + ... + ar^n + ar^{n+1} = \frac{ar^{n+2} - a}{r - 1}$$

$\therefore p(n + 1)$ is true. Hence $p(n)$ is true, $\forall\ n \geq$ o.

Example 1.28 Prove that a finite set A with n element has 2^n elements.

Solution Let $p(n)$ be the statement that a set A with n elements has 2^n elements.

Basis step: $p(0)$ is true, since a set with zero elements, the empty set, has exactly $2^0 = 1$ subset.

Inductive step: Assume that $p(n)$ is true. Under this assumption, we have to prove that $p(n)$ is true. i.e., every set with $n + 1$ element has 2^{n+1} subsets. Let B be a set with $n + 1$ elements. Now we can write $B = AU\{x\}$, where x is one of the elements of B and $A = b - \{x\}$. The subsets of B can be obtained as: For every subset of S of A, there are exactly two subsets of B, (i) S & (ii) $S \cup \{x\}$. These constitute all the subsets of B and are all distinct. Since tshere are 2^n subsets of s, there are $2.2^n = 2^{n+1}$ subsets of B.

$\therefore$ $p(n + 1)$ is true. Hence $p(n + 1)$ is true for all positive integers.

Example 1.29 Prove that the sum of the first n odd positive integers is n^2.

Solution Let $p(n)$ be the proposition that the sum of the first n odd positive integers is n^2.

Basis step $p(1)$ is true, since the sum of the first odd positive integer is 1.

Inductive step Assume that $p(n)$ is true, $n \in Z+$.

i.e $1+3+5+...+(2n-1)=n^2$. (1)

Claim $p(n+1)$ is true, assuming that $p(n)$ is true.

Now $1+3+5+...+(2n-1)+(2n+1)=[1+3+5+.....+(2n-1)]+(2n+1)$

$=n^2+(2n+1)$ [by (1)]

$=n^2+2n+1=(n+1)^2$

$\therefore p(n+1)$ is true. Hence $p(n)$ is true for all positive integers.

EXERCISE VII

Prove the following by Mathematical inducton.

1) Prove that $1^2+2^2+...+n^2 = \frac{1}{6}[n[n+1)(2n+1)], \forall n>0$.
2) Prove that $1^2+3^2+5^2+...+(2n+1)^2=(n+1)(2n+1)(2n+3)/3, \forall n \geq 0$.
3) Prove that $3^n < n! \forall n > 6$.
4) Prove that $1+\frac{1}{4}+\frac{1}{9}+...+\frac{1}{n^2}<2-\frac{1}{n}, \forall n>1$.
5) Prove that if $A_1, A_2,....,A_n$ and B are sets, then $(A_1 \cup A_2 \cup .... \cup A_n) \cap B=(A_1 \cap B) \cup (A_2 \cap B) \cup .... \cup (A_n \cap B)$
6) Let $a_1, a_2,....,a_n$ be positive real numbers. The arithmetic mean of these numbers is defined by $\bar{x}=\frac{1}{n}(a_1+a_2+...+a_n)$ and the geometric mean $G = (a_1a_2....a_n)^{\frac{1}{n}}$. By mathematical induction prove that $\bar{x} \geq G$.
7) Prove that 21 divides $4^{n+1}+5^{2n-1}, \forall n>0$.

SET RELATIONS AND FUNCTIONS

Recursive Functions

In this section we first define a class of functions inductively and show that any such function can be evaluated in a purely mechanical manner. We shall restrict ourselves to only those functions whose arguments and values are natural numbers.

Total and partial functions Any function $f: N^m = N \times N \times ...... \rightarrow N$ is called total if it is defined for every n-tuple in N^m. If $f: D \rightarrow N$ is defined, where $D \leq N^m$, then f is called partial.

Example If $f: N \times N \rightarrow$ by $f(x, y) = x + y$, then f is a total function. If $g: N \times N \rightarrow N$ by $g(xy) = x-y$, then g is a partial function, since it is defined only for x, y when $x, y \in N$.

Zero function A function $Z: N \rightarrow N$ by $Z(x) = 0$, for all $x \in N$, is called a zero function.

Successor function A function $S: N \rightarrow N$ by $S(x) = x + 1$ for all $x \in N$, is called successor function.

Projection function A function $U_i^n : N^n \rightarrow N$ by

$U_i^n (x_1, x_2, ..., x_i, ... x_n) = x_i$, for all $x_i \in N$, $i = 1, 2, ... n$, is called projection function.

These three functions are known as initial functions.

Now we extend the definition of composition of functions for more than one variable.

Let $f_1 : N \times N \to N, f_2: N \times N \to N$ and $g: N \times N \to N$ be any three functions. When $h: N \times N \to N$, the composition of g with f_1 and f_2 is given by

$h(x, y) = g(f_1(x, y), f_2(x, y))$ for all $x, y \in N$.

Here we assume that $R_{f1} \times R_{f2} \leq D_g$ and $D_h = D_{f1} \cap D_{f2}$

Example Define $f_1: N \times N \to N$ by $f_1(x, y) = x + y$,

$f_2: N \times N \to N$ by $f_2(x,y) = xy + y^2$ and $g: N \times N \to N$ by $g(x,y) = xy$. Then $h: N \times N \to N$ is

$$h(x, y) = g(f_1(x, y)), (f_2(x, y))$$
$$= g(x + y, x, y + y^2) = (x + y)(x, y + y^2)$$

similarly, we can extend this for more variables.

Given a function $f(x_1, x_2, \ldots, x_n)$ of n-variables, it is convenient to consider $n - 1$ of the variables as fixed and vary only the remaining variable over the set of N. For example, suppose $f: N \times N \to N$ is a function defined by $f(x, y) = x + y$. To compute $f(x, y)$, we treat x as fixed and vary f. Suppose $f(2, 0) = 2$ is given, to find $f(2, 3)$,

$$f(2, 3) = [(f(2, 0) + 1) + 1] + 1$$
$$= [(2 + 1) + 1] + 1 = [3 + 1] + 1$$
$$= 4 + 1 = 5$$

In general, given a function $f(x_1, x_2, \ldots, x_n)$ of n-variables, consider $n - 1$ of the variables as fixed and vary only the remaining variable over N. The fixed $n - 1$ variables are called parameters.

Recursive function Consider the known functions $g(x_1, x_2, \ldots x_n)$ and $h(x_1, x_2, \ldots, x_n, y, z)$ of n and $n + 2$ variables. Define a function $f(x_1, x_2, \ldots, x_n, y)$ of $n + 1$ variables by $f(x_1, x_2, \ldots, x_n, 0) = g(x_1, x_2, \ldots, x_n)$.

and $f(x_1, x_2, \ldots, x_n, y + 1) = h(x_1, x_2, \ldots, x_n, y, f(x_1, x_2, \ldots, x_n, y))$.

Then f is called a recursive function or recursion.

Note: Here the value of f at $y + 1$ is expressed in terms of the value of f at y. The variables $x_1, x_2 \ldots, x_n$ are treated as parameters.

Primitive recursive A function f is called primitive recursive if it can he obtained from the initial functions by a finite number of operations of composition and recursion.

Example 1.30 Show that the function $f(x, y) = x + y$ is primitive recursive. Use it to compute $f(2, 3)$.

Solution Since

$$x + (y + 1) = (x + y) + 1,$$
$$f(x, y+1) = f(x, y) + 1 = S(f(x, y))$$

Also, $f(x, 0) = x$.

Now we define $f(x,y)$ as

$$f(x, 0) = x = \cup_1'(x)$$
$$f(x, y + 1) = S(\cup_3^3(x, y, f(x, y))$$

Therefore, f is primitive recursive. Now

$$f(2, 3) = S(f(2, 2))$$
$$= S(S(f(2, 1))$$

$= S(S(S(f(2, 0))))$

$= S(S(S(2)))$ Since $f(2, 0) = 2$.

$= S(S(3))$

$= S(4)$

$= 5$

Example 1.31 Using recursion, define the multiplication function '$*$' given by $g(x, y) = x * y$.

Solution We have

$$g(x, 0) = 0$$
$$g(x, y+1) = x * (y+1)$$
$$= (x * y) + x = g(x, y) + x$$

So we define,

$$g(x, 0) = 0 = z(x)$$
$$g(x, y+1) = f(\cup_3^3(x, y, g(x, y)), \cup_1^3(x, y, g(x, y))),$$

where $f(x, y) = x + y$ is primitive recursive. (see previous example).

Example 1.32 Show that the function $f(x) = k$, where k a constant, is primitive recursive.

Solution Suppose $k = 0$, Then.

$$f(x) = 0 = z(x)$$

Otherwise,

$$f(x+1) = k = f(x)$$
$$= \cup_2^2(x, f(x)).$$

so $f(x)$ is primitive recursive.

Example 1.33 Show that the function $x!$ is primitive recursive, where $0! = 1$ and $n! = n^* (n-1)!$.

Solution Let $f(x) = x!$ Then

$$f(0) = 0! = 1 = S(0) \text{ and } f(x+1) = (x+1)!$$
$$= (x+1) * x! = (x+1) * f(x)$$
$$= (x * f(x) + f(x)$$
$$= \cup_1^2(x, f(x)) * = \cup_2^2(x, f(x)) + = \cup_2^2(x, f(x)$$

Since the sum (+) and product ($*$) are primitive recursive, $f(x)$ is primitive recursive.

The following are some of the primitive recursive functions which are used frequently.

Sign function (S_g)

$S_g(0) = 0, \quad S_g(y+1) = 1$

(Since $S_g(0) = Z(0), S_g(y+1) = S(Z\cup_2^2(y, S_g(4)))$

Zero test function ($\overline{S_g}$)

$\overline{S_g}(0) = 1, S_g(y+1) = 0$

(Since $\overline{S_g}(0) = s(0), \overline{S_g}(y+1) = Z(\cup_2^2(Sg(y))$

Predecessor function (P)

$P(o) = 0; P(y+1) = y = \cup_1^2(y, p(y))$

(Since $P(0) = 0, P(1) = 0, P(2) = 1, P(3) = 2, \ldots$)

Odd and even parity function (Pr)

$Pr(0) = 0, Pr(y+1) = \overline{S_g}(\cup_2^2(y, Pr(y)))$

(Since $Pr(0) = 0, Pr(1) = 1, Pr(2) = 0, Pr(3) = 1, \ldots$)

Proper subtraction function (∸)

$x \dot{-} 0 = x, x \dot{-} (y+1) = P(x \dot{-} y)$

(Since $x \dot{-} y = 0$ for $x<y$ and $x \dot{-} y = x - y$ for $x >, y$).

Absolute value function (||)

$|x - y| = (x \dot{-} y) + (y \dot{-} x)$

Minimum of x and y (Min (x,y))

$\text{Min}(x, y) = x \dot{-} (x \dot{-} y).$

Maximum of x and y (Max (x,y))

$\text{Max}(x, y) = y + (x \dot{-} y)$

Square function (y^2)

$f(y) = y^2 = \cup_1'(y) * \cup_1'(y)$

Example 1.34 Show that $f(x, y) = x^y$ is a primitive recursive function.

Solution Note that, $x^0 = \begin{cases} 1 & \text{for } x \neq 0. \\ 0 & \text{for } x = 0. \end{cases}$

Also, $x^{y+1} = x^y * x$

$= f(x, y) * x$

$f(x, 0) = S_g(x)$

$f(x, y+1) = \cup_3^3(x, y, f(x, y) * \cup_1^3(x, y, f(x,y)))$. Therefore, $f(x, y)$ is primitive recursive.

Example 1.35 Show that the function $\left[\frac{x}{2}\right]$ which is equal to the greatest integer is primitive recursive.

Solution Note that,

$$\left[\frac{x}{2}\right] = \frac{x}{2} \text{ when } x \text{ is even}$$

$$= \frac{x-1}{2} \text{ when } x \text{ is odd.}$$

So we define, $[0/2] = 0 = z(x)$

$\left[\frac{y+1}{2}\right] = \left[\frac{y}{2}\right] + p_r(y)$. Therefore $\left[\frac{x}{2}\right]$ is primitive recursive.

Example 1.36 Show that the function $f(x) = \begin{cases} \frac{x}{2} \text{ when } x \text{ is even} \\ \frac{x-1}{2} \text{ when } x \text{ is odd} \end{cases}$ is primitive recursive. (M.E. May 99)

Solution Since

$f(0) = 0 = z(x)$

$f(y+1) = f(y) + P_r(y)$

f is primitive recursive. (see the previous example).

Example 1.37 Show that $f(x, y) = x^y + y^x$ is primitive recursive. (M.E, May 98, 2000)

Solution We know that x^y is primitive recursive (refer Example). Similarly y^x is also primitive recursive, by Example, the sum $x^y + y^x$ is also a primitive recursive function.

Example 1.38 Show that if $f(x + y)$ defines the remainder upon devision of y by x, then it is a primitive recursive function.

Solution Consider the following case for $f(x, y)$.

$f(5, 11) = \text{Remainder of} \left(\frac{11}{5}\right) = 1$

Similarly, $f(5,12) = 2$

$f(5,13) = 3$

$f(5,14) = 4$

$f(5,15) = 0$

For $y = 0$, $f(x,0) = 0$ and the value of $f(x, y)$ increases by 1 when y is increased by 1, until the value becomes equal to x; in which case it is put equal to 0, and the process continuous. We therefore build a function which increases by 1 each time y increases by 1, i.e. $S(f(x,y))$. Now we multiply this function by another primitive recursive function which becomes 0 whenever $S(f(x, y))'\ x$, but $S(f(x, y))$ is always $\leq x$ and hence, such a function is

$S_g(x \dot{-} S(f(x, y))$

Therefore, $f(x, 0) = 0$

$f(x, y+1) = S(f(x, y)) * S_g(x \dot{-} S(f(x, y)))$

Hence $f(x, y)$ is a primitive recursive function.

Characteristic function The characteristic function of a relation R is defined as

$$X_R(x_1, x_2, ..., x_n) = \begin{cases} 1 \text{ if } (x_1, x_2, ... x_n) \in R \\ 0 \text{ if } (x_1, x_2, ... x_n) \notin R \end{cases}$$

Here X_R is a function from R to N and $R \leq N^n$, an n-ary relation.

Primitive recursive A relation R is said to be primitive recursive if its characteristic function is primitive recursive.

Example 1.39 Show that $\{x, x) / x \in N\}$ which defines that the relation of equality is primitive recursive.

Solution Let $R = \{(x, x)/x \in N\}$. Here, we have to find a function $f(x, y)$ such that $f(x, y) = 1\ y(x, y) \in R$ and 0 if $(x, y) \notin R$. Such a function is

$f(x, y) = \overline{S_g}\ (|x - y|)$

for $x = y$, $\overline{S_g}\ (|x-x|) = \overline{S_g}\ (0)=1$ and for $x \neq y$, $\overline{S_g}(|x-y|)= 0$. Therefore $f(x, y)$ is the required characteristic function, which is primitive recursive. Hence R is primitive recursive.

Example 1.40 Show that for any fixed k the relation given by $R = \{(k, y)/y>k\}$ is primitive recursive.

Solution We have to find a function such that it is equal to 1 if $y > k$ and 0 if $y \leq k$. Such a function is $S_g(y \dot{-} k)$, since $y \dot{-} k = 0$ for $y \leq k$ and $y \dot{-} k = y - k$ for $y > k$. Therefore, of $R(x, y) = S_g(y \dot{-} k)$ is primitive recursive and hence R is primitive recursive.

Example 1.41 Show that the function $f(x_1, x_2, y)$ defined by $f(x_1, x_2, y) = \begin{cases} x_2 \text{ for } x_1 > y \\ (x_1 * y) + x \text{ for } x_1 \leq y \end{cases}$

is primitive recursive.

Solution Here we have to show that

$$f(x_1, x_2, y) = x_2 + (x_1 * y) * (a \text{ function})$$

and that function is zero when $x_1 > y$ and 1 when $x_1 \leq y$. Such a function is $\overline{S_g}\ (x_1 \dot{-} y)$, since $x_1 \dot{-} y = 0$ for, $x_1 \leq y$ and $x_1 - y$ for $x_1 > y \dot{-}$

Hence $f(x_1, x_2, y) = x_2 + (x_1 * y) * \overline{S_g}\ (x_1 \dot{-} y)$ is primitive recursive.

Regular function Let $g(x_1, x_2,...,x_n,y)$ be a total function. If there exists atleast one value of y, say $\bar{y} \in N$ such that the function $g\ (x_1,x_2,...,x_n, \bar{y}) = 0$ for all $(x_1, x_2,...,x_n) \in N^n$, then g is called a regular function.

Example Consider the function $g(x, y) = y \dot{-} x$. Since $g(x, 0) = 0 \dot{-} x = 0$ for all x, $g\ (x, y)$ is regular.

Example Consider the total function $g(x, y) = |y^2 - x|$. Here $g(x, y) = 0$ for only those values of x which are perfect squares. Therefore, no values of y in N is such that $|y^2 - x| = 0$ for all x. Hence g is not regular.

Total function A function $f(x_1, x_2,...,x_n)$ is said to be defined a total function $g(x_1,x_2,...,x_n, y)$ by minimization if

$$f(x_1, x_2,..., x_n) = \begin{cases} \because_y (g(x_1,x_2,..., x_n,y) = 0) \\ \text{undefined} \end{cases} \quad \text{Otherwise, if there exists such a } y.$$

where $\because_y$ means the least y greater than or equal to zero.

Recursive function A function is said to be recursive if it can be obtained from the initial functions by a finite number of applications of the operations of composition, recursion and minimization over regular function.

Partial recursive function A function is said to be partial recursive if it can be obtained from the initial functions by a finite number of applications of the operations of composition, recursion and minimization.

Example 1.42 Show that the function $f(x)= \dfrac{x}{2}$ is a partial recursive function.

Solution Let $g\ (x, y) = |2y - x|$. It is not regular because $|2y - x| = 0$ only for even values of x.

Define $f(x) = \mu_y(|2y - x| = 0)$

$$= \frac{x}{2} \text{ for } x \text{ is even.}$$

Here, the smallest y for which $|2y - x| = 0$ is $\dfrac{x}{2}$

Therefore, g is partial recursive.

Example 1.43 Let $\left[\sqrt{x}\right]$ be the greatest integer $\leq \sqrt{x}$. Show that $\left[\sqrt{x}\right]$ is primitive recursive.

Solution We know that

$$(y+1)^2 \dot{-} x = \begin{cases} 0 \text{ for } (y+1)^2 < x \\ \text{not zero for } (y+1)^2 \geq x. \end{cases}$$

$$\text{Therefore } \overline{s_g}((y+1)^2 \dot{-} x = \begin{cases} 1 \text{ if } (y+1)^2 < x \\ 0 \text{ if } (y+1)^2 \geq x. \end{cases}$$

The smallest value of y for which $(y+1)^2 \geq x$ is $\sqrt{x}$,

Since $\left([\sqrt{x},+1]^2 > x\right)$. Hence $[\sqrt{x},] = \mu_y(\overline{S_g}\ (y+1)^2 \dot{-} x) = 0)$

and since $\left[\sqrt{x}\right]$ is defined for all x, $\left[\sqrt{x}\right]$ is a recursive function.

Characteristic function Let U be an universal set and A be a subset of U. The function $Y_A : \cup \rightarrow [0,1]$ is defined by $Y_A(x) = \begin{cases} 1 \text{ if } x \in A \\ 0 \text{ if } x \notin A \end{cases}$ is called the characterstic function of the se t A.

Recursive Any set A is called recursive (partial recursive) if its characteristic function Y_A is recursive (partial recursive).

Note: If a set A is recursive, then Y_A is recursive. Since $Y_{\overline{A}} = 1 \dot{-} Y_A = \overline{S_g}\ (Y_A)$, $\overline{A}$ is also recursive. If A and B are recursive, then $A \cap B$ and $A \cup B$ are also recursive since, $Y_{A \cap B} = Y_A * Y_B$ and $Y_{A \cup B} = (Y_A + Y_B) \dot{-} Y_{A \cap B}$

Example 1.44 Show that the set of even and odd natural numbers are both primitive recursive. (M.E. May 98)

Solution Let E be the set of odd natural numbers.

$$\text{since } Y_E(x) = \begin{cases} 1 \text{ if } x \in E \\ 0 \text{ if } x \notin E \end{cases} = P_r(x)$$

$Y_E(x)$ is the required characteristic function which is primitive recursive. Hence the set E is primitive recursive. Since $\overline{E}$ is the set of even natural numbers, it is also primitive recursive.

Example 1.45 Show that the set of divisors of a positive integer n is recursive. (M.E. May 99)

Solution If x divides n then $\dfrac{n}{x} = i, 1 \leq i \leq n$.

i.e. $x * i = n$. So $x \leq n$ is a division of n if $|x * i - n| = 0$ for one fixed value of i, and if x is not a divisor, $|x * i - n| \neq 0$ for all $1 \leq i \leq n$. Let B be the set of divisors of n. consider

$$\sum_{i=1}^{n} \overline{S_g}\ |x * i - n| = \begin{cases} \overline{S_g}\ (0) = 1 \text{ if } x \text{ is a divisor} \\ \overline{S_g}\ (\text{a non zero}) = 0 \text{ if } x \text{ is not a divisor} \end{cases}$$

$$= \begin{cases} 1 \text{ if } x \in B \\ 0 \text{ if } x \notin B \end{cases}$$

$$= Y_B(x)$$

where B is the set of divisors of n. Hence the set B is recursive.

Example 1.46 Let $D(x)$ devote "number of divisors of x". Show that $D(x)$ is primitive recursive.

Solution We know that, the function which defines the remainder upon division of y by x is primitive recursive. we shall denote such a function by $rm(x, y)$. If a number x divides y, then $rm(x, y) = 0$ and $\overline{S_g}\ rm(x, y) = 1$. So that the number of divisors for y is given by

$$D(y) = \sum_{x=1}^{y} \overline{S_g}\left(rm(x, y)\right)$$

Therefore, $D(y)$ is primitive recursive.

Example 1.47 Show that the predicate"x is a prime" is primitive recursive.

Solution A number $x(\neq 0, \neq 1)$ is a prime if it has only two divisors 1 and x. Let $\overline{p_r}$ be the set containing non primes. Since

$$S_g(D(x) \dot{-} 2) + \overline{S_g}\,(|x-1|) + \overline{S_g}\,(|x-0|)$$

$$= \begin{cases} 1 \text{ if } x \in \overline{pr} \\ 0 \text{ if } x \notin \overline{pr} \end{cases}$$

$$= Y_{\overline{\text{Pr}}}(x),$$

$Y_{\overline{\text{P}_r}}$ is primitive recursive. But $Y_{\overline{\text{P}_r}}(x) = 1 \dot{-} Y_{\overline{\text{Pr}}}(x)$,

$Y_{\overline{\text{P}_r}}(x)$ is also primitive recursive. Therfore P_r is primitive recursive.

Ackermann's function The function $A(x, y)$ defined by

$$A(0, y) = y + 1$$

$$A(x+1, 0) = A(x, 1)$$

$$A(x+1, y+1) = A(x)A(x+1, y))$$

is known as Ackermann's function.

Note $A(x, y)$ is well defined and total. Further, $A(x, y)$ is not primitive recursive but recursive.

Now we demonstrate, how he above definition can be used in finding the value of $A(3, 3.)$

$A(3, 3) = A(2, A\,(3, 2))$

$A(3, 2) = A(2, A\,(3, 1))$

$A(3, 1) = A(2, A\,(3, 0))$

$\quad = A(2,\ A(2, 1)) = A(2, A(0, 4)) = A(2, 5)$

Since $A(2, 1) = A(1, A\,(2, 0))$

$\quad = A(1, A\,(1, 1))$

$A(1, 1) = A(0, A(1, 0))$

$\quad = A(0, A(0, 1)) = A(0,\ 2) = 3$

Now, $A(2, 1) = A\,(1, 3)$

$\quad = A(0, A\,(1, 2))$

$A(1, 2) = A(0, A\,(1, 1))$

$\quad = A(0, 3) = 4$

so, $A(2, 1) = A(0, 4) = 5$

Again, $A(2,2)=A(1,5)=A(0,A(1,4))$

$A(1,4)=A(0,A(1,3))$

$A(1,3)=A(0,A(1,2))=A(0,4)=5$

$A(1,4)=A(0,5)=6$

$A(2,2)=A(0,6)=7$

Again, $A(2,5)=A(1,A(2,4))$

$A(2,4)=A(1,A(2,3))$

$A(2,3)=A(1,A(2,2))=A(1,7)$

$A(1,7)=A(0,A(1,6))$

$A(1,6)=A(0,A(1,5))$

$A(1,5)=A(0,A(1,4))=A(0,6)=7$

$A(1,6)=A(0,7)=8$

$A(1,7)=A(0,8)=9$

so, $A(2,3)=9$

Similarly,

$A(2,4)=11$

$A(2,5)=13$

so, $A(3,1)=13$

Now, $A(3,2)=A(2,13)=29$ by similar method.

Finally, $A(3,3)=A(2,29)=61$

Since $A(2,n)=2n+3$.

Note that $A(0,n)=n+1$

$A(1,n)=n+2$

$A(2,n)=2n+3$

EXERCISE VIII

1. Prove that the function $f(x,y)=x-y$ is partial recursive
2. Show that the quotient function $g(x,y)$ = quotient upon division of y by x is primitive recursive.
3. If $\pi(x)$ = number of primes $\le x$, show that $\pi(x)$ = is primitive recursive.
4. Show that every finite set is primitive recursive.
5. Prove that Ackermann's function $A(x,y)$ is not primitive recursive, but only recursive.

PERMUTATIONS AND COMBINATIONS

Permutations

1. If there are things represented by α, β, γ, then the selections that can be made from these, taken two at a time are $\beta\gamma, \gamma\alpha, \alpha\beta$.

   If however we take into account, the order or the arrangement in each of the selections we have $\beta\gamma, \gamma\beta, \gamma\alpha, \alpha\gamma, \alpha\beta, \beta\alpha$ as the different arrangements of three things taken two at a time.

The different selections that can be obtained by taking the same number r of things from a given collection of n different things, without any regard to the order of the things, are called the combination of n things taken r at a time and the number of such combination is denoted by nC_r. The different arrangements that can be obtained by taking the same number r of things from a given collection of n different things, are called the Permutations of n things taken r at a time, and the number of such permutations is denoted by nP_r.

2. The number of ways of selecting one thing at a time from n different things is n and hence $nC_1 = n$.

   The number of ways of selecting all the n things is 1 and hence $nC_n = 1$.

   The number of ways of arranging one thing at a time from n different things is n, and hence $nP_1 = n$.

Note: In nC_r and nP_r, r must be necessarily less than n and its maximum value is n.

3. If a certain operation can be performed in any one of the m ways and then, a second operation can be performed in anyone of the n ways, then both the operations can be performed in anyone of the mn ways.
4. The number of permutations of n different things taken r at a time is $n(n-1)(n-2)\ldots\ldots\ldots(n-r+1)$.

   The number of permutations of n things taken r at a time is the same as the number of ways in which r blank spaces can be filled up by the n things.
5. We have $nP_n = n(n-1)(n-2)\ldots\ldots\ldots$ to n factors $= n(n-1)(n-2)\ldots 3\cdot 2\cdot 1 = n!$

   Also, we have

$$nP_r = n(n-1)(n-2)\ldots\ldots\ldots(n-r+1)$$

$$= \frac{n(n-1)(n-2)\ldots\ldots(n-r+1).(n-r)\ldots\ldots\ldots 2\cdot 1}{(n-r)(n-r-1)\ldots.2\cdot 1} = \frac{n!}{(n-r)!}$$

Example 1.48 How many numbers between 1,000 and 10,000 can be formed with the digits 1, 3, 5, 7, 9, each digit being used only once in each number ?

Solution Each number should consist of 4 digits and the required number is the same as the number of permutations of 5 different things taken 4 at a time = $5P_4$=120.

Example 1.49 Eleven papers are set for the engineering examination of which two are in Mathematics. In how many ways can the papers be arranged if the two Mathematics papers do not come together ?

Solution We shall find the total number of ways of arranging all the papers if there is no restriction and subtract from this the number of ways in which the two Mathematics papers come together.

The total number of ways in which all the papers can be arranged if there is no restriction = 11!

To find the number of ways in which Mathematics papers come together, consider that the two papers are bound together ; this can be done in 2! ways. Now the number of ways in which the resulting 10 papers can be arranged is 10! hence the total number of ways in which the Mathematics papers come together is $2! \times 10!$. Hence the number of ways in which the Mathematics papers do not come together is $11! - 2! \times 10! = 9! \times 10!$

Example 1.50 There are 35 micro computers in a computer centre. Each microcomputer has 18 ports. How many different ports are there in the centre?

Solution The procedure for choosing a port consists of two jobs, first picking a microcomputer and then picking, a port on this microcomputer. Since there are 35 ways to choose the microcomputer and 18 ways to choose the port no matter which microcomputer has been choosen, by rule 3, there are $35 \times 18 = 630$ ports.

Example 1.51 How many functions are there from a set with P elements to one with Q elements ?

Solution A function corresponds to a choice of one of the Q elements in the co-domain for each of the P elements in the domain. Hence by rule 3, there are QP functions from a set with P elements to one with Q elements.

Example 1.52 In how many ways can the letters of the word EDINBURGH be arranged.

i. With the vowels only in the odd places;
ii. Beginning and ending with vowels;
iii. Beginning and ending with consonants;

Solution

i. There are five odd places and as the three vowels should be in these places only, they can be first arranged $5P_3 = 60$ ways.

When the vowels have been arranged in any one way as shown below

1 2 3 4 5 6 7 8 9
u *e*

the remaining six places are to be filled up by the six consonants and this can be done in $6! = 720$ ways. Hence the total number of arrangements is $60 \times 720 = 43{,}200$ ways.

ii. The first and last places should be occupied by vowels and this can be done in $3P_2 = 6$ ways. Further for each of these ways the other 7 letters can be arranged in $7! = 5040$ ways. Hence the total number of arrangements is $= 6 \times 5040 = 30240$

iii. The total number of ways $= 6P_2 \times 7! = 30 \times 5040 = 151200$

Example 1.53 How many numbers between 5,000 and 10,000 can be formed from the digits 1, 2, 3, 4, 5, 6, 7, 8, 9, each digit not appearing more than once in each number?

Solution The first digit from the left may be 5 or 6 or 7 or 8 or 9; and so the first place from the left can be filled in 5 different ways; as the number should consist of 4 digits, the remaining 3 digits can be arranged in $8P_3 = 336$ ways. Hence the total number of numbers that can be formed is $5 \times 336 = 1{,}680$.

Circular Permutations

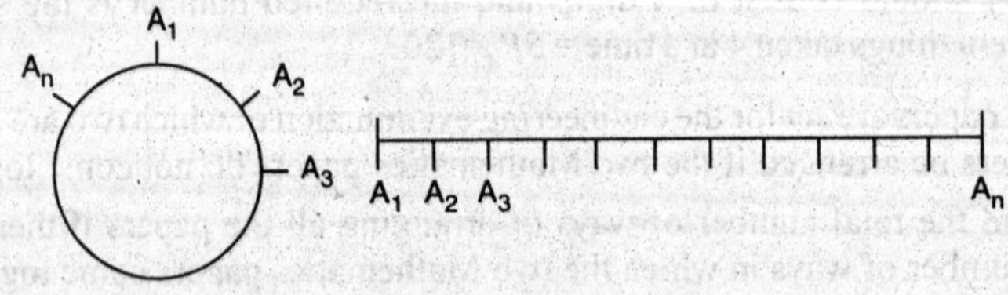

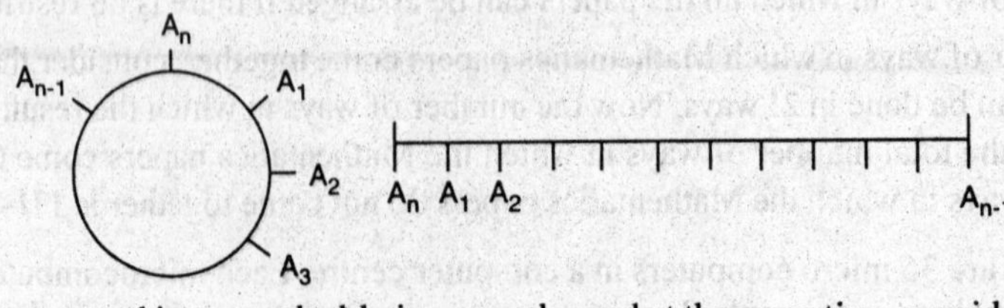

Consider n persons seated in a round table in any order and at the same time consider them arranged in the same order in a line, as shown above.

The number of circular arrangements of n persons is $\frac{n!}{n} = (n-1)!$

Example 1.54 In how many ways can 6 different beads be strung together to form a necklace?

Solution The number of circular permutations of 6 different things is 5!

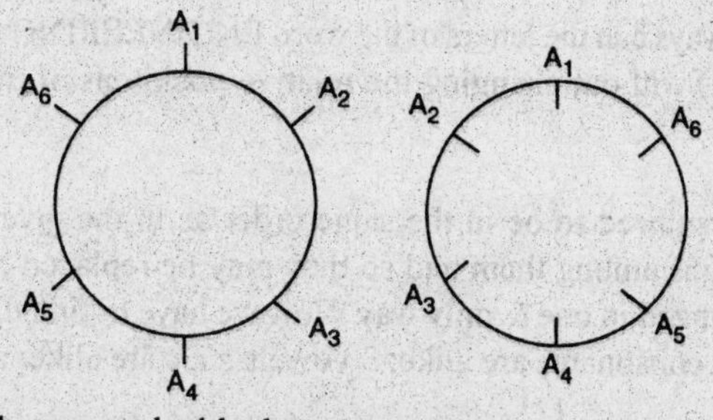

When 6 persons are seated at a round table the two arrangements shown above will have to be considered different; but in the case of the necklace, the above arrangements can be obtained by simply turning over the first arrangement same and so must be considered identical.

Hence the 5! circular permutation gives us only $\frac{1}{2}.5! = 60$ ways in which a necklace of 6 beads can be formed.

Permutations with Repetitions

The number of permutations of n different things taken r at a time when the things can be repeated any number of times is n^r.

Note: The total number of permutations of n things taken 1,2,3,....., r at a time when the things may be repeated any number of times is $n + n^2 + ...+ n^r = n.\dfrac{n^r - 1}{n-1}$

Example 1.55 How many numbers of four digits can be formed with the digits 1, 2, 3 ? Find the sum of all such numbers.

Solution The number of four digit numbers $= 3 \times 3 \times 3 \times 3 = 81$

If the numbers are all written down we find that any one of the digits 1, 2, 3 occurs in the units place in $\dfrac{81}{3} = 27$ times. Hence the total sum of the digits in the units place is $27 (1 + 2 + 3) = 162$. This is also the sum of all digits in the tens, hundreds and thousands places.

Hence the required sum is $162 + 162 \times 10 + 162 \times 162 \times 1000 = 179982$.

Permutations when all the Things are not Different

The number of permutations of n things taken together when all the things are not different is given by $\dfrac{n!}{p!q!r!...}$

Here, among the n things, p things of them are of one kind; q of them are of second kind and so on.

$$(p + q + r + ... = n)$$

Example 1.56 How many different words can be made out of the letters which form the word ALLAHABAD?

Solution There are 9 letters of which 4 letters are of one sort (A, A, A, A); 2 are of second sort (L, L); 1 is of third sort (H); 1 is of fourth sort (B); and 1 is of different sort (D).

The required number $= \dfrac{9!}{4!2!}$

Example 1.57 **In how many ways can the letters of the word ENGINEERING be arranged (i) without changing the order of the consonants (ii) without changing the relative posititons of the vowels and the consonants.**

Solution

i. The consonants are required to be in the same order as in the given word and so there can be no interchange of positions among them and so they may be replaced by letters say *c,c,c,c,c,c*. For all these 6 c's can be arranged in one & only way. Now we have to find the number of permutations of 11 letters of which the 6 consonants are alike; 3 vowels *e,e,e* are alike; and the two vowels *i, i* are alike.

Hence the required number $= \dfrac{11!}{6!3!2!} = 4{,}620.$

ii. The places originally occupied by vowels must always be occupied by vowels and those occupied by consonants, always by consonants. The vowels *e, e, e, i, i* can be arranged among themselves in $\dfrac{5!}{3!2!} = 10$ ways and the consonants *n, n, n, g, g, r* can be arranged among themselves in $\dfrac{6!}{3!2!} = 60$ ways.

Hence the required number is $10 \times 60 = 600$ ways.

EXERCISE IX

1. How many numbers of 3 digits can be formed with 0, 2, 4, 6, 8 ?
2. Numbers are formed by using all the digits 1, 2, 5, 6, 9. Each digit appearing only once in each number, how many of these will be
   a. divisible by 5
   b. divisible by 25
   c. odd
   d. even
   e. greater than 50,000
3. A family consisting of 7 members are to be arranged for a photograph. In how many of these arrangements
   a. The 3 sisters come together
   b. The three sisters not coming together
   c. The 4 brothers come together
   d. Four brothers not coming together
   e. The sisters occupy the middle seats
   f. The sisters occupy the odd seats
   g. The sisters occupy the even seats?
4. Find the number of different arrangements that can be made out of the 7 rainbow colours, so that the blue and green colours not shall come together.
5. In how many ways can 7 gentlemen and 7 ladies sit together at a round table so that
   a. no two ladies may be together
   b. the ladies are together.
6. In how many ways can 8 different flowers be strung together to form a garland?
7. Find the number of different signals that can be made with 6 flags, 2 of which are white, 2 are black and 2 are red. Six flags hoisted side by side to be used for each signal.
8. In how many ways can 6 different question papers be distributed for 6 days so that the English paper may always be given before the Mathematics paper and the Mathematics paper be given before the Vernacular papers? In how many of these shall be Vernacular papers immediately follow from the mathematics paper.
9. Find the number of permutations taking together all the letters of the following words.
   a. ARRANGEMENT b. EXAMINATION c. MATHEMATICS d. VEROMOUSNESS.
10. How many different numbers, each of six digits, can be expressed by means of the digits of the number 121202?

Combinations

The number of combinations of n different things taken r at a time is $nc_r = \dfrac{n(n-1)(n-2)......(n-r+1)}{1.2.3....r}$

Example 1.58 How many diagonals are there in a polygon of n sides ?

Solution The number of diagonals is the number of straight lines joining any two consecutive vertices excepted; and hence the required number $= nC_2 - n = \dfrac{n(n-1)}{2} - n = \dfrac{n^2 - 3n}{2}$.

Example 1.59 There are n stations on a Railway. Prove that a train can stop at three of these in $\frac{1}{6}(n-2)(n-3)(n-4)$ ways, no two of the stopping stations being consecutive.

Solution Given that there are 3 stopping stations and $(n-3)$ green flags and 3 red flags such that no two red flags come together

First, the $(n-3)$ greeen flags are arranged on a line. This can be achieved in one way. Now there are $(n-2)$ spaces where the 3 red flags can be put such that no two red flags are together. This can be done in $(n-2)\,C_3$ ways.

i.e. $(n-2)\,C_3 = \frac{1}{6}(n-2)(n-3)(n-4)$ ways.

Example 1.60 The English language consists of 21 consonants and 5 vowels. How many 5 lettered words, consisting of atleast a vowel and two consonants, can be formed from them ?

Solution Each 5 - lettered word might consists of

i. 1 Vowel & 4 consonants
ii. 2 Vowels & 3 consonants
iii. 3 Vowels & 2 consonants

Now

i. gives rises to $5C_1 \times 21C_4 \times 5!$
ii. gives rises to $5C_2 \times 21C_3 \times 5!$
iii. gives rises to $5C_3 \times 21C_2 \times 5!$

$\therefore$ The required number of words is 5,439,000.

Example 1.61 Find the number of permutations when six letters at a time are taken from the word RAMAYANAM.

Solution There are 9 letters of 5 sorts, namely, *aaaa ; mm; r ; y ; n*

The combination & permutation can be grouped. Then:

i. 4 alike ; 2 alike

The number of combinations = 1.1 = 1

The number of permutations $= \dfrac{6!}{4!2!} = 15$

ii. 4 alike ; 2 different

The number of combinations $= 1.4\,C_2 = 6$

The number of permutations $= 6\dfrac{6!}{4!} = 180$

iii. 3 alike ; 2 alike ; 1 different

The number of combinations = $1.1.3\,C_1 = 3$

The number of permutations = $3.\dfrac{6!}{3!2!} = 180$

iv. 3 alike ; 3 different

The number of combinations = $1.4C_3 = 4$

The number of pemutations = $4.\dfrac{6!}{3!} = 480$

v. 2 alike ; 2 alike ; 2 different

The number of combinations = $1.1.3\,C_2 = 3$

The number of permutations = $3\dfrac{6.1}{2!2!} = 540.$

vi. 2 alike; 4 different

The number of combinations = $2C_1\,4C_4 = 2$

The number of permutations = $2.\dfrac{6!}{2!} = 720$

The total number of permutations = $15 + 180 + 180 + 480 + 540 + 720 = 2.115.$

Example 1.62 I have 4 friends : in how many ways can I invite them for dinner.

Solution The required number is $4C_1 + 4C_2 + 4C_3 + 4C_4 = 4 + 6 + 4 + 1 = 15$

EXERCISE X

1. A father with 8 children takes them three at a time to M.G.M, as often as he can without taking the same three children together more than once. How often will he go and how often will each child go ?
2. If n straight lines be drawn in a plane, no two being parallel and no three concurrent, except p which meets in another point, how many other points of intersection will be there ?
3. Find the total number of ways in which 10 different cakes can be divided into two parcels each containing the same number ?
4. A person has 15 acquaintances of whom 5 are relatives. In how many ways may he invite 12 guests from among them so that 3 of them may be relatives?
5. The Tamil alphabet consists of 12 vowels and 18 consonants. How many seven lettered words can be formed from them, if each word is to include atleast 3 vowels and 2 consonants ?
6. The crew for a ten-oared boat are to chosen from 16 men; if 3 of them can row on one side only, 4 of them on the other side only and the rest on both sides, in how many ways can the crew be selected ?
7. How many combinations and permutations can be made with the letters of the word PARABOLA, taken three at a time?
8. How many six-lettered word can be made from the letters of the word ALLITERATION ?
9. How many different colours can be formed by mixing together, in the same proportion, the seven rainbow colours ?
10. Find the number of factors of 2, 268.
11. A man wants to host as many parties as possible for his 15 friends; how many should he invite at a time and in how many of them will a particular man be found ?

CHAPTER

2

Matrix Algebra

Matrix Algebra was developed systematically by A.Cayley (1821-1895), a French Mathematician, the discoverer, J.J. Sylvester (1814–1897), S.B. Frobenious (1849–1917), Jordan, Jacobi and Hermite. Today, the 'theory of matrices' is one of the most important and powerful tools not only in the study of higher mathematics (pure and applied) but also in the study of engineering, physical and social sciences such as computer science, discrete mathematics, network analysis, nuclear physics, quantum mechanics, mathematical economics, sociology, probability and statistics. So it is essential for one to have the knowledge of the subject. In this chapter we study the subject from the fundamental concept onwards.

Definition of a Matrix

A set of *mn* numbers (real or complex) arranged in the form of a rectangular array, having an ordered set of *m* rows (horizontal lines) and *n* columns (vertical lines) being enclosed by the brackets [ ] or ( ) or || || is called an $m \times n$ matrix (to be read as '*m* by *n*' matrix or a matrix of order $m \times n$).

Each of the above *mn* numbers is called an element. A matrix is usually denoted by capital letters *A* or *B* or *C*.

A $m \times n$ matrix is usually written as

$$A = \begin{pmatrix} a_{11} & a_{12} \ldots a_{1j} \ldots a_{1n} \\ a_{21} & a_{22} \ldots a_{2j} \ldots a_{2n} \\ \ldots & \ldots\ldots\ldots\ldots \\ a_{i1} & a_{i2} \ldots a_{ij} \ldots a_{in} \\ \ldots & \ldots\ldots\ldots\ldots \\ a_{m1} & a_{m2} \; a_{mj} \; a_{mn} \end{pmatrix} \quad \text{or} \quad A = [a_{ij}];\ 1 \le i \ge m,\ 1 \le j \ge n \quad \text{or} \quad A = [a_{ij}]_{m \times n}$$

In the notation used, each of the element has two suffixes. The first indicates the row and the second indicates the column in which the element lies. Thus the element a_{ij} is at the intersection of *i*-th row and *j*-th column.

Note: A matrix is a system of numbers. It has no numerical value.

TYPES OF MATRICES

Let $A = [a_{ij}]_{m \times n}$ be the general matrix.

Rectangular and Square Matrices

If $m \neq n$ in (1) then the matrix A is a rectangular matrix of order $m \times n$ and if $m = n$ in (1) then the matrix A is a square matrix of order n.

If $A = [a_{ij}]_{n \times n}$ (square matrix) then the diagonal consisting the elements $a_{ij} \forall i = j$ is called principal or leading diagonal, and the elements are known as principal diagonal elements. The sum of principal diagonal elements of a square matrix is known as trace of the matrix.

$$\therefore \quad \text{Trace of } A = \sum_{i=1}^{n} a_{ii} = a_{11} + a_{22} + - - - + a_{nn}$$

$$\begin{pmatrix} 1 & 2 & 3 & -4 \\ 3 & 6 & 7 & 5 \\ 3 & 1 & -2 & 6 \end{pmatrix}_{3\times4} \text{ and } \begin{pmatrix} 1 & 2 & 3 \\ 2 & 3 & 4 \\ 5 & 6 & 7 \end{pmatrix}_{3\times3}$$

are examples of rectangular and square matrices. For the square matrix, trace $= 1 + 3 + 7 = 11$.

Row and Column Matrices

If $m = 1$ in (1) then the matrix A is a $1 \times n$ matrix, called row matrix (or row vector) and if $n = 1$ in (1) then the matrix A is a $m \times 1$ matrix called column matrix (or column vector).

Diagonal Matrix

If $m = n$, $a_{ij} = 0 \ \forall i \neq j$ and $a_{ij} \neq 0 \ \forall i = j$ in (1) then the matrix A is called a diagonal matrix of order n. A diagonal matrix of order n with diagonal elements $a_{11}, a_{22} ..., a_{nn}$ is denoted by $A = \text{diag}\,[a_{11}, a_{22}, ..., a_{nn}]$

Scalar Matrix

If $m = n$, $a_{ij} = 0 \ \forall i \neq j$ and $a_{ij} = k \forall i = j$ in (1), where k is some scalar, then the matrix A is called a scalar matrix of order n. *Example*

$$\begin{pmatrix} 2 & 0 & 6 \\ 0 & -3 & 0 \\ 0 & 0 & 6 \end{pmatrix} \text{ and } \begin{pmatrix} 5 & 0 & 0 \\ 0 & 5 & 0 \\ 0 & 0 & 5 \end{pmatrix}$$ are the diagonal and scalar matrices of order 3.

Unit Matrix or Identity Matrix

If $m = n$, $a_{ij} = 0$, $\forall \neq j$ and $a_{ij} = 1, \forall = j$ in (1), then the matrix A is called a unit matrix of order n and generally denoted by I_n. For example:

$$I_1 = (1), I_2, \begin{pmatrix} 1 & 0 \\ 0 & 1 \end{pmatrix}, I_3 = \begin{pmatrix} 1 & 0 & 0 \\ 0 & 1 & 0 \\ 0 & 0 & 1 \end{pmatrix}$$ are unit matrices of order 1, 2, 3 etc.

Zero or Null matrix

If $a_{ij} = 0 \ \forall \ i \neq j$ and $i = j$, then the matrix A is called a null matrix of order $m \times n$ and is denoted by $0_{m \times n}$.

Example $$0_{3\times4} = \begin{pmatrix} 0 & 0 & 0 & 0 \\ 0 & 0 & 0 & 0 \\ 0 & 0 & 0 & 0 \end{pmatrix}$$ is a null matrix of order 3×4.

Upper and Lower Triangular Matrix

If $a_{ij} = 0 \ \forall \ i < j$ in (1) then the matrix A is said to be lower triangular matrix and if $a_{ij} = 0, \forall \ i > j$ in (1), then the matrix A is said to be upper triangular matrix.

$$\begin{pmatrix} 3 & 0 & 0 & 0 \\ -2 & 3 & 0 & 0 \\ 1 & 1 & 2 & 0 \end{pmatrix} \text{ and } \begin{pmatrix} 2 & 3 & 4 \\ 0 & 5 & 6 \\ 0 & 0 & 7 \end{pmatrix} \text{ are examples of lower and upper triangular matrices.}$$

Note:

1. Triangular matrix need not be square.
2. A square matrix which is both upper and lower triangular is a diagonal matrix.
3. If $m = n$; $a_{ij} = 0, \forall \ i \le j$ in (1) then A is strictly lower triangular matrix and if $m = n$, $a_{ij} = 0, \forall \ i \ge j$ in (1) then A is strictly upper triangular matrix.

MATRICES ASSOCIATED WITH A GIVEN MATRIX

Transpose of a Matrix

If $A = [a_{ij}]_{m \times n}$, then the matrix obtained by interchanging the rows and columns of A is called Transpose of A and is denoted by A^T or A^1. Evidently the transpose of $m \times n$ matrix is an $n \times m$ matrix.

$\therefore \quad A^T = [a^1_{ij}]$ where $a^1_{ij} = a_{ji}$ i.e. (j, i)th element of A^T is the (i, j) th element of A.

Example If $A = \begin{pmatrix} 2 & 3 & 7 & 4 \\ 4 & 5 & 6 & 0 \end{pmatrix}$ then $A^T = \begin{pmatrix} 2 & 4 \\ 3 & 5 \\ 7 & 6 \\ 4 & 0 \end{pmatrix}$

and also $(A^T)^T = A$, i.e. the transpose of A coincides with A.

Conjugate of a Matrix

If A is a $m \times n$ matrix, then the $m \times n$ matrix obtained by replacing each element of A by its complex conjugate is called the conjugate matrix of A and is denoted by $\overline{A}$. Thus if $A = [a_{ij}]_{m \times n}$ then $\overline{A} = [\overline{a}_{ij}]_{m \times n}$ where $\overline{a}_{ij}$ is the complex conjugate of a_{ij}. For example:

$$\text{If } A = \begin{pmatrix} 2+i & -2 & -3+i \\ 4-i & -2-i & 3 \end{pmatrix} \text{ then } \overline{A} = \begin{pmatrix} 2-i & -2 & -3-i \\ 4+i & -2+i & 3 \end{pmatrix}$$

Note:

1. If A is real i.e. all the elements are over the field of real numbers, then $\overline{A} = A$ i.e. the conjugate of A coincides with A.
2. $(\overline{\overline{A}}) = A$ i.e. the conjugate of $\overline{A}$ coincides with A.

Conjugate Transpose of a Matrix

The conjugate of the transpose of a matrix A is called the conjugate transpose of A and is denoted by A^*. Thus if $A = [a_{ij}]_{m \times n}$ then $A^* = (\overline{A^T})$.

Example If $A = \begin{pmatrix} 1+i & 2-i \\ -1-i & 3-i \end{pmatrix}$ then $A^* = \begin{pmatrix} 1-i & -1+i \\ 2+i & 3+i \end{pmatrix}$

Note:

1. The conjugate of the transpose of A coincides with the transpose of the conjugate of A i.e. $A^* = \overline{(A^T)} = (\bar{A})^T$.
2. $(A^*)^* = A$

Symmetric and Skew-Symmetric Matrix

A square matrix $A = [a_{ij}]_{m\times n}$ is said to be symmetric if $a_{ij} = a_{ji}\ \forall\ i, j$ and is said to be skew-symmetric if $a_{ij} = -a_{ji} \forall i, j$ i.e. symmetric if $A^T = A$ and skew-symmetric if $A^T = -A$ (or $A = -A^T$)

In case of skew-symmetric, all the principal diagonal elements are necessarily to be zero ($\because$ we have $a_{ij} = -a_{ji} \Leftrightarrow a_{ii} = -a_{ii}$ if $j = i$; $\therefore a_{ii} = 0$.)

Example $\begin{pmatrix} 1 & 2 & 3 \\ 2 & 5 & 6 \\ 3 & 6 & 7 \end{pmatrix}$ and $\begin{pmatrix} 0 & 8 & 5 \\ -8 & 0 & 6 \\ -5 & -6 & 0 \end{pmatrix}$ are symmetric and skew - symmetric matrices.

Note:

1. If A is a symmetric matrix of order n, then the number of independent elements $= \frac{1}{2}n(n+1)$.
2. If A is a skew-symmetric matrix of order n, then the number of independent elements $= \frac{1}{2}n(n-1)$.

Hermitian and Skew-Hermitian Matrices

A square matrix $A = [a_{ij}]_{n\times n}$ is said to be Hermitian if $a_{ij} = \bar{a}_{ji}$ and is said to be skew-Hermitian if $a_{ij} = -\bar{a}_{ji}$ i.e. Hermitian if $A^* = A$, and skew-Hermitian if $A^* = -A$ (or $A = -A^*$).

$\therefore$ It is clear from above that the principal diagonal elements of a Hermitian matrix are purely real ($\because$ when $j = i$, $a_{ii} = \bar{a}_{ii} \Rightarrow a_{ii}$ is real) and the principal diagonal elements of skew-Hermitian are purely imaginary or zero.

Example

$\begin{pmatrix} 2 & 3-i & 4+3i \\ 3+i & 5 & 6i \\ 4-3i & -6i & 1 \end{pmatrix}$ and $\begin{pmatrix} -2i & 4+3i & -3-i \\ -4+3i & 0 & 6i \\ -3+i & -6i & 4i \end{pmatrix}$ are Hermitian and skew-Hermitian matrices.

SUB-MATRIX

The matrix obtained on deleting any number of rows and columns of the given matrix A is called the sub-matrix *of A.*

If $A = \begin{pmatrix} 1 & 2 & 3 & 4 \\ 7 & 2 & 1 & 0 \end{pmatrix}$ then $B = \begin{pmatrix} 1 & 3 \\ 7 & 1 \end{pmatrix}$ and $C = \begin{pmatrix} 2 & 3 & 4 \\ 2 & 1 & 0 \end{pmatrix}$ are submatrices of A.

EQUALITY OF MATRICES

Two matrices A and B are said to be equal and is written as $A = B$ iff (i) they are of the same order and (ii) each element of A is equal to the corresponding element of B. If only the condition (i) is true of A and B then they are said to be comparable matrices.

ADDITION AND SUBTRACTION OF MATRICES

Two matrices A and B are said to conformable for either addition or subtraction if they are of the same order.

Then their sum $A + B$ is defined as the matrix, each element of which is the sum of the corresponding elements of A and B. Similarly $A - B$ is defined as the matrix, each element of which is the difference of the corresponding elements of A and B.

$$\text{Thus}\begin{pmatrix} a & b & c \\ d & e & f \\ g & h & f \end{pmatrix} \pm \begin{pmatrix} 1 & 2 & 3 \\ 4 & 5 & 6 \\ 7 & 8 & 9 \end{pmatrix} = \begin{pmatrix} a\pm1 & b\pm2 & c\pm3 \\ d\pm4 & e\pm5 & f\pm6 \\ g\pm7 & h\pm8 & f\pm9 \end{pmatrix}$$

SCALAR MULTIPLE OF A MATRIX

If A is a matrix and α is a number then αA is defined as the matrix, each element of which is α times the corresponding element of the matrix A.

$$\text{Thus if } A = \begin{pmatrix} a & b \\ 3 & 4 \end{pmatrix} \text{ then } \alpha A = \begin{pmatrix} \alpha a & \alpha b \\ 3\alpha & 4\alpha \end{pmatrix}$$

Note:

1. $\alpha A = A\alpha$
2. $\alpha A = -A$ if $\alpha = -1$, in such case $-A$ is called negative of A.

PROPERTIES OF MATRIX ADDITION

Let M be the set of all matrices of order $m \times n$. Then the following laws hold.

i. ***Commutative law*** $A + B = B + A \ \forall A, B \in M$

ii. ***Associative law*** $A + (B + C) = (A + B) + C \ \forall A, B, C \in M$

iii. ***Identity law*** $O \in M$ such that $A + O = A = O + A \ \forall A \in M$, here O (null matrix) is called additive identity of A.

iv. ***Inverse law*** We can always find a matrix $X \in M$ such that $A + X = O = X + A \ \forall A \in M$ where $O \in M$ is a null matrix. Here is called additive inverse of A and is equal to $-A$ if, the above had unique solution.

v. ***Distributive law*** $\alpha(A + B) = \alpha A + \alpha B$

$$(A + B)^T = A^T + B^T$$

MULTIPLICATION OF MATRICES

If A and B are two matrices such that the number of columns in A is equal to the number of rows in B, then product AB is defined. (in other words we say that A and B are conformable for multiplication.)

$\therefore$ $A = [a_{ij}]_{m\times n}$ and $B = [b_{ij}]_{n\times p}$ then the product AB is a matrix of order $m \times p$ and is defined as

$$AB = [a_{ij}]_{m\times n}[b_{ij}]_{n\times p} = \left(\sum_{k=1}^{n} a_{ik} b_{kj} \right)_{m\times p}$$

Example

$$\text{If } A = \begin{pmatrix} a_{11} & a_{12} & a_{13} \\ a_{21} & a_{22} & a_{23} \end{pmatrix}_{2\times3} \text{ and } B = \begin{pmatrix} b_{11} \\ b_{21} \\ b_{31} \end{pmatrix}_{3\times1} \text{ then the product}$$

$$AB = \begin{pmatrix} a_{11}b_{11} + a_{12}b_{21} + a_{13}b_{31} \\ a_{21}b_{11} + a_{22}b_{21} + a_{23}b_{31} \end{pmatrix}_{2\times1}$$

Note:

1. If the matrices A and B are square matrices of the same order then both the products AB and BA are defined and each is of the same order as A and B.
2. If the product AB exists, then the matrix A is known as the pre-factor (pre-multiplier) and the matrix B is known as the post -factor (post-multiplier).

PROPERTIES OF MATRIX MULTIPLICATION

i. Matrix multiplication is not commutative Suppose that $A = [a_{ij}]_{m\times n}$ and $B = [b_{ij}]_{n\times p}$ then only the product AB exists and BA does not. If $A = [a_{ij}]_{m\times n}$ and $B = [b_{ij}]_{n\times m}$ and both AB and BA exists, but not equal since they are of size $m \times m$ and $n \times n$. Even when AB and BA both exist and are matrices of same order it is not necessary that $AB \neq BA$.

The necessary condition for commutative law to hold good is that both AB and BA should exist and are to be of the same order.

These conditions are not sufficient for commutative property.

$$\text{If } A = \begin{pmatrix} 1 & 2 & 1 \\ 3 & 4 & 2 \\ 1 & 3 & 2 \end{pmatrix};\ B = \begin{pmatrix} 10 & -4 & -1 \\ -11 & 5 & 0 \\ 9 & -5 & 1 \end{pmatrix} \text{ then } AB = BA$$

$$\text{and if } A = \begin{pmatrix} 1 & 2 & 3 \\ 2 & 3 & 4 \\ -1 & 1 & 2 \end{pmatrix};\ B = \begin{pmatrix} 0 & 2 & -1 \\ 1 & 3 & 4 \\ 0 & -2 & -3 \end{pmatrix} \text{ then } AB \neq BA$$

Note:

1. If $AB = -BA$ then the matrices A and B are said to be anti-commute.
2. The product of two diagonal matrices of order n is another diagonal matrix of the same order and commute always.

ii. Matrix multiplication is associative If A, B and C are any three matrices conformable for multiplication then $(AB)C = A(BC)$.

Let $A = [a_{ij}]_{m\times n}, B = [b_{ij}]_{n\times p}$ and $C = [c_{ij}]_{p\times q}$ then

$$AB = \left(\sum_{k=1}^{n} a_{ik} b_{kj}\right)_{m\times p} \quad \text{and } (AB)C = \left(\sum_{s=1}^{n}\left(\sum_{k=1}^{n} a_{ik} b_{ks}\right) c_{si}\right)_{m\times q}$$

Since the order of summation is arbitrary in a finite sum, we have

$$(AB)C = \left(\sum_{k=1}^{n} a_{ik}\left(\sum_{s=1}^{n} b_{ks} c_{sj}\right)\right)_{m\times q} = A(BC). \text{ Hence } (AB)C = A(BC)$$

iii. Matrix multiplication is distributive $A(B \pm C) = AB \pm AC$ and $(B \pm C)D = BD \pm CD$ where A, B, C and D are matrices conformable for the operations involved.

iv. If $AB = 0$ it does not imply that either $A = 0$ or $B = 0$ even if $A \neq 0$ and $B \neq 0$, still AB may be a null matrix.

Example

$$\text{If } A = \begin{pmatrix} 0 & -3 & 1 \\ 2 & -1 & 1 \\ 2 & -1 & 1 \end{pmatrix} \text{ and } B = \begin{pmatrix} 0 & -1 & 1 \\ 0 & 1 & -1 \\ 0 & 3 & 3 \end{pmatrix} \text{ then } AB = 0$$

v. Cancellation law This law does not hold good in matrix equation : i.e. if $A \neq 0$ $AB = AC$, or $BA = CA$ then it does not imply $B = C$.

vi. Existence of identity If A is a matrix of order $m \times n$, then there exists unit matrices of order m and n always such that $AI_n = A = I_mA$. Here I_n is called right multiplicative identity and I_m is called left multiplicative identity. If A is a square matrix of order n, then there exists unit matrix of order n such that $IA = A = AI$. Here I is called multiplicative identity.

vii. If A and B are any two matrices conformable for multiplication then $(AB)^T = B^TA^T$. Let $A = [a_{ij}]_{m\times n}$ and $B = [b_{ij}]_{n\times p}$ then $A^T = [a^1_{ij}]_{n\times m}$ and $B^T = [b^1_{ij}]_{p\times n}$ where $a^1_{ij} = a_{ji}$ and $b^1_{ji} = b_{ji}$

$\therefore$ $(AB)^T$ and B^TA^T have same dimensions.

Now let $AB = [c_{ij}]_{m\times p}$ where $c_{ij} = \sum_{k=1}^{n} a_{ik}b_{kj}; i = 1, 2, ..., m; j = 1, 2, ..., p$

$$\therefore (i, j) \text{ th element of } (AB)^T = c^1_{ij} = c_{ji} = \sum_{k=1}^{n} a_{jk}b_{ki}$$

$$= \sum_{k=1}^{n} a^1_{kj}b^1_{ik} = \sum_{k=1}^{n} b^1_{ik}a^1_{kj}$$

$= (i, j)$ th element of B^TA^T. Hence the result.

Similarly we can show that

$(A_1, A_2, ..., A_{n-1}, A_n)^T = A^T_n, A^T_{n-1}, ..., A^T_2, A^T_1$ where $A_1, A_2 ..., A_{n-1}$ and A_n are conformable for matrix multiplication.

DETERMINANT OF A SQUARE MATRIX

Determinants occur only for square matrices and plays an important role in solving a system of linear equations. It is purely a number. If A is any square matrix we denote determinant of A as det A or $|A|$ or Δ.

For a matrix of order 1 say for $A = [a]$, det $A = a$

For a matrix of order 2×2 say for $A = \begin{pmatrix} a & b \\ c & d \end{pmatrix}$, $\Delta = (ad - bc)$.

To define it in general, we need the definitions of minor and co-factor. Let us consider $A = [a_{ij}]_{n\times n}$. The minor of a_{ij} is denoted by M_{ij} and is defined as the determinant of the matrix obtained after deleting the i-th row and j-th column passing through a_{ij} in A. The co-factor of a_{ij} is denoted by C_{ij} and is equal to $(-1)^{i+j} M_{ij}$.

For example let us consider $A = \begin{pmatrix} a_{11} & a_{12} & a_{13} \\ a_{21} & a_{22} & a_{23} \\ a_{31} & a_{32} & a_{33} \end{pmatrix}$

$$M_{11} \text{ (minor of } A_{11}) = \begin{vmatrix} a_{22} & a_{23} \\ a_{32} & a_{33} \end{vmatrix} = (a_{22}a_{33} - a_{32}a_{23})$$

$$M_{23} \text{ (minor of } A_{23}) = \begin{vmatrix} a_{11} & a_{12} \\ a_{31} & a_{32} \end{vmatrix} = (a_{11}a_{32} - a_{31}a_{12}) \text{ etc.}$$

and C_{11} (Co-factor of a_{11}) $= (-1)^{1+1}M_{11} = +(a_{22}a_{33} - a_{32}a_{23})$

C_{23} (Co-factor of a_{23}) $= (-1)^{2+3} M_{23} = -(a_{11}a_{32} - a_{31}a_{12})$

Definition Let $A = [a_{ij}]_{n\times n}$. Then the determinant of A is defined as the sum of the products of the elements of any row (or column) by their corresponding co-factors. So for the above matrix A of order 3, the determinant when expanded along the first row is given by

$$|A| = a_{11}C_{11} + a_{12}C_{12} + a_{13}C_{13}$$

$$\text{or } |A| = a_{11}\begin{vmatrix} a_{22} & a_{23} \\ a_{32} & a_{33} \end{vmatrix} - a_{12}\begin{vmatrix} a_{21} & a_{23} \\ a_{31} & a_{33} \end{vmatrix} + a_{13}\begin{vmatrix} a_{21} & a_{22} \\ a_{31} & a_{32} \end{vmatrix}$$

$$= a_{11}(a_{22}a_{33} - a_{23}a_{32}) - a_{12}(a_{21}a_{33} - a_{31}a_{23}) + a_{13}(a_{21}a_{32} - a_{31}a_{22})$$

Similarly by expanding along the 2nd, 3rd rows, 1st, 2nd and 3rd columns we obtain the same value of the determinant. So the determinant of a 3 × 3 matrix can be obtained in the following way.

$$\Delta = \sum_{j=1}^{3} a_{ij}c_{ij}; i = 1,2,3 = \sum_{i=1}^{3} a_{ij}c_{ij}; j = 1,2,3$$

In the same way, we can write for $A = [a_{ij}]_{n\times n}$

$$\Delta = \sum_{j=1}^{n} a_{ij}c_{ij};\ i = 1, 2, ..., n = \sum_{i=1}^{n} a_{ij}c_{ij};\ j = 1, 2, ..., n$$

PROPERTIES OF DETERMINANT OF MATRIX

These properties hold good for any order.

i. The sum of the products of the elements of any row (or column) by the co-factors of another row (or column) is zero.
ii. $|A^T| = |A|$
iii. If a matrix B is formed from the matrix A by the interchange of two rows (or columns) then $|B| = -|A|$
iv. If two rows (or columns) of a square matrix are identical or if all the elements of a row (or column) of a square matrix are k times the elements of another row (or column) then its determinant is zero.
v. The determinant of a square matrix with a zero row (or zero column) is zero.
vi. If B is the matrix obtained from a matrix A, by multiplying the elements of any row (or column) by a scalar quantity k then $|B| = k|A|$. If A is any square matrix of order n then $|kA| = k^n|A|$.
vii. The determinant of a triangular matrix is the product of the principle diagonal elements.
viii. If A and B are any two square matrices of equal order then $|AB| = |A|\,|B|$.
ix. If a matrix B is obtained from a matrix A, by adding a multiple of a row (or column) to another row (or column) then $|B| = |A|$.
x. If each element of a row is a sum of m terms, then the determinant can be expressed as the sum of m determinants.

SINGULAR AND NON-SINGULAR MATRICES

A matrix $A = [a_{ij}]_{n\times n}$ is said to be singular if $|A| = 0$, otherwise non-singular (ie $|A| \neq 0$).

Note: If A and B are two non-singular matrices of the same order, them AB is also a non-singular of the same order.

ADJOINT OF A SQUARE MATRIX

If the elements of a square matrix A are replaced by corresponding co-factors of elements, then the transpose of the resulting matrix is called the adjoint of the matrix A and is denoted by adj A. If $A = [a_{ij}]_{n\times n}$ then adj $A = [c_{ij}]^T_{n\times n}$ where c_{ij} is the co-factor of a_{ij}.

PROPERTIES OF ADJOINT OF A MATRIX

The following are the few properties which can be verified easily.

i. $A\,(\text{adj}\,A) = |A|\,I = (\text{adj}\,A)\,A$, iff $|A| \neq 0$

ii. $A\left\{\dfrac{(\text{adj}\,A)}{|A|}\right\} = I = \left\{\dfrac{(\text{adj}\,A)}{|A|}\right\}A$, iff $|A| \neq 0$

iii. $\text{adj}\,(AB) = (\text{adj}\,A)\,(\text{adj}\,B)$, iff $|A| \neq 0$; $|B| \neq 0$.

INVERSE OF A MATRIX

If for a matrix A, there exists another matrix B such that $AB = BA = I$, then B is called the inverse of A and is denoted by A^{-1} i.e. $B = A^{-1}$. The products AB and BA should be defined and be equal. For it is necessary that A and B, should both be square matrices of the same order. Thus rectangular matrices cannot have an inverse. It follows from the definition (i) $A = B^{-1}$ (ii) $AA^{-1} = I = A^{-1}A$.

Now it follows from the property (ii) of adjoint matrix, we have for a square matrix A, inverse of A as

$$A^{-1} = \frac{\text{adj}\,A}{|A|};\ |A| \neq 0$$

PROPERTIES OF INVERSE OF A MATRIX

i. The necessary and sufficient condition for a square matrix A to possess the inverse is that, A is to be non-singular.

ii. If A is non-singular then
(a) $(A^{-1})^{-1} = A$ (b) $(A^{-1})^T = (A^T)^{-1}$

iii. The inverse of a matrix if exists is unique.

Proof If possible let B and C be the two inverses of the matrix A then $AB = BA = I$ and $AC = CA = I$

Hence $CAB = C(AB) = CI = C$

and $CAB = (CA)\,B = IB = B$

$\therefore B = C$ i.e. the inverse is unique.

iv. *If* A and B are two non-singular matrices of the same order then $(AB)^{-1} = B^{-1}A^{-1}$

Proof $(AB)\,B^{-1}A^{-1} = A(BB^{-1})A^{-1}$(associative law)

$= AIA^{-1} = (AI)A^{-1} = AA^{-1} = I$

similarly $B^{-1}A^{-1}(AB) = I$ $\qquad \therefore (AB)^{-1} = B^{-1}A^{-1}$

In general if $A_1, A_2, \ldots, A_n$ are non-singular matrices 'then $(A_1A_2\ldots A_n)^{-1} = A_n^{-1}\ldots A_2^{-1}A_1^{-1}$

SPECIAL SQUARE MATRICES

Involutory matrix A square matrix A is said to be involutory if $A^2 = I$

Idempotent matrix A square matrix A is said to be idempotent if $A^2 = A$.

Nilpotent matrix A square matrix A is said to be nilpotent if there exists $k \in N$ such that $A^k = 0$. If k is the least of N such that $A^k = 0$, then k is called the index of A.

Periodic matrix A square matrix A is said to be periodic if $A^{k+1} = A \; \forall \; k \in N$. If k is the least of N such that $A^{k+1} = A$ then k is called the period of A.

Note Period of idempotent matrix is one.

POSITIVE INTEGRAL POWERS OF MATRICES

If A is a square matrix of order n, we write products $AA = A^2$, $A(AA) = (AA)A = AAA = A^3$; AA.....p times $= A^p$; $A^p A^q = A^{p+q}$ and $(A^p)^q = A^{pq}$, where p and q are positive integers.

$\therefore$ For all positive integral powers of p, A^p is a square matrix of order n. In particular we define $A^0 = I_n$.

NEGATIVE INTEGRAL POWERS OF A NON-SINGULAR MATRIX

If A be a non-singular matrix and p is any positive integer, then we define $A^{-p} = (A_p)^{-1}$, so that A^{-p} is the inverse of A^p. Also $A^{-p} = (A^p)^{-1} = (A^{-1})^p$.

ORTHOGONAL MATRIX

A square matrix A is called an orthogonal matrix if $AA^T = I = A^T A$, where I is the identity matrix.

Note:

1. If A is an orthogonal then A^T and A^{-1} are also orthogonal.
2. If A is an orthogonal matrix then $|A| = \pm 1$.

UNITARY MATRIX

A square matrix A is called an unitary matrix if $AA^* = I = A^*A$ where A^* is the transposed conjugate of A and I is the unit matrix.

Note: If A is unitary then A^T and A^{-1} are also unitary.

Example 2.1 If $A = \begin{pmatrix} 2 & -1 & 0 \\ 0 & -2 & 1 \\ 1 & 0 & 1 \end{pmatrix}$; $B = \begin{pmatrix} -2 & 1 & -1 \\ 1 & 2 & -2 \\ 2 & -1 & -4 \end{pmatrix}$ and $C = \begin{pmatrix} 1 & 1 & -1 \\ 2 & -3 & 4 \\ 3 & -2 & 3 \end{pmatrix}$ evaluate $2A + 3B - 4C$, AB, BA, BC, $A(BC)$ and $(AB)C$

Solution

$$2A+3B-4C = \begin{pmatrix} 2(2)+3(-2)-4(1) & 2(-1)+3(1)-4(1) & 2(0)+3(-1)-4(-1) \\ 2(0)+3(1)-4(2) & 2(-2)+3(2)-4(-3) & 2(1)+3(-2)-4(4) \\ 2(1)+3(2)-4(3) & 2(0)+3(-1)-4(-2) & 2(1)+3(-4)-4(3) \end{pmatrix}$$

$$= \begin{pmatrix} -6 & -3 & 1 \\ -5 & 14 & -20 \\ -4 & 5 & -22 \end{pmatrix}$$

$$AB = \begin{pmatrix} 2 & -1 & 0 \\ 0 & -2 & 1 \\ 1 & 0 & 1 \end{pmatrix} \begin{pmatrix} -2 & 1 & -1 \\ 1 & 2 & -2 \\ 2 & -1 & -4 \end{pmatrix} = \begin{pmatrix} -4-1+0 & 2-2+0 & -2+2-0 \\ 0-2+2 & 0-4-1 & 0 \; 4-4 \\ -2+0+2 & 1+0-1 & -1+0-4 \end{pmatrix}$$

$$= \begin{pmatrix} -5 & 0 & 0 \\ 0 & -5 & 0 \\ 0 & 0 & -5 \end{pmatrix}$$

Similarly $BA=\begin{pmatrix}-5 & 0 & 0\\ 0 & -5 & 0\\ 0 & 0 & -5\end{pmatrix}; BC=\begin{pmatrix}-3 & -3 & 3\\ -1 & -1 & 1\\ -12 & 13 & -18\end{pmatrix}$

$$A(BC)=\begin{pmatrix}2 & -1 & 0\\ 0 & -2 & 1\\ 1 & 0 & 1\end{pmatrix}\begin{pmatrix}-3 & -3 & 3\\ -1 & -1 & 1\\ -12 & 13 & -18\end{pmatrix}=\begin{pmatrix}-5 & -5 & 5\\ -10 & 15 & -20\\ -15 & 10 & -15\end{pmatrix}$$

$$\text{and } (AB)C=\begin{pmatrix}-5 & 0 & 0\\ 0 & -5 & 0\\ 0 & 0 & -5\end{pmatrix}\begin{pmatrix}1 & 1 & -1\\ 2 & -3 & 4\\ 3 & -2 & 3\end{pmatrix}=\begin{pmatrix}-5 & -5 & 5\\ -10 & 15 & -20\\ -15 & 10 & -15\end{pmatrix}$$

Hence in this problem $AB = BA$ and $A(BC) = (AB)C$.

Example 2.2 Prove that $A^3 - 4A^2 - 3A + 11I = 0$ where $A=\begin{pmatrix}1 & 3 & 2\\ 2 & 0 & -1\\ 1 & 2 & 3\end{pmatrix}$ and I is the unit matrix of order 3.

Solution Given $A=\begin{pmatrix}1 & 3 & 2\\ 2 & 0 & -1\\ 1 & 2 & 3\end{pmatrix}$

$$A^2 = A.A=\begin{pmatrix}1 & 3 & 2\\ 2 & 0 & -1\\ 1 & 2 & 3\end{pmatrix}\begin{pmatrix}1 & 3 & 2\\ 2 & 0 & -1\\ 1 & 2 & 3\end{pmatrix}=\begin{pmatrix}9 & 7 & 5\\ 1 & 4 & 1\\ 8 & 9 & 9\end{pmatrix}$$

$$A^3 = A^2.A=\begin{pmatrix}9 & 7 & 5\\ 1 & 4 & 1\\ 8 & 9 & 9\end{pmatrix}\begin{pmatrix}1 & 3 & 2\\ 2 & 0 & -1\\ 1 & 2 & 3\end{pmatrix}=\begin{pmatrix}28 & 37 & 20\\ 10 & 5 & 1\\ 35 & 42 & 34\end{pmatrix}$$

$$\therefore\ A^3-4A^2-3A+11I=\begin{pmatrix}28 & 37 & 26\\ 10 & 5 & 1\\ 35 & 42 & 34\end{pmatrix}-\begin{pmatrix}36 & 28 & 20\\ 4 & 16 & 4\\ 32 & 36 & 36\end{pmatrix}-\begin{pmatrix}3 & 9 & 6\\ 6 & 0 & -3\\ 3 & 6 & 9\end{pmatrix}+\begin{pmatrix}11 & 0 & 0\\ 0 & 11 & 0\\ 0 & 0 & 11\end{pmatrix}=\begin{pmatrix}0 & 0 & 0\\ 0 & 0 & 0\\ 0 & 0 & 0\end{pmatrix}$$

$\therefore\ A^4A^2-3A+11I=0$

Example 2.3 If $A=\begin{pmatrix}1 & -2 & 3\\ 2 & 0 & -1\\ -3 & 1 & 0\end{pmatrix}$ and $B=\begin{pmatrix}4 & -1 & 2\\ 0 & 4 & 0\\ 2 & 3 & 1\end{pmatrix}$ show that $(A+B)^T = A^T + B^T$ and $(AB)^T = B^T A^T$.

Solution Given $A=\begin{pmatrix}1 & -2 & 3\\ 2 & 0 & -1\\ -3 & 1 & 0\end{pmatrix}$ and $B=\begin{pmatrix}4 & -1 & 2\\ 0 & 4 & 0\\ 2 & 3 & 1\end{pmatrix}$

$$\therefore\quad A^T=\begin{pmatrix}1 & 2 & -3\\ -2 & 0 & 1\\ 3 & -1 & 0\end{pmatrix}\text{ and } B^T=\begin{pmatrix}4 & 0 & 2\\ -1 & 4 & 3\\ 2 & 0 & 1\end{pmatrix}$$

$$(A+B)^T=\begin{pmatrix}5 & -3 & 5\\ 2 & 4 & -1\\ -1 & 4 & 1\end{pmatrix}^T=\begin{pmatrix}5 & 2 & -1\\ -3 & 4 & 4\\ 5 & -1 & 1\end{pmatrix}$$

$$\therefore\ A^T+B^T=\begin{pmatrix}5 & 2 & -1\\ -3 & 4 & 4\\ 5 & -1 & 1\end{pmatrix}=(A+B)^T.$$

$$\text{Now } (AB)^T=\begin{pmatrix}4 & 0 & 5\\ 8 & -5 & 3\\ -12 & 7 & -6\end{pmatrix}^T=\begin{pmatrix}4 & 8 & -12\\ 0 & -5 & 7\\ 5 & 3 & -6\end{pmatrix}$$

$$B^T A^T = \begin{pmatrix} 4 & 0 & 2 \\ -1 & 4 & 3 \\ 2 & 0 & 1 \end{pmatrix} \begin{pmatrix} 1 & 2 & -3 \\ -2 & 0 & 1 \\ 3 & -1 & 0 \end{pmatrix} = \begin{pmatrix} 4 & 8 & -12 \\ 0 & -5 & 7 \\ 5 & 3 & -6 \end{pmatrix} = (AB)^T$$

Example 2.4 If $A = \begin{pmatrix} 1 & 1 & 1 \\ 1 & 2 & 3 \\ 1 & 4 & 9 \end{pmatrix}$ and $B = \begin{pmatrix} 2 & 5 & 3 \\ 3 & 1 & 2 \\ 1 & 2 & 1 \end{pmatrix}$ find A^{-1}, B^{-1} and hence show that $(AB)^{-1} = B^{-1}A^{-1}$.

Solution Given $A = \begin{pmatrix} 1 & 1 & 1 \\ 1 & 2 & 3 \\ 1 & 4 & 9 \end{pmatrix}$; $B = \begin{pmatrix} 2 & 5 & 3 \\ 3 & 1 & 2 \\ 1 & 2 & 1 \end{pmatrix}$

$$|A| = 1(18-12) - 1(9-3) + 1(4-2) = 2 \neq 0$$
$$|B| = 2(1-4) - 5(3-2) + 3(6-1) = 4 \neq 0$$

Now $A^{-1} = \dfrac{\text{adj}\, A}{|A|}$ and $B^{-1} = \dfrac{\text{adj}\, B}{|B|}$, where

$$\text{adj}\, A = \begin{pmatrix} +(18-12) & -(9-3) & +(4-2) \\ -(9-4) & +(9-1) & -(4-1) \\ +(3-2) & -(3-1) & +(2-1) \end{pmatrix}^T \text{ and } \text{adj}\, B = \begin{pmatrix} +(1-4) & -(3-2) & +(6-1) \\ -(5-6) & +(2-3) & -(4-6) \\ +(10-3) & -(4-9) & +(2-15) \end{pmatrix}^T$$

$$\text{or adj}\, A = \begin{pmatrix} 6 & -6 & 2 \\ -5 & 8 & -3 \\ 2 & -3 & 1 \end{pmatrix} \text{ and adj}\, B = \begin{pmatrix} -3 & 1 & 7 \\ -1 & -1 & 5 \\ 5 & 1 & -13 \end{pmatrix}$$

$$\therefore \quad A^{-1} = \frac{1}{2}\begin{pmatrix} 6 & -6 & 2 \\ -5 & 8 & -3 \\ 2 & -3 & 1 \end{pmatrix} \text{ and } B^{-1} = \frac{1}{4}\begin{pmatrix} -3 & 1 & 7 \\ -1 & -1 & 5 \\ 5 & 1 & -13 \end{pmatrix}$$

$$\text{Again } AB = \begin{pmatrix} 1 & 1 & 1 \\ 1 & 2 & 3 \\ 1 & 4 & 9 \end{pmatrix}\begin{pmatrix} 2 & 5 & 3 \\ 3 & 1 & 2 \\ 1 & 2 & 1 \end{pmatrix} = \begin{pmatrix} 6 & 8 & 6 \\ 11 & 13 & 10 \\ 23 & 27 & 20 \end{pmatrix}$$

$$|AB| = 6(260-270) - 8(220-230) + 6(297-299) = 8$$

$$(AB)^{-1} = \frac{\text{adj}(AB)}{|AB|}$$

$$\text{adj}(AB) = \begin{pmatrix} (260-270) & -(220-230) & (297-299) \\ -(160-162) & (120-138) & -(162-184) \\ (80-78) & -(60-66) & (78-88) \end{pmatrix}^T$$

$$\therefore \text{adj}(AB) = \begin{pmatrix} -10 & 2 & -2 \\ 10 & -18 & 6 \\ -2 & 22 & -10 \end{pmatrix} \text{ and } (AB)^{-1} = \frac{1}{8}\begin{pmatrix} -10 & 2 & -2 \\ 10 & -18 & 6 \\ -2 & 22 & -10 \end{pmatrix}$$

$$\text{Now } B^{-1}A^{-1} = \frac{1}{2}\begin{pmatrix} 6 & -6 & 2 \\ -5 & 8 & -3 \\ 2 & -3 & 1 \end{pmatrix} \frac{1}{4}\begin{pmatrix} -3 & 1 & 7 \\ -1 & -1 & 5 \\ 5 & 1 & -13 \end{pmatrix}$$

$$= \frac{1}{8}\begin{pmatrix} -10 & 2 & -2 \\ 10 & -18 & 6 \\ -2 & 22 & -10 \end{pmatrix}$$

Hence $(AB)^{-1} = B^{-1}A^{-1}$.

Example 2.5 Prove that every square matrix can be uniquely expressed as the sum of symmetric and skew-symmetric matrices. Give an example.

Solution Let A be a square matrix and A^T its transpose. Then we have $A=\frac{1}{2}(A+A^T)+\frac{1}{2}(A-A^T)$

$$= B + C \text{ (say)} \tag{1}$$

$$\text{Where } B=\frac{1}{2}(A+A^T) \text{ and } C=\frac{1}{2}(A-A^T) \tag{2}$$

Now $B^T=\frac{1}{2}(A+A^T)^T=\frac{1}{2}(A^T+A)=B$

and $C^T=\frac{1}{2}(A-A^T)^T=\frac{1}{2}(A^T-A)=-C$ $(\because\ (A^T)^T=A)$

Thus B is a symmetric matrix and C is a skew-symmetric matrix.

Hence from (1) we conclude that a square matrix can be expressed as the sum of a symmetric and skew-symmetric matrices.

Uniqueness *of* (1)

$$\text{Let } A = D + E \tag{3}$$

where D is a symmetric and E is skew symmetric matrix so that $D^T = D$ and $E^T = -E$.

$$\text{Then } A^T=\frac{1}{2}(D+E)^T=\frac{1}{2}(D^T+E^T)=D.E \tag{4}$$

On adding and subtracting (3) and (4) respectively we get $D=\frac{1}{2}(A+A^T)=B$ and $E=\frac{1}{2}(A-A^T)=C$ which proves the uniqueness of (1).

Example Let $A=\begin{pmatrix}1&2&4\\0&1&2\\1&2&0\end{pmatrix}$ then $A^T=\begin{pmatrix}1&0&1\\2&1&2\\4&2&0\end{pmatrix}$

$B=$ Symmetric part of $A=\frac{1}{2}(A+A^T)=\frac{1}{2}\begin{pmatrix}2&2&5\\2&2&4\\5&4&0\end{pmatrix}$

$C=$ Skew-Symmetric part of $A=\frac{1}{2}(A-A^T)=\frac{1}{2}\begin{pmatrix}0&2&3\\-2&0&0\\-3&0&0\end{pmatrix}$

Example 2.6 If A and B are idempotent then AB is also idempotent if $AB = BA$.

Solution It is given that $AB = BA$. Also $A^2 = A$ and $B^2 = B$

$$(AB)^2 = (AB)(AB) = A(BA)B$$
$$= A(AB)B$$
$$= (AA)(BB)$$

$= A^2B^2 = AB$. Hence AB is idempotent.

Example 2.7 If A and B are Hermitian matrices then show that $AB + BA$ is Hermitian and $AB - BA$ is skew-Hermitian matrix.

Solution Since A and B are Hermitian matrices, we have $A^* = A$ and $B^* = B$ (1)

Now $(AB+BA)^* = (AB)^*+(BA)^* = B^*A^*+A^*B^*$

$$= BA + AB = (AB + BA)$$

$\Rightarrow$ $(AB + BA)$ is Hermitian matrix

Again $(AB - BA)^* = (AB)^* = (BA)^* = B^*A^* - A^*B^*$

$$= BA - AB = (AB - BA)$$

$\Rightarrow$ $(AB - BA)$ is skew-Hermitian matrix.

Example 2.8 If a non-singular matrix A is symmetric, then A^{-1} is also symmetric.

Solution If A is symmetric $A = A^T$ (1)

If A is non-singular then $A^{-1}A = I = I^T$

$$= (AA^{-1})^T \quad (\because I = A^{-1}A = AA^{-1})$$

$$= (A^{-1})^T A^T \quad (\because (AB)^T = B^T A^T$$

$\therefore$ $A^{-1}A = (A^{-1})^T A$ $(\because A^T = A)$ or $A^{-1} = (A^{-1})^T$, by right cancellation law

Hence A^{-1} is symmetric by definition.

Example 2.9 Prove that if A is an orthogonal matrix then $|A| = \pm 1$.

Solution Since A is orthogonal

$$AA^T = I \Rightarrow |AA^T| = |I|; \text{ But } |I| = 1$$

$$\Rightarrow |A|\,|A^T| = |I| \Rightarrow |A|.|A| = 1 \; (\because |A| = |A^T| \;\&\; |I| = 1)$$

$$\Rightarrow |A|^2 = 1 \;\therefore\; |A| = \pm 1$$

Example 2.10 Show that $\frac{1}{\sqrt{3}}\begin{pmatrix} 1 & 1-i \\ 1+i & -1 \end{pmatrix}$ is unitary.

Solution Let A be the given matrix then

$$A^* = \frac{1}{\sqrt{3}}\begin{pmatrix} 1 & 1-i \\ 1+i & -1 \end{pmatrix}^T \quad \therefore A^* = \frac{1}{\sqrt{3}}\begin{pmatrix} 1 & 1+i \\ 1-i & -1 \end{pmatrix}$$

Now $A^*A = \frac{1}{\sqrt{3}}\begin{pmatrix} 1 & 1+i \\ 1-i & -1 \end{pmatrix} \frac{1}{\sqrt{3}}\begin{pmatrix} 1 & 1+i \\ 1-i & -1 \end{pmatrix} = \frac{1}{3}\begin{pmatrix} 1+1-i^2 & 0 \\ 0 & 1-i^2+1 \end{pmatrix} = \begin{pmatrix} 1 & 0 \\ 0 & 1 \end{pmatrix} = I$

Similarly $AA^* = I$. Hence A is unitary.

EXERCISE I

1. If $A = \begin{pmatrix} -2 & 3 & -1 \\ -1 & 2 & -1 \\ -6 & 9 & 4 \end{pmatrix}$ and $B = \begin{pmatrix} 1 & 3 & -1 \\ 2 & 2 & -1 \\ 3 & 0 & -1 \end{pmatrix}$ find $2A + 4B$, $A - B$, AB and BA.

2. Express $\begin{pmatrix} 1 & 2 & 6 \\ 3 & 4 & 5 \\ 3 & 1 & 7 \end{pmatrix}$ as (i) the sum of lower triangular and upper triangular matrix with zero leading diagonal (ii) the sum of symmetric and skew-symmetric matrices.

3. If $A = \begin{pmatrix} 1 & 1 & -1 \\ 2 & -3 & 4 \\ 3 & -2 & 3 \end{pmatrix}$, $B = \begin{pmatrix} -1 & -2 & -1 \\ 6 & 12 & 6 \\ 5 & 10 & 6 \end{pmatrix}$ and $C = \begin{pmatrix} -1 & -1 & 1 \\ 2 & 2 & -2 \\ -3 & -3 & 3 \end{pmatrix}$ show that AB and CA are null matrices where as BA and AC are not.

4. If $A=\begin{pmatrix}2 & -3 & 1\\ 3 & 1 & 3\\ -5 & 2 & -4\end{pmatrix}$ show that $A(A-I)(A+2I)=0$.

5. If $E=\begin{pmatrix}0 & 1 & 0\\ 0 & 0 & 1\\ 0 & 0 & 0\end{pmatrix}$ and $F=\begin{pmatrix}0 & 0 & 0\\ 1 & 0 & 0\\ 0 & 1 & 0\end{pmatrix}$ then calculate EF and FE and show that $EF^2+F^2E=E$.

6. If $A=\begin{pmatrix}2 & -5 & 7\\ -3 & 4 & 6\\ 1 & 8 & 5\end{pmatrix}$; $B=\begin{pmatrix}8 & 1 & 3\\ 2 & -9 & 2\\ -4 & 5 & 6\end{pmatrix}$ and $C=\begin{pmatrix}-7 & 10 & 13\\ 1 & 6 & -3\\ 9 & 2 & 0\end{pmatrix}$ then verify that

i. $A(B+C)=AB+AC$

ii. $A(BC)=(AB)C$

iii. $(A+B+C)^T=A^T+B^T+C^T$

iv. $(AB)^T=B^TA^T$

v. $(ABC)^T=C^TB^TA^T$

vi. $(AB)^{-1}=B^{-1}A^{-1}$

vii. $(BC)^{-1}=C^{-1}B^{-1}$

viii $(ABC)^{-1}=C^{-1}B^{-1}A^{-1}$

7. Show that every square matrix can be expressed uniquely as the sum of Hermitian and a skew-Hermitian matrix.

8. If A is a Hermitian (skew-Hermitian) matrix then prove that iA is skew-Hermitian matrix.

9. If $A=\begin{pmatrix}2 & -1 & 1\\ -2 & 3 & -2\\ -4 & 4 & -3\end{pmatrix}$; $B=\begin{pmatrix}1 & 2 & 3\\ 1 & 2 & 3\\ 1 & -2 & -3\end{pmatrix}$; $C=\begin{pmatrix}-5 & -8 & 0\\ 3 & 5 & 0\\ 1 & 2 & -1\end{pmatrix}$ and $D=\begin{pmatrix}1 & -2 & -6\\ -3 & 2 & 9\\ 2 & 0 & -3\end{pmatrix}$

Show that (i) A is idempotent (ii) B is nilpotent of order 2 (iii) C is involutory (iv) D is periodic of order 2.

10. If A and B are two square matrices of the same order which commute then prove that

i. $(A\pm B)^2=A^2\pm 2AB+B^2$

ii. $(A\pm B)(A^2\mp AB+B^2)=A^3\pm B^3$

iii. $(A\pm B)^3=A^3\pm 3A^2B+3AB^2\pm B^3$

11. Show that the matrix $\frac{1}{3}\begin{pmatrix}1 & 2 & 2\\ 2 & 1 & -2\\ 2 & -2 & 1\end{pmatrix}$ is orthogonal.

12. Prove that $\begin{bmatrix}\frac{1+i}{2\sqrt{2}} & \frac{\sqrt{3}(1-i)}{2\sqrt{2}}\\ \frac{\sqrt{2+i}}{2} & \frac{i\sqrt{2}-1}{2\sqrt{3}}\end{bmatrix}$ is unitary.

13. If $D=\text{diag }[d_1, d_2, \ldots, d_n]$, $d_1, d_2, \ldots, d_n \neq 0$ prove that $D^{-1}=\text{diag }[d_1^{-1}, d_2^{-1}, \ldots, d_n^{-1}]$

ANSWERS

1. $2A+4B=\begin{pmatrix}0 & 18 & -6\\ 6 & 12 & -6\\ 0 & 18 & 4\end{pmatrix}$ $A-B=\begin{pmatrix}-3 & 0 & 0\\ -3 & 0 & 0\\ -9 & 9 & 5\end{pmatrix}$ $AB=\begin{pmatrix}1 & 0 & 0\\ 0 & 1 & 0\\ 24 & 0 & -7\end{pmatrix}$ $BA=\begin{pmatrix}1 & 0 & 0\\ 0 & 1 & -8\\ 0 & 0 & -7\end{pmatrix}$

2. $\text{L.T.M}=\begin{pmatrix}1 & 0 & 0\\ 3 & 4 & 0\\ 3 & 1 & 7\end{pmatrix}$ $\text{U.T.M}=\begin{pmatrix}0 & 2 & 6\\ 0 & 0 & 5\\ 0 & 0 & 0\end{pmatrix}$ $\text{Sy.M}=\frac{1}{2}\begin{pmatrix}2 & 5 & 9\\ 5 & 8 & 6\\ 9 & 6 & 14\end{pmatrix}$ $\text{Skew Sy.M}=\begin{pmatrix}0 & -1 & 3\\ 1 & 0 & 4\\ -3 & -4 & 0\end{pmatrix}$

RANK OF MATRIX

The number $r \in N$ is called the rank of the matrix A, if

i. there exists atleast one minor of order r of A which does not vanish.

ii. every minor of order $(r + 1)$ of A, if any, vanishes.

In other words, the rank of a matrix is defined as the largest order of all the non-vanishing minors of the matrix.

From property (ii), every minor of order greater than r will vanish. The rank of the matrix A is denoted by $\rho(A)$.

Note:

i. $\rho(A) = 0$ when A is a null matrix.

ii. $\rho(A) \geq 1$ when A is a non-zero matrix.

iii. $\rho(A) \geq r$ when there exists a non-zero minor of order r.

iv. $\rho(A) \leq r$ when every minor of order $(r + 1)$ vanishes.

v. $\rho(A) \leq m$ when A is any $m \times n$ matrix such that $m \leq n$.

vi. $\rho(A) \leq n$ when A is any $m \times n$ matrix such that $m \geq n$.

vii. $\rho(A) = n$ when A is any $n \times n$ non-singular matrix.

viii. If I_n is unit matrix of order n, then $\rho(I_n) = n$.

ix. If A is a diagonal matrix of order n or upper (or lower) triangular matrix with non-zero diagonal elements, then $\rho(A) = n$.

x. $\rho(A^T) = \rho(A)$.

Example 2.11 Find the rank of the following matrices.

i. $\begin{pmatrix} 1 & 2 & 3 \\ 4 & 5 & 6 \\ 7 & 8 & 9 \end{pmatrix}$ *(M.U. 1997)* ii. $\begin{pmatrix} 1 & 2 & 3 & -1 \\ 3 & 6 & 9 & -3 \\ 2 & 4 & 6 & -2 \end{pmatrix}$ *(M.U. 1994, 1995)*

iii. $\begin{pmatrix} 4 & 2 & 2 & 3 \\ 0 & 4 & 2 & 3 \\ 0 & 0 & 0 & 5 \\ 0 & 0 & 4 & 5 \end{pmatrix}$ *(M.U. 1995)* iv. $\begin{pmatrix} 1 & 1 & -1 & 1 \\ 1 & -1 & 2 & -1 \\ 3 & 1 & 0 & 1 \end{pmatrix}$ *(M.U. 1997)*

Solution

i. Consider $\begin{pmatrix} 1 & 2 & 3 \\ 4 & 5 & 6 \\ 7 & 8 & 9 \end{pmatrix} = A(\text{say})$

Now $|A| = 1(45-48) - 2(36-42) + 3(32-35) = -3 - 2(-6) + 3(-3) = 0$

Since the largest minor of A is its determinant itself and its value is zero, $\rho(A) \neq 3$.

$\therefore \rho(A) < 3$

Consider any minor of order 2×2 of A, say, $\begin{vmatrix} 1 & 2 \\ 4 & 5 \end{vmatrix} = 5 - 8 = -3 \neq 0$. Therefore, $\rho(A) = 2$

ii. Let $A = \begin{pmatrix} 1 & 2 & 3 & -1 \\ 3 & 6 & 9 & -3 \\ 2 & 4 & 6 & -2 \end{pmatrix}$

$$\begin{matrix} R_2 \to R_2 - 3R_1 \\ R_3 \to R_3 - 2R_1 \end{matrix} \sim \begin{pmatrix} 1 & 2 & 3 & -1 \\ 0 & 0 & 0 & 0 \\ 0 & 0 & 0 & 0 \end{pmatrix}$$

$\Rightarrow$ There occurs only minors of order 1 × 1. Therefore, $\rho(A) = 1$

iii. Let $A = \begin{pmatrix} 4 & 2 & 2 & 3 \\ 0 & 4 & 2 & 3 \\ 0 & 0 & 0 & 5 \\ 0 & 0 & 4 & 5 \end{pmatrix}; R(3,4) \sim \begin{pmatrix} 4 & 2 & 2 & 3 \\ 0 & 4 & 2 & 3 \\ 0 & 0 & 4 & 5 \\ 0 & 0 & 0 & 5 \end{pmatrix}$

$\therefore \rho(A) = 4$ ($\because$ upper triangular matrix have non-zero diagonal elements).

iv. Let $A = \begin{pmatrix} 1 & 1 & -1 & 1 \\ 1 & -1 & 2 & -1 \\ 3 & 1 & 0 & 1 \end{pmatrix}$

It is of order 3 × 4. Largest minor is of order 3 × 3.

Consider $\begin{vmatrix} 1 & 1 & -1 \\ 1 & -1 & 2 \\ 3 & 1 & 0 \end{vmatrix} = 1(0-2) - 1(0-6) - 1(1+3) = 0.$

Similarly, if we consider all the other possible 3 × 3 minors i.e.,

$\begin{vmatrix} 1 & 1 & 1 \\ 1 & -1 & -1 \\ 3 & 1 & 1 \end{vmatrix}, \begin{vmatrix} 1 & -1 & 1 \\ 1 & 2 & -1 \\ 3 & 0 & 1 \end{vmatrix}, \begin{vmatrix} 1 & -1 & 1 \\ -1 & 2 & -1 \\ 1 & 0 & 1 \end{vmatrix}$ they will all vanish.

$\therefore \rho(A) \neq 3$

Now consider the next lower minor i.e., of order 2 × 2.

$\begin{vmatrix} 1 & 1 \\ 1 & -1 \end{vmatrix} = -1 - 1 = -2 \neq 0.$ Therefore, $\rho(A) = 2.$

Aliter:

Let $A = \begin{pmatrix} 1 & 1 & -1 & 1 \\ 1 & -1 & 2 & -1 \\ 3 & 1 & 0 & 1 \end{pmatrix}$

$$\begin{matrix} R_2 \to R_2 - R_1 \\ R_3 \to R_3 - R_1 \end{matrix} \sim \begin{pmatrix} 1 & 1 & -1 & 1 \\ 0 & -2 & 3 & -2 \\ 0 & -2 & 3 & -2 \end{pmatrix}$$

$$R_3 \to R_3 - R_2 \sim \begin{pmatrix} 1 & 1 & -1 & 1 \\ 0 & -2 & 3 & -2 \\ 0 & 0 & 0 & 0 \end{pmatrix}$$

It is clear that the largest non-vanishing minor is of order 2.

Therefore, $\rho(A) = 2$.

Note: An elementary operation (or an elementary transformation) is an operation of any one of the following three types:

i. Interchange of any two rows (or columns) Symbolically, $R(i, j)$ stands for the interchange of the ith and j^{th} rows [$C(i, j)$ for columns].
ii. Multiplication of the elements of any row (or column) by any non-zero number Symbolically, $k\,R_i$ stands for the multiplication of the elements of the ith row by the non-zero k [$k\,C_i$ for columns].
iii. Addition of the elements of any row (or column) with the corresponding elements of any other row (or column) multiplied by any number Symbolically, $R_i + kR_j$ stands for the addition of k times the elements of the jth row to the corresponding elements of the ith row [$C_i + kC_j$ for columns].

An elementary transformation is called a row transformation or a column transformation in accordance with its application to rows or columns. These elementary operations are oftenly known as E-operations.

Rank of a matrix is invariant under elementary operations.

NORMAL FORM OF A MATRIX

Every non-zero matrix of rank r can be reduced to one of the forms $\begin{pmatrix} I_r & \vdots & 0 \\ \cdots & \cdots & \cdots \\ 0 & \vdots & 0 \end{pmatrix}$, $[I_r : 0]$, $\begin{pmatrix} I_r \\ \cdots \\ 0 \end{pmatrix}$ or I_r by a finite numbers of elementary transformations, where I_r denotes the identity matrix of order r which is called as normal or canonical form of the given non-zero matrix.

Now it is obvious from the definition that rank of the normal form of a given matrix A is the same as the rank of A.

Example 2.12 Find the rank of the matrix $A = \begin{pmatrix} 1 & 2 & 1 & 2 \\ 1 & 3 & 2 & 2 \\ 2 & 4 & 3 & 4 \\ 3 & 7 & 4 & 6 \end{pmatrix}$

Solution Given $A = \begin{pmatrix} 1 & 2 & 1 & 2 \\ 1 & 3 & 2 & 2 \\ 2 & 4 & 3 & 4 \\ 3 & 7 & 4 & 6 \end{pmatrix}$

Reduce A to normal form.

$$C_4 \rightarrow C_4 - 2C_1 \sim \begin{pmatrix} 1 & 2 & 1 & 0 \\ 1 & 3 & 2 & 0 \\ 2 & 4 & 3 & 0 \\ 3 & 7 & 4 & 0 \end{pmatrix}$$

$$\begin{matrix} R_1 \rightarrow R_2 - R_1 \\ R_3 \rightarrow R_3 - 2R_1 \\ R_4 \rightarrow R_4 - 3R_1 \end{matrix} \sim \begin{pmatrix} 1 & 2 & 1 & 0 \\ 0 & 1 & 1 & 0 \\ 0 & 0 & 1 & 0 \\ 0 & 1 & 1 & 0 \end{pmatrix}$$

$$\begin{matrix} R_4 \to R_4 - R_2 \\ R_2 \to R_2 - R_3 \end{matrix} \sim \begin{pmatrix} 1 & 2 & 1 & 0 \\ 0 & 1 & 0 & 0 \\ 0 & 0 & 1 & 0 \\ 0 & 0 & 0 & 0 \end{pmatrix}$$

$$\begin{matrix} C_4 \to C_2 - 2C_1 \\ C_3 \to C_3 - C_1 \end{matrix} \sim \begin{pmatrix} 1 & 0 & 0 & 0 \\ 0 & 1 & 0 & 0 \\ 0 & 0 & 1 & 0 \\ 0 & 0 & 0 & 0 \end{pmatrix} = \begin{pmatrix} I_3 & \vdots & 0 \\ \cdots & \cdots & \cdots \\ 0 & \vdots & 0 \end{pmatrix}$$

Therefore $\rho(A) = 3$.

Example 2.13 Find the rank of the matrix $A = \begin{pmatrix} 2 & 2 & 4 & 6 \\ 3 & 3 & 7 & 10 \\ 5 & 7 & 11 & 17 \\ 6 & 8 & 13 & 16 \\ 3 & 4 & 6 & 6 \end{pmatrix}$.

Solution Given $A = \begin{pmatrix} 2 & 2 & 4 & 6 \\ 3 & 3 & 7 & 10 \\ 5 & 7 & 11 & 17 \\ 6 & 8 & 13 & 16 \\ 3 & 4 & 6 & 6 \end{pmatrix}$

Reduce A to normal form.

$$R_1 \to \frac{1}{2}R_1 \sim \begin{pmatrix} 1 & 1 & 2 & 3 \\ 3 & 3 & 7 & 10 \\ 5 & 7 & 11 & 17 \\ 6 & 8 & 13 & 16 \\ 3 & 4 & 6 & 6 \end{pmatrix}$$

$$\begin{matrix} R_2 \to R_2 - 3R_1 \\ R_3 \to R_3 - 5R_1 \\ R_4 \to R_4 - 6R_1 \\ R_5 \to R_5 - 3R_1 \end{matrix} \sim \begin{pmatrix} 1 & 1 & 2 & 3 \\ 0 & 1 & 1 & 1 \\ 0 & 2 & 1 & 2 \\ 0 & 2 & 1 & -2 \\ 0 & 1 & 0 & -3 \end{pmatrix}$$

$$\begin{matrix} R_1 \to R_1 - R_2 \\ R_3 \to R_3 - 2R_2 \\ R_4 \to R_4 - 2R_2 \\ R_5 \to R_5 - R_4 \end{matrix} \sim \begin{pmatrix} 1 & 0 & 1 & 2 \\ 0 & 1 & 1 & 1 \\ 0 & 0 & -1 & 0 \\ 0 & 0 & -1 & -4 \\ 0 & 0 & -1 & -4 \end{pmatrix}$$

$$\begin{matrix} R_1 \to R_1 + R_3 \\ R_2 \to R_2 + R_3 \\ R_5 \to R_5 - R_4 \end{matrix} \sim \begin{pmatrix} 1 & 0 & 0 & 2 \\ 0 & 1 & 0 & 1 \\ 0 & 0 & -1 & 0 \\ 0 & 0 & -1 & -4 \\ 0 & 0 & 0 & 0 \end{pmatrix}$$

$$\begin{matrix} R_3 \to -R_3 \\ R_4 \to -\frac{1}{4}(R_4 + R_3) \\ R_1 \to R_1 - 2R_4 \\ R_2 \to R_2 - R_4 \end{matrix} \sim \begin{pmatrix} 1 & 0 & 0 & 0 \\ 0 & 1 & 0 & 0 \\ 0 & 0 & 1 & 0 \\ 0 & 0 & 0 & 1 \\ 0 & 0 & 0 & 0 \end{pmatrix} = \begin{pmatrix} I_4 \\ \cdots \\ 0 \end{pmatrix}$$

Therefore, $\rho(A) = 4$.

CONSISTENCY OF A SYSTEM OF ALGEBRAIC LINEAR EQUATIONS

Consider the system of m linear equation in n unknowns $x_1, x_2, \ldots, x_n$ i.e.,

$$\begin{aligned} a_{11}x_1 + a_{12}x_2 + \cdots + a_{1n}x_n &= b_1 \\ a_{21}x_1 + a_{22}x_2 + \cdots + a_{2n}x_n &= b_2 \\ a_{m1}x_1 + a_{m2}x_2 + \cdots + a_{mn}x_n &= b_n \end{aligned} \qquad (1)$$

'In matrix notation, the above system (1) can be written as $AX = B$ where

$$A = \begin{pmatrix} a_{11} & a_{12}\cdots & a_{1n} \\ a_{21} & a_{22}\cdots & a_{2n} \\ \cdots & \cdots & \cdots \\ a_{m1} & a_{m2}\cdots & a_{mn} \end{pmatrix}; X = \begin{pmatrix} x_1 \\ x_2 \\ \vdots \\ x_n \end{pmatrix} \text{ and } B = \begin{pmatrix} b_1 \\ b_2 \\ \vdots \\ b_n \end{pmatrix}$$

If $B = 0$ i.e., zero column matrix, then the system (1) reduces to $AX = 0$. Such a system is called system of homogeneous linear equations. If $B \neq 0$, then system (1) is called system of non-homogeneous linear equations.

Solution to System of Non-homogeneous Linear Equations *AX* = *B*

To determine whether the system (1) possesses a solution or not, we consider the matrix A so called coefficient matrix, and augmented matrix $(A : B)$.

$$\text{Here } (A:B) = \begin{bmatrix} a_{11} & a_{12} & a_{1n} & b_1 \\ a_{21} & a_{22} & a_{n2} & b_2 \\ \cdots & \cdots & \cdots & \cdots \\ a_{m1} & a_{m2} & a_{mn} & b_n \end{bmatrix}$$

A system of equations $AX = B$ need not always have a solution. If there is no solution then the system is said to be inconsistent. If the system possesses one or more solutions then the system is said to be *consistent*. This can be tested easily using the rank concept for both coefficient matrix and augmented matrix with the help of unknown quantities.

Working procedure Let $AX = B$ be the system as in (1) then

i. If $\rho(A) = \rho(A{:}B) =$ number of unknowns, then the system is consistent and possesses a unique solution.

ii. If $\rho(A) = \rho(A{:}B) <$ number of unknowns, then the system is consistent and possesses many solutions.

   In this case, if both ranks are equal to k (say) and n is the number of unknowns then there will be $(n - k)$ parameters.

iii. If $\rho(A) \neq \rho(A{:}B)$ then the system is inconsistent.

Note: Use only elementary row operations.

Example 2.14 Test for consistency and solve the following system.

$$x + y - z = 1,\ 3x + y - 6z = -2,\ 2x - 3y + z = 0 \qquad (M.U.\ 1994)$$

Solution The augmented matrix $(A{:}\ B)$ of the given system is

$$(A{:}B) = \left(\begin{array}{ccc|c} 1 & 1 & -1 & 1 \\ 3 & 1 & -6 & -2 \\ 2 & -3 & 1 & 0 \end{array}\right)$$

$$R_2 \to R_2 - 3R_1 \quad R_3 \to R_3 - 2R_1 \sim \left(\begin{array}{ccc|c} 1 & 1 & -1 & 1 \\ 0 & -2 & -3 & -5 \\ 0 & -5 & 3 & -2 \end{array}\right)$$

$$R_2 \to -(3R_2 - R_3) \sim \left(\begin{array}{ccc|c} 1 & 1 & -1 & 1 \\ 0 & 1 & 12 & 13 \\ 0 & -5 & 3 & -2 \end{array}\right)$$

$$R_1 \to R_1 - R_2 \quad R_3 \to \frac{1}{63}(R_3 - 5R_2) \sim \left(\begin{array}{ccc|c} 1 & 0 & -13 & -12 \\ 0 & 1 & 12 & 13 \\ 0 & 0 & 1 & 1 \end{array}\right)$$

$$R_1 \to R_1 + 13R_3 \quad R_2 \to R_2 - 12R_3 \sim \left(\begin{array}{ccc|c} 1 & 0 & 0 & 1 \\ 0 & 1 & 0 & 1 \\ 0 & 0 & 0 & 1 \end{array}\right)$$

$$\therefore (A{:}B) \sim \begin{pmatrix} 1 & 0 & 0 & 1 \\ 0 & 1 & 0 & 1 \\ 0 & 0 & 1 & 1 \end{pmatrix}; A \sim \begin{pmatrix} 1 & 0 & 0 \\ 0 & 1 & 0 \\ 0 & 0 & 1 \end{pmatrix}$$

$\Rightarrow \rho(A:B) = 3$ and $\rho(A) = 3$

$\because \rho(A:B) = \rho(A) = 3 =$ number of unknowns.

Therefore the system is consistent and has a unique solution. Now the given system in matrix form reduces to

$$\begin{pmatrix} 1 & 0 & 0 \\ 0 & 1 & 0 \\ 0 & 0 & 1 \end{pmatrix}\begin{pmatrix} x \\ y \\ z \end{pmatrix} = \begin{pmatrix} 1 \\ 1 \\ 1 \end{pmatrix} \Rightarrow x = y = z = 1$$

Example 2.15 Test for consistency and solve

$$x + 2y - z = 3,\ 3x - y + 2z = 1,\ 2x - 2y + 3z = 2,\ x - y + z = -1.$$ *(M.U. 1995)*

Solution The augmented matrix of the system is

$$(A:B) = \left(\begin{array}{ccc|c} 1 & 2 & -1 & 3 \\ 3 & -1 & 2 & 1 \\ 2 & -2 & 3 & 2 \\ 1 & -1 & 1 & -1 \end{array}\right)$$

$$R_2 \to R_2 - 3R_1 \quad R_3 \to R_3 - 2R_1 \quad R_4 \to R_4 - R_1 \sim \left(\begin{array}{ccc|c} 1 & 2 & -1 & 3 \\ 0 & -7 & 5 & 4 \\ 0 & -6 & 5 & -4 \\ 0 & -3 & 2 & -4 \end{array}\right)$$

$$R_2 \to -(R_2 - R_3) \sim \left(\begin{array}{ccc|c} 1 & 2 & -1 & 3 \\ 0 & 1 & 0 & 4 \\ 0 & -6 & 5 & -4 \\ 0 & -3 & 2 & -4 \end{array}\right)$$

$$R_1 \to R_1 - 2R_2 \quad R_3 \to \frac{1}{5}(R_3 + 6R_2) \quad R_4 \to \frac{1}{2}(R_4 + 3R_2) \sim \left(\begin{array}{ccc|c} 1 & 0 & -1 & -5 \\ 0 & 1 & 0 & 4 \\ 0 & 0 & 1 & 4 \\ 0 & 0 & 1 & 4 \end{array}\right)$$

$$R_1 \to R_1 + R_3 \quad R_4 \to R_4 - R_3 \sim \left(\begin{array}{ccc|c} 1 & 0 & 0 & -1 \\ 0 & 1 & 0 & 4 \\ 0 & 0 & 1 & 4 \\ 0 & 0 & 0 & 0 \end{array}\right)$$

$$\therefore (A:B) \sim \begin{pmatrix} 1 & 0 & 0 & -1 \\ 0 & 1 & 0 & 4 \\ 0 & 0 & 1 & 4 \\ 0 & 0 & 0 & 0 \end{pmatrix} \text{ and } A \sim \begin{pmatrix} 1 & 0 & 0 \\ 0 & 1 & 0 \\ 0 & 0 & 1 \\ 0 & 0 & 0 \end{pmatrix}$$

$\Rightarrow \rho(A:B) = 3$ and $\rho(A) = 3$

$\because \rho(A:B) = \rho(A) = 3 =$ number of unknowns

Therefore, the system is consistent and has a unique solution. Now the given system in matrix form reduces to

$$\begin{pmatrix} 1 & 0 & 0 \\ 0 & 1 & 0 \\ 0 & 0 & 1 \\ 0 & 0 & 0 \end{pmatrix} \begin{pmatrix} x \\ y \\ z \end{pmatrix} = \begin{pmatrix} -1 \\ 4 \\ 4 \\ 0 \end{pmatrix} \Rightarrow x = -1, y = 4, z = 4$$

Example 2.16 Test for consistency and solve

$$3x - 4y + 5z = 0,\ 5x + 3y - 4z = 15,\ 4x - 5y - 3z = 19. \qquad (M.U.\ 1997)$$

Solution The augmented matrix of the given system is

$$(A:B) = \left(\begin{array}{ccc|c} 3 & -4 & 5 & 0 \\ 5 & 3 & -4 & 15 \\ 4 & -5 & -3 & 19 \end{array}\right)$$

$$R_1 \to -(R_1 - R_3) \sim \left(\begin{array}{ccc|c} 1 & -1 & -8 & 19 \\ 5 & 3 & -4 & 15 \\ 4 & -5 & -3 & 19 \end{array}\right)$$

$$\begin{array}{l} R_2 \to \frac{1}{4}(R_2 - 5R_1) \\ R_3 \to R_3 - 4R_1 \end{array} \sim \left(\begin{array}{ccc|c} 1 & -1 & -8 & 19 \\ 0 & 2 & 9 & -20 \\ 0 & -1 & 29 & -57 \end{array}\right)$$

$$R_2 \to R_2 + R_3 \sim \left(\begin{array}{ccc|c} 1 & -1 & -8 & 19 \\ 0 & 1 & 38 & -77 \\ 0 & -1 & 29 & -57 \end{array}\right)$$

$$\begin{array}{l} R_1 \to R_1 + R_2 \\ R_3 \to \frac{1}{67}(R_3 + R_2) \end{array} \sim \left(\begin{array}{ccc|c} 1 & 0 & 30 & -58 \\ 0 & 1 & 38 & -77 \\ 0 & 0 & 1 & -2 \end{array}\right)$$

$$\begin{array}{l} R_1 \to R_1 - 30R_3 \\ R_2 \to R_2 - 38R_3 \end{array} \sim \left(\begin{array}{ccc|c} 1 & 0 & 0 & 2 \\ 0 & 1 & 0 & -1 \\ 0 & 0 & 1 & -2 \end{array}\right)$$

$$\therefore (A:B) = \begin{pmatrix} 1 & 0 & 0 & 2 \\ 0 & 1 & 0 & -1 \\ 0 & 0 & 1 & -2 \end{pmatrix};\ A = \begin{pmatrix} 1 & 0 & 0 \\ 0 & 1 & 0 \\ 0 & 0 & 1 \end{pmatrix}$$

i.e. $\rho(A:B) = 3$ and $\rho(A) = 3$ $\because \rho(A:B) = \rho(A) = 3 =$ number of unknowns

Therefore the system is consistent and has a unique solution. Now the given system in matrix form reduces to

$$\begin{pmatrix} 1 & 0 & 0 \\ 0 & 1 & 0 \\ 0 & 0 & 1 \end{pmatrix} \begin{pmatrix} x \\ y \\ z \end{pmatrix} = \begin{pmatrix} 2 \\ -1 \\ -2 \end{pmatrix} \Rightarrow x = 2, y = -1, z = -2.$$

Example 2. 17 Test for consistency and solve

$2x_1 + x_2 + 2x_3 + x_4 = 6,\ 6x_1 - 6x_2 + 6x_3 + 12x_4 = 36,$
$4x_1 + 3x_2 + 3x_3 - 3x_4 = -1,\ 2x_1 + 2x_2 - x_3 + x_4 = 10.$

Solution The augmented matrix of the given system is

$$(A:B) = \left(\begin{array}{cccc|c} 2 & 1 & 2 & 1 & 6 \\ 6 & -6 & 6 & 12 & 36 \\ 4 & 3 & 3 & -3 & -1 \\ 2 & 2 & -1 & 1 & 10 \end{array}\right)$$

$$R_2 \to \frac{1}{6}R_2 \sim \left(\begin{array}{cccc|c} 2 & 1 & 2 & 1 & 6 \\ 1 & -1 & 1 & 2 & 6 \\ 4 & 3 & 3 & -3 & -1 \\ 2 & 2 & -1 & 1 & 10 \end{array}\right)$$

$$R_1 \to R_1 - R_2 \sim \left(\begin{array}{cccc|c} 1 & 2 & 1 & -1 & 0 \\ 1 & -1 & 1 & 2 & 6 \\ 4 & 3 & 3 & -3 & -1 \\ 2 & 2 & -1 & 1 & 10 \end{array}\right)$$

$$\begin{array}{l} R_2 \to \frac{-1}{3}(R_2 - R_1) \\ R_3 \to R_3 - 4R_1 \\ R_4 \to R_4 - 2R_1 \end{array} \sim \left(\begin{array}{cccc|c} 1 & 2 & 1 & -1 & 0 \\ 0 & 1 & 0 & -1 & -2 \\ 0 & -5 & -1 & 1 & -1 \\ 0 & -2 & -3 & 3 & 10 \end{array}\right)$$

$$\begin{array}{l} R_1 \to - R_1 - 2R_2 \\ R_3 \to (R_3 + 5R_2) \\ R_4 \to R_4 + 2R_2 \end{array} \sim \left(\begin{array}{cccc|c} 1 & 0 & 1 & 1 & 4 \\ 0 & 1 & 0 & -1 & -2 \\ 0 & 0 & 1 & 4 & 11 \\ 0 & 0 & -3 & 1 & 6 \end{array}\right)$$

$$\begin{array}{l} R_1 \to R_1 - R_3 \\ R_3 \to \frac{1}{3}(R_3 + 3R_1) \end{array} \sim \left(\begin{array}{cccc|c} 1 & 0 & 0 & -3 & -7 \\ 0 & 1 & 0 & -1 & -2 \\ 0 & 0 & 1 & 4 & 11 \\ 0 & 0 & 0 & 1 & 3 \end{array}\right)$$

$$\begin{array}{l} R_1 \to R_1 + 3R_4 \\ R_2 \to R_2 + R_4 \\ R_3 \to R_3 - 4R_4 \end{array} \sim \left(\begin{array}{cccc|c} 1 & 0 & 0 & 0 & 2 \\ 0 & 1 & 0 & 0 & 1 \\ 0 & 0 & 1 & 0 & -1 \\ 0 & 0 & 0 & 1 & 3 \end{array}\right)$$

$$\therefore (A:B) = \begin{pmatrix} 1 & 0 & 0 & 0 & 2 \\ 0 & 1 & 0 & 0 & 1 \\ 0 & 0 & 1 & 0 & -1 \\ 0 & 0 & 0 & 1 & 3 \end{pmatrix} \quad A = \begin{pmatrix} 1 & 0 & 0 & 0 \\ 0 & 1 & 0 & 0 \\ 0 & 0 & 1 & 0 \\ 0 & 0 & 0 & 1 \end{pmatrix}$$

$\Rightarrow \rho(A:B) = 4$ and $\rho(A) = 4$

$\because \rho(A:B) = \rho(A) = 4 =$ number of unknowns

Therefore the system is consistent and has a unique solution. Now the given system in matrix form reduces to

$$\begin{pmatrix} 1 & 0 & 0 & 0 \\ 0 & 1 & 0 & 0 \\ 0 & 0 & 1 & 0 \\ 0 & 0 & 0 & 1 \end{pmatrix} \begin{pmatrix} x_1 \\ x_2 \\ x_3 \\ x_4 \end{pmatrix} = \begin{pmatrix} 2 \\ 1 \\ -1 \\ 3 \end{pmatrix} \Rightarrow x_1 = 2,\ x_2 = 1,\ x_3 = -1,\ x_4 = 3$$

Example 2. 18 Test for consistency and solve $x-4y+7z=14,\ 3x+8y-2z=13,\ 7x-8y+26z=5.$

Solution The augmented matrix is given by

$$(A:B)=\left(\begin{array}{ccc|c}1&-4&7&14\\3&8&-2&13\\7&-8&26&5\end{array}\right)$$

$$\begin{array}{l}R_2\to R_2-R_1\\R_3\to R_3-7R_1\end{array}\sim\left(\begin{array}{ccc|c}1&-4&7&14\\0&20&-23&-29\\0&20&-23&-93\end{array}\right)$$

$$R_3\to-\frac{1}{64}(R_3-R_2)\sim\left(\begin{array}{ccc|c}1&-4&7&14\\0&20&-23&-29\\0&0&0&1\end{array}\right)$$

$$\therefore\ (A:B)=\begin{pmatrix}1&-4&7&14\\0&20&-23&-29\\0&0&0&1\end{pmatrix}\Rightarrow A=\begin{pmatrix}1&-4&7\\0&20&-23\\0&0&0\end{pmatrix}$$

$\Rightarrow\rho(A:B)=3$ and $\rho(A)=2$

i.e., $\rho(A:B)\neq\rho(A)\Rightarrow$ the system is inconsistent.

Example 2. 19 Test for consistency of the system $x+y+z=1, 2x-2y+3z=1, x-y+2z=5,\ 3x+y+z=2.$

Solution The augmented matrix of the given system is

$$(A:B)=\left(\begin{array}{ccc|c}1&1&1&1\\2&-2&3&1\\1&-1&2&5\\3&1&1&2\end{array}\right)$$

$$\begin{array}{r}R_2\to R_2-2R_1\\R_3\to R_3-R_1\\R_4\to R_4-3R_1\end{array}\sim\left(\begin{array}{ccc|c}1&1&1&1\\0&-4&1&-1\\0&-2&1&4\\0&-2&-2&-1\end{array}\right)$$

$$R_2\to R_2/(-4)\sim\left(\begin{array}{ccc|c}1&1&1&1\\0&1&-1/4&1/4\\0&-2&1&4\\0&-2&-2&-1\end{array}\right)$$

$$\begin{array}{l}R_1\to R_1-R_2\\R_3\to\frac{1}{2}(R_3+2R_2)\\R_4\to\frac{1}{2}(R_4+2R_2)\end{array}\sim\left(\begin{array}{ccc|c}1&0&5/4&3/4\\0&1&-1/4&1/4\\0&0&1&9\\0&0&-5&-1\end{array}\right)$$

$$R_4\to\frac{1}{44}(R_4+5R_3)\sim\left(\begin{array}{ccc|c}1&0&5/4&3/4\\0&1&-1/4&1/4\\0&0&1&9\\0&0&0&1\end{array}\right)$$

$$\therefore\ (A:B)=\left(\begin{array}{ccc|c}1&0&5/4&3/4\\0&1&-1/4&1/4\\0&0&1&9\\0&0&0&1\end{array}\right);\qquad A=\begin{pmatrix}1&0&5/4\\0&1&-1/4\\0&0&1\\0&0&0\end{pmatrix}$$

$\Rightarrow\rho(A:B)=4$ and $\rho(A)=3$; i.e., $\rho(A:B)\neq\rho(A)\Rightarrow$ the system is inconsistent.

Example 2. 20 Show that the system $x+y+z=3,\ 2x-y+3z=4,\ 5x-y+7z=11$ has one parameter family of solution. *(M.U. 1997)*

Solution The augmented matrix of the given system is

$$(A:B)=\begin{pmatrix}1 & 1 & 1 & 3\\ 2 & -1 & 3 & 4\\ 5 & -1 & 7 & 11\end{pmatrix}$$

$$\begin{matrix}R_2\to R_2-2R_1\\ R_3\to R_3-5R_1\end{matrix}\sim\begin{pmatrix}1 & 1 & 1 & 3\\ 0 & -3 & 1 & -2\\ 0 & -6 & 2 & -4\end{pmatrix}$$

$$R_3\to R_3-2R_2\sim\begin{pmatrix}1 & 1 & 1 & 3\\ 0 & -3 & 1 & -2\\ 0 & 0 & 0 & 0\end{pmatrix}$$

$$\therefore\ (A:B)=\begin{pmatrix}1 & 1 & 1 & 3\\ 0 & -3 & 1 & -2\\ 0 & 0 & 0 & 0\end{pmatrix}$$

$$A=\begin{pmatrix}1 & 1 & 1\\ 1 & -3 & 1\\ 0 & 0 & 0\end{pmatrix}$$

$\Rightarrow \rho(A:B)=2$ and $\rho(A)=2$

i.e., $\rho(A:B)=\rho(A)=2<3$ (number of unknowns)

Therefore the system is consistent and has one parameter family of solution. Now the given system in matrix form is

$$\begin{pmatrix}1 & 1 & 1\\ 0 & -3 & 1\\ 0 & 0 & 0\end{pmatrix}\begin{pmatrix}x\\ y\\ z\end{pmatrix}=\begin{pmatrix}3\\ -2\\ 0\end{pmatrix}$$

$\therefore\quad x+y+z=3$ and (1)

$-3y+z=-2$ (2)

Let $z=\lambda$

then from eqn.(2): $y=\frac{1}{3}(2+\lambda)$ and

from eqn.(1): $x=3-\frac{1}{3}(2+\lambda)-\lambda=\frac{1}{3}(7-4\lambda)$

$\therefore\ x=\frac{1}{3}(7-4\lambda),\ y=\frac{1}{3}(2+\lambda),\ z=\lambda$ is the required one parameter family of solution.

Example 2. 21 Test for consistency and solve the system

$$5x+3y+7z=4,\ 3x+26y+2z=9,\ 7x+2y+10z=5.$$

Solution The augmented matrix of the given system is

$$(A:B)=\left(\begin{array}{ccc|c}5 & 3 & 7 & 4\\ 3 & 26 & 2 & 9\\ 7 & 2 & 10 & 5\end{array}\right)$$

$$R_1\to 2R_1-3R_2\sim\left(\begin{array}{ccc|c}1 & -72 & 8 & -19\\ 3 & 26 & 2 & 9\\ 7 & 2 & 10 & 5\end{array}\right)$$

$$\begin{array}{l} R_2 \to \frac{1}{22}(R_2 - 3R_1) \\ R_3 \to R_3 - 7R_1 \end{array} \sim \left(\begin{array}{ccc|c} 1 & -72 & 8 & -19 \\ 0 & 11 & -1 & 3 \\ 0 & 11 & -1 & 3 \end{array}\right)$$

$$R_3 \to R_3 - R_2 \sim \left(\begin{array}{ccc|c} 1 & -72 & 8 & -19 \\ 0 & 11 & -1 & 3 \\ 0 & 0 & 0 & 0 \end{array}\right)$$

$$\therefore\ (A:B) = \begin{pmatrix} 1 & -72 & 8 & -19 \\ 0 & 11 & -1 & 3 \\ 0 & 0 & 0 & 0 \end{pmatrix}$$

$$A = \begin{pmatrix} 1 & -72 & 8 \\ 0 & 11 & -1 \\ 0 & 0 & 0 \end{pmatrix}$$

$\Rightarrow \rho(A:B) = 2$ and $\rho(A) = 2$

i.e., $\rho(A:B) = \rho(A) = 2 < 3$ (number of unknowns).

Hence the given system is consistent and has (3 – 2 = 1) one parameter family of solution.

Now the given system in matrix form is

$$\begin{pmatrix} 1 & -72 & 8 \\ 0 & 11 & -1 \\ 0 & 0 & 0 \end{pmatrix} \begin{pmatrix} x \\ y \\ z \end{pmatrix} = \begin{pmatrix} -19 \\ 3 \\ 0 \end{pmatrix}$$

i.e., $x - 72y + 8z = 19$ (1)

and $11y - z = 3$ (2)

Let $z = \alpha$ then

from eqn.(2): $y = \frac{1}{11}(3 + \alpha)$

from eqn.(1): $x = -19 + 72\left\{\frac{1}{11}(3+\alpha)\right\} - 8\alpha = \frac{1}{11}\{-16\alpha + 7\}$

$\therefore\ x = \frac{1}{11}(-16\alpha + 7),\ y = \frac{1}{11}(3 + \alpha),\ z = \alpha$ and produces infinite number of solutions for different values of α (parameter).

Example 2. 22 Test for consistency and solve the system

$$2x - y + 3z = 3,\ x + 2y - z - 5w = 4,\ x + 3y - 2z - 7w = 5.$$

Solution The augmented matrix of the system is

$$(A:B) = \left(\begin{array}{cccc|c} 2 & -1 & 3 & 0 & 3 \\ 1 & 2 & -1 & -5 & 4 \\ 1 & 3 & -2 & -7 & 5 \end{array}\right)$$

$$R(1,2) \sim \left(\begin{array}{cccc|c} 1 & 2 & -1 & -5 & 4 \\ 2 & -1 & 3 & 0 & 3 \\ 1 & 3 & -2 & -7 & 5 \end{array}\right)$$

$$\begin{array}{l} R_2 \to -\frac{1}{5}(R_2 - 2R_1) \\ R_3 \to R_3 - R_1 \end{array} \sim \left(\begin{array}{cccc|c} 1 & 2 & -1 & -5 & 4 \\ 0 & 1 & -1 & -2 & 1 \\ 0 & 1 & -1 & -2 & 1 \end{array}\right)$$

$$R_3 \to R_3 - R_2 \atop R_1 \to R_1 - 2R_2 \sim \left(\begin{array}{cccc|c} 1 & 0 & 1 & -1 & 2 \\ 0 & 1 & -1 & -2 & 1 \\ 0 & 0 & 0 & 0 & 0 \end{array}\right)$$

$$\therefore (A:B) \sim \begin{pmatrix} 1 & 0 & 1 & -1 & 2 \\ 0 & 1 & -1 & -2 & 1 \\ 0 & 0 & 0 & 0 & 0 \end{pmatrix}$$

$$A = \begin{pmatrix} 1 & 0 & 1 & -1 \\ 0 & 1 & -1 & -2 \\ 0 & 0 & 0 & 0 \end{pmatrix}$$

$\Rightarrow \rho(A:B) = 2$ and $\rho(A) = 2$

$\therefore \rho(A:B) = \rho(A) = 2 < 4$ (number of unknowns).

Hence the system is consistent and has infinite number of solutions with $(4-2=2)$ parameters. Now the matrix form of the system is

$$\begin{pmatrix} 1 & 0 & 1 & -1 \\ 0 & 1 & -1 & -2 \\ 0 & 0 & 0 & 0 \end{pmatrix} \begin{pmatrix} x \\ y \\ z \\ w \end{pmatrix} = \begin{pmatrix} 2 \\ 1 \\ 0 \end{pmatrix}$$

i.e., $x + z - w = 2$ and (1)

$y - z - 2w = 1$ (2)

Let $z = \alpha$, $w = \beta$. Then from eqns.(1) and (2)

$x = 2 - \alpha + \beta$; $y = 1 + \alpha + 2\beta$.

For different values of α and β we get infinite number of solutions.

Example 2. 23 For what values of a and b do the equations $x + 2y + 3z = 6$, $x + 3y + 5z = 9$ and $2x + 5y + az = b$ have (i) no solution (ii) unique solution (iii) infinite number of solutions. *(M.U. 1996)*

Solution The augmented matrix of the given system is

$$(A:B) = \left(\begin{array}{ccc|c} 1 & 2 & 3 & 6 \\ 1 & 3 & 5 & 9 \\ 2 & 5 & a & b \end{array}\right)$$

$$R_2 \to R_2 - R_1 \atop R_3 \to R_3 - 2R_1 \sim \left(\begin{array}{ccc|c} 1 & 2 & 3 & 6 \\ 0 & 1 & 2 & 3 \\ 0 & 1 & a-6 & b-12 \end{array}\right)$$

$$R_3 \to R_3 - R_2 \sim \left(\begin{array}{ccc|c} 1 & 2 & 3 & 6 \\ 0 & 1 & 2 & 3 \\ 0 & 0 & a-8 & b-15 \end{array}\right)$$

Case (i) If $a - 8 = 0$; $b - 15 \neq 0$ i.e., $a = 8$ and $b \neq 15$ then

$$(A:B) = \begin{pmatrix} 1 & 2 & 3 & 6 \\ 0 & 1 & 2 & 3 \\ 0 & 0 & 0 & * \end{pmatrix}; \qquad A = \begin{pmatrix} 1 & 2 & 3 \\ 0 & 1 & 2 \\ 0 & 0 & 0 \end{pmatrix}$$

$\therefore\ \rho(A:B) = 3$ and $\rho(A) = 2 \Rightarrow \rho(A:B) \neq \rho(A)$

i.e., the system is inconsistent.

Therefore the system has no solution if $a = 8,\ b \neq 15$.

Case (ii) If $a - 8 \neq 0$ and b be any value, then

$$(A:B) = \begin{pmatrix} 1 & 2 & 3 & 6 \\ 0 & 1 & 2 & 3 \\ 0 & 0 & * & * \end{pmatrix}$$

$$A = \begin{pmatrix} 1 & 2 & 3 \\ 0 & 1 & 2 \\ 0 & 0 & * \end{pmatrix}$$

$$\therefore\ \rho(A:B) = 3;\ \rho(A) = 3$$

i.e., $\rho(A:B) = \rho(A) = 3 =$ number of unknowns.

$\Rightarrow$ The system is consistent and has a unique solution.

Case (iii) If $a - 8 = 0;\ b - 15 = 0$ i.e., $a = 8,\ b = 15$ then

$$(A:B) = \begin{pmatrix} 1 & 2 & 3 & 6 \\ 0 & 1 & 2 & 3 \\ 0 & 0 & 0 & 0 \end{pmatrix};\ A = \begin{pmatrix} 1 & 2 & 3 \\ 0 & 1 & 2 \\ 0 & 0 & 0 \end{pmatrix}$$

$$\therefore \rho(A:B) = 2 \text{ and } \rho(A) = 2$$

i.e., $\rho(A:B) = \rho(A) = 2 < 3$ (number of unknowns).

Hence the system is consistent and has infinite number of solutions for $a = 8,\ b = 15$.

Example 2.24 For what values of λ, the system $x + y + z = 1,\ x + 2y + 4z = \lambda, x + 4y + 10z = \lambda^2$ has a solution and solves completely.

Solution The augmented matrix of the system is

$$(A:B) = \begin{pmatrix} 1 & 1 & 1 & 1 \\ 1 & 2 & 4 & \lambda \\ 1 & 4 & 10 & \lambda^2 \end{pmatrix}$$

$$\begin{matrix} R_2 \to R_2 - R_1 \\ R_3 \to R_3 - R_1 \end{matrix} \sim \left(\begin{array}{ccc|c} 1 & 1 & 1 & 1 \\ 0 & 1 & 3 & \lambda - 1 \\ 0 & 3 & 9 & \lambda^2 - 1 \end{array}\right)$$

$$\begin{matrix} R_1 \to R_1 - R_2 \\ R_3 \to R_3 - 3R_2 \end{matrix} \sim \left(\begin{array}{ccc|c} 1 & 0 & -2 & 2 - \lambda \\ 0 & 1 & 3 & \lambda - 1 \\ 0 & 0 & 0 & \lambda^2 - 3\lambda + 2 \end{array}\right)$$

Case (i) If $\lambda^2 - 3\lambda + 2 \neq 0$ i.e., $(\lambda - 2)(\lambda - 1) \neq 0$ or $\lambda \neq 1, 2$ then,

we get $\rho(A) = 2,\ \rho(A:B) = 3$ and $\rho(A:B) \neq \rho(A) \Rightarrow$ the system is inconsistent.

Case (ii) If $\lambda^2 - 3\lambda + 2 = 0$ i.e., $\lambda = 1$ or 2

[then $\rho(A:B) = 2,\ \rho(A) = 2$ i.e., $\rho(A:B) = \rho(A) = 2 < 3$ (number of unknowns)]

$\Rightarrow$ The system is consistent and has many solutions with one parameter.

Case (iii) There is no possibility for unique solution $\because \rho(A)=2$

Now the matrix form of the system is

$$\begin{pmatrix} 1 & 0 & -2 \\ 0 & 1 & 3 \\ 0 & 0 & 0 \end{pmatrix}\begin{pmatrix} x \\ y \\ z \end{pmatrix}=\begin{pmatrix} 2-\lambda \\ \lambda-1 \\ \lambda^2-3\lambda+2 \end{pmatrix} \quad (1)$$

i. when $\lambda=1$ the system (1) is equivalent to $x-2z=1;\ y+3z=0 \therefore x=1+2\alpha, y=-3\alpha, z=\alpha$ is the solution with α as parameter.

ii. when $\lambda=2$ the system (1) is equivalent to $x-2z=0,\ y+3z=1 \therefore x=2\beta,\ y=1-3\beta,\ z=\beta$ is the solution with β as parameter.

Solution to System of Homogeneous Linear Equations AX = 0

A system of homogeneous linear equations has always a solution $X=0$ which is also called as null solution or the trivial solution. If there are many solutions, is such case $X=0$ in one of the solution. Hence the system $AX=0$ is always consistent.

The system possesses only trivial solution if $\rho(A)$ = number of unknowns. The system possesses infinite number of solutions if $\rho(A)$ < number of unknowns.

Example 2. 25 Test for consistency and solve $x+y+z=0,\ 2x-5y+3z=0,\ 2x+5y+7z=0.$

Solution The system in matrix form is

$$\begin{pmatrix} 1 & 1 & 1 \\ 2 & -5 & 3 \\ 2 & 5 & 7 \end{pmatrix}\begin{pmatrix} x \\ y \\ z \end{pmatrix}=\begin{pmatrix} 0 \\ 0 \\ 0 \end{pmatrix} \text{ i.e., } AX=0 \quad (1)$$

$$A=\begin{pmatrix} 1 & 1 & 1 \\ 2 & -5 & 3 \\ 2 & 5 & 7 \end{pmatrix}$$

$$\begin{matrix} R_2 \to R_2-2R_1 \\ R_3 \to R_3-2R_1 \end{matrix} \sim \begin{pmatrix} 1 & 1 & 1 \\ 0 & -7 & 1 \\ 0 & 3 & 5 \end{pmatrix}$$

$$R_2 \to -(R_2+2R_3) \sim \begin{pmatrix} 1 & 1 & 1 \\ 0 & 1 & -11 \\ 0 & 3 & 5 \end{pmatrix}$$

$$\begin{matrix} R_1 \to R_1-R_2 \\ R_3 \to \frac{1}{38}(R_3-3R_2) \end{matrix} \sim \begin{pmatrix} 1 & 0 & 12 \\ 0 & 1 & -11 \\ 0 & 0 & 1 \end{pmatrix}$$

$$\begin{matrix} R_1 \to R_1-12R_3 \\ R_2 \to R_2+11R_3 \end{matrix} \sim \begin{pmatrix} 1 & 0 & 0 \\ 0 & 1 & 0 \\ 0 & 0 & 1 \end{pmatrix}$$

$\Rightarrow \rho(A)=3=$ number of unknowns. Hence the system is consistent and has only a trivial solution. i.e $x=0,\ y=0, z=0.$

Example 2.26 Test for consistency and solve $x-y+2z-3t=0,\ 3x+2y-4z+t=0,\ 5x-3y+2z+6t=0.$

Solution The given system in matrix form can be written as

$$\begin{pmatrix} 1 & -1 & 2 & -3 \\ 3 & 2 & -4 & 1 \\ 5 & -3 & 2 & 6 \end{pmatrix} \begin{pmatrix} x \\ y \\ z \\ w \end{pmatrix} = \begin{pmatrix} 0 \\ 0 \\ 0 \\ 0 \end{pmatrix} \text{ i.e., } AX = 0 \qquad (1)$$

Here

$$A = \begin{pmatrix} 1 & -1 & 2 & -3 \\ 3 & 2 & -4 & 1 \\ 5 & -3 & 2 & 6 \end{pmatrix}$$

$$\begin{matrix} R_2 \to \frac{1}{5}(R_2 - 3R_1) \\ R_3 \to R_3 - 5R_1 \end{matrix} \sim \begin{pmatrix} 1 & -1 & 2 & -3 \\ 0 & 1 & -2 & 2 \\ 0 & 2 & -8 & 21 \end{pmatrix}$$

$$R_3 \to -\frac{1}{4}(R_3 - 2R_2) \sim \begin{pmatrix} 1 & -1 & 2 & -3 \\ 0 & 1 & -2 & 2 \\ 0 & 0 & 1 & -17/4 \end{pmatrix}$$

$\Rightarrow \rho(A) = 3 < 4$ (number of unknowns).

$\therefore$ The system is consistent and has many solutions with one parameter. i.e., from above and (1) we get

$$x - y + 2z - 3w = 0,\ y - 2z + 2w = 0;\ z - \frac{17}{4}w = 0$$

$\therefore$ If $w = k$, then $z = \frac{17}{4}k;\ y = 2z - 2w = \frac{13}{2}k;\ x = y - 2z + 3w = k$

$\therefore\ x = k,\ y = \frac{13}{2}k,\ z = \frac{17}{4}k$ and $w = k$ in particular if $k = 0$ then

$x = y = z = w = 0$ which is a trivial solution.

EXERCISE II

Test for consistency and solve the following system of linear equations.

1. $9x+7y+3z=6,\ 5x+y+4z=1,\ 6x+8y+2z=4$
2. $x+2y+3z=14,\ 3x+y+2z=11,\ 2x+3y+z=11$
3. $2x-y+3z=9,\ x+y+z=6,\ x-y+z=2$
4. $2x+y-z=0,\ 2x+5y+7z=52,\ x+y+z=9$
5. $x+y+z=9,\ 2x+5y+7z=50,\ 2x+y-z=2$
6. $4x+3y+6z=25,\ x+5y+7z=13,\ 2x+9y+z=1$
7. $x-2y+3z=11, 3x+y-z=2, 5x+3y+2z=3$
8. $2x+5y+3z=9,\ 3x+y+2z=3,\ x+2y-z=6$
9. $5x-6y+4z=15,\ 7x+4y-3z=19,\ 2x+y+6z=46$

10. $x+2y-z=1, 3x+y+z=2,\ x-y+z=0,\ 2x+y-z=0$

11. $x-y+2z=3,\ 2x+z=1,\ 3x+2y+z=4,\ 2x+y+2z=6$

12. $x+y+z=6,\ x-y+z=2,\ 2x+y-z=1,\ 4x+3y-2z=4$

13. $2x+y+5z+w=5,\ x+y+3z-4w+1=0,\ 3x+6y-2z+w=8,\ 2x+2y+2z-3w=2$

14. $x+2y-3z-4w=6,\ x+3y+z-2w=4,\ 2x+5y-2z-5w=10$

15. $2x-4y-z=9,\ 3x-y+5z=5,\ 8x+2y+9z=12$

16. $2x+y+z=2,\ \ y+z=0,\ \ x+2y+2z=1,\ \ 3x+2y+2z=3$

17. $5x+3y+7z=4,\ 3x+26y+2z=9,\ 7x+2y+10z=5$

18. $x+y+z=6,\ x+2y+3z=14,\ x+4y+7z=30$

19. $5x-3y-7z+w=10,\ -x+2y+6z-3w=-3,\ x+y+4z-5w=0$

20. $2x-y+z=4,\ 3x-y+z=6,\ 4x-y+2z=7,\ -x+y-z=9$

21. $x+y+2z+w=5,\ 2x+3y-z-2w=2,\ 4x+5y+3z=7$

22. $x-2y+z-w=-1,\ 3x-2z+3w=-4,\ 5x-4y+w=-3$

23. $2x-y+3z=3,\ x+2y-z-5w=4,\ x+3y-2z-7w=5$

24. $x+y+z=4,\ 2x+5y-2z=3,\ x-7y-7z=15$

25. For what values of λ and μ the system of equations $x+y+z=6,\ x+2y+3z=10,\ x+2y+\lambda z=\mu$ has (a) a unique solution (b) no solution (c) an infinite number of solutions. *(B.U. 1994, M.U. 1996)*

26. For what values of a and b the system of equations $x+2y+z=8,\ 2x+y+3z=13,\ 3x+4y-az=b$ has (a) no solution (b) a unique solution (c) many solutions.

27. Find the value of λ such that the system $ax+hy+g=0,\ hx+by+f=0,\ gx+by+c=0$ is consistent.

28. Show that the system $-2x+y+z=a,\ x-2y+z=b,\ x+y-2z=c$ have no solution unless $a+b+c=0$ in which case they have infinitely many solutions. Find the solutions when $a=1,\ b=1,\ c=-2$.

29. Solve the equations with the help of matrices considering specially in the case when $\lambda=2$. $\lambda x+2y-2z=1,\ 4x+2\lambda y-z=2,\ 6x+6y+\lambda z=3$.

30. Solve the equations given below:
    (a) $2x-y+z=0,\ 3x+2y+z=0,\ x-3y+5z=0$.
    (b) $x+2y-z=0,\ 3x+y-z=0,\ 2x-y=0$.
    (c) $3x-y-z+2w=0,\ 6x+2y-2z-4w=0,\ 12x-3y+6z+6w=0$.
    (d) $x-2y+z-w=0,\ x+y-2z+3w=0,\ 4x+y-5z+8w=0,\ 5x-7y+2z-w=0$.

31. Find the real value of λ for which the system $x+2y+3z=\lambda x,\ 3x+y+2z=\lambda y,\ 2x+2y+z=\lambda x$ has a non-zero solution.

32. Prove that the system $x=\lambda y+z,\ y=z+\lambda x,\ z=x+y$ has a non-zero solution if $\lambda+1=0$.

ANSWERS

1. $x=4,\ y=2,\ z=1$	2. $x=1,\ y=2,\ z=3$	3. $x=1,\ y=2,\ z=3$
4. $x=1,\ y=3,\ z=5$	5. $x=1,\ y=4,\ z=4$	6. $x=4,\ y=-1,\ z=2$

7. $x=2,\ y=-3,\ z=1$ 8. $x=1,\ y=2,\ z=-1$ 9. $x=3,\ y=4,\ z=6$

10. $x=0,\ y=1,\ z=1$ 11. $x=-1,\ y=2,\ z=3$ 12. $x=1,\ y=2,\ z=3$

13. $x=2,\ y=\frac{1}{5},\ z=0,\ w=4/5$

14. Many solutions, $x=11k+10,\ y=-4k-2,\ z=k,\ w=0$

15. Many solutions, $x=-\frac{1}{14}(19k-29),\ y=\frac{1}{14}(13k+17)z=k$

16. Many solutions $x=1,\ y=-k,\ z=k$

17. Many solutions, $x=\frac{1}{11}(7-16k),\ y=\frac{1}{11}(k+3),\ z=k$

18. Many solutions $x=-2+k,\ y=8-2k, z=k$

19. Many solutions, $x=5-7k,\ y=19-44k,\ z=k,\ w=-6+14k$

20. inconsistent 21. inconsistent 22. inconsistent

23. Many solutions, $x=2-\lambda+\mu, y=1+\lambda+2\mu, z=\lambda, w=\mu$ 24. inconsistent

25. Unique solution when $\lambda \neq 3$ whatever μ may be. No solution when, $\lambda=3,\ \mu\neq 10$ and many solutions when $\lambda=3,\ \mu\neq 10$.

26. Unique solution when $a\neq -11/3$ whatever b may be. No solution when $a=-11/3,\ b\neq 22$ and many solutions when $a=-11/3,\ b=22$.

27. $(abc+2fgh-af^2-bg^2-ch^2)/(ab-h^2)$

28. $x=k-1,\ y=k-1,\ z=k$ 29. $x=-k+\frac{1}{2},\ y=k,\ z=0$

30. *a.* $x=y=z=0$ *b.* $x=k,\ y=2k,\ z=5k$

c. $x=\lambda/3,\ y=2\mu,\ w=\mu$ *d.* $x=\lambda-\frac{5\mu}{3},\ y=\lambda-\frac{4}{3}\mu,\ z=\lambda,\ w=\mu$

32. $\lambda=6$

EIGEN VALUES AND EIGEN VECTORS

Eigen values and eigen vectors are also known as characteristic values and characteristic vectors respectively.

Consider the linear transformation $Y=AX$ which transforms a column vector X into a column vector Y by means of the square matrix A.

Let $Y=AX$ (1)

Where $A=\begin{bmatrix} a_{11} & a_{12} & \cdots & a_{1n} \\ a_{21} & a_{22} & \cdots & a_{2n} \\ \cdots & \cdots & \cdots & \cdots \\ \cdots & \cdots & \cdots & \cdots \\ a_{n1} & a_{n2} & \cdots & a_{nn} \end{bmatrix}$ $X=\begin{bmatrix} x_1 \\ x_2 \\ . \\ . \\ x_n \end{bmatrix}$ and $Y=\begin{bmatrix} y_1 \\ y_2 \\ . \\ . \\ y_n \end{bmatrix}$

If a vector X is transformed into a scalar multiple of the same vector i.e., X is transformed to λX then

$$Y=\lambda X=AX \Rightarrow AX=\lambda X \text{ or } (A-\lambda I)X=0$$

where I is the unit matrix of order n. The system will have non-trivial solutions if

$$|A-\lambda I|=0 \quad (2)$$

$$\text{i.e} \quad \begin{vmatrix} a_{11}-\lambda & a_{12} & \cdots & a_{1n} \\ a_{22} & a_{22}-\lambda & \cdots & a_{2n} \\ \cdots & \cdots & \cdots & \cdots \\ \cdots & \cdots & \cdots & \cdots \\ a_{n1} & a_{n2} & \cdots & a_{nn}-\lambda \end{vmatrix} = 0 \qquad (3)$$

(2) or (3) is said to be characteristic equation of the transformation or the characteristic equation of the matrix A. The determinant $|A-\lambda I|$ is a polynomial of nth degree in λ and it is called the characteristic polynomial of the transformation or the characteristic polynomial of the matrix A. Now $|A-\lambda I|=0$, possess n values of λ which are called as characteristic roots (values) or eigen values or latent roots of the matrix A. Let $\lambda_i\,(i=1,2,\ldots)$ be the n roots (eigen values) of $|A-\lambda I|=0$. Corresponding to each value of λ, the equation $AX=\lambda X$ has a non-zero solution vector X. If X_i be the non-zero solution vector satisfying $AX=\lambda X$ when $\lambda=\lambda_i$, X_i is said to be the characteristic vector or eigen vector or latent vector of the matrix A corresponding to λ_i. *For* values of λ other than the characteristic values of A, the equation $AX=\lambda X$ will possess only trivial solution $X=0$.

Note: Since $AX=\lambda X$ is a system of homogeneous linear equations if X is a solution, then αX is also a solution of it where α is a non-zero scalar.

$\therefore$ Characteristic vector corresponding to a characteristic value is not unique.

IMPORTANT POINTS

(a) Characteristic values

1. If $A=(a_{ij})$ is a square matrix of order n, then the sum of its characteristic values is equal to trace of A and product of its characteristic values is equal to $|A|$.
2. Zero is the characteristic value of a square matrix A iff $|A|=0$ i.e., A is singular and if zero is not one of the characteristic value then A is non-singular i.e., $|A|\neq 0$.
3. The characteristic values of a triangular matrix are precisely the diagonal elements of the matrix.
4. The characteristic values of a diagonal matrix are just the elements in the leading diagonal.
5. If A and B are non-singular square matrices then AB and BA have the same characteristic values.
6. If A and B are square matrices of order n and if A is non-singular then matrices $A^{-1}B$ and BA^{-1} have the same characteristic values.
7. If λ is a characteristic value of a non-singular matrix A, then $|A|/\lambda$ is a characteristic value of matrix adj A if $\lambda\neq 0$.
8. If $\lambda_i\,(i=1,2,\ldots)$ are characteristic values of A then $\lambda_i^p\,(i=1,2,\ldots)$ are characteristic values of A^p where p is a positive integer.
9. If $\lambda_i\,(i=1,2,\ldots)$ are characteristic values of A then $\lambda_i^{-1}\,(i=1,2,\ldots)$ are characteristic values of A^{-1}, if $\lambda_i\neq 0$.
10. The characteristic values of A and A^T are equal.
11. If λ is the characteristic value of A then $\alpha\lambda$ is the characteristic value of αA.
12. The characteristic values of a Hermitian matrix are all real.
13. The characteristic values of a real symmetric matrix are all real.

14. The characteristic roots of a skew-Hermitian matrix is either zero or purely imaginary.
15. The modulus of each characteristic value of an orthogonal or unitary (since every orthogonal matrix is unitary) is unity.

(b) Characteristic vectors

16. If all the eigen values λ_i $(i=1,2,...)$ of a matrix A of order n are distinct then the corresponding eigen vectors $X_i (i=1,2,...)$ are all linearly independent.
17. If two or more eigen values are equal then the eigen vectors may be linearly independent or dependent.
18. The eigen vectors corresponding to distinct eigen values of (i) real symmetric matrix (ii) Hermitian matrix and (iii) Unitary matrix are orthogonal.

Example 2.27 Find the eigen values of $A=\begin{pmatrix}4 & 1\\ 3 & 2\end{pmatrix}$. *(M.U. 1994)*

Solution The eigen values are given by $|A-\lambda I|=0$

$$\text{i.e., } \begin{vmatrix}4-\lambda & 1\\ 3 & 2-\lambda\end{vmatrix}=0$$

$$\text{or } (4-\lambda)(2-\lambda)-3 \text{ or } \lambda^2-6\lambda+5=0$$

$\therefore$ $(\lambda-1)(\lambda-5)=0 \Rightarrow \lambda=1, 5$ are the eigen values.

Example 2.28 Find the sum and product of the eigen values of the matrix $\begin{pmatrix}3 & -4 & 4\\ 1 & -2 & 4\\ 1 & -1 & 3\end{pmatrix}$ without finding the eigen values. *(M.U. 1997)*

Solution Let A be the given matrix. We know that,

i. Sum of the eigen values = Trace of the matrix A
ii. Product of the eigen values = Determinant of A

Now trace of A = sum of principal diagonal elements $=3+(-2)+3=4$

$$|A|=\begin{vmatrix}3 & -4 & 4\\ 1 & -2 & 4\\ 1 & -1 & 3\end{vmatrix}=3(-6+4)+4(3-4)+4(-1+2)=-6$$

$\therefore$ Sum of eigen values = 4, Product of eigen values = -6

Example 2.29 If 3 and 15 are the eigen values of $\begin{pmatrix}8 & -6 & 2\\ -6 & 7 & -4\\ 2 & -4 & 3\end{pmatrix}$, what is the third eigen value. *(M.U. 1996)*

Solution Let α be the third eigen value.

Sum of eigen values $=\lambda+3+15=\lambda+18$. (1)

Trace of the given matrix $=8+7+3=18$ (2)

But eqn.(1) = eqn.(2) $\Rightarrow \lambda+18=18 \Leftrightarrow \lambda=0$.

i.e., the third eigen value is 0.

Example 2.30 If $A = \begin{pmatrix} 2 & 2 & 2 \\ 0 & 5 & 2 \\ 0 & 0 & 4 \end{pmatrix}$ find the eigen values of A, A^2, A^{-1}, A^{100}.

Solution Since A is upper triangular matrix, the eigen values are principal diagonal elements only.

$\therefore$ eigen values of $A = 2, 5, 4$

Now eigen values of $A^2 = 2^2, 5^2, 4^2$ i.e., 4, 25, 16

eigen value of $A^{-1} = \frac{1}{2}, \frac{1}{5}, \frac{1}{4}$

eigen values of $A^{100} = 2^{100}, 5^{100}, 4^{100}$

Example 2.31 Find the eigen values and eigen vectors of $A = \begin{pmatrix} 3 & -1 & 1 \\ -1 & 5 & -1 \\ 1 & -1 & 3 \end{pmatrix}$. *(M.U. 1997)*

Solution The characteristic equation is given by $|A - \lambda I| = 0$.

i.e. $\begin{vmatrix} 3-\lambda & -1 & 1 \\ -1 & 5-\lambda & -1 \\ 1 & -1 & 3-\lambda \end{vmatrix} = 0$

i.e. $(3-\lambda)\{(5-\lambda)(3-\lambda)-1\}+1\{(-1)(3-\lambda)+1\}+1\{1-(5-\lambda)\}=0$

or $\lambda^3 - 11\lambda^2 + 36\lambda - 36 = 0$ or $(\lambda-2)(\lambda-3)(\lambda-6)=0$ $\therefore$ $\lambda = 2, 3, 6$

i.e., the eigen values are 2, 3, 6

eigen vector for $\lambda = 2$ is given by $[A - 2I]X_1 = 0$

$$\begin{pmatrix} 3-2 & -1 & 1 \\ -1 & 5-2 & -1 \\ 1 & -1 & 3-2 \end{pmatrix} \begin{pmatrix} x_1 \\ x_2 \\ x_3 \end{pmatrix} \begin{pmatrix} 0 \\ 0 \\ 0 \end{pmatrix}$$

$\Rightarrow x_1 - x_2 + x_3 = 0;\ -x_1 + 3x_2 - x_3 = 0;\ x_1 - x_2 + x_3 = 0$

Solving the first two equations by cross multiplication

$$\begin{array}{cccc} & x_1 & x_2 & x_3 & \\ -1 & 1 & 1 & -1 \\ 3 & -1 & -1 & 3 \end{array}$$

$$\frac{x_1}{1-3} = \frac{x_2}{-1+1} = \frac{x_3}{3-1} \quad \text{or} \quad \frac{x_1}{1} = \frac{x_2}{0} = \frac{x_3}{-1}$$

$$\therefore \quad X_1 = \begin{pmatrix} x_1 \\ x_2 \\ x_3 \end{pmatrix} = \begin{pmatrix} 1 \\ 0 \\ -1 \end{pmatrix}$$

eigen vector for $\lambda = 3$ is given by $[A - 3I]X_2 = 0$

$$\begin{pmatrix} 3-3 & -1 & 1 \\ -1 & 5-3 & -1 \\ 1 & -1 & 3-3 \end{pmatrix} \begin{pmatrix} x_1 \\ x_2 \\ x_3 \end{pmatrix} = \begin{pmatrix} 0 \\ 0 \\ 0 \end{pmatrix}$$

$\Rightarrow -x_2 + x_3 = 0;\ -x_1 + 2x_2 - x_3 = 0;\ x_1 - x_2 = 0 \Leftrightarrow x_1 = x_2 = x_3 = 1$

$$\therefore \quad X_2 = \begin{pmatrix} x_1 \\ x_2 \\ x_3 \end{pmatrix} = \begin{pmatrix} 1 \\ 1 \\ 1 \end{pmatrix}$$

eigen vector for $\lambda = 6$ is given by $[A - 6I]\, X_3 = 0$

$$\text{i.e.,} \quad \begin{pmatrix} 3-6 & -1 & 1 \\ -1 & 5-6 & -1 \\ 1 & -1 & 3-6 \end{pmatrix} \begin{pmatrix} x_1 \\ x_2 \\ x_3 \end{pmatrix} = \begin{pmatrix} 0 \\ 0 \\ 0 \end{pmatrix}$$

$$\Rightarrow -3x_1 - x_2 + x_3 = 0;\ -x_1 - x_2 - x_3 = 0;\ x_1 - x_2 - 3x_3 = 0$$

Solving any two equations by cross-multiplication method we get

$$\frac{x_1}{2} = \frac{x_2}{-4} = \frac{x_3}{2} \quad \text{or} \quad \frac{x_1}{1} = \frac{x_2}{-2} = \frac{x_3}{1}$$

$$\therefore \quad X_3 = \begin{pmatrix} x_1 \\ x_2 \\ x_3 \end{pmatrix} = \begin{pmatrix} 1 \\ -2 \\ 1 \end{pmatrix}$$

Example 2.32 Find the eigen values and eigen vectors of $A = \begin{pmatrix} 7 & -2 & -2 \\ -2 & 1 & 4 \\ -2 & 4 & 1 \end{pmatrix}$. *(M.U. 1995, 1996)*

Solution The characteristic equation is given by $|A - \lambda I| = 0$

$$\text{i.e.} \quad \begin{vmatrix} 7-\lambda & -2 & -2 \\ -2 & 1-\lambda & 4 \\ -2 & 4 & 1-\lambda \end{vmatrix} = 0$$

$$\text{or} \quad (7-\lambda)\{(1-\lambda)^2 - 16\} + 2\{-2(1-\lambda) + 8\} - 2\{-8 + 2(1-\lambda)\} = 0$$

$$\text{or} \quad \lambda^3 - 9\lambda^2 - 9\lambda + 81 = 0 \quad \text{or} \quad (\lambda - 3)(\lambda + 3)(\lambda - 9) = 0$$

$\therefore \lambda = 3, -3, 9$ are the eigen values.

eigen vector for $\lambda = 3$ is given by $[A - 3I]X_1 = 0$

$$\text{i.e.,} \quad \begin{pmatrix} 7-3 & -2 & -2 \\ -2 & 1-3 & 4 \\ -2 & 4 & 1-3 \end{pmatrix} \begin{pmatrix} x_1 \\ x_2 \\ x_3 \end{pmatrix} = \begin{pmatrix} 0 \\ 0 \\ 0 \end{pmatrix}$$

$$\Rightarrow 4x_1 - 2x_2 - 2x_3 = 0;\ -2x_1 - 2x_2 + 4x_3 = 0;\ -2x_1 + 4x_2 - 2x_3 = 0$$

On solving any two equations by cross-multiplication method we get

$$\frac{x_1}{1} = \frac{x_2}{1} = \frac{x_3}{1} \quad \therefore \quad X_1 = [1\ 1\ 1]^T$$

eigen vector for $\lambda = -3$ is given by $[A + 3I]\ X_2 = 0$

$$\text{i.e.,} \quad \begin{pmatrix} 7+3 & -2 & -2 \\ -2 & 1+3 & 4 \\ -2 & 4 & 1+3 \end{pmatrix} \begin{pmatrix} x_1 \\ x_2 \\ x_3 \end{pmatrix} = \begin{pmatrix} 0 \\ 0 \\ 0 \end{pmatrix}$$

$$\Rightarrow 10x_1 - 2x_2 - 2x_3 = 0;\ -2x_1 + 4x_2 + 4x_3 = 0;\ -2x_1 + 4x_2 + 4x_3 = 0$$

On solving any two equations by cross-multiplication method, we get

$$\frac{x_1}{0}=\frac{x_2}{1}=\frac{x_3}{-1} \quad \therefore\ X_2=[0\ 1\ -1]^T$$

eigen vector for $\lambda = 9$ is given by $[A-9I]\ X_3 = 0$

$$\text{i.e.,}\quad \begin{pmatrix} 7-9 & -2 & -2 \\ -2 & 1-9 & 4 \\ -2 & 4 & 1-9 \end{pmatrix}\begin{pmatrix} x_1 \\ x_2 \\ x_3 \end{pmatrix}=\begin{pmatrix} 0 \\ 0 \\ 0 \end{pmatrix}$$

$$\Rightarrow -2x_1-2x_2-2x_3=0;\ -2x_1-8x_2+4x_3=0;\ -2x_1+4x_2-8x_3=0$$

On solving any two equations by cross-multiplication method we get

$$\frac{x_1}{2}=\frac{x_2}{-1}=\frac{x_3}{1} \quad \therefore\ X_3=[2\ -1\ 1]^T$$

Example 2.33 Find the eigen values and eigen vectors of $A=\begin{pmatrix} 1 & 2 & 3 \\ 0 & -4 & 2 \\ 0 & 0 & 7 \end{pmatrix}$.

Solution Given A is upper triangular matrix

$\therefore$ The eigen values are $\lambda = 1, -4, 7$.

The eigen vector corresponding to $\lambda = 1$ is given by $[A-(1)I]\ X_1 = 0$

$$\text{i.e.,}\quad \begin{pmatrix} (1-1) & 2 & 3 \\ 0 & (-4-1) & 2 \\ 0 & 0 & (7-1) \end{pmatrix}\begin{pmatrix} x_1 \\ x_2 \\ x_3 \end{pmatrix}=\begin{pmatrix} 0 \\ 0 \\ 0 \end{pmatrix} \quad \text{or} \quad \begin{pmatrix} 0 & 2 & 3 \\ 0 & -5 & 2 \\ 0 & 0 & 6 \end{pmatrix}\begin{pmatrix} x_1 \\ x_2 \\ x_3 \end{pmatrix}=\begin{pmatrix} 0 \\ 0 \\ 0 \end{pmatrix}$$

Clearly matrix A is singular i.e., $|A| = 0$ and hence it has a non- trivial solution. The equivalent system of equations are $2x_2+3x_3=0;\ -5x_2+2x_3=0;\ 6x_3=0$. Here rank of coefficient matrix = 2 < 3 (no. of unknowns) $\therefore$ There will be one parameter. From the above equations it is clear that $x_2 = x_3 = 0\ \therefore$ let $x_1 = k$

$\therefore\ X_1=[k, 0, 0]^T$, in particular if $k = 1$ then $X_1 = [1\ 0\ 0]^T$

the eigen vector for $\lambda = -4$ is given by $[A+4I]\ X_2 = 0$

$$\text{i.e.,}\quad \begin{pmatrix} (1+4) & 2 & 3 \\ 0 & (-4+4) & 2 \\ 0 & 0 & (7+4) \end{pmatrix}\begin{pmatrix} x_1 \\ x_2 \\ x_3 \end{pmatrix}=\begin{pmatrix} 0 \\ 0 \\ 0 \end{pmatrix} \quad \text{or} \quad \begin{pmatrix} 5 & 2 & 3 \\ 0 & 0 & 2 \\ 0 & 0 & 11 \end{pmatrix}\begin{pmatrix} x_1 \\ x_2 \\ x_3 \end{pmatrix}=\begin{pmatrix} 0 \\ 0 \\ 0 \end{pmatrix}$$

The rank of coefficient matrix = 2 < 3 (no. of unknowns).

$\therefore$ There will be one parameter. The equivalent system is

$5x_1+2x_2+3x_3=0;\ 2x_3=0;\ 11x_3=0$

$$\therefore\ x_3=0 \Rightarrow 5x_1+2x_2=0 \ \text{ or } \ \frac{x_1}{-2}=\frac{x_2}{5}=k \ \text{ or } \ \frac{x_1}{2}=\frac{x_2}{-5}=k$$

$\therefore\ x_1 - 2k;\ x_2 = 5k$ or $x_1 = 2k;\ x_2 = -5k$

$$X_2=[-2k\ 5k\ 0]^T \ \text{ or } \ X_2=[2k\ -5k\ 0]^T$$

In particular if $k = 1$; $X_2 = [-2\ 5\ 0]^T$ or $[2\ -5\ \ 0]^T$.

The eigen vector corresponding to $\lambda = 7$ is given by: $[A-7I]\ X_3 = 0$

$$\text{i.e.,}\quad \begin{pmatrix} 1-7 & 2 & 3 \\ 0 & -4-7 & 2 \\ 0 & 0 & 7-7 \end{pmatrix}\begin{pmatrix} x_1 \\ x_2 \\ x_3 \end{pmatrix} = \begin{pmatrix} 0 \\ 0 \\ 0 \end{pmatrix} \quad \text{or} \quad \begin{pmatrix} -6 & 2 & 3 \\ 0 & -11 & 2 \\ 0 & 0 & 0 \end{pmatrix}\begin{pmatrix} x_1 \\ x_2 \\ x_3 \end{pmatrix} = \begin{pmatrix} 0 \\ 0 \\ 0 \end{pmatrix}$$

The rank of coefficient matrix is 2 < 3 (no. of unknowns).

∴ There will be one parameter. The equivalent system is $-6x_1 + 2x_2 + 3x_3 = 0;\ -11x_2 + 2x_3 = 0;\ 0 = 0$

$$\therefore\quad x_2 = \frac{2x_3}{11},\ x_1 = \frac{37}{66}x_3.$$

If $x_3 = k$ then $X_3 = \left(\frac{37}{66}k\ \ \frac{2}{11}k\right)^T$, $k = 1$ and $X_3 = \left(\frac{37}{66}\ \frac{2}{11}\right)^T$

Example 2.34 Find the eigen values and eigen vectors of the matrix.

$$A = \begin{pmatrix} 6 & -2 & 2 \\ -2 & 3 & -1 \\ 2 & -1 & 3 \end{pmatrix}$$

Solution The eigen values of A are given by $|A - \lambda I| = 0$

$$\text{i.e.,}\quad \begin{pmatrix} 6-\lambda & -2 & 2 \\ -2 & 3-\lambda & -1 \\ 2 & -1 & 3-\lambda \end{pmatrix} = 0$$

i.e. $-\lambda^3 + 12\lambda^2 - 36\lambda + 32 = 0$ or $\lambda^3 - 12\lambda^2 + 36\lambda - 32 = 0$

By trial $\lambda = 2$ is a root.

∴ $(\lambda-2)(\lambda-2)(\lambda-8)$ or $\lambda = 2, 2, 8$.

The eigen vector for $\lambda = 2$ is given by $[A - 2I]\,X = 0$

$$\text{i.e.,}\quad \begin{pmatrix} 4 & -2 & 2 \\ -2 & 1 & -1 \\ 2 & -1 & 1 \end{pmatrix}\begin{pmatrix} x_1 \\ x_2 \\ x_3 \end{pmatrix} = \begin{pmatrix} 0 \\ 0 \\ 0 \end{pmatrix}$$

Now the system of equations reduces to a single equation $2x_1 - x_2 + x_3 = 0$.

⇒ The rank of the coefficient matrix is one, less than the number of unknowns by two.

∴ Independent characteristic vectors can be obtained by giving arbitrary values of two of the quantities x_1, x_2, x_3.

Putting $x_2 = 0$ we get $2x_1 + x_3 = 0$ or $x_3 = -2x_1$.

If $x_1 = k$ than $x_3 = -2k$

∴ One vector is $X_1 = [k\ 0\ -2k]^T$ or $X_1 = [1\ 0\ -2]^T$ if $k = 1$.

Putting $x_1 = 0$, we get $x_3 - x_2 = 0$ or $x_3 = x_2$

If $x_2 = k$ then $x_3 = k$ and hence the other vector is $X_2 = [0\ k\ k]^T$ or $X_2 = [0\ 1\ 1]^T$ if k = 1

Now eigen vector *for* $\lambda = 8$ is given by $[A-8I]X = 0$

$$\text{i.e.,} \begin{pmatrix} -2 & -2 & 2 \\ -2 & -5 & -1 \\ 2 & -1 & -5 \end{pmatrix} \begin{pmatrix} x_1 \\ x_2 \\ x_3 \end{pmatrix} = \begin{pmatrix} 0 \\ 0 \\ 0 \end{pmatrix}$$

i.e., $-2x_1 - 2x_2 + 2x_3 = 0;\ -2x_1 - 5x_2 - x_3 = 0;\ 2x_1 - x_2 - 5x_3 = 0$

Solving any two of the equations, we get the corresponding characteristic vector as

$$X_3 = [2k - k \quad k]^T \text{ or } x_3 = [2 -1 \quad 1]^T \text{ if } k = 1.$$

Example 2.35 Show that the characteristic values of an orthogonal matrix are ± 1.

Solution Let A be an orthogonal matrix.

$\therefore \quad AA^T = A^T = I$

Let λ be a characteristic value and X be its characteristic vector then

$AX = \lambda X \quad \ldots$ (1)

or $\quad X^T A^T = \lambda X^T \ldots$ (2)

Now $(X^T A^T)(AX) = \lambda^2 X^T X$

$X^T(A^T A) = \lambda^2 X^T X \Rightarrow X^T X(1-\lambda^2) = 0 \quad (\because \ A^T A = I)$

$X^T X \neq 0 \Rightarrow 1 - \lambda^2 = 0 \ \therefore \ \lambda = \pm 1.$

CAYLEY–HAMILTON THEOREM

Every square matrix satisfies its own characteristic equation (*M.U. 1995*)

$\therefore$ If $|A - \lambda I| = a_0\lambda^n + a_1\lambda^{n-1} + a_2\lambda^{n-2} + \cdots + a_{n-1}\lambda + a_n = 0$ is the characteristic equation of A then

$a_0 A^n + a_1 A^{n-1} + a_2 A^{n-2} + \cdots + a_{n-1}A + a_n I = 0$ (1)

Inverse of *A* Cayley-Hamilton theorem may also be used to obtain A^{-1} if A is non-singular.

From (1)

$$-a_n I = a_0 A^n + a_1 A^{n-1} + a_2 A^{n-2} + \cdots + a_{n-1}A.$$

Premultiplying by A^{-1} and transposing, we get

$$A^{-1} = -\frac{1}{a_n}[a_0 A^{n-1} + a_1 A^{n-2} + a_2 A^{n-3} + \cdots + a_{n-1} I]$$

Example 2.36 Verify Cayley-Hamilton theorem for the matrix $\begin{pmatrix} 11 & -4 & -7 \\ 7 & -2 & -5 \\ 10 & -4 & -6 \end{pmatrix}$. Also find its inverse if possible.

Solution Let A be the given matrix. Then the characteristic equation of A is given by

$$|A - \lambda I| = \begin{vmatrix} 11-\lambda & -4 & -7 \\ 7 & -2-\lambda & -5 \\ 10 & -4 & -6-\lambda \end{vmatrix} = 0.$$

$\Rightarrow -\lambda^3 + 3\lambda^2 - 2\lambda = 0$ or $\lambda^3 - 3\lambda^2 + 2\lambda = 0$ (1)

Now by Cayley-Hamilton theorem we have from (1)

$$A^3 - 3A^2 + 2A = 0 \tag{2}$$

Verification

$$A^2 = A \cdot A = \begin{pmatrix} 23 & -8 & -5 \\ 13 & -4 & -9 \\ 22 & -8 & -14 \end{pmatrix}; \; A^3 = A^2 \cdot A = \begin{pmatrix} 47 & -16 & -31 \\ 25 & -8 & 17 \\ 46 & -16 & -30 \end{pmatrix}$$

$$\therefore \; A^3 - 3A^2 + 2A = \begin{pmatrix} 47 & -16 & -31 \\ 25 & -8 & 17 \\ 46 & -16 & -30 \end{pmatrix} - 3\begin{pmatrix} 23 & -8 & -5 \\ 13 & -4 & -9 \\ 22 & -8 & -14 \end{pmatrix} + 2\begin{pmatrix} 11 & -4 & -7 \\ 7 & -2 & -5 \\ 10 & -4 & -6 \end{pmatrix} = \begin{pmatrix} 0 & 0 & 0 \\ 0 & 0 & 0 \\ 0 & 0 & 0 \end{pmatrix} = 0$$

Hence the theorem is verified.

In this problem $\lambda = 0$ is one of the characteristic value $\Rightarrow A$ is a singular matrix.

$\therefore \; A^{-1}$ is not possible using C-H theorem.

Example 2.37 Find the characteristic equation of the matrix $A = \begin{pmatrix} 2 & 1 & 1 \\ -1 & 2 & -1 \\ 1 & -1 & 2 \end{pmatrix}$ and verify that it has satisfied A. Also compute A^{-1} and A^4.

Solution The characteristic equation of the matrix A is given by

$$|A - \lambda I| = \begin{vmatrix} 2-\lambda & -1 & 1 \\ -1 & 2-\lambda & -1 \\ 1 & -1 & 2-\lambda \end{vmatrix} = 0$$

$$\Rightarrow -\lambda^3 + 6\lambda^2 - 9\lambda + 4 = 0 \tag{1}$$

Consider the matrix equation $-A^3 + 6A^2 - 9A + 4I = 0$ (2)

Now

$$A^2 = A \cdot A = \begin{pmatrix} 6 & -5 & 5 \\ -5 & 6 & -5 \\ 5 & -5 & 6 \end{pmatrix}; \; A^3 = A^2 \cdot A = \begin{pmatrix} 22 & -21 & 21 \\ -21 & 22 & -21 \\ 21 & -21 & 22 \end{pmatrix}$$

$$\therefore -A^3 + 6A - 9A + 4I = -\begin{pmatrix} 22 & -21 & 21 \\ -21 & 22 & -21 \\ 21 & -21 & 22 \end{pmatrix} + 6\begin{pmatrix} 6 & -5 & 5 \\ -5 & 6 & -5 \\ 5 & -5 & 6 \end{pmatrix}$$

$$-9\begin{pmatrix} 2 & -1 & 1 \\ -1 & 2 & -1 \\ 1 & -1 & 2 \end{pmatrix} + 4\begin{pmatrix} 1 & 0 & 0 \\ 0 & 1 & 0 \\ 0 & 0 & 1 \end{pmatrix} = \begin{pmatrix} 0 & 0 & 0 \\ 0 & 0 & 0 \\ 0 & 0 & 0 \end{pmatrix} = 0$$

Hence $-A^3 + 6A - 9A + 4I = 0$ i.e., the characteristic equation of the matrix A i.e., eqn.(1) is satisfied by A. [Cayley Hamilton theorem verification]

Computing A^{-1}: Pre-multiplying eqn.(2) by A^{-1}, we get

$$4A^{-1} = A^2 - 6A + 9I \tag{3}$$

Substituting for A^2, A and I in eqn.(3), we get

$$4^{A-1} = \begin{pmatrix} 6 & -5 & 5 \\ -5 & 6 & -5 \\ 5 & -5 & 6 \end{pmatrix} - 6\begin{pmatrix} 2 & -1 & 1 \\ -1 & 2 & -1 \\ 1 & -1 & 2 \end{pmatrix} + 9\begin{pmatrix} 1 & 0 & 0 \\ 0 & 1 & 0 \\ 0 & 0 & 1 \end{pmatrix} \text{ or } A^{-1} = \frac{1}{4}\begin{pmatrix} 3 & 1 & -1 \\ 1 & 3 & 1 \\ -1 & 1 & 3 \end{pmatrix}$$

Computing A^4: Pre-multiplying eqn.(2) by A, we get

$$-A^4+6A^3-9A^2+4A=0 \text{ or } A^4=-6A^3+9A^2-4A.$$

$$\therefore\ A^4=-6\begin{pmatrix}22 & -21 & 21\\ -21 & 22 & -21\\ 21 & -21 & 22\end{pmatrix}+9\begin{pmatrix}6 & -5 & 5\\ -5 & 6 & -5\\ 5 & -5 & 6\end{pmatrix}-4\begin{pmatrix}2 & -1 & 1\\ -1 & 2 & -1\\ 1 & -1 & 2\end{pmatrix}$$

$$\therefore\ A^4=\begin{pmatrix}-86 & 85 & -85\\ 85 & -86 & 85\\ -85 & 85 & -86\end{pmatrix}$$

DIAGONALISATION OF A MATRIX

A square matrix A with distinct characteristic values may be diagonalised by a similarity transformation $D=P^{-1}AP$ where P is the matrix whose columns are the characteristic vectors of A. The elements of the diagonal matrix D are characteristic values of A.

Proof Let A be a square matrix of order n. Let $\lambda_i\,(i=1, 2, ...)$ be its n distinct characteristic values and $X_i\,(i=1, 2, ...)$ be the corresponding characteristic vectors. $\therefore\ AX_i=\lambda_i X_i$ for $i=1,2,...,n$

Let $P=[X_1 X_2 ... X_n]$

$$\therefore\ AP=[AX_1\ AX_2 ... AX_n]=[\lambda_1 X_1\ \lambda_2 X_2 ..., \lambda_n X_n]$$

$$\text{or}\quad AP=[X_1\ X_2 ... X_n]\begin{bmatrix}\lambda_1 & 0 & \cdots & 0\\ 0 & \lambda_2 & \cdots & 0\\ \cdots & \cdots & \cdots & \cdots\\ \cdots & \cdots & \cdots & \cdots\\ 0 & 0 & \cdots & \lambda_n\end{bmatrix}=PD$$

i.e., $AP=PD$ $\therefore$ $P^{-1}AP=P^{-1}PD=D$.

$\therefore$ If A is non-singular and its characteristic values are distinct, then $P^{-1}AP=D$.

Note:

1. Since $A \sim D$, A and D should have the same characteristic values.
2. This process of finding P, the model matrix, such that $P^{-1}APP=D$ is called diagonalising the matrix.

COMPUTATION OF POWERS OF THE MATRIX *A*

Consider $D=P^{-1}AP$ (From above process)

$$\therefore\quad A=PDP^{-1}$$

$$\Rightarrow A^k=(PDP^{-1})(PDP^{-1})...k \text{ times } (PDP^{-1})$$

$$=PD(P^{-1}P)D(P^{-1}P)...D(P^{-1}P)DP^{-1}=PD^kP^{-1}$$

$$\therefore\quad A^k=PD^kP^{-1}=P\begin{bmatrix}\lambda_1^k & 0 & 0 & \cdots & 0\\ 0 & \lambda_2^k & 0 & \cdots & 0\\ \cdots & \cdots & \cdots & & \cdots\\ \cdots & \cdots & \cdots & & \cdots\\ 0 & 0 & 0 & \cdots & \lambda_n^k\end{bmatrix}P^{-1}$$

Example 2.38 Given the matrix $A=\begin{pmatrix}3&2&4\\2&0&2\\4&2&3\end{pmatrix}$ find matrix P such that $P^{-1}AP$ is a diagonal matrix. (diagonalise A).

Solution The characteristic values of A are given by

$$|A-\lambda I|=\begin{vmatrix}3-\lambda&2&4\\2&-\lambda&2\\4&2&3-\lambda\end{vmatrix}=0$$

$$\Rightarrow \lambda^3-6\lambda^2-15\lambda-8=0$$

Sum of alternative coefficients is equal.

i.e. $[1+(-15)=(-6)+(-8)]$

$\therefore\ \lambda=-1$ is a factor.

By synthetic division the other two values are $\lambda=-1$ and $\lambda=8$.

$\therefore\quad \lambda=-1,-1,8$.

Characteristic vector *for* $\lambda=-1$: It is given by

$$[A-(-1)I]X=0 \text{ i.e., } [A+I]\ X=0$$

$$\text{i.e., }\begin{pmatrix}4&2&4\\2&1&2\\4&2&4\end{pmatrix}\begin{pmatrix}x_1\\x_2\\x_3\end{pmatrix}=\begin{pmatrix}0\\0\\0\end{pmatrix}$$

The system reduces to a single equation $2x_1+x_2+2x_3=0$ (1)

Putting $x_2=0$ in eqn.(1) we get

$$2x_1+2x_3=0 \text{ or } x_1+x_3=0\Rightarrow \frac{x_1}{-1}=\frac{x_3}{1}$$

$$\therefore\quad \frac{x_1}{-1}=\frac{x_2}{0}=\frac{x_3}{1}=k \text{ (say)}$$

Hence $X_1=[-k\ 0\ k]^T$ or $X_1=[-1\ 0\ 1]^T$ If k = 1

Putting $x_3=0$ in eqn.(1) we get $2x_1+x_2=0$

$$\Rightarrow \frac{x_1}{-1}=\frac{x_2}{2} \Leftrightarrow \frac{x_1}{-1}=\frac{x_2}{2}=\frac{x_3}{0}=k \text{ (say)}$$

Hence $X_2=[-k2k0]^T$ or $X_2=[-120]^T$

Characteristic Vector *for* $\lambda=8$: It is given by $[A-8I]\ X=0$

$$\text{i.e., }\begin{pmatrix}-5&2&4\\2&-8&2\\4&2&-5\end{pmatrix}\begin{pmatrix}x_1\\x_2\\x_3\end{pmatrix}=\begin{pmatrix}0\\0\\0\end{pmatrix} \text{ which gives}$$

$$-5x_1+2x_2+4x_3=0;\ 2x_1-8x_2+2x_3=0 \text{ and } 4x_1+2x_2-5x_3=0.$$

Solving any two of the equations we get

$$\frac{x_1}{2}=\frac{x_2}{1}=\frac{x_3}{2}=k \text{ (say)}$$

$X_3=[2k\ k\ 2k]^T$ or $X_3=[2\ 1\ 2]^T$ if $k=1$

Let $P=\underset{\text{(modal matrix)}}{[X_1 X_2 X_3]}=\begin{pmatrix}-1 & -1 & 2\\ 0 & 2 & 1\\ 1 & 0 & 2\end{pmatrix}$ $\therefore\ P^{-1}=-\frac{1}{9}\begin{pmatrix}4 & 2 & -5\\ 1 & -4 & 1\\ -2 & -1 & -2\end{pmatrix}$

$$AP=\begin{pmatrix}3 & 2 & 4\\ 2 & 0 & 2\\ 4 & 2 & 3\end{pmatrix}\begin{pmatrix}-1 & -1 & 2\\ 0 & 2 & 1\\ 1 & 0 & 2\end{pmatrix}=\begin{pmatrix}1 & 1 & 16\\ 0 & -2 & 8\\ -1 & 0 & 16\end{pmatrix}$$

$$P^{-1}(AP)=-\frac{1}{9}\begin{pmatrix}4 & 2 & -5\\ 1 & -4 & 1\\ -2 & -1 & -2\end{pmatrix}\begin{pmatrix}1 & 1 & 16\\ 0 & -2 & 8\\ -1 & 0 & 16\end{pmatrix}=-\frac{1}{9}\begin{pmatrix}9 & 0 & 0\\ 0 & 9 & 0\\ 0 & 0 & -72\end{pmatrix}=\begin{pmatrix}-1 & 0 & 0\\ 0 & -1 & 0\\ 0 & 0 & 8\end{pmatrix}=D$$

$$\therefore P=\underset{\text{(modal matrix)}}{[X_1 X_2 X_3]}=\begin{pmatrix}-1 & -1 & 2\\ 0 & 2 & 1\\ 1 & 0 & 2\end{pmatrix}$$

Example 2.39 Find a matrix P such that $P^{-1}AP=D$ where $A=\begin{pmatrix}3 & 1 & 4\\ 2 & 2 & 4\\ 4 & 1 & 3\end{pmatrix}$ and D is a diagonal matrix whose non-zero elements are the eigen values of A.

Solution The eigen or characteristic values of the matrix A are given by

$$|A-\lambda I|=\begin{vmatrix}3-\lambda & 1 & 1\\ 2 & 2-\lambda & 4\\ 4 & 1 & 3-\lambda\end{vmatrix}=0$$

$\Rightarrow -\lambda^3+8\lambda^2+\lambda-8=0$

or $\quad -(\lambda+1)(\lambda-1)(\lambda-8)=0$ or $\lambda=-1,1,8$

eigen vector for $\lambda=-1$: It is given by $[A-(-1)I]\ X=0$ or $[A+I]\ X=0$

$$\text{i.e., }\begin{pmatrix}4 & 1 & 4\\ 2 & 3 & 4\\ 4 & 1 & 4\end{pmatrix}\begin{pmatrix}x_1\\ x_2\\ x_3\end{pmatrix}=\begin{pmatrix}0\\ 0\\ 0\end{pmatrix}\Rightarrow \begin{matrix}4x_1+x_2+4x_3=0\\ 2x_1+3x_2+4x_3=0\end{matrix}$$

Solving the system $\dfrac{x_1}{-4}=\dfrac{x_2}{-4}=\dfrac{x_3}{5}=k$ (say)

$\therefore\ X_1=[-4k -4k\ \ 5k]^T$ or $X_1=[-4\ -4\ \ 5]^T$ if $k=1$

eigen vector for $\lambda=1$: It is given by $[A-(1)I]\ X=0$ or $[A-I]\ X=0$

$$\text{i.e., }\begin{pmatrix}2 & 1 & 4\\ 2 & 1 & 4\\ 4 & 1 & 2\end{pmatrix}\begin{pmatrix}x_1\\ x_2\\ x_3\end{pmatrix}=\begin{pmatrix}0\\ 0\\ 0\end{pmatrix}\Rightarrow \begin{matrix}2x_1+x_2+4x_3=0\\ 4x_1+x_2+2x_3=0\end{matrix}$$

Solving the system $\dfrac{x_1}{1}=\dfrac{x_2}{-6}=\dfrac{x_3}{1}=k$ (say)

$\therefore \quad X_1 = [k \ -6k \ \ k]^T$ or $X_2 = [1 \ -6 \ 1]^T$ if $k = 1$

eigen vector for $\lambda = 8$: It is given by $[A - 8I] \ X = 0$

$$\text{i.e.,} \begin{pmatrix} -5 & 1 & 4 \\ 2 & -6 & 4 \\ 4 & 1 & -5 \end{pmatrix} \begin{pmatrix} x_1 \\ x_2 \\ x_3 \end{pmatrix} = \begin{pmatrix} 0 \\ 0 \\ 0 \end{pmatrix} \Rightarrow \begin{matrix} -5x_1 + x_2 + 4x_3 = 0 \\ 2x_1 - 6x_2 + 4x_3 = 0 \\ 4x_1 + x_2 - 5x_3 = 0 \end{matrix}$$

Solving the system $\dfrac{x_1}{1} = \dfrac{x_2}{1} = \dfrac{x_3}{1} = k$ (say)

$\therefore \ X_3 = [k \ k \ k]^T$ or $X_3 = [1 \ 1 \ 1]$ if $k = 1$

$$\text{Let } P = \underset{\text{(modal matrix)}}{[X_1 X_2 X_3]} = \begin{pmatrix} -4 & 1 & 1 \\ -4 & -6 & 1 \\ 5 & 1 & 1 \end{pmatrix} \therefore \ P^{-1} = \frac{1}{63} \begin{pmatrix} 7 & 0 & 7 \\ 9 & -9 & 0 \\ 26 & 9 & 28 \end{pmatrix}$$

$$\therefore AP = \begin{pmatrix} 3 & 1 & 4 \\ 2 & 2 & 4 \\ 4 & 1 & 3 \end{pmatrix} \begin{pmatrix} -4 & 1 & 1 \\ -4 & -6 & 1 \\ 5 & 1 & 1 \end{pmatrix} = \begin{pmatrix} 4 & 1 & 8 \\ 4 & -6 & 8 \\ -5 & 1 & 8 \end{pmatrix}$$

$$P^{-1}(AP) = \frac{1}{63} \begin{pmatrix} -7 & 0 & 7 \\ 9 & -9 & 0 \\ 26 & 9 & 28 \end{pmatrix} \begin{pmatrix} 4 & 1 & 8 \\ 4 & -6 & 8 \\ -5 & 1 & 8 \end{pmatrix} = \frac{1}{63} \begin{pmatrix} -63 & 0 & 0 \\ 0 & 63 & 0 \\ 0 & 0 & 504 \end{pmatrix}$$

$$\therefore \quad P^{-1} AP = \begin{pmatrix} -1 & 0 & 0 \\ 0 & 1 & 0 \\ 0 & 0 & 8 \end{pmatrix} = D$$

Hence the matrix P such that $P^{-1} AP = D$ is

$$P = \begin{pmatrix} -4 & 1 & 1 \\ -4 & -6 & 1 \\ 5 & 1 & 1 \end{pmatrix}$$

EXERCISE III

1. Find the eigen values and eigen vectors of the following matrices.

a. $\begin{pmatrix} 1 & -2 \\ -5 & 4 \end{pmatrix}$ b. $\begin{pmatrix} 2 & -1 \\ -8 & 4 \end{pmatrix}$ c. $\begin{pmatrix} 13 & -3 & 5 \\ 0 & 4 & 0 \\ -15 & 9 & -7 \end{pmatrix}$

d. $\begin{pmatrix} 3 & 1 & 4 \\ 0 & 2 & 6 \\ 0 & 0 & 5 \end{pmatrix}$ e. $\begin{pmatrix} 2 & 2 & 1 \\ 1 & 3 & 1 \\ 1 & 2 & 2 \end{pmatrix}$ f. $\begin{pmatrix} 1 & 2 & 2 \\ 0 & 2 & 2 \\ -1 & 1 & 3 \end{pmatrix}$

g. $\begin{pmatrix} 8 & -6 & 2 \\ -6 & 7 & -4 \\ 2 & -4 & 3 \end{pmatrix}$ h. $\begin{pmatrix} 3 & 1 & 0 \\ 0 & 3 & 1 \\ 0 & 0 & 3 \end{pmatrix}$ i. $\begin{pmatrix} 1 & -1 & 0 \\ 1 & 2 & 1 \\ -2 & 1 & -1 \end{pmatrix}$

j. $\begin{pmatrix} 10 & -2 & -5 \\ -2 & 2 & 3 \\ -5 & 3 & 5 \end{pmatrix}$ k. $\begin{pmatrix} -2 & 2 & -3 \\ 2 & 1 & -6 \\ -1 & 2 & 0 \end{pmatrix}$ i. $\begin{pmatrix} 2 & 1 & 0 \\ 1 & 3 & 0 \\ 0 & 0 & 2 \end{pmatrix}$

m. $\begin{pmatrix} 2 & 1 & 0 \\ 0 & 2 & 1 \\ 0 & 0 & 2 \end{pmatrix}$ n. $\begin{pmatrix} 2 & 1 & -1 \\ 0 & 3 & -2 \\ 2 & 4 & -3 \end{pmatrix}$ o. $\begin{pmatrix} 2 & 0 & 1 \\ 0 & 2 & 0 \\ 1 & 0 & 2 \end{pmatrix}$

2. Without finding the eigen values find the sum and product of the eigen values of

a. $\begin{pmatrix} 1 & 1 & 1 \\ 1 & 2 & 2 \\ 1 & 2 & 3 \end{pmatrix}$ *(M.U. 1994)* b. $\begin{pmatrix} 1 & 2 & -1 \\ -2 & 0 & 0 \\ 4 & 5 & 0 \end{pmatrix}$ *(M.U. 1997)*

3. If 1, 5 are the two eigen values of $\begin{pmatrix} 2 & 2 & 1 \\ 1 & 3 & 1 \\ 1 & 2 & 2 \end{pmatrix}$ what is the third eigen value.

4. If $A = \begin{pmatrix} 8 & -6 & 2 \\ -6 & 7 & -4 \\ 2 & -4 & 3 \end{pmatrix}$ find the eigen values of A, A^2, A^4, and A^{-1}.

5. Verify Cayley-Hamilton theorem for the matrix $\begin{pmatrix} 1 & 0 & -2 \\ 2 & 2 & 4 \\ 0 & 0 & 2 \end{pmatrix}$ *(M.U. 1994)*

6. Verify Cayley-Hamilton theorem for the following matrices and find the inverse.

a. $\begin{pmatrix} 1 & 2 & 4 \\ 2 & 1 & 2 \\ 4 & 2 & 1 \end{pmatrix}$ b. $\begin{pmatrix} 1 & 1 & 2 \\ 3 & 1 & 1 \\ 2 & 3 & 1 \end{pmatrix}$

7. Diagonalise the following matrices.

a. $\begin{pmatrix} 8 & -6 & 2 \\ -6 & 7 & -4 \\ 2 & -4 & 3 \end{pmatrix}$ b. $\begin{pmatrix} 1 & 2 & 2 \\ 0 & 2 & 2 \\ -1 & 1 & 3 \end{pmatrix}$ c. $\begin{pmatrix} -2 & 2 & -3 \\ 2 & 1 & -6 \\ -1 & 2 & 0 \end{pmatrix}$

ANSWERS

1. a. $6, -1, [2 -5]^T, [5\ 2]^T$ b. $6, 6, [4\ 1]^T, [2 -1]^T$

c. $-2, 4, 8, [1\ 6\ 3]^T, [\frac{-1}{2}\ 1\ \frac{3}{2}]^T, [1\ 0\ -1]^T$ d. $2, 3, 5, [1 -1\ 0]^T, [1\ 0\ 0]^T, [3\ 2\ 1]^T$

e. $5, 1, 1, [1\ 2\ 1]^T, [1\ 0\ -1]^T, [1 -1\ 0]^T$ f. $1, 2, 3[1\ -1\ 0]^T, [2 -1\ 2]^T, [1 -1\ 2]^T$

g. $0, 3, 15[1\ 2\ 2]^T, [2\ 1 -2]^T, [2 -2\ 1]^T$ h. $3, 3, 3\ [1\ 0\ 0]^T$

i. 1, 2, $-1[3\ 2\ 1]^T$, $[1\ 3\ 1]^T$, $[1\ 0\ 1]^T$

j. 0, 3, $4[1 -5\ 4]^T$, $[1\ 1\ 1]^T$, $[-3\ 1\ 2]^T$

k. 5, -3, $-3[1\ 0\ 0]^T$, $[-2\ 1\ 0]^T$, $[3\ 0\ 1]^T$

l. 1, 3, $3[1\ 0 -1]^T$, $[1\ 0\ 1]^T$, $[0\ 1\ 0]^T$

m. 2, 2, $2[1\ 0\ 0]^T$

n. 0, 1, 1, $\left[\frac{1}{6}\ \frac{2}{3}\ 1\right]^T$, $[011]^T$

o. 1, 2, 3, $[1\ 0 -1]^T$, $[0\ 1\ 0]^T$, $[1\ 0\ 1]^T$

2. a. 6

   b. 1, 10

3. 1

4. 0, 3, 15; 0, 3^2, 15^2; 0, 3^4, 15^4; No eigen values for A^{-1}.

7. a. $\frac{1}{9}\begin{pmatrix} -3 & 6 & 0 \\ 6 & -15 & 6 \\ 0 & 6 & -3 \end{pmatrix}$

   b. $\frac{1}{11}\begin{pmatrix} -2 & 5 & -1 \\ -1 & -3 & 5 \\ 7 & -1 & -2 \end{pmatrix}$

CHAPTER

3

Mathematical Logic

INTRODUCTION

This chapter deals with mathematical logic. One of the main aims of mathematical logic is to provide rules. The rules of logic give precise meaning to mathematical statements and distinguish between valid and invalid mathematical arguments. In addition, logic has numerous applications in computer science. These rules are used in the design of computer circuits, construction of computer programs, verification of the correctness of programs, and in many other ways.

PROPOSITIONS AND LOGICAL OPERATORS

Proposition **A proposition is a statement to which only one of the terms, true or false, can be meaningfully applied.**

The value of a proposition if true is denoted by 1 and false if denoted by 0. Occasionally they are also denoted by the symbols *T* and *F*.

The following are propositions.

a. $4+2=6$

b. **4 is an even integer and 5 is not**

c. **5 is a prime number**

d. **New Delhi is the capital of India**

e. $2 \in \{1, 3, 5, 7\}$

f. $42 \geq 51$

g. **Paris is in England.**

Of the above propositions, (a)–(d) are true whereas (e)–(g) are false.

The following are *not* propositions.

a. **Where are you going?**

b. $x+2=5$

c. $x+y<z$

d. **Beware of dogs.**

The expressions (a) and (d) are not propositions since neither is true or false. The expression (b) and (c) are not propositions, since the variables in these expressions have not been assigned values and hence they are neither true or false.

Note: Letters are used to denote propositions just as letters are used to denote variables. The conventional letters used for this purpose are $p, q, r, s, \ldots$

Logical Operators

There are several ways in which we commonly combine simple statements into compound ones. The words *or*, and, not, if... then and if and only if can be added to one or more propositions to create a new proposition. New propositions are called compound propositions. Logical operators are used to form new propositions or compound propositions. These logical operators are also called connectives.

Truth Table

The truth table of a logical operator specifies how the truth value of a proposition using that operator is determined by the truth values of the propositions. A truth table lists all possible combination of truth values of the propositions in the left most columns and the truth value of the resulting propositions in the right most column.

Negation (NOT) If p is a proposition, its negation not p is another proposition called the negation p. The negation of p is denoted by $\sim p$. The proposition $\sim p$ is read "not p".
Alternate symbols used in the literature are $\neg$ p, $\bar{p}$ and "not p".

Note that a negation is called a connective although it only modifies a statement. In this sense, negation is the only operator that acts on a single proposition.

***Table 3.1** Truth table for negation*

p	~p
0	1
1	0

Example 3.1 Find the negation of the propositions

a. It is cold.
b. Today is Sunday.
c. Ravi is poor.

Solution The negation of the proposition are

a. It is not cold.
b. Today is not Sunday.
c. Ravi is not poor.

Conjunction (AND) If p and q are propositions, then the propositions "p and q", denoted by $p \wedge q$, is true when both p and q are true and is false otherwise. The proposition $p \wedge q$ is called the conjunction of p and q.

The truth table for $p \wedge q$ is shown in Table 3.2. Note that there are four rows in this truth table, one row for each possible combination of truth values of the propositions p and q.

Table 3.2 Truth table for conjunction

p	q	$p \wedge q$
0	0	0
0	1	0
1	0	0
1	1	1

Example 3.2 Find the conjunction of the propositions p and q where p is the proposition "Today is Sunday" and q is the proposition "It is raining today".

Solution The conjunction of these propositions $p \wedge q$ is two proposition. "Today is Sunday and it is raining today"

Example 3.3 Let p be "Ravi is rich" and let q be "Ravi is happy". Write each of the following in symbolic form.

a. Ravi is poor but happy.
b. Ravi is neither rich nor happy.
c. Ravi is rich and unhappy.

Solution

a. $\sim p \wedge q$
b. $\sim p \wedge \sim q$
c. $p \wedge \sim q$

Disjunction (OR) If p and q are propositions, then disjunction p or q, denoted $p \vee q$, is false when p and q are both false and true otherwise. The proposition $p \vee q$ is called the disjunction of p and q.

Note that in the table 3.3, the connectives ~ and $\wedge$ defined earlier have the same meaning as the words "not" and "and" in general. However, the connective $\vee$ is not always the same as the word "or" because of the fact that the word "or" in English is commonly used both as an "exclusive or" and as an "inclusive or". For example, consider the following statements:

a. I shall watch the movie on TV or go to cinema.
b. There is something wrong with the fan or with the switch.
c. Ten or twenty people were killed in the fire today.

In statement (a), the connective "or" is used in the exclusive sense; that is to say, one or the other possibility exists but not both. In (b) the intended meaning is clearly one or the other or both. The connective "or" used in (b) is the "inclusive or". In (c) the "or" is used for indicating an approximate numbr of peoples, and it is not used as a connective. From the definition of disjunction it is clear that $\vee$ is " inclusive or".

Table 3.3 Truth table for disjunction

p	q	$p \vee q$
0	0	0
0	1	1
1	0	1
1	1	1

Example 3.4 **Find the disjunction of the propositions p and q where p is the proposition "Today is Sunday" and q is the proposition "It is raining today".**

Solution The disjunction of p or q, $p \vee q$, is the proposition "Today is Sunday or it is raining today".

Example 3.5 **Let p be "Ravi is tall" and let q be "Ravi is handsome". Write each of the following statements in symbolic form.**

a. Ravi is short or not handsome.
b. Ravi is tall or handsome.
c. It is not true that Ravi is short or not handsome.

Solution

a. $\sim p \vee \sim q$
b. $p \vee q$
c. $\sim(\sim p \vee \sim q)$

Example 3.6 **Let p be "Ravi speaks Tamil" and let q be "Ravi speaks Hindi". Give a simple verbal sentence which describes each of the following.**

a. $p \vee q$ b. $p \wedge q$ c. $p \wedge \sim q$ d. $\sim p \vee \sim q$ e. $\sim(\sim p)$.

Solution

a. Ravi speaks Tamil or Hindi
b. Ravi speaks Tamil and Hindi.
c. Ravi speaks Tamil but not Hindi.
d. Ravi does not speak Tamil or he does not speak Hindi.
e. It is not true that Ravi does not speak Tamil.

Exclusive OR Let p and q be propositions. The exclusive or of p and q, denoted by $p \oplus q$, is true only when one of p or q (but not both) is true and is false otherwise.

Table 3.4 Truth table for Exclusive OR

p	q	$p \oplus q$
0	0	0
0	1	1
1	0	1
1	1	0

Conditional operator (if ... then) Let p and q be propositions. The implication $p \rightarrow q$ is false when p is true and q is false and true otherwise. In the implication p is called the *premise* (or hypothesis) and q is called the *consequence* (or conclusion).

Table 3.5 Truth table for if.... then

p	q	$p \rightarrow q$
0	0	1
0	1	1
1	0	0
1	1	1

Because implications arise in many places in mathematical argument, a wide variety of terminology is used to express $p \to q$. Some of the more common ways of expressing this implication are

a. p implies q.
b. if p, then q.
c. q if p.
d. p only if q.
e. p is sufficient for q.
f. q whenever p.
g. q is necessary for p.

We shall avoid the word "implies" since it might be used in different contexts.

Example 3.7 Let p denote "It is below freezing" and let q denote "It is snowing". Write the following statements in a symbolic form.

a. If it is below freezing, it is also snowing.
b. It is not snowing if it is below freezing.
c. It is below freezing is a necessary condition for it to be snowing.

Solution Recall that $p \to q$ can be read " if p, then q" or "p only if q" or "q is necessary for p". Then (a) $p \to q$ (b) $p \to \sim q$ (c) $q \to p$

Example 3.8 Let p and q be the propositions

p: You drive over 80 kms per hour.
q: You get a speeding ticket.

Write the following propositions in symbolic form.

a. You will get a speeding ticket if you drive over 80 kms per hour.
b. If you do not drive over 80 kms per hour, then you will not get a speeding ticket.
c. Driving over 80 kms per hour is sufficient for getting a speeding ticket.
d. Whenever you get a speeding ticket, you are driving over 80 kms per hour.

Solution (a) $p \to q$ (b) $\sim p \to \sim q$ (c) $p \to q$ (d) $q \to p$

Example 3.9 Write each of the following statements in the form "if p then q".

a. To get tenure as a professor, it is sufficient to be world-famous.
b. It shows whenever the wind blows from the north-east.
c. The apple trees will bloom if it stays warm for a week.

Solution

a. If you are world-famous, then you will get tenure as a professor.
b. If the wind blows from the north-east, then it snows.
c. If it stays warm for a week, then the apple trees will bloom.

Example 3.10 Determine the truth value of each of the following statements:

a. If Calcutta is in India, then $1 + 1 = 2$.
b. If Calcutta is in Sri Lanka, then $1 + 1 = 2$.
c. If Calcutta is in India, then $1 + 1 = 3$.
d. If Calcutta is in Sri Lanka, then $1 + 1 = 3$.

Solution Since the statement "if p then q" is false only when p is true and q is false. only (c) is false.

Biconditional operator (if and only if) Let p and q be propositions. The biconditional $p \leftrightarrow q$ is true when p and q have the same truth values and is false otherwise.

Note that the biconditional $p \leftrightarrow q$ is true when both the implications $p \rightarrow q$ and $q \rightarrow p$ are true. So "p if and only if q" is used for biconditional. Other common ways of expressing the proposition $p \leftrightarrow q$ or $p = q$ are "p is necessary and sufficient for q" and " if p then q, and conversely".

Table 3.6 Truth table for if and only if

p	q	$p \leftrightarrow q$
0	0	1
0	1	0
1	0	0
1	1	1

Each of the following theorems is well known, and each can be symbolized in the form $p \leftrightarrow q$.

a. Two lines are parallel if and only if they have the same slope.
b. Two triangles are congruent if and only if all three sets of corresponding sides are congruent.

Example 3.11 Determine the truth value of each of the following statements.

a. Calcutta is in India if and only if $1 + 1 = 2$.
b. Calcutta is in Sri Lanka if and only if $1 + 1 = 2$.
c. Calcutta is in India if and only if $1 + 1 = 3$.
d. Calcutta is in Sri Lanka if and only if $1 + 1 = 3$.

Solution (a) and (d) are true since the substatements are both true in (a) and both false in (d). On the other hand, (b) and (c) are false since the substatements have different truth values.

Example 3.12 Let p denote " He is poor" and let q denote "He is happy". Write each statement in symbolic form using p and q.

a. To be poor is to be unhappy.
b. He is rich if and only if he is unhappy.
c. Being rich is a necessary and sufficient condition to being happy.

Solution

a. $p \leftrightarrow q$
b. $\sim p \leftrightarrow q$
c. $\sim p \leftrightarrow q$

CONSTRUCTION OF TRUTH TABLES

Truth tables have already been introduced in the definitions of the connectives. Our basic concern is to determine the truth table of a proposition for each possible combination of the truth values of the compound propositions. A table showing all such truth values is called truth table of the formula. In general, if there are n distinct components in a proposition or formula, we need to consider 2^n possible combinations of truth values in order to obtain the truth table.

Two methods of constructing truth table are shown in the following examples.

Example 3.13 Construct the truth table for the statement formula $\sim p \wedge q$.

Solution It is necessary to consider all possible values of p and q (for the variables, as here, four rows are necessary). These values are entered in two first two columns of table for both methods.

Method 1 In this method, the truth values of $\sim$ p are entered in the third column, and the truth values of $\sim p \wedge q$ are entered in the fourth column.

Truth Table (Method 1)

p	q	$\sim p$	$\sim p \wedge q$
0	0	1	0
0	1	1	1
1	0	0	0
1	1	0	0

Truth Table (Method 2)

p	q	p	$\sim$	$\wedge$	q
0	0	0	1	0	0
0	1	0	1	1	1
1	0	1	0	0	0
1	1	1	0	0	1
Step number		1	2	3	1

Method 2 In this method, a column is drawn for each statement as well as for the connectives that appear. The truth values are entered step by step. The step numbers at the bottom of the table shows the sequence followed in arriving at the final step.

Example 3.14 Construct the truth table for $p \wedge \sim p$.

Solution

Method 1

p	$\sim p$	$p \wedge \sim p$
0	1	0
1	0	0

Method 2

p	p	$\wedge$	$\sim$	p
0	0	0	1	0
1	1	0	0	1
Step number	1	3	2	1

Example 3.15 Construct the truth table for $(p \vee q) \vee \sim p$.

Solution

Method 1

p	q	$p \vee q$	$\sim p$	$(p \vee q) \vee \sim p$
0	0	0	1	1
0	1	1	1	1
1	0	1	0	1
1	1	1	0	1

Method 2

p	q	p	$\vee$	q	$\vee$	$\sim$	p
0	0	0	0	0	1	1	0
0	1	0	1	1	1	1	0
1	0	1	1	0	1	0	1
1	1	1	1	1	1	0	1
Step number		1	2	1	3	2	1

Example 3.16 Construct truth tables for

a. $\sim (p \wedge q)$ b. $p \wedge (\sim q)$ c. $(p \wedge q) \wedge r$
d. $(p \wedge q) \vee (q \wedge r) \vee (r \wedge p)$ e. $(\sim p) \vee (\sim q)$ f. $(p \vee q) \vee r \vee s$

Solution

a. $\sim (p \wedge q)$

p	q	$p \wedge q$	$\sim (p \wedge q)$
0	0	0	1
0	1	0	1
1	0	0	1
1	1	1	0

b. $p \wedge (\sim q)$

p	q	$\sim q$	$p \wedge (\sim q)$
0	0	1	0
0	1	0	0
1	0	1	1
1	1	0	0

c. $(p \wedge q) \wedge r$

p	q	r	$p \wedge q$	$(p \wedge q) \wedge r$
0	0	0	0	0
0	0	1	0	0
0	1	0	0	0
0	1	1	0	0
1	0	0	0	0
1	0	1	0	0
1	1	0	1	0
1	1	1	1	1

d. $(p \wedge q) \vee (q \wedge r) \vee (r \wedge p)$

p	q	r	$p \wedge q$	$q \wedge r$	$r \wedge p$	$(p \wedge q) \vee (q \wedge r)$	$(p \wedge q) \vee (q \wedge r) \vee (r \wedge p)$
0	0	0	0	0	0	0	0
0	0	1	0	0	0	0	0
0	1	0	0	0	0	0	0
0	1	1	0	1	0	1	1
1	0	0	0	0	0	0	0
1	0	1	0	0	1	0	1
1	1	0	1	0	0	1	1
1	1	1	1	1	1	1	1

e. $(\sim p) \vee (\sim q)$

p	q	$\sim p$	$\sim q$	$(\sim p) \vee (\sim q)$
0	0	1	1	1
0	1	1	0	1
1	0	0	1	1
1	1	0	0	0

f. $(p \vee q) \vee r \vee s$

p	q	r	s	$p \vee q$	$p \vee q \vee r$	$p \vee q \vee r \vee s$
0	0	0	0	0	0	0
0	0	0	0	0	0	1
0	0	1	0	0	1	1
0	1	0	1	1	1	1
0	1	0	1	1	1	1
0	1	0	1	1	1	1
0	0	1	0	0	1	1
0	1	1	1	1	1	1
1	0	1	1	1	1	1
1	0	0	1	1	1	1
1	0	0	1	1	1	1
1	1	1	1	1	1	1
1	1	0	1	1	1	1
1	1	1	1	1	1	1
1	0	1	1	1	1	1
1	1	1	1	1	1	1

Example 3.17 Construct the truth table for

a. $\sim(\sim p \vee \sim q)$

b. $\sim(\sim p \wedge \sim q)$

c. $p \wedge (p \vee q)$

d. $p \wedge (q \wedge p)$.

Solution

a. $\sim(\sim p \vee \sim q)$

p	q	$\sim p$	$\sim q$	$(\sim p \vee \sim q)$	$\sim(\sim p \vee \sim q)$
0	0	1	1	1	0
0	1	1	0	1	0
1	0	0	1	1	0
1	1	0	0	0	1

b. $\sim(\sim p \wedge \sim q)$

p	q	$\sim p$	$\sim q$	$(\sim p \wedge \sim q)$	$\sim(\sim p \wedge \sim q)$
0	0	1	1	1	0
0	1	1	0	0	1
1	0	0	1	0	1
1	1	0	0	0	1

c. $p \wedge (p \vee q)$

p	q	$p \vee q$	$p \wedge (p \vee q)$
0	0	0	0
0	1	1	0
1	0	1	1
1	1	1	1

d. $p \wedge (q \wedge p)$

p	q	$q \wedge p$	$p \wedge (q \wedge p)$
0	0	0	0
0	1	0	0
1	0	0	0
1	1	1	1

Example 3.18 **Construct the truth table for**

a. $(p \rightarrow q) \rightarrow (p \wedge q)$
b. $\sim p \rightarrow (q \rightarrow p)$
c. $(p \wedge q) \rightarrow (p \vee q)$
d. $p \rightarrow (q \wedge r)$

Solution

a. $(p \to q) \to (p \wedge q)$

p	q	$p \to q$	$p \wedge q$	$(p \to q) \to (p \wedge q)$
0	0	1	0	0
0	1	1	0	0
1	0	0	0	1
1	1	1	1	1

b. $\sim p \to (q \to p)$

p	q	$\sim p$	$q \to p$	$\sim p \to (q \to p)$
0	0	1	1	1
0	1	1	0	0
1	0	0	1	1
1	1	0	1	1

c. $(p \wedge q) \to (p \vee q)$

p	q	$p \wedge q$	$p \wedge q$	$(p \wedge q) \to (p \vee q)$
0	0	0	0	1
0	1	0	1	1
1	0	0	1	1
1	1	1	1	1

d. $p \to (q \wedge r)$

p	q	r	$q \wedge r$	$p \to (q \wedge r)$
0	0	0	0	1
0	0	1	0	1
0	1	0	0	1
0	1	1	1	1
1	0	0	0	0
1	0	1	0	0
1	1	0	0	0
1	1	1	1	1

EXERCISE I

1. Which of the following sentences are proposition? What are the truth values of those that are propositions?
   a. Calcutta is the capital of India.
   b. Answer this question.
   c. What time is it?
   d. $x+y=y+x$ for every pair of real number x and y.
   e. $S+6=10$
   f. $X+3=3$
   g. $X+2=5$ if $x=1$
   h. Do not go.
2. What is the negation of each of the following propositions.
   a. Tomorrow is Monday.
   b. There is no pollution in New Delhi.
   c. $3+2=5$.
3. Let p, q and r be the propositions
   p: you have the flu
   q: you miss the final examination
   r: you pass the course.
   Express the following in statements.
   a. $p \to q$
   b. $\sim q \leftrightarrow r$
   c. $q \to \sim r$
   d. $p \vee q \vee r$
   e. $(p \to \sim r) \vee (q \to \sim r)$
   f. $(p \wedge q) \vee (\sim q \wedge r)$
4. Let p be "Ravi reads The Hindu", let q be "Ravi reads The Indian Express", and let r be "Ravi reads The Times of India". Write each of the following in symbolic form.
   a. Ravi reads The Hindu or The Indian Express, but not The Times of India.
   b. Ravi reads The Hindu and The Indian Express, or he does not read The Hindu and The Times of India.
   c. It is not true that Ravi reads The Hindu but not The Times of India.
   d. It is not true that Ravi reads The Times of India or The Indian Express but not The Hindu.
5. Given the truth values of p and q as 1 and those of r and s as 0, find the truth values of the following:
   a. $p \vee (q \wedge r)$
   b. $(p \wedge (q \wedge r)) \vee \sim ((\sim p \vee q) \wedge (r \vee s))$
   c. $(\sim (p \wedge q) \vee \sim r) \vee (((\sim p \wedge q) \vee \sim p) \wedge s)$
6. Translate into symbolic form:
   a. Jack and Jill went up the hill.
   b. I was absent from my class yesterday.
   c. Ravi is poor or he is both rich and unhappy.
   d. If either Jerry takes Calculus or Jack takes Sociology, then Larry will take English.
   e. The crop will be destroyed if there is a flood.
   f. For you to get an 'O' Grade in this course, it is necessary and sufficient that you learn how to solve discrete mathematics problems.
   g. If you read the newspaper every day, you will be informed, and conversely.
   h. A positive integer is a prime only if it has no divisors other than 1 and itself.
7. Construct a truth table for the following propositions.
   a. $p \vee \sim p$
   b. $\sim (p \wedge \sim q)$
   c. $p \wedge (q \vee r)$
   d. $(p \wedge q) \wedge (p \wedge r)$
   e. $(p \wedge q) \vee (p \wedge r)$
   f. $(p \to q) \wedge (q \to p)$

g. $p \vee \sim(p \wedge q)$ h. $\sim(\sim p \vee \sim q)$ i. $p \oplus \sim p$

j. $\sim p \oplus \sim q$ k. $\sim p \leftrightarrow q$ l. $(p \to q) \vee (\sim p \to q)$

m. $\sim p \leftrightarrow \sim q$ n. $(p \to q)\ (p \to r)$ o. $(p \leftrightarrow q) \leftrightarrow (p \wedge q) \vee (\sim p \wedge \sim q)$

(M.C.A Apr. 98)

ANSWERS

1. a. Yes, 0 b. No c. No d. Yes, 1

 e. Yes, 0 f. Yes, 0 g. No.

4. a. $(p \vee q) \wedge \sim r$ b. $(p \wedge q) \vee \sim(p \wedge r)$ c. $\sim(p \wedge \sim r)$ d. $\sim[(r \vee q) \wedge \sim p]$

5. a. 1 b. 1 c. 1

TAUTOLOGIES AND CONTRADICTIONS

We know how to construct the truth table of a given formula. In general, the final column of a truth table of a given formula contains both 1 and 0. There are some formulae whose truth values are always 1 or always 0 regardless of the truth value assignments to the variables. Consider for example the statement formula $p \vee \sim p$ and $p \wedge \sim p$ in Table 3.7

Table 3.7 Tautology and Contradiction

p	$\sim p$	$p \vee \sim p$	$p \wedge \sim p$
0	1	1	0
1	0	1	0

The truth values of $p \vee \sim p$ and $p \wedge \sim p$, which are 1 and 0 respectively, are independent of the statement by which the variable p may be replaced.

Taulotogy A statement formula which is true regardless of the truth values of the statements which replace the variables in it is called a tautology or a logical truth or a universally valid formula.

Contradiction A statement formula which is false regardless of the truth values of the statements which replace the variables in it is called a contradiction.

Contingency A statement formula that is neither a tautology nor a contradiction is called a *contingency*.

Note: A straight forward method to determine whether a given formula is a tautology is to construct its truth table . We may say that a statement formula which is a tautology is identically true (their truth tables consist of a column of ones) and a formula which is a contradiction is identically false (their truth tables consist of a column of zeros). Obviously, the negation of a contradiction is a tautology.

Example. 3.19 Verify if the following propositions are tautologies.

a. $(p \wedge q) \to p$.
b. $q \to (p \vee q)$.
c. $(p \vee q) \leftrightarrow (q \vee p)$
d. $p \vee \sim(p \wedge q)$
e. $\sim(p \wedge q) \leftrightarrow (\sim p) \vee (\sim q)$.

Solution Construct the truth table of the propositions.

p	q	$\sim p$	$\sim q$	$p \vee q$	$p \wedge q$	$q \vee p$	$\sim(p \wedge q)$	(a)	(b)	(c)	(d)	(e)
0	0	1	1	0	0	0	1	1	1	1	1	1
0	1	1	0	1	0	1	1	1	1	1	1	1
1	0	0	1	1	0	1	1	1	1	1	1	1
1	1	0	0	1	1	1	0	1	1	1	1	1

Since the truth value of all propositions is 1 for all values of p and q, the given propositions are tautologies.

Example. 3.20 Verify if the proposition $(p \wedge q) \wedge \sim(p \vee q)$ is a contradiction.

Solution

p	q	$p \wedge q$	$p \vee q$	$\sim(p \vee q)$	$(p \wedge q) \wedge \sim(p \vee q)$
0	0	0	0	1	0
0	1	0	1	0	0
1	0	0	1	0	0
1	1	1	1	0	0

Since the truth value of $(p \wedge q) \wedge \sim(p \vee q)$ is 0 for all values of p and q, the proposition is a contradiction.

Example 3.21 Show that the conjunction of two tautologies is also a tautology.

Solution Let us denote A and B by two statement formulae which are tautologies. If we assign any truth values to the variables of A and B, then the truth values of both A and B will be 1. Thus the truth value of $A \wedge B$ will be 1, so that $A \wedge B$ will be a tautology.

Example. 3.22 From the formulae given below, indicate if they are tautologies or contradictions.

a. $p \to (p \vee q)$

b. $(p \to \sim p) \to \sim p$

c. $(\sim q \wedge p) \wedge q$

d. $(p \vee q) \wedge (\sim p \wedge \sim q)$

Solution

p	q	$\sim p$	$\sim q$	$p \vee q$	$\sim p \wedge \sim q$	$p \to \sim p$	$\sim q \wedge p$	(a)	(b)	(c)	(d)
0	0	1	1	0	1	1	0	1	1	0	0
0	1	1	0	1	0	1	0	1	1	0	0
1	0	0	1	1	0	0	1	1	1	0	0
1	1	0	0	1	0	0	0	1	1	0	0

Since the truth values of (a) and (b) are 1 for all values of p and q, (a) and (b) are tautologies. Since the truth values of (c) and (d) are 0 for all values of p and q, (e) and (d) are contradictions.

Substitution instance A formula A is called a substitution instance of another formula B if A can be obtained from B by substituting formulae for some variables of B, with the condition that the same formula is substituted for the same variables each time it occurs.

Note: Suppose A (p, q, ...) is a tautology. Then it does not depend upon the particular truth values of its variables $p,q,\ldots$ so we can substitute P for p, Q for $q\ldots$ for any propositions P, $Q,\ldots$ in the tautology $A(p,q\ldots)$ and still have a tautology $A(P, Q,\ldots)$

Example 3.23 Verify that the proposition $(p \wedge \sim q) \vee \sim (p \wedge \sim q)$ is a tautology. (*B.E.* Apr. 98)

Solution This proposition has the form $p \vee \sim p$ where $p = p \wedge \sim q$. Since $p \vee \sim p$ is a tautology, by the above Note, $(p \wedge \sim p) \vee \sim (p \wedge \sim q)$ is also a tautology.

Note: It is possible to substitute more than one variable by other formulae provided all substitutions are considered to occur simultaneously.

Example 3.24 Produce the substitution instance of the following formulae for the given substitutions.

a. $(((p \to q) \to p) \to p)$; substitute $(p \to q)$ for p and $(p \wedge q \to r)$ for q.

b. $((p \to q) \to (q \to p)$; substitute q for p and $(p \wedge \sim p)$ for q.

Solution

a. Substitute $(p \to q)$ for p and $(p \wedge q \to r)$ for q simultaneously, we get

$$(((p \to q) \to (p \wedge q \to r)\ (p \to q) \to (p \to q))$$

b. Substitute q for p and $(p \wedge \sim p)$ for q simultaneously in

$((p \to q) \to \ (q \to p))$, we get $(q \to (p \wedge \sim p)) \to ((p \wedge \sim p) \to q))$.

Note: Note that in constructing substitution instances of a formula, substitutions are made for the simple proposition (without connectives) and never for the compound proposition. For example, $p \to q$ is not a substitution instance of $p\ \sim r$, because it must be replaced by r and not $\sim r$

Example. 3. 25 Verify if the proposition $((p \vee \sim q) \to r) \leftrightarrow s \vee \sim (((p \vee \sim q) \to r) \leftrightarrow s)$ is a tautology.

Solution This proposition has the form $p \vee \sim p$ where $p = (((p \vee \sim q) \to r) \leftrightarrow s)$.

EQUIVALENCE AND IMPLICATION

An important step used in mathematical argument is the replacement of a statement with another statement with the same truth value. Because of this, methods that produce propositions with the same truth value as a given compound proposition are used extensively in the construction of mathematical arguments.

Equivalence Two propositions are logically equivalent or simply equivalent if they have exactly the same truth values under all circumstances.

We can also define this notion as follows

The propositions p and q are called logically equivalent if $p \leftrightarrow q$ is a tautology. The equivalence of p and q is denoted by $p \Leftrightarrow q$.

Note 1: One way to determine whether two propositions are equivalent is to use a truth table. In particular, the propositions p and q are logically equivalent if and only if the columns giving their truth values agree.

Note 2: Whenever we find logically equivalent statements, we can substitute one for another as we wish, since this action will not change the truth value of any statement

Example 3.26 Show that $\sim (p \wedge q)$ and $\sim p \vee \sim q$ are logically equivalent.

Solution Construct the truth table of these propositions as shown in Truth Table. Since the truth values are same for all combinations, it follows that these propositions are logically equivalent.

p	q	$\sim p$	$\sim q$	$p \wedge q$	$\sim(p \wedge q)$	$\sim p \vee \sim q$
0	0	1	1	0	1	1
0	1	1	0	0	1	1
1	0	0	1	0	1	1
1	1	0	0	1	0	0

Example 3.27 **Show that two propositions $p \rightarrow q$ and $\sim p \vee \sim q$ are logically equivalent.**

Solution Construct the required truth table. Since the truth values of $p \rightarrow q$ and $\sim p \vee q$ agree,

$$p \rightarrow q \Leftrightarrow \sim p \vee q.$$

p	q	$\sim p$	$p \rightarrow q$	$\sim p \vee q$
0	0	1	1	1
0	1	1	1	1
1	0	0	0	0
1	1	0	1	1

Example 3.28 Show that $p \rightarrow (q \rightarrow r) \Leftrightarrow p \rightarrow (\sim q \vee r) \Leftrightarrow (p \wedge q) \rightarrow r$.

Solution Recall from Example 3.27 that $q \rightarrow r \Leftrightarrow \sim q \vee r$.

Replacing $q \rightarrow r$ by $\sim q \vee r$ (see Note 2), we get

$$\begin{aligned} p \rightarrow (q \rightarrow r) &\Leftrightarrow p \rightarrow (\sim q \vee r) \\ &\Leftrightarrow \sim p \vee (\sim q \vee r) \text{ [by Example 3.27]} \\ &\Leftrightarrow (\sim p \vee \sim q) \vee r \text{ using associativity of } \vee \\ &\Leftrightarrow \sim (p \wedge q) \vee r \text{ [by Example 3.26]} \\ &\Leftrightarrow (p \wedge q) \rightarrow r \text{ [by Example 3.27]} \end{aligned}$$

Table 3.8 contains some important equivalences. In these equivalences, 1 denotes any proposition that is a tautology, and 0 denotes any proposition that is a contradiction. The symbol p, q, r represent arbitrary propositions. Most of the equivalences in this table have straightforward intuitive interpretations and all of them can be verified by constructing truth tables.

Table 3.8 *Logical Equivalences*

Equivalence	Name
1. $p \Leftrightarrow (p \vee p)$	idempotents of $\vee$
2. $p \Leftrightarrow (p \wedge p)$	idempotents of $\wedge$
3. $(p \wedge q) \Leftrightarrow (q \vee p)$	commutativity of $\vee$
4. $(p \wedge q) \Leftrightarrow (q \wedge p)$	communtativity of $\wedge$
5. $(p \vee q) \vee r \Leftrightarrow p \vee (q \vee r)$	associativity of $\vee$
6. $(p \wedge q) \wedge r \Leftrightarrow p \wedge (q \wedge r)$	associativity of $\wedge$
7. $\sim (p \vee q) \Leftrightarrow \sim p \wedge \sim q$	De Morgan's law 1
8. $\sim (p \wedge q) \Leftrightarrow \sim p \vee \sim q$	De Morgan's law 2
9. $p \wedge (q \vee r) \Leftrightarrow (p \wedge q) \vee (p \wedge r)$	distributive of $\wedge$ over $\vee$
10. $p \vee (q \wedge r) \Leftrightarrow (p \vee q) \wedge (p \vee r)$	distributive of $\vee$ over $\wedge$
11. $p \vee 1 \Leftrightarrow 1$	(null) or domination law 1
12. $p \wedge 0 \Leftrightarrow 0$	(null) or domination law 2
13. $p \wedge 1 \Leftrightarrow p$	identity law 1
14. $p \vee 0 \Leftrightarrow p$	identity law 2
15. $p \vee \sim p \Leftrightarrow 1$	negation law 1
16. $p \wedge \sim p \Leftrightarrow 0$	negation law 2
17. $\sim (\sim p) \Leftrightarrow p$	double negation law (involution)
18. $p \rightarrow q \Leftrightarrow \sim p \vee q$	implication law
19. $p \leftrightarrow q \Leftrightarrow (p \rightarrow q) \wedge (q \rightarrow p)$	equivalence law
20. $(p \wedge q) \rightarrow r \Leftrightarrow p \rightarrow (q \rightarrow r)$	exportation law
21. $(p \rightarrow q) \wedge (p \rightarrow \sim q) \Leftrightarrow \sim p$	absurdity law
22. $p \rightarrow q \Leftrightarrow \sim q \rightarrow \sim p$	contrapositive law
23. $p \vee (p \wedge q) \Leftrightarrow p$	absorption law 1
24. $p \wedge (p \vee q) \Leftrightarrow p$	absorption law 2
25. $p \leftrightarrow q \Leftrightarrow (p \wedge q) \vee (\sim p \wedge \sim q)$	biconditional law

Example. 3.29 Show that $\sim (p \vee (\sim p \wedge q))$ and $\sim p \wedge \sim q$ are logically equivalent.

Note: Instead of a truth table we shall establish equivalence by developing a series of logical equivalences using Table 3.8 at a starting with $\sim (p \vee (\sim p \wedge q))$ and ending with $\sim p \wedge \sim q$.

$\sim(p \vee (\sim p \wedge q)) \Leftrightarrow \sim p \wedge \sim(\sim p \wedge q)$ by De Morgan's law 1

$\Leftrightarrow \sim p \wedge (\sim(\sim p) \vee \sim q)$ by De Morgan's law 2

$\Leftrightarrow \sim p \wedge (p \vee \sim q)$ by involution law.

$\Leftrightarrow (\sim p \wedge p) \vee (\sim p \wedge \sim q)$ by distributive law of $\wedge$ over $\vee$

$\Leftrightarrow 0 \vee (\sim p \wedge \sim q)$ by negation law 2

$\Leftrightarrow (\sim p \wedge \sim q) \vee 0$ by commutative law of $\vee$

$\Leftrightarrow \sim p \wedge \sim q$ by identity law 2

Hence, $\sim(p \vee (\sim p \wedge q))$ and $\sim p \wedge \sim q$ are logically equivalent.

Example. 3.30 Show that $(p \wedge q) \rightarrow (p \vee q)$ is a tautology.

Solution

$(p \wedge q) \rightarrow (p \vee q) \Leftrightarrow \sim(p \wedge q) \vee (p \vee q)$ by implication law

$\Leftrightarrow (\sim p \vee \sim q) \vee (p \vee q)$ by De Morgan's law 2.

$\Leftrightarrow \sim p \vee (\sim q \vee p) \vee q$ by associative law of $\vee$

$\Leftrightarrow \sim p \vee (p \vee \sim q) \vee q$ by commutative law of $\vee$

$\Leftrightarrow (\sim p \vee p) \vee (\sim q \vee q)$ by associative law of $\vee$

$\Leftrightarrow 1 \vee 1$ by negation law 1

$\Leftrightarrow 1$ by domination law 1

Hence the result.

Note: This could also be done using a truth table.

Example 3.31 Show that $\sim(p \wedge (\sim q \wedge r)) \vee (q \wedge r) \vee (p \wedge r) \Leftrightarrow r$. *(M.C.A. Nov 96, May 97)*

Solution

$(\sim p \wedge (\sim q \wedge r)) \vee (q \wedge r) \vee (p \wedge r)$

$\Leftrightarrow (\sim p \wedge (\sim q \wedge r)) \vee ((q \vee p) \wedge r)$ by distributive law of $\wedge$ over $\vee$

$\Leftrightarrow ((\sim p \wedge \sim q) \wedge r) \vee ((q \vee p) \wedge r)$ by associative law of $\wedge$

$\Leftrightarrow ((\sim p \wedge \sim q) \vee (q \vee p) \wedge r$ by distributive law of $\wedge$ on $\vee$

$\Leftrightarrow (\sim(p \vee q) \vee (p \vee q) \wedge r$ by De Morgan's law 1 and commutative law of $\vee$

$\Leftrightarrow 1 \wedge r$ by negation law 1

$\Leftrightarrow r$ by identity law 1

Hence the result.

Example 3.32 Show that $((p \vee q) \wedge \sim(\sim p \wedge (\sim q \vee \sim r))) \vee (\sim p \wedge \sim q) \vee (\sim p \wedge \sim r)$ is a tautology.
(*M.C.A.* Nov. 96, May 97)

Solution Consider

$(\sim p \wedge \sim q) \vee (\sim p \wedge \sim r) \Leftrightarrow \sim(p \vee q) \vee \sim(p \vee r)$ by De Morgan's law 1

$\Leftrightarrow \sim((p \vee q) \wedge (p \vee r))$ by De Morgan's law 2

$\sim(\sim p \wedge (\sim q \vee \sim r)) \Leftrightarrow \sim(\sim p \wedge \sim(q \wedge r))$ by De Morgan's law 2

$\Leftrightarrow \sim(\sim p) \vee \sim(\sim(q \wedge r))$ by De Morgan's law 2

$\Leftrightarrow p \vee (q \wedge r)$ by involution law

$\Leftrightarrow (p \vee q) \wedge (p \vee r)$ by distributive law of $\vee$ over $\wedge$

Hence the given formula is equivalent to

$$((p \vee q) \wedge (p \vee r)) \vee \sim((p \vee q) \wedge (p \vee r))$$

If we substitute $(p \vee q) \wedge (p \vee r)$ for p in $p \vee \sim p$, we get the above formula. But $p \vee \sim p \Leftrightarrow 1$, the given formula is a tautology.

Example. 3.33 Show the following equivalences.

a. $p \to (q \to p) \Leftrightarrow \sim p \to (p \to q)$ (*B.E.* Apr.97)

b. $p \to (q \vee p) \Leftrightarrow (p \to q) \vee (p \to r)$

c. $(p \to q) \wedge (r \to q) \Leftrightarrow (p \vee r) \to q$

d. $\sim (p \leftrightarrow q) \Leftrightarrow (p \vee q) \wedge \sim (p \vee q) \wedge (p \wedge q)$

Solution

a. Consider

$p \to (q \to p) \Leftrightarrow \sim p \vee (\sim q \vee p)$ by implication law

$\Leftrightarrow (\sim p \vee p) \sim q$ by associative and commutative laws of $\vee$

$\Leftrightarrow 1 \vee \sim q$ by negation law 1

$\Leftrightarrow 1$ by null law 1

Again, $\sim p \to (p \to q) \Leftrightarrow \sim(\sim p) \vee (\sim p \vee q)$ by implication law

$\Leftrightarrow p \vee (\sim p \vee q)$ by involution law

$\Leftrightarrow (p \vee \sim p) \vee q$ by associative law of $\vee$

$\Leftrightarrow 1 \vee q$ by negation law 1

$\Leftrightarrow 1$ by null law 1

Hence $p \to (q \to p) \Leftrightarrow \sim p \to (p \to q)$.

b. $p \to (q \vee r) \Leftrightarrow \sim p \vee (q \vee r)$ by implication law

$\Leftrightarrow \sim p \vee \sim p \vee (q \vee r)$ since $\sim p \Leftrightarrow \sim p \vee \sim p$

$\Leftrightarrow (\sim p \vee q) \vee (\sim p \vee r)$ by associative and commutative laws of $\vee$.

$\Leftrightarrow (p \to q) \vee (p \to r)$ by implication law.

c. $(p \to q) \wedge (r \to q) \Leftrightarrow (\sim p \wedge q) \wedge (\sim r \vee q)$ by implication law

$\Leftrightarrow (\sim p \wedge \sim r) \vee q$ by distributive of $\vee$ over $\wedge$

$\Leftrightarrow \sim (p \vee r) \vee q$ by De Morgan's law 1

$\Leftrightarrow (p \vee q) \to q$ by implication law

d. $\sim (p \to q) \Leftrightarrow \sim ((p \wedge q) \vee \sim (p \vee q))$ by biconditional and De Morgan's law 1

$\Leftrightarrow \sim (p \wedge q) \wedge \sim (\sim (p \vee q))$ by De Morgan's law 1

$\Leftrightarrow \sim (p \wedge q) \wedge (p \vee q)$ by involution law

$\Leftrightarrow (p \wedge q) \wedge \sim (p \wedge q)$ by commutative law of $\wedge$

Example 3.34 Write an equivalent formula for $p \wedge (q \leftrightarrow r) \vee (r \leftrightarrow p)$ which does not contain the biconditional.

Solution Since $p \leftrightarrow q \Leftrightarrow (p \to q) \wedge (q \to p), p \wedge (q \leftrightarrow r) \vee (r \leftrightarrow p) \Leftrightarrow p \wedge (q \to r) \wedge (r \to q)) \vee ((r \to p) \wedge (p \to r))$.

Example 3.35 Write an equivalent formula for $p \wedge (q \leftrightarrow r)$ which contains neither the biconditional nor the conditional.

Solution Since $p \leftrightarrow q \Leftrightarrow (p \to q) \wedge (q \to p)$ and $p \to q \Leftrightarrow \sim p \vee q$.

$$p \wedge (q \leftrightarrow r) \Leftrightarrow p \wedge ((q \to r) \wedge (r \to q))$$
$$\Leftrightarrow p \wedge ((\sim q \vee r) \wedge (\sim r \vee q))$$

Duality Two formulae or propositions are said to be *dual* of each other if either one can be obtained from the other by replacing $\wedge$ by $\vee$ and $\vee$ by $\wedge$.

Note 1: The connectives $\wedge$ and $\vee$ are called duals. If the proposition contains the special variables 1 or 0, then its dual, is obtained by repacing 1 by 0 and 0 by 1 in addition to the above-mentioned interchanges.

Note 2: If any two formulae are equivalent, then their duals are also equivalent to each others.

Example 3.36 Write the duals of

a. $(p \vee q) \wedge r$

b. $(p \wedge q) \vee 1$

c. $\sim (p \vee q) \wedge (p \vee \sim (q \wedge \sim s))$ *(M.C.A. May 97)*

Solution The duals are

a. $(p \wedge q) \vee r$

b. $(p \vee q) \wedge 0$

c. $\sim (p \wedge q) \vee (p \wedge \sim (q \vee \sim s))$

Example 3.37 Show that

a. $\sim (p \wedge q) \to (\sim p \vee (\sim p \vee q)) \Leftrightarrow (\sim p \vee q)$

b. $(p \vee q) \wedge (\sim p \wedge (\sim p \wedge q)) \Leftrightarrow (\sim p \wedge q)$ *(M.C.A. Nov. 96)*

Solution

a. $\sim (p \wedge q) \to (\sim p \vee (\sim p \vee q))$

$\Leftrightarrow (p \wedge q) \vee (\sim p \vee (\sim p \vee q))$ by implication and involution laws

$\Leftrightarrow (p \wedge q) \vee (\sim p \vee q)$ by associative law

$\Leftrightarrow ((p \wedge q) \vee \sim p) \vee q$ by associative law

$\Leftrightarrow ((p \vee \sim p) \wedge (q \vee \sim p)) \vee q$ by distributive law

$\Leftrightarrow (1 \wedge (q \vee \sim p)) \vee q$ negation law

$\Leftrightarrow (q \vee \sim p) \vee q$ identity law 1

$\Leftrightarrow \sim p \vee q$ by associative and commutative laws

b. In (a) we have proved that

$$(p \wedge q) \vee (\sim p \vee (\sim p \vee q)) \Leftrightarrow (\sim p \vee q).$$

By Note 2,

$$(p \vee q) \wedge (\sim p \wedge (\sim p \wedge q)) \Leftrightarrow (\sim p \wedge q)$$

Implication A proposition p is said to logically imply or tautologically imply or simply imply a proposition q if q is true whenever p is true.

We can also define this notion as follows.

A proposition p is said to logically imply a proposition q if $p \to q$ is a tautology. The implication of p to q is denoted by $p \Rightarrow q$.

Note 1: One can determine whether $p \Rightarrow q$ by constructing the truth table of p and q in the same manner as was done in the determination of $p \Leftrightarrow q$.

Note 2: Since $(p \to q) \Leftrightarrow (p \to q) \wedge (q \to p)$ (equivalence law), it is easy to see from the definition of equivalence and implication that $p \Leftrightarrow q$ if and only if $p \Rightarrow q$ and if $q \Rightarrow p$. This statement is an alternative definition of the equivalence as two propositions.

The implications in Table 3.9 have important applications. All of them can be proved by truth table or by other methods.

Table 3.9 Logical Implications.

Implication	Name
1. $p \wedge q \Rightarrow p$	conjunctive simplification
2. $p \wedge q \Rightarrow q$	conjunctive simplification
3. $p \Rightarrow p \vee q$	disjunctive addition
4. $p \wedge (p \to q) \Rightarrow q$	detachment
5. $\sim q \wedge (p \to q) \Rightarrow \sim p$	contra positive
6. $(p \vee q) \wedge \sim q \Rightarrow p$	disjunctive simplification
7. $(p \vee q) \wedge \sim p \Rightarrow p$	disjunctive simplification
8. $(p \to q) \wedge (q \to r) \Rightarrow (p \to r)$	chain rule
9. $\sim p \Rightarrow p \to q$	
10. $q \Rightarrow p \to q$	
11. $\sim (p \to q) \Rightarrow p$	
12. $\sim (p \to q) \Rightarrow \sim q$	
13. $(p \vee q) \wedge (p \to q) \wedge (q \to r) \Rightarrow r$	dilemma
14. $p \wedge q \Rightarrow p \leftrightarrow q$	
15. $p \to (q \to r) \Rightarrow (p \to q) \to (p \to r)$	

Example 3.38 Show that $p \leftrightarrow q$ logically implies $p \to q$.

Solution Consider the truth table of $p \leftrightarrow q$ and $p \to q$. Now $p \to q$ is true in line 1 and 4, and in these cases $p \leftrightarrow q$ is also true. Hence $p \leftrightarrow q$ implies $p \to q$.

p	q	$p \leftrightarrow q$	$p \to q$
0	0	1	1
0	1	0	1
1	0	0	0
1	1	1	1

Now we will introduce two methods to show $p \Rightarrow q$.

Method 1 To show the implication $p \Rightarrow q$, we assume that p has the truth value 1 and then show that this assumption leads to q having the value 1. Then $p \to q$ must have the value 1.

Method 2 To show $p \Rightarrow q$, we assume that q has the truth value 0 and then show that this assumption leads to p having the value 0. Then $p \to q$ must have the truth value 1.

Example 3.39 Consider the implication from the Table 3.9 $(p \vee q) \wedge (p \to r) \wedge (q \to r) \Leftrightarrow r$

Solution Now by method 1, we assume that $(p \vee q) \wedge (p \to r) \wedge (q \to r)$ is true. This assumption means that $p \vee q, p \to r$ and $q \to r$ are true. Since $p \vee q$ is true, atleast one p or q is true. If p is true then r must be true since $p \to r$ is true. If q is true then, r must be true since $q \to r$ is true. So r is true. Hence the result (13) in Table 3.12.

Example 3.40 Consider the implication from the Table 3.11 $\sim q \wedge (p \to q) \sim p$

Solution Now by method 2, assume that $\sim p$ is false, so p is true. If q is true then $\sim q$ is false. If q is false then $p \to q$ is false. So in both cases, $\sim q \wedge (p \to q)$ is false. Hence the implication (5) in Table 3.12.

Example 3.41 Show the following implications.

a. $(p \wedge q) \Rightarrow p \to q$

b. $p \Rightarrow q \to p$

c. $p \to (q \to r) \Rightarrow (p \to q) \to (p \to r)$ *(B.E. Oct. 97)*

Solution

a. Assume that $p \wedge q$ is true. This means that p and q are true. So $p \to q$ must also be true. By method 1, the result follows.

b. Assume that p is true. Then for all possible truth values of q, $q \to p$ is true. Hence, by method 1 the result follows.

c. Assume that $(p \to q) \to (p \to r)$ is false. Then $p \to r$ is false and $p \to q$ is true. This means that p and q are true since r is false. So $q \to r$ is false and hence $p \to (q \to r)$ must also be false. Hence, by method 2, the implication follows.

Example 3.42 Show the following implications without constructing the truth tables.

a. $p \to q \Rightarrow \to p\,(p \wedge q)$

b. $(p \to q) \to q \Rightarrow p \wedge q$

c. $((p \to \vee \sim p) \to q) \to ((p \vee \sim p) \to r) \Rightarrow q \to r$

d. $(q \to (p \vee \sim p)) \to (r \to (p \vee \sim p)) \Rightarrow r \to q$

Solution

a. Assume that $p \to (p \wedge q)$ is false. Then p is true and $p \wedge$ q is false. So q is false. Hence $p \to q$ is false. So by method 2, the implication follows.

b. Assume that $p \vee q$ is false. This means that both p and q are false. So $p \rightarrow q$ is true and hence $(p \rightarrow q) \rightarrow q$ is false. Again by method 2, the result follows.

c. Suppose that $q \rightarrow r$ is false. Then q is true and r is false. This implies that $(p \vee \sim p) \rightarrow q$ is true and $p \vee \sim q \rightarrow r$ is false. Hence $(p \vee \sim p) \rightarrow q ((p \vee \sim p) \rightarrow$ r) is false. By method 2, the result follows.

d. Assume that $r \rightarrow q$ is false. Then r is true and q is false. So $q \rightarrow (($ p $\wedge \sim$ p) is true and $r \nrightarrow (p \wedge \sim p)$ is false. Hence $(q \rightarrow (p \wedge \sim p)) \rightarrow r (p \wedge \sim p))$ is false and the result follows.

Note: Note that $p \vee \sim p$ is always true and $p \wedge \sim p$ is always false.

Example 3.43 If $p_1, p_2, \ldots, p_n$ and p imply q, then $p_1, p_2, \ldots, p_n$ imply $p \rightarrow q$.

Solution Let us assume that $(p_1 \wedge p_2 \wedge \ldots \wedge p_n \wedge p) \Rightarrow q$

This means $(p_1 \wedge p_2 \wedge \ldots \wedge p_n \wedge p) \rightarrow q$ is a tautology

But we know that (Example 3.28) $(p \wedge q) \rightarrow$ r $\Leftrightarrow p \rightarrow (q \rightarrow r)$

Using this, we get $(p_1 \wedge p_2 \wedge \ldots \wedge p_n) \rightarrow (p \rightarrow q)$ is a tautology.

Hence, $(p_1 \wedge p_2 \wedge \ldots \wedge p_n) \Rightarrow (p \rightarrow q)$

NAND AND NOR

NAND and NOR are connectives which have useful applications in the design of computers.

NAND If p and q are propositions, then the proposition "p NAND q" denoted by $p \uparrow q$ is false when both p and q are true and is true otherwise. That is the word "NAND" is a combination of "NOT" and "AND" so,

$$p \uparrow q \Leftrightarrow \sim (p \wedge q)$$

NOR If p and q propositions, then the proposition "p NOR q" denoted by $p \downarrow q$ is true when p and q are both false and true otherwise. That is, the word "NOR" is a combination of "NOT" and "OR" so

$$p \downarrow q \Leftrightarrow \sim (p \vee q)$$

***Table 3.10** Truth Table for NAND and NOR*

p	q	$p \uparrow q$	$p \downarrow q$
0	0	1	0
0	1	1	1
1	0	1	1
1	1	0	1

Note 1: $\uparrow$ and $\downarrow$ are duals of each other (see Table 3.14). Therefore, in order to obtain the dual of a formula which includes $\uparrow$ or $\downarrow$, we should interchange $\uparrow$ and $\downarrow$ in addition to making the other interchanges mentioned earlier.

Note 2: The connection $\uparrow$ and $\downarrow$ have been defined in terms of the connectives $\wedge$, $\vee$ and $\sim$. Therefore, for any formula containing the connection $\uparrow$ or $\downarrow$, one can obtain an equivalent formula containing the connectives $\wedge$, $\vee$ and $\sim$ only.

Example. 3.44 If $A(p, q, r)$ is given by $p \uparrow (q \wedge \sim (r \downarrow p))$, find its dual $A^*(p, q, r)$. Also find formulas which are equivalent to A and A^*, but which contain the connectives $\wedge$, $\vee$ and $\sim$ only.

Solution By Note 1, if $A(p, q, r) = p \uparrow (q \wedge \sim (r \downarrow p))$

Then its dual, $A^*(p, q, r) = p \downarrow (q \vee \sim (r \uparrow p))$

Now

$$A(p, q, r) = p \uparrow (q \wedge \sim (r \downarrow p))$$
$$\Leftrightarrow \sim (p \wedge (q \wedge \sim (r \downarrow p))) \text{ since } p \uparrow q \Leftrightarrow \sim (p \wedge q)$$
$$\Leftrightarrow \sim (p \wedge (q \wedge \sim (\sim (r \vee p)))) \text{ since } p \downarrow q \Leftrightarrow \sim (p \vee q))$$
$$\Leftrightarrow \sim p \vee \sim (q \wedge (r \vee p)) \text{ since by } \sim p \Leftrightarrow p \text{ and De Morgan's Law.}$$
$$\Leftrightarrow \sim p \vee \sim q \vee \sim (r \vee p) \text{ by De Morgan's Law.}$$
$$\Leftrightarrow \sim p \vee \sim q \vee (\sim r \wedge \sim p) \text{ by De Morgan's Law.}$$

Similarly,

$$A^*(p, q, r) = p \downarrow (q \vee \sim (r \uparrow p))$$
$$\Leftrightarrow \sim (p \vee (q \vee \sim (\sim (r \wedge p)))) \text{ since } p \uparrow q \Leftrightarrow \sim (p \wedge q) \text{ and } p \downarrow q \Leftrightarrow \sim (p \vee q)$$
$$\Leftrightarrow \sim p \wedge \sim q \wedge \sim (r \wedge p) \text{ by De Morgan's Law and } \sim \sim p \Leftrightarrow p$$
$$\Leftrightarrow \sim p \wedge \sim q \wedge (\sim r \vee \sim p) \text{ by De Morgan's law.}$$

Example 3.45 Show the following

a. $\sim (p \uparrow q) \Leftrightarrow \sim p \downarrow \sim q$ b. $\sim (p \downarrow q) \Leftrightarrow \sim p \uparrow \sim q$

Solution

a. $\sim (p \uparrow q) \Leftrightarrow \sim (\sim (p \wedge q))$ by $p \uparrow q \Leftrightarrow \sim (p \wedge q)$

$\Leftrightarrow p \wedge q$ since $p \Leftrightarrow \sim \sim p$

$\Leftrightarrow \sim (\sim p) \wedge \sim (\sim q)$

$\Leftrightarrow \sim (\sim p \vee \sim q)$ by De Morgan's Law

$\Leftrightarrow \sim p \downarrow \sim q$ by $p \downarrow q \Leftrightarrow \sim (p \vee q)$

b. $\sim (p \downarrow q) \Leftrightarrow \sim (\sim p \vee q)$ since $p \downarrow q \Leftrightarrow \sim (p \vee q)$

$\Leftrightarrow p \vee q$ since $p \Leftrightarrow \sim \sim p$

$\Leftrightarrow \sim (\sim p) \vee \sim (\sim q)$ since $p \Leftrightarrow \sim \sim p$

$\Leftrightarrow \sim (\sim p \wedge \sim q)$ by De Morgan's Law

$\Leftrightarrow \sim p \uparrow \sim q$ since $p \uparrow q \Leftrightarrow \sim (p \wedge q)$

Example 3.46 Prove the following

a. $(p \uparrow q) \Leftrightarrow (\sim p \wedge q) \vee (p \wedge \sim q) \vee (\sim p \wedge \sim q)$ b. $(p \downarrow q) \Leftrightarrow (\sim p \vee q) \wedge (p \vee \sim q) \wedge (\sim p \vee \sim q)$

Solution

a. $p \uparrow q \Leftrightarrow (\sim p \wedge q) \Leftrightarrow \sim p \vee \sim q$

$\Leftrightarrow (\sim p \wedge (q \vee \sim q)) \vee (\sim q \wedge (p \vee \sim p))$ since $p \wedge 1 \Leftrightarrow p$ and $p \vee \sim p \Leftrightarrow 1$

$\Leftrightarrow (\sim p \wedge q) \vee (\sim p \wedge \sim q) \vee (p \wedge \sim q)$ by distributive law and commutative law

Dually (b) follows.

Example 3.47 Prove that the connectives NAND and NOR are commutative but not associative.

Solution Since $p \uparrow q \Leftrightarrow \sim(p \wedge q) \Leftrightarrow \sim(q \wedge p) \Leftrightarrow q \uparrow p$ and

$p \downarrow q \Leftrightarrow \sim(p \vee q) \Leftrightarrow \sim(q \vee q) \Leftrightarrow q \downarrow p$. $\uparrow$ and $\downarrow$ are commutative. Now

$$p \uparrow (q \uparrow r) \Leftrightarrow p \uparrow \sim(q \wedge r)$$
$$\Leftrightarrow \sim(p \wedge \sim(q \wedge r))$$
$$\Leftrightarrow \sim p \vee (q \vee r) \text{ by De Morgan's Law}$$

and

$$(p \uparrow q) \uparrow r \Leftrightarrow \sim(p \wedge q) \uparrow r$$
$$\Leftrightarrow \sim(\sim(p \wedge q) \wedge r)$$
$$\Leftrightarrow (p \wedge q) \vee \sim r \text{ by De Morgan's Law}$$

implies that the connective $\uparrow$ is not associative. Similarly, $\downarrow$ is not associative.

Note: Note that $p \uparrow q \uparrow r$ is not equivalent to $p \uparrow (q \uparrow r)$ or to $q \uparrow (p \uparrow r)$. This non-associativity of the connective NAND and NOR creates some difficulty in using them.

FUNCTIONALLY COMPLETE SETS

Any set of connectives in which every formula can be expressed in terms of an equivalent formula containing the connectives from $\wedge, \vee, \sim, \rightarrow, \leftrightarrow, \uparrow$ and $\downarrow$ is called a functionally complete set of connectives.

Example 3.48 Prove that $\{\wedge, \sim\}$ and $\{\vee, \sim\}$ are functionally complete.

Solution Since $p \leftrightarrow q \Leftrightarrow (p \rightarrow q) \wedge (q \rightarrow p)$ and $p \rightarrow q \Leftrightarrow \sim p \vee q$

We can replace the part containing $\leftrightarrow$ and $\rightarrow$ by an equivalent formula NOR containing $\leftrightarrow$ and $\rightarrow$. By De Morgan's law,

$$p \wedge q \Leftrightarrow \sim(\sim p \vee \sim q) \quad \text{and} \quad p \vee q \Leftrightarrow \sim(\sim p \wedge \sim q)$$

So we can obtain a formula which is equivalent to a given formula which contains either the negation and disjunction only or the negation and conjunction only. This means that $\{\wedge, \sim\}$ and $\{\vee, \sim\}$ are functionally complete sets of connectives.

Example 3.49 Write formulae which are equivalent to the formulas given below and which contain the connectives $\wedge$ and $\sim$ only.

a. $\sim(p \leftrightarrow (q \rightarrow (r \vee p)))$ b. $((p \vee q) \wedge r) \rightarrow (p \vee r)$

Solution Since $p \leftrightarrow q \Leftrightarrow (p \rightarrow q) \wedge (q \rightarrow q)$

$$p \rightarrow q \Leftrightarrow \sim p \vee q$$
$$p \vee q \Leftrightarrow \sim(\sim p \wedge \sim q)$$

a. $\sim(p \leftrightarrow (q \rightarrow (r \vee q))) \Leftrightarrow \sim(p \rightarrow (q \rightarrow (r \vee p))) \wedge ((q \rightarrow (r \vee q)) \rightarrow p))$

$\Leftrightarrow \sim((\sim p \vee \sim q \vee (r \vee p))) \wedge (\sim(\sim q \vee (r \vee p)) \vee p)$

$\Leftrightarrow \sim(\sim(p \wedge \sim(q \wedge \sim(r \vee p)) \wedge ((q \wedge \sim(r \vee p) \vee p))$

$\Leftrightarrow \sim(\sim(p \wedge \sim(q \wedge (\sim r \wedge \sim p)) \wedge ((q \wedge \sim r \wedge \sim p) \vee p)$

$\Leftrightarrow \sim(\sim p \wedge \sim(q \wedge (\sim r \wedge \sim p))) \wedge \sim((\sim(q \wedge \sim r \wedge \sim p) \wedge \sim p))$

b. $((p \vee q) \wedge r) \to (p \vee r)$

$\Leftrightarrow \sim ((p \vee q) \wedge r) \wedge (p \vee r)$

$\Leftrightarrow \sim (((p \vee q) \wedge r) \wedge \sim (p \vee r))$

$\Leftrightarrow \sim (((\sim (\sim p \wedge \sim q) \wedge r) \wedge (\sim p \wedge \sim r)))$

Example 3.50 **Show that the connectives $\uparrow$ and $\downarrow$ are functionally complete.**

Solution It is sufficient to show that the sets of connectives $\{\wedge, \sim\}$ and $\{\vee, \sim\}$ can be expressed either in terms of $\uparrow$ alone or in terms of $\downarrow$ alone. Since

$$p \uparrow p \Leftrightarrow \sim (p \wedge p) \Leftrightarrow \sim p \vee \sim p \Leftrightarrow \sim p \quad (1)$$

$$\begin{aligned}(p \uparrow q) \uparrow (p \uparrow q) &\Leftrightarrow \sim ((p \uparrow q) \wedge (p \uparrow q)) \\ &\Leftrightarrow \sim (p \uparrow q) \\ &\Leftrightarrow \sim (\sim (p \wedge q)) \Leftrightarrow p \wedge q \quad (2)\end{aligned}$$

$$\begin{aligned}(p \uparrow p) \uparrow (q \uparrow q) &\Leftrightarrow \sim (p \wedge p) \uparrow \sim (q \wedge q) \\ &\Leftrightarrow \sim p \uparrow \sim q \\ &\Leftrightarrow \sim (\sim p \wedge \sim q) \\ &\Leftrightarrow p \wedge q \quad (3)\end{aligned}$$

$\uparrow$ is functionally complete.

$$\text{Dually, } \sim p \Leftrightarrow p \downarrow p \quad (4)$$

$$p \vee q \Leftrightarrow (p \downarrow q) \downarrow (p \downarrow q) \quad (5)$$

$$(p \wedge q) \Leftrightarrow (p \downarrow p) \downarrow (q \downarrow q) \quad (6)$$

implies that $\downarrow$ is functionally complete.

Example 3.51 Express $p \uparrow q$ in terms of $\downarrow$ only.

Solution $p \uparrow q \Leftrightarrow \sim (p \wedge q)$

$\Leftrightarrow \sim ((p \downarrow p) \downarrow (q \downarrow q))$ (by [6])

$\Leftrightarrow \sim ((p \downarrow p) \downarrow (q \downarrow q)) \downarrow ((p \downarrow p) \downarrow (q \downarrow q))$ (by [4])

Example 3.52 Express $p \to (\sim p \to q)$ in terms of $\uparrow$ only. Express the same formula in terms of $\downarrow$ only.

Solution $p \to (\sim p \to q)$

$\Leftrightarrow \sim p \vee (\sim p \to q)$ since $p \to q \Leftrightarrow \sim p \vee q$

$\Leftrightarrow \sim p \vee (p \vee q)$ since $p \to q$

$\Leftrightarrow \sim p \vee q$ and $\sim \sim p \Leftrightarrow p$

$\Leftrightarrow (\sim p \uparrow \sim p) \uparrow ((p \vee q) \uparrow (p \vee q))$ (by [3])

$\Leftrightarrow ((p \uparrow p) \uparrow (p \uparrow p))\ (((p \uparrow p) \uparrow (q \uparrow q)) \uparrow ((p \uparrow p) \uparrow (q \uparrow q))$ (by [3])

Again,

$p \to (\sim p \to q) \Leftrightarrow \sim p \vee (p \vee q)$

$\Leftrightarrow (\sim p \downarrow (p \vee q)) \downarrow (\sim p \downarrow (p \vee q))$ (by [5])

$\Leftrightarrow (\sim p \downarrow ((p \downarrow q)) \downarrow (p \downarrow q))) \downarrow (\sim p \downarrow ((p \downarrow q) \downarrow (p \downarrow q)))$

$\Leftrightarrow ((p \downarrow q)) \downarrow ((p \downarrow q) \downarrow (p \downarrow q))) \downarrow ((p \downarrow p) \downarrow ((p \downarrow q) \downarrow (p \downarrow q)))$.

Example 3.53 Write a formula which is equivalent to the formula $p \wedge (q \leftrightarrow r)$ and contains the connective ~ and ($\uparrow$) only.

Solution $p \wedge (q \leftrightarrow r)$

$$\Leftrightarrow p \wedge ((q \to r) \wedge (r \to q))$$
$$\Leftrightarrow p \wedge ((\sim q \vee r) \wedge (\sim r \vee q))$$
$$\Leftrightarrow (p \uparrow ((\sim q \vee r) \wedge (\sim r \vee q))) \uparrow (p ((\sim q \vee r) \wedge (\sim r \vee q)))$$
$$\Leftrightarrow (p \uparrow (((\sim q \uparrow \sim q) \uparrow (r \uparrow r)) \wedge ((\sim r \uparrow \sim r) \uparrow (q \uparrow q)))) \uparrow$$
$$(p \uparrow (((\sim q \uparrow \sim q) \uparrow (r \uparrow r)) \wedge ((\sim r \to \sim r)\ (q \uparrow q))))$$
$$\Leftrightarrow (p \uparrow ((((q \uparrow \sim q) \uparrow (q \uparrow q)) \uparrow (r \uparrow r)) \wedge (((r \uparrow r) \uparrow (r \uparrow r)) \uparrow (q \uparrow q))))) \uparrow$$
$$\Leftrightarrow (A \uparrow A) \uparrow (A \uparrow A)$$

TWO-STATE DEVICES AND STATEMENT LOGIC

We know about two valued logic, i.e. statements that have truth values true or false. Now, we shall discuss about two-state devices. Two-state devices are devices having two states "on" or "off" For example, a transistor either in conducting state or non-conducting state] is a two-state device.

Let us consider an electric lamp controlled by a mechanical switch as shown

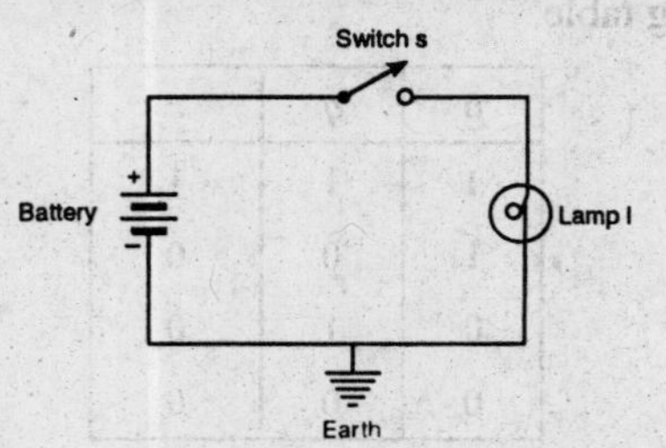

When switch s is open, no current flows through the circuit and hence the lamp is off. When switch s is closed the lamp is on. It can be represented by the following table of combinations.

State of s	State of l
Closed	on
Open	off

If p and q are represented by statements

p: The switch s is closed.

q: The lamp l is "on".

Then we can have the following table.

p (s)	q (l)
1	1
0	0

Here, we represent the truth value of p and q by 1 and 0 respectively in place of "on" and "off".

Next, consider a circuit in which we have two switches p and q connected in series

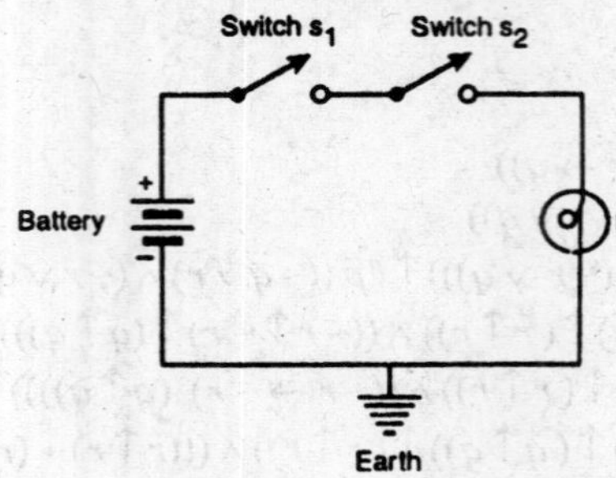

The lamp will be on only when both s_1 and s_2 are closed, otherwise it will be in the off state.

If p, q, r are statements represented by

p: The switch s_1 is closed

q: The switch s_2 is closed

r: The lamp l is on,

Thus we can have the following table.

p	q	r
1	1	1
1	0	0
0	1	0
0	0	0

Similarly, consider another circuit in which switch p_1 and p_2 are connected in parallel. The circuit and truth table is shown below.

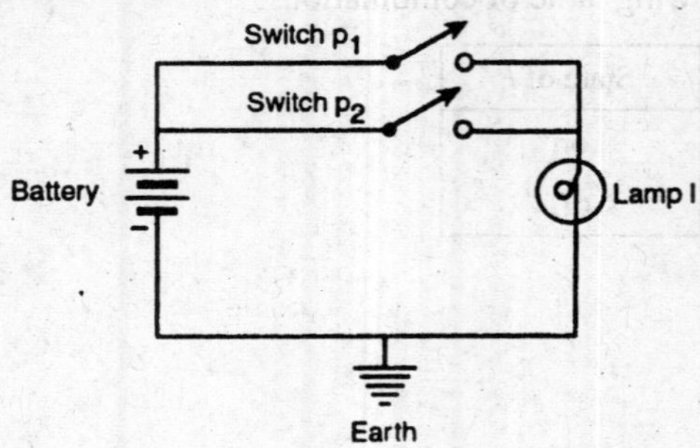

p	q	r
1	1	1
1	0	1
0	1	1
0	0	0

Hence the connectives $\wedge$ and $\vee$ correspond to switch connected in series and parallel respectively. Instead of representing $\wedge$, $\vee$ and $\sim$ by the circuits, we generally represent them using gates.

A gate is a logic circuit having one or more inputs and only one output and is also called a module. The block diagram symbol for OR, and NOT along with their truth tables are shown below. Here, we denote $\wedge$, $\vee$ and $\sim$ by, + and – respectively.

These modules or gates can be interconnected to realize various logical expressions. These systems of modules are known as logic or combinational networks.

The following block-diagram symbols show for an AND, OR and NOT gates as well as its associated table of combinations.

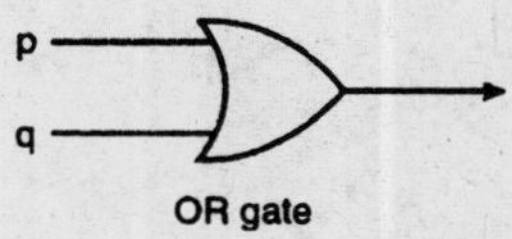

p	q	$p+q$
1	1	1
1	0	0
0	1	0
0	0	0

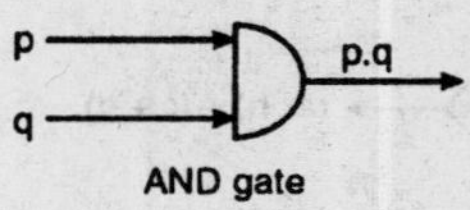

p	q	$p.q$
1	1	1
1	0	0
0	1	0
0	0	0

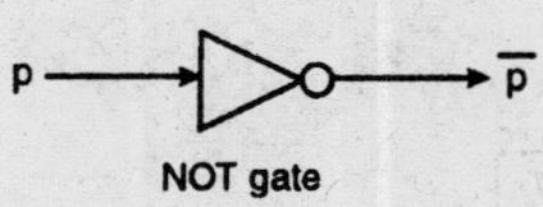

p	$\overline{p}$
1	0
0	1

Example 3.54 Draw logical networks for

a. $(a+b).c$ b. $(a.\overline{b})+(\overline{a}.b)$ c. $(a+b).(\overline{a}+\overline{b})$

d. $(a.b)+(c.d)$e. $(a+b).(c+d)$

Solution

a.

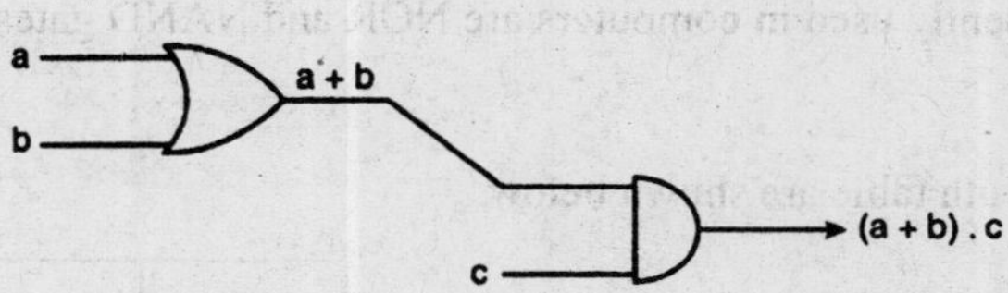

b.

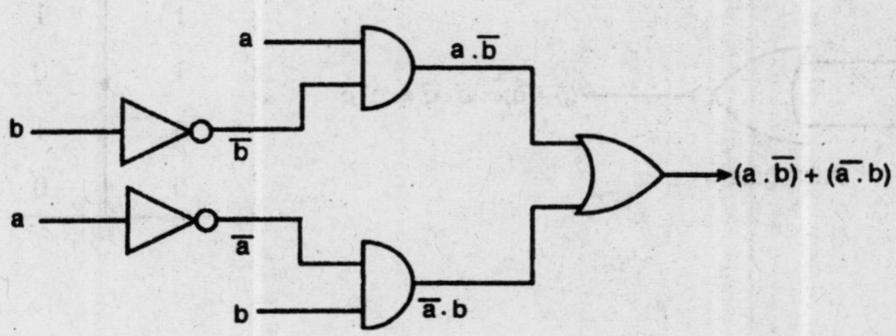

c.

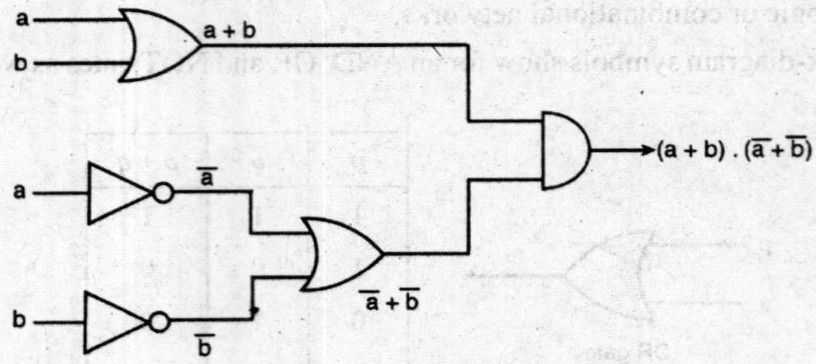

d.

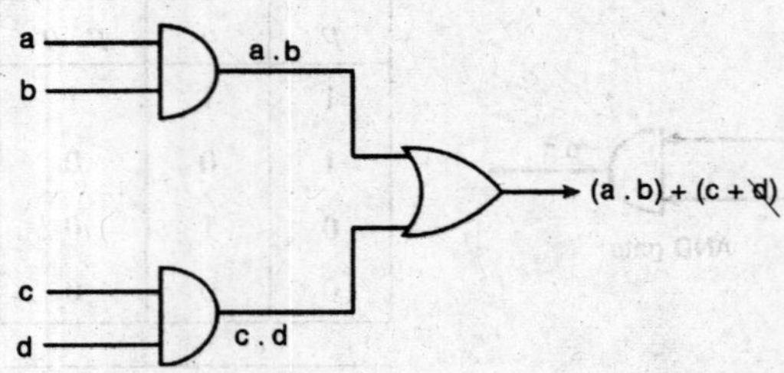

e.

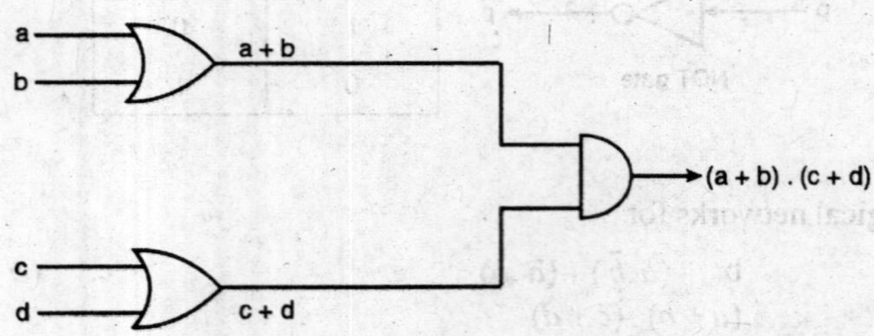

Other types of gates frequently used in computers are NOR and NAND gates.

NOR and NAND gates

The block diagrams and truth table are shown below.

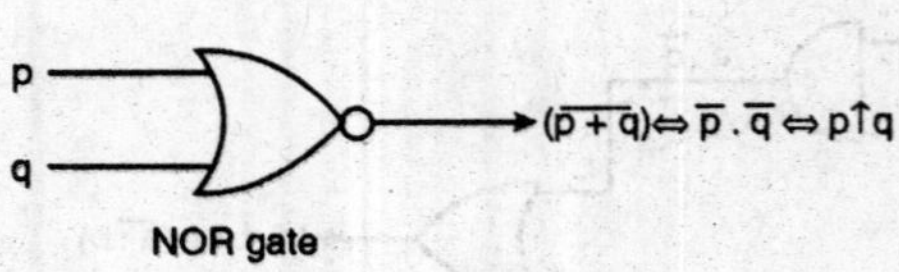

p	q	$p \downarrow q$
1	1	0
1	0	0
0	1	0
0	0	1

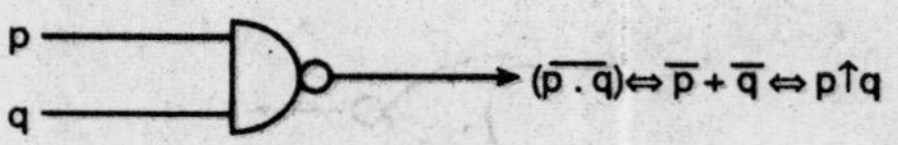

NAND gate

p	q	$p \uparrow q$
1	1	0
1	0	1
0	1	1
0	0	1

We know that ↑ and ↓ are functionally complete, i.e., they can be used to obtain AND, OR and NOT operators. These operators accomplished using NAND gates are shown below.

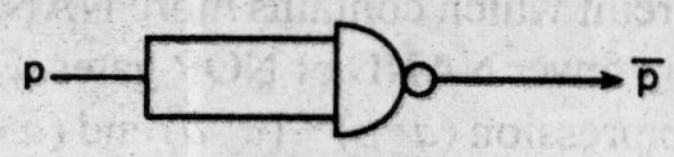

NOR gate

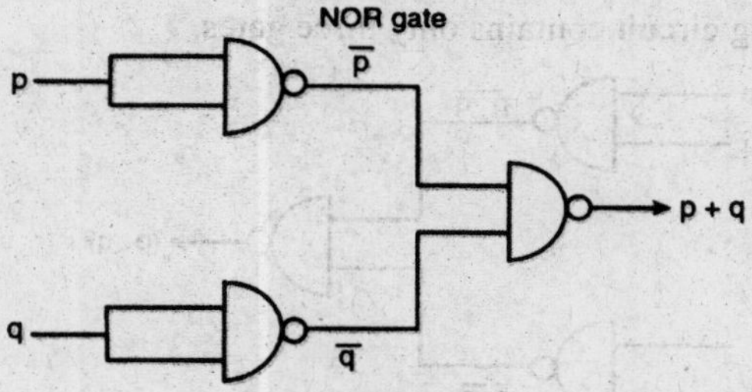

OR gate

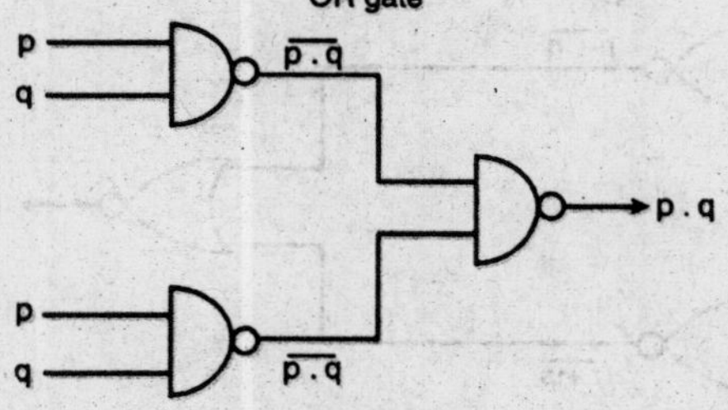

AND gate

The above expression can also be obtained using NOR gates as follows.

NOR gate

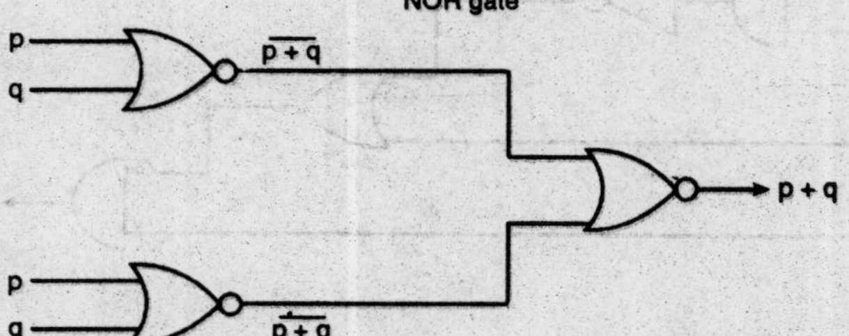

OR gate

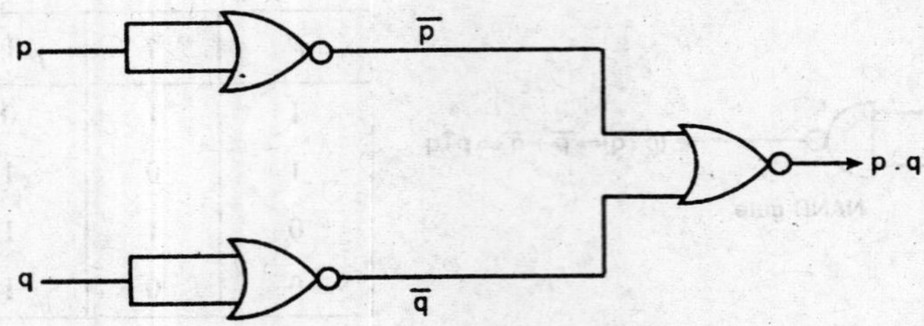

AND gate

But this replacement results in a circuit which contains more NAND or NOR gates. There are techniques to generate equivalent circuits containing fewer NAND or NOR gates. For example, if AND and OR operations are replaced by NAND or NOR in the expression $(a \,.\, b) + (c \,.\, d)$ and $(a + b) \,.\, (c + d)$, it will require nine NAND or NOR gates. But the following circuit contains only three gates.

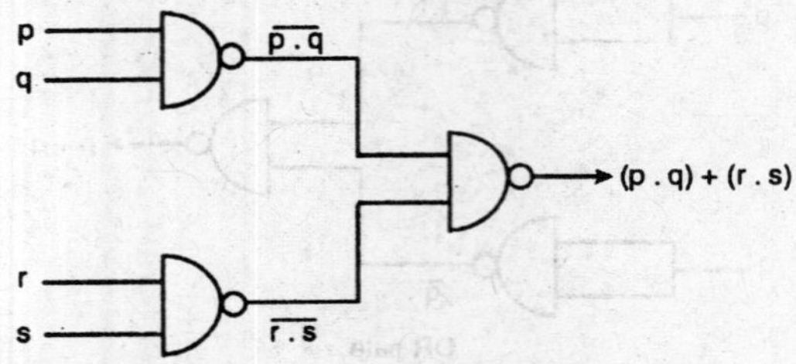

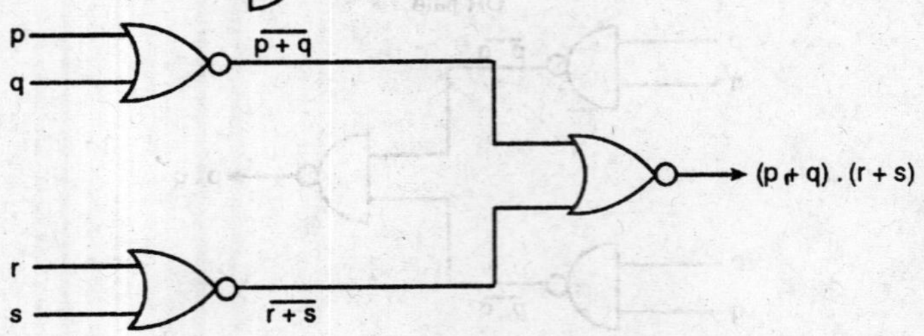

Thus, if a formula containing connectives of statement logic is given, we can physically realize a circuit that corresponds to the formula by replacing the connectives by the appropriate gates.

Example 3.55 Write expressions corresponding to the following logic diagrams.

a.

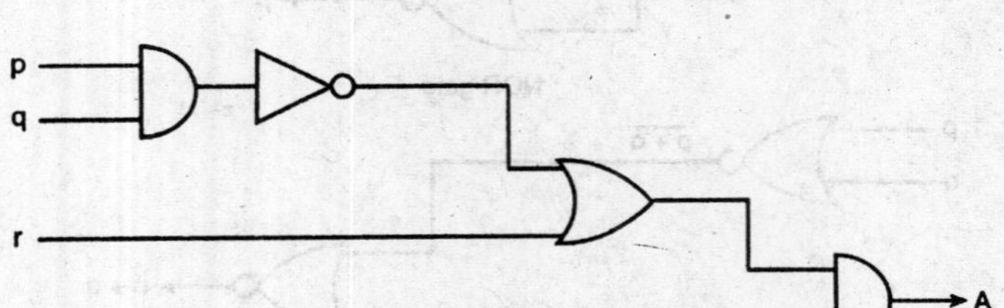

b.

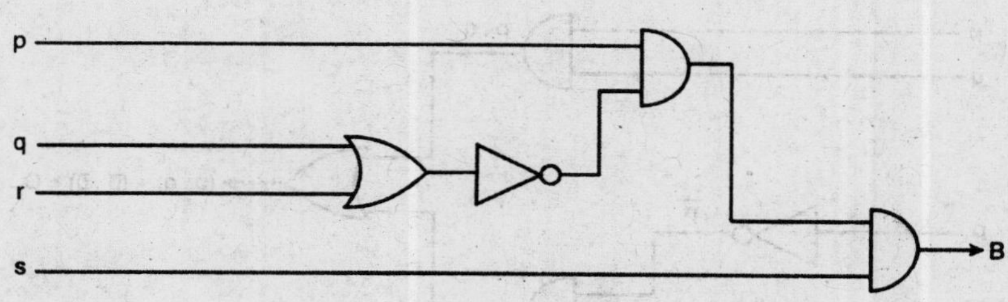

Solution

a. Here the first operator is $p\ .\ q$; this quantity is then complemented; then "OR" with r; and finally "AND" with s. This sequence of operation is equivalent to the expression

$$A = ((\overline{p.q})+r).s$$

b. Here the first operation is $q + r$, which is then complemented; then "AND" with p; then "AND" with s.So

$$B = ((\overline{q+r}).p).s$$

Note: Both of these functions involve two ANDs, one OR and one NOT, but, they are quite different.

Example 3.56 Draw circuit diagrams to represent the following equations.

a. $A = (p\ .\ q)\ .\ r$

b. $B = (\overline{\overline{p}.q}) + (p.q)$

c. $C = (p.q) + (\overline{p}.\overline{q})$

Solution

a.

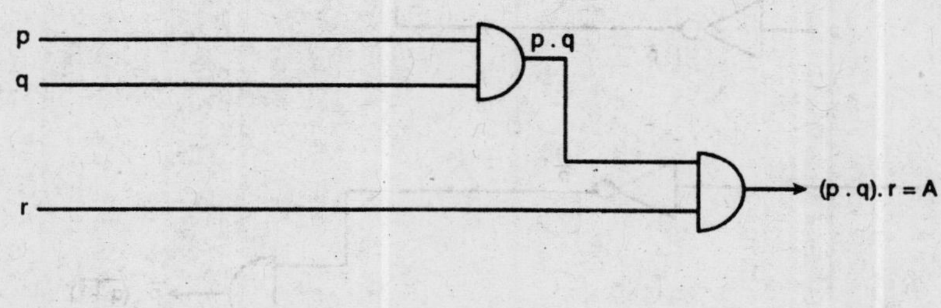

b.

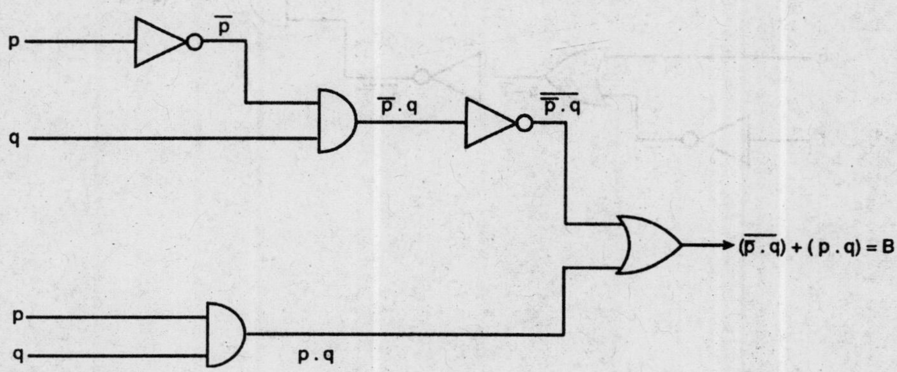

c.

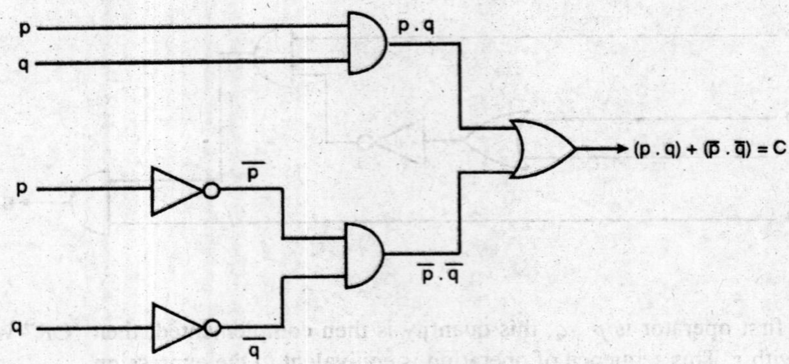

Example 3.57 Construct circuits that produce the following outputs.

a. $(p+q).\overline{p}$ b. $\overline{p}.(\overline{q+\overline{r}})$ c. $(p+q).(\overline{p}.\overline{q})$

Solution

a.

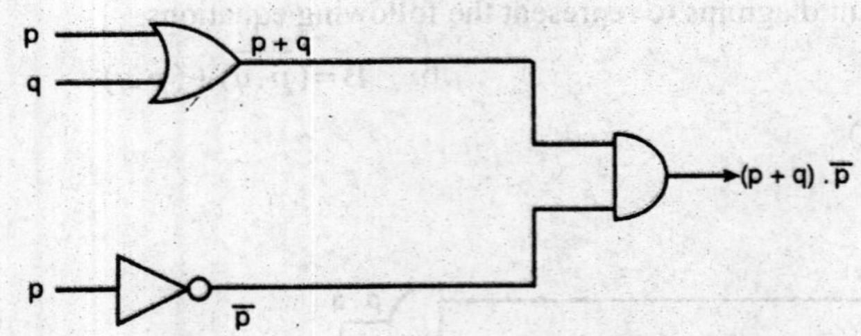

b.

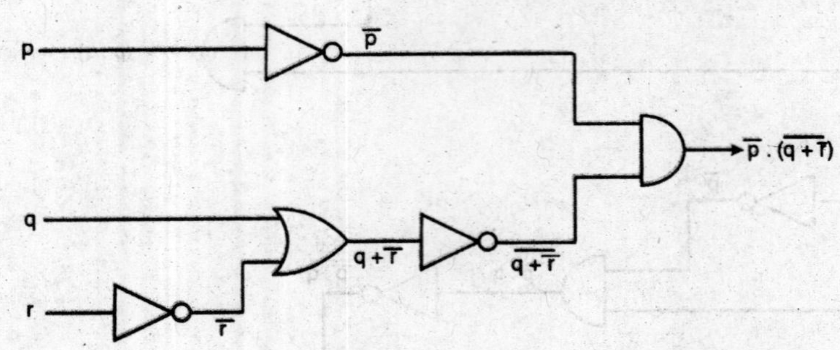

c.

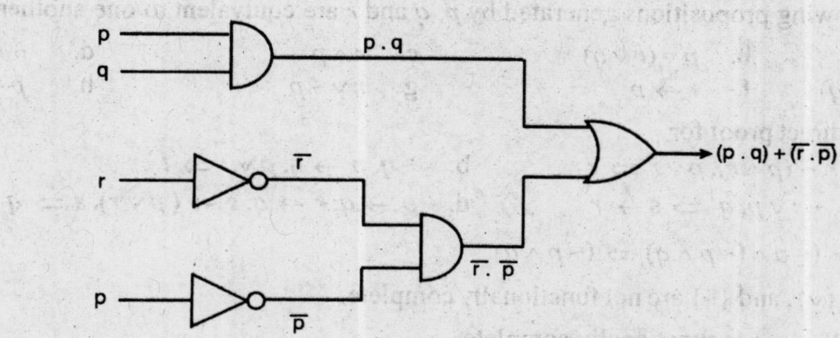

Example 3.58 For the formula $(p \wedge q) \vee (\sim r \wedge \sim p)$, draw a corresponding circuit diagram using

a. NOT, AND, and OR gates;

b. NAND gates only. *(MCA, Nov. 96, May 97, Apr. 98)*

Solution The given expression is same as $(p \,.\, q) + (\bar{r} \,.\, \bar{p})$

a.

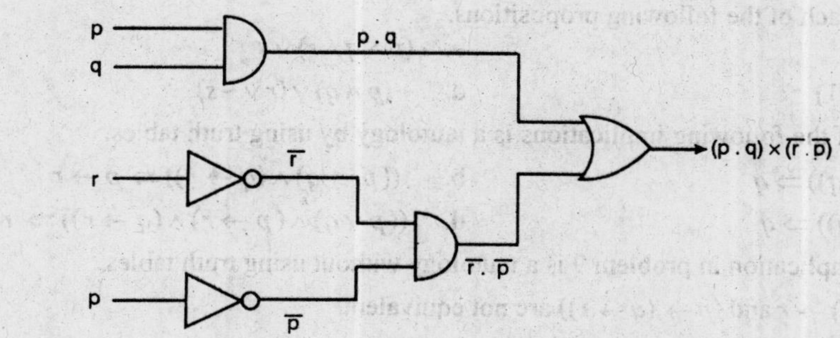

Using NOT, AND, and OR gates

b.

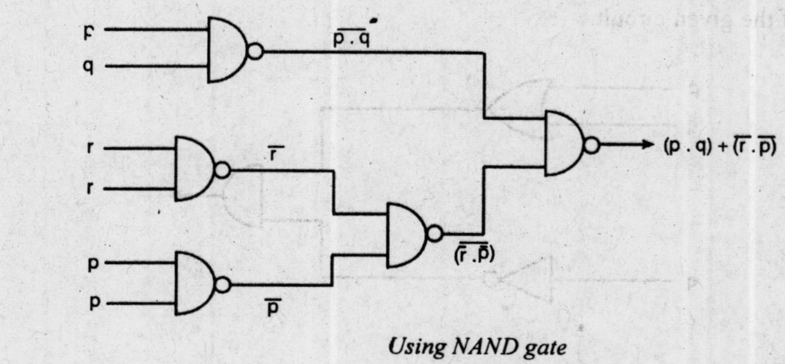

Using NAND gate

EXERCISE II

1. Show that p is equivalent to the following formulae.

$\sim\sim p \quad p \wedge p \quad p \vee p \quad p \vee (p \wedge p) \quad p \wedge (p \vee q) \quad (p \wedge q) \vee (p \wedge \sim q) \quad (p \vee q) \wedge (p \vee \sim q).$

2. Which of the following propositions generated by p, q and r are equivalent to one another?

   a. $(p \wedge r) \vee q$ b. $p \vee (r \vee q)$ c. $r \wedge p$ d. $\sim r \vee q$
   e. $(p \vee q) \wedge (r \vee q)$ f. $r \rightarrow p$ g. $r \vee \sim p$ h. $p \rightarrow r$

3. Give direct and indirect proof for

   a. $p \rightarrow q, q \rightarrow r, \sim(p \wedge r), p \vee r \Rightarrow r$
   b. $\sim q, p \rightarrow q, p \vee t \Rightarrow t$
   c. $p \rightarrow (q \rightarrow r), \sim s \vee p, q \Rightarrow s \rightarrow r$
   d. $p \rightarrow q, r \rightarrow q, s \rightarrow (p \vee r), s \Rightarrow q$

4. Show that $(p \vee q) \wedge (\sim p \wedge (\sim p \wedge q) \Rightarrow (\sim p \wedge q)$.

5. Show that $\{\wedge, \vee\}$, $\{\vee\}$, and $\{\sim\}$ are not functionally complete.

6. Show that the set $\{\sim, \rightarrow\}$ is functionally complete.

7. Show the following equivalences.

   a. $p \rightarrow (q \vee r) \Leftrightarrow (p \wedge \sim q) \rightarrow r$
   b. $(p \rightarrow q) \wedge (r \rightarrow q) \Leftrightarrow (p \vee r) \rightarrow q$
   c. $((p \wedge q) \rightarrow r) \wedge (q \rightarrow (s \vee r)) \Leftrightarrow (q \wedge (s \rightarrow p)) \rightarrow r$
   d. $((p \wedge q \wedge r) \rightarrow s) \wedge (r \rightarrow (p \vee q \vee s) \Leftrightarrow (r \wedge (p \leftrightarrow q)) \rightarrow s)$

8. Find the dual of each of the following propositions.

   a. $p \wedge \sim q \wedge \sim r$
   b. $(p \wedge q \wedge r) \vee s$
   c. $(p \vee 0) \wedge (q \vee 1)$
   d. $\sim(p \wedge q) \vee (r \vee \sim s)$

9. Show that each of the following implications is a tautology by using truth tables.

   a. $(\sim p \wedge (p \vee q)) \Rightarrow q$
   b. $((p \rightarrow q) \wedge (q \rightarrow r)) \Rightarrow p \rightarrow r$
   c. $(p \wedge (p \rightarrow q)) \Rightarrow q$
   d. $((p \vee q) \wedge (p \rightarrow r) \wedge (q \rightarrow r)) \Rightarrow r$

10. Show that each implication in problem 9 is a tautology without using truth tables.

11. Show that $(p \rightarrow q) \rightarrow r$ and $(p \rightarrow (q \rightarrow r))$ are not equivalent.

12. Write a formula which is equivalent to the formula $p \wedge (q \leftrightarrow r)$ and contains the connective $\uparrow$ only. Obtain an equivalent formula which contains the connective $\downarrow$ only.

13. Find the output of the given circuit.

    a.

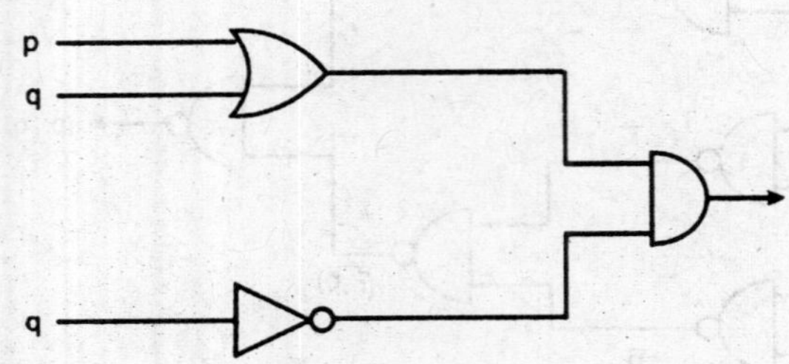

    b.

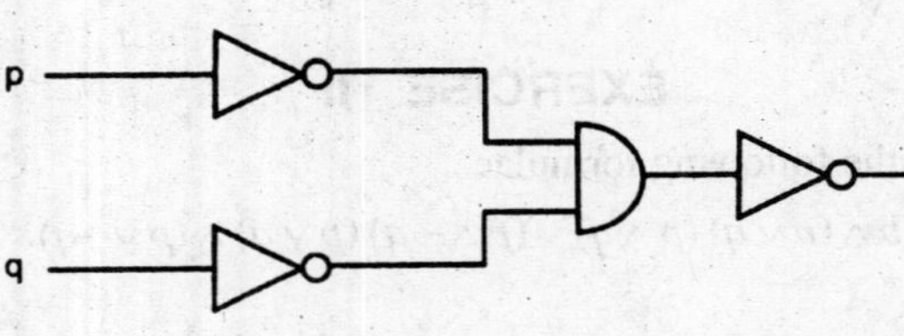

c.

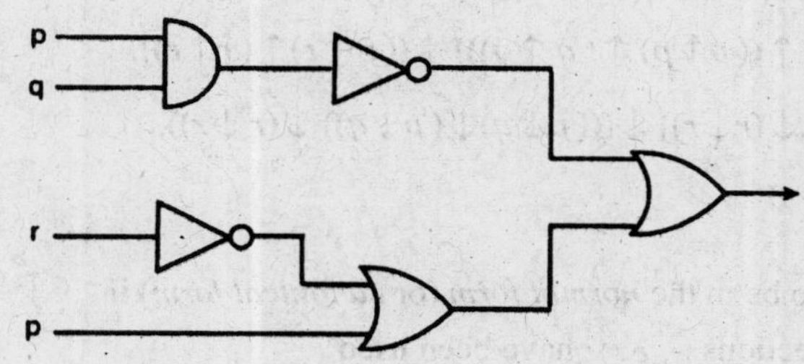

d.

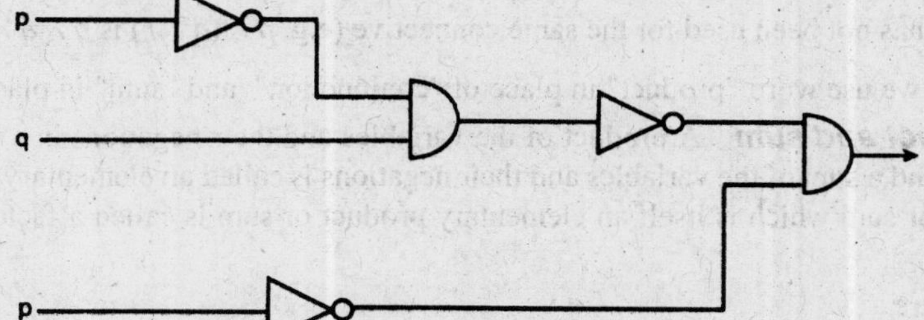

14. Construct circuits to produce the following outputs.

a. $(\bar{x}+y).\bar{y}$ b. $\overline{(\bar{x}+z).(y+\bar{z})}$

c. $\overline{(x+y).x}$ d. $\overline{((x.y)+y)}.(x+y)$

15. Show that the following circuits are equivalent.

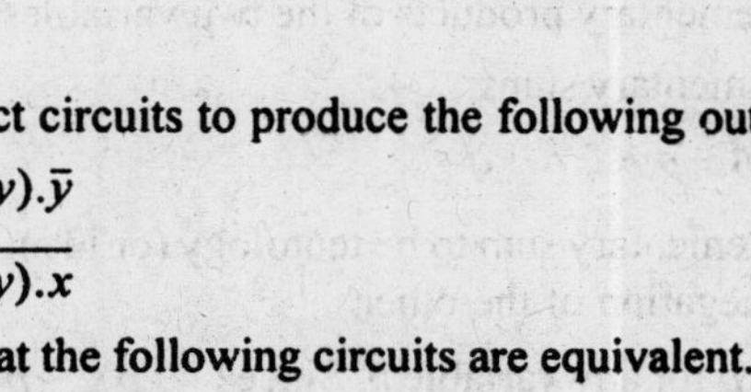

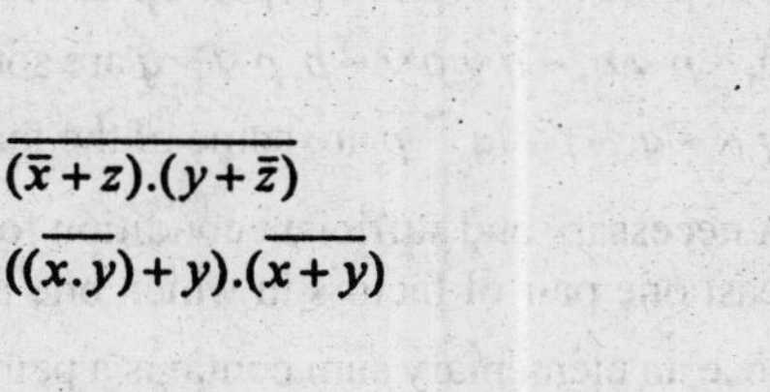

16. a. Express $p \vee q \wedge \sim r$ by use of the NAND operator only.

b. Express $p \vee q \vee \sim r$ by use of the NOR operator only.

17. Find whether $q \vee (p \wedge \sim q) \vee (\sim p \wedge \sim q)$ is a tautology or not.

ANSWERS

2. (a) ⇔ (e), (d) ⇔ (f), (g) ⇔ (h)

8. a. $p \vee \sim q \vee \sim r$ b. $(p \vee q \vee r) \wedge s$ c. $(p \wedge 1) \vee (q \wedge 0)$ d. $\sim(p \vee q) \wedge (r \wedge \sim s)$

13. a. $(p+q).\bar{q}$ b. $(\overline{p.q})$ c. $(p.\bar{q}.\bar{p})$

15. The first circuit is $(p + q) . r$. The second is $(p . r) + (q . r)$. They are equivalent by distributive property.

16. a. $(((p \uparrow p) \uparrow (q \uparrow q)) \uparrow ((p \uparrow p) \uparrow (q \uparrow q))) \uparrow ((r \uparrow r) \uparrow (r \uparrow r))$.

b. $(((p \downarrow q) \downarrow (p \downarrow q)) \downarrow (r \downarrow r)) \downarrow (((p \downarrow q) \downarrow (p \downarrow q)) \downarrow (r \downarrow r))$.

NORMAL FORMS

A statement formula is said to be in the *normal form* (or *canonical form*) if

i. only the three connections $\sim, \wedge, \vee$ have been used
ii. negation has not been used for a group of letters
iii. distributive law has been applied.
iv. paranthesis has not been used for the same connective (e.g. $p \wedge (q \wedge r)$ is $p \wedge q \wedge r$).

Note: In this section we use word "product" in place of "conjunction" and "sum" in place of "disjunction".

Elementary product and sum A product of the variables and their negations in a formula is called an elementary product and a sum of the variables and their negations is called an elementary sum. Any part of an elementary product or sum which is itself an elementary product or sum is called a factor of the elementary product or sum.

Some examples are.

1. $p, \sim p, \sim p \wedge q, \sim q \wedge p \wedge \sim p, p \wedge \sim p$ are some elementary products of the two variables p and q.
2. $p, \sim p, \sim p \vee q, \sim q \vee p \vee \sim p, p \vee \sim q$ are some elementary sums.
3. $\sim p, q \wedge \sim q, \sim p \wedge q, \sim p$ are some of the factors of $\sim p \wedge q \wedge \sim q$.

Theorem A necessary and sufficient condition for an elementary sum to be tautology (or identically true) is it contains atleast one pair of factors in which one is the negation of the other.

Proof Suppose an elementary sum contains a pair $p \vee \sim p$ for any variable p. Since $p \vee \sim p \Leftrightarrow 1$ and $1 \vee p \Leftrightarrow 1$ the sum is a tautology. Conversely, if an elementary sum is a tautology and does not contain atleast one pair of factors of type $p \vee \sim p$, we can assign truth values 1 and 0 to variables and negated variables respectively that appear in the sum. This mean that the elementary sum has the truth value 0. But it is contrary to our assumption. Hence the proof.

We can prove the following theorems in a similar way.

Theorem A necessary and sufficient condition for an elementary product to be a contradiction (or identically false) is that it contains atleast one pair of factors in which one is the negation of the other.

Disjunction normal A formula which is equivalent to a given formula and which consists of a sum of elementary products is called a disjunctive normal form of the given formula.

Conjunctive normal A formula which is equivalent to a given formula and which consists of a product of elementary sums is called a conjunctive normal form of the given formula.

Note: The disjunctive normal form or conjunctive normal form of a given formula is not unique. In fact, we can obtain different disjunctive (or conjunctive) normal forms for a given formula if the distributive laws are applied in different ways. For example, the formula $p \vee (q \wedge r)$ is in disjunctive normal form. However,

$$p \wedge (q \wedge r) \Leftrightarrow (p \vee q) \wedge (p \vee r)$$
$$\Leftrightarrow (p \wedge p) \vee (p \wedge q) \vee (p \wedge r) \vee (q \wedge r)$$

Example 3.59 **Obtain the disjunctive normal form and conjunctive normal forms of**

a. $p \wedge (p \rightarrow q)$ b. $\sim (p \vee q) \Leftrightarrow (p \wedge q)$

Solution Distributive normal form

a. $p \wedge (p \rightarrow q) \Leftrightarrow p \wedge (\sim p \vee q)$

b.

$\sim (p \vee q) \leftrightarrow (p \wedge q)$

$\Leftrightarrow (\sim (p \vee q) \wedge (p \wedge q) \vee ((p \vee q) \wedge \sim (p \wedge q))$ since $p \leftrightarrow q \Leftrightarrow (p \wedge q) \vee (\sim p \wedge \sim q)$

$\Leftrightarrow (\sim p \wedge \sim q \wedge (p \wedge q)) \vee ((p \vee q) \wedge (\sim p \vee \sim q))$ by De Morgan's Law

$\Leftrightarrow (\sim p \wedge \sim q \wedge p \wedge q)((p \vee q) \wedge \sim p)) \vee ((p \vee q) \vee \sim q)$ by distributive law)

$\Leftrightarrow (\sim p \wedge \sim q \wedge p \wedge q) \vee (p \wedge \sim q) \vee (q \wedge \sim p) \vee (p \wedge \sim q) \vee (q \wedge \sim q)$

Conjunctive normal form

a. $p \wedge (p \rightarrow q) \Leftrightarrow p \wedge (\sim p \vee q)$

b. $\sim (p \vee q) \leftrightarrow (p \wedge q)$

$\Leftrightarrow (\sim (p \vee q) \rightarrow (p \vee q) \wedge ((p \wedge q) \sim (p \wedge q))$ since $p \leftrightarrow \Leftrightarrow (p \rightarrow q) \wedge (q \rightarrow p)$

$\Leftrightarrow ((p \vee q) \vee (p \wedge q) \wedge (\sim (p \wedge q) \wedge \sim (p \vee q))$ since $p \rightarrow q \Leftrightarrow \sim p \vee q$

$\Leftrightarrow (p \vee q \vee p) \wedge (p \vee q \vee q) \wedge (\sim (p \wedge q) \vee \sim (p \vee q))$ by distributive law)

$\Leftrightarrow (p \vee q \vee p) \wedge (p \vee q \vee q) \wedge (\sim (p \vee \sim q) \wedge (\sim p \wedge \sim q))$ by De Morgan's law

$\Leftrightarrow (p \vee q) \wedge (\sim p \vee q \vee \sim p) \wedge (\sim p \vee \sim q \vee \sim q)$

$\Leftrightarrow (p \vee q) \wedge (\sim p \vee \sim q)$

Note: A formula is a tautology if every elementary sum appearing in its conjunctive normal form is identically true or a tautology. For the assumption to be true, every elementary sum should have atleast two factors, of which one is the negation of the other.

Example 3.60 Show that the formula $q \vee (p \wedge \sim q) \vee (\sim p \wedge \sim q)$ is a tautology.

Solution First we reduce the given formula into a conjunctive normal form.

$q \vee (p \wedge \sim q) \vee (\sim p \wedge \sim q)$

$\Leftrightarrow q \vee ((p \vee \sim p) \wedge \sim q)$ [distributive law]

$\Leftrightarrow (q \vee (p \vee \sim p)) \wedge (q \vee \sim q)$

$\Leftrightarrow (q \vee p \vee \sim p) \wedge (q \vee \sim q)$

Since both the elementary sum is a tautology, the given formula is a tautology.

Minterm For a given number of variables, the minterm consists of conjunctions in which each variable or its negation, but not both, appears only once.

For example, for two variables p and q, $p \wedge q$, $p \wedge \sim q$, $\sim p \wedge q$ are called minterms. From the truth table of these minterms, it is clear that no two minterms are equivalent. Each minterm has the truth value 1 for exactly one combination of the true values of p and q.

Table 3.11

p	q	$p \wedge q$	$p \wedge \sim q$	$\sim p \wedge q$	$\sim p \wedge \sim q$
0	0	0	0	0	1
0	1	0	0	1	0
1	0	0	1	0	0
1	1	1	0	0	0

Principal disjunctive normal For a given formula, an equivalent formula consisting of disjunctions of only minterms is known as its principal disjunctive normal form. It is also known as the sum of products caononical form.

Note: For every truth value 1 in the truth table of the given formula, select the minterm which also has the value 1 for the same combination of the truth table of p and q. The disjunction of these minterms will then be the equivalent to the given formula.

Example 3.61 Obtain the principal disjunctive normal forms of

a. $p \to q$ b. $p \vee q$ c. $\sim(p \wedge q)$

Solution The truth table for $p \to q, p \vee q$ and $\sim(p \wedge q)$ are given in Table 3.12.

Table 3.12

p	q	$p \to q$	$p \vee q$	$\sim(p \wedge q)$
0	0	1	0	1
0	1	1	1	1
1	0	0	1	1
1	1	1	1	0

Using Table 3.11,

$$p \to q \Leftrightarrow (p \wedge q) \vee (\sim p \wedge q) \vee (\sim p \wedge q)$$

$$p \vee q \Leftrightarrow (p \wedge q) \vee (p \wedge \sim q) \vee (\sim p \wedge q)$$

$$\sim(p \wedge q) \Leftrightarrow (p \wedge \sim q) \vee (\sim p \wedge q) \vee (\sim p \wedge \sim q)$$

Note: If a formula is a tautology, then clearly all the minterms appear in its principal disjunctive normal form.

The method to obtain the principal disjunctive normal form of a given formula without constructing its truth table is given below.

1. **Replace $\leftrightarrow$ and $\to$ by $\wedge, \vee$ and $\sim$.**
2. **Use De Morgan's law and distributive law.**
3. **Drop any elementary product which is a contradiction.**
4. **Introduce the missing factors to obtain minterms.**
5. **Delete the identical minterms.**

Example 3.62 Obtain the principal disjunctive normal forms of

a. $\sim p \vee q$ b. $(p \wedge q) \vee (\sim p \wedge r) \vee (q \wedge r)$ *(MCA May 97)*

Solution

a. $\sim p \vee q \Leftrightarrow (\sim p \wedge (q \wedge \sim q)) \vee (q \wedge (p \vee \sim p))$ **(since $p \wedge 1 \Leftrightarrow p$.)**

$\Leftrightarrow (\sim p \wedge q) \vee (\sim p \wedge \sim q) \vee (q \wedge p) \vee (q \wedge \sim p)$ **by distributive law.**

$\Leftrightarrow (\sim p \wedge q) \vee (\sim p \wedge \sim q) \vee (p \wedge q)$ **by communtative law and $p \vee p \Leftrightarrow p$**

b. $(p \wedge q) \vee (\sim p \wedge r) \vee (q \wedge r)$ **(three variables)**

$\Leftrightarrow (p \wedge q \wedge (r \vee \sim r)) \vee (\sim p \wedge r \wedge (q \vee \sim q)) \vee (q \wedge r \wedge (p \vee \sim p))$ **since $p \wedge 1 \Leftrightarrow p$**

$\Leftrightarrow (p \wedge q \wedge r) \vee (p \wedge q \wedge \sim r) \vee (\sim p \wedge r \wedge q)$

$\vee (\sim p \wedge r \wedge r \wedge \sim q) \vee (q \wedge r \wedge p) \vee (q \wedge r \wedge \sim p)$ **by distributive law**

$\Leftrightarrow (p \wedge q \wedge r) \vee (p \wedge q \wedge \sim r) \vee (\sim p \wedge r \wedge r) \vee (\sim p \wedge \sim q \wedge r)$ **by $p \vee p \Leftrightarrow p$**

Example 3.63 Show that the following are equivalent formulae

a. $p \vee (p \wedge q) \Leftrightarrow p$ b. $p \vee (\sim p \wedge q) \Leftrightarrow p \vee q$

Solution We write the principal disjunctive normal form of each formula and compare these formulae

a. $p \vee (p \wedge q) \Leftrightarrow (p \wedge (q \vee \sim q)) \vee (p \wedge q)$

$\Leftrightarrow (p \wedge q) \vee (p \wedge \sim q) \vee (p \wedge q)$

$\Leftrightarrow (p \wedge q) \vee (p \wedge \sim q)$ (1)

$p \Leftrightarrow p \wedge (q \vee \sim q)$

$\Leftrightarrow (p \wedge q) \vee (p \vee \sim q)$ (2)

From (1) and (2), $p \vee (p \wedge 1) \Leftrightarrow p$

b. $p \vee (\sim p \wedge q) \Leftrightarrow (p \wedge (q \vee \sim q)) \vee (\sim p \wedge q)$

$\Leftrightarrow (p \wedge q) \vee (p \wedge \sim q) \vee (\sim p \wedge q)$ (3)

$p \vee q \Leftrightarrow (p \wedge (q \vee \sim q)) \vee (q \wedge (p \vee \sim p))$

$\Leftrightarrow (p \wedge q) \vee (p \wedge \sim q) \vee (q \wedge p) \vee (q \vee \sim q)$

$\Leftrightarrow (p \wedge q) \vee (p \wedge \sim q) \vee (\sim p \wedge q)$ (4)

From (3) and (4), $p \vee (\sim p \wedge q) \Leftrightarrow p \vee q$

Example 3.64 Obtain the principal disjunctive normal form of $p \rightarrow ((p \rightarrow q) \wedge \sim(\sim q \vee \sim p))$ *(MCA, Nov. 96)*

Solution $p \rightarrow ((p \rightarrow q) \wedge \sim(\sim q \vee \sim p))$

$\Leftrightarrow \sim p \vee ((\sim p \vee q) \wedge \sim(\sim q \vee \sim p))$ since $p \rightarrow q \Leftrightarrow \sim p \vee q$

$\Leftrightarrow \sim p \vee ((\sim p \wedge q) \wedge (q \wedge p))$ by De Morgan's law

$\Leftrightarrow \sim p \vee (\sim p \wedge q \wedge p) \vee (q \wedge q \wedge p))$ by distributive law

$\Leftrightarrow \sim p \vee (q \wedge p)$ since $p \wedge \sim p \Leftrightarrow 0$

$\Leftrightarrow (\sim p \wedge (q \vee \sim q)) \vee (q \wedge p)$

$\Leftrightarrow (\sim p \wedge q) \vee (\sim p \wedge \sim q) \vee (q \wedge p)$ by distributive law

$\Leftrightarrow (\sim p \wedge q) \vee (\sim p \wedge \sim q) \vee (p \wedge q)$ by commutative law

Maxterm For a given number of variables, the *maxterm* consists of disjunctions in which each variable or its negation, but not both, appears only once.

Note: The maxterms are the duals of minterms. Either from the duality principle or directly from the truth tables, it can be ascertained that each of the maxterms has the truth value 0 for exactly one combination of the truth values of the variables.

Principal conjunctive normal For a given formula, an equivalent formula consisting of conjunctions of the maxterms only is known as its principal conjunctive normal form. It is also called the product-of-sums canonical forms.

Note 1: The method for obtaining the principal conjunctive normal form for a given formula is similar to the principal disjunctive normal form.

Note 2: If the principal disjunctive normal form of a given formula A containing n variables is known, then the principal disjunctive normal form of $\sim A$ will consist of the disjunction of the remaining minterms which do not appear in the principal disjunctive normal form of A. From $A \Leftrightarrow \sim\sim A$, one can obtain the principal conjunctive normal form of A by repeated applications of De Morgan's laws to the principal disjunctive normal of $\sim A$.

Example 3.65 **Obtain the principal conjunctive and disjunctive normal of the formula S given by**

$$(\sim p \rightarrow r) \wedge (q \leftrightarrow p) \qquad \textit{(MCA, May 97)}$$

Solution

$S = (\sim p \rightarrow r) \vee (q \leftrightarrow p)$

$\Leftrightarrow (p \vee r) \wedge ((q \rightarrow p) \wedge (p \rightarrow q))$

$\Leftrightarrow (p \vee r) \wedge ((\sim q \vee p) \wedge (\sim p \vee q))$

$\Leftrightarrow (p \vee r \vee (q \wedge \sim q)) \wedge (\sim q \vee p \vee (r \wedge \sim r)) \wedge (\sim p \vee q \vee (r \wedge \sim r))$ since $p \vee 0 \Leftrightarrow p$.

$\Leftrightarrow (p \vee r \vee q) \wedge (p \vee r \vee \sim q) \wedge (\sim q \vee p \vee r) \wedge (\sim q \vee p \vee \sim r) \wedge (\sim p \vee q \vee r) \wedge (\sim p \vee q \vee \sim r)$

$\Leftrightarrow (p \vee q \vee r) \wedge (p \vee \sim q \vee r) \wedge (p \vee \sim q \vee r) \wedge (\sim p \vee q \vee r) \wedge (\sim p \vee q \vee \sim r)$

which is the required principal conjunctive normal form.

Now $\sim S \Leftrightarrow (p \vee q \vee \sim r) \wedge (\sim p \vee \sim q \vee r) \wedge (\sim p \vee \sim q \vee \sim r)$ (remaining maxterms) is also a principal conjunctive form of $\sim S$. So

$S \Leftrightarrow \sim\sim S \Leftrightarrow \sim((p \vee q \vee \sim r) \wedge (\sim p \vee \sim q \vee r) \wedge (\sim p \vee \sim q \vee \sim r))$

$\Leftrightarrow \sim (p \vee q \vee \sim r) \vee (\sim p \vee \sim q \vee r) \wedge \sim (\sim p \vee \sim q \vee \sim r)$

$\Leftrightarrow (\sim p \wedge \sim q \wedge r) \vee (p \wedge q \wedge \sim r) \vee (p \wedge q \wedge r)$ **by De Morgan's law**

which is the principal disjunctive normal of S.

Example 3.66 **Obtain the principal conjunctive normal forms (or products of sums of canonical forms) for**

a. $(p \wedge q \wedge r) \vee (\sim p \wedge r \wedge q) \vee (\sim p \wedge \sim q \wedge \sim r)$ b. $(p \wedge q) \vee (\sim p \wedge q) \vee (p \wedge \sim q)$

c. $(p \wedge q) \vee (\sim p \wedge q \wedge r)$

Solution

a. **Let** $A = (p \wedge q \wedge r) \vee (\sim p \wedge r \wedge q) \vee (\sim p \wedge \sim q \wedge \sim r)$

$\sim A \Leftrightarrow (p \wedge \sim q \wedge r) \vee (p \wedge q \wedge \sim r) \vee (p \wedge \sim q \wedge \sim r)$

$\vee (\sim p \wedge \sim q \wedge \sim r) \vee (\sim p \wedge q \wedge r)$ **(remaining minterms)**

Now

$A \Leftrightarrow \sim\sim A \Leftrightarrow \sim (p \wedge \sim q \wedge r) \wedge \sim (p \wedge q \wedge \sim r) \wedge \sim (p \wedge \sim q \wedge \sim r)$

$\wedge \sim (\sim p \wedge \sim q \wedge \sim r) \wedge \sim (\sim p \wedge \sim q \wedge \sim r)$ **by De Morgan's law.**

$\Leftrightarrow (\sim p \vee q \vee \sim r) \wedge (\sim p \vee \sim q \vee \vee)$

$\wedge (\sim p \vee q \vee r) \wedge (p \vee q \vee \sim r) \wedge (p \vee \sim q \vee r)$ **by De Morgan's law**

b. **Let** $A = (p \wedge q) \vee (\sim p \wedge q) \vee (p \wedge \sim q)$

$\sim A \Leftrightarrow (\sim p \wedge \sim q)$ **(remaining minterms)**

So $A \Leftrightarrow \sim\sim A \Leftrightarrow \sim (\sim p \wedge \sim q) \Leftrightarrow (p \vee q)$

c. **Let** $A = (p \wedge q) \vee (\sim p \wedge q \wedge r)$

$\Leftrightarrow (p \wedge q \wedge (r \vee \sim r) \vee (\sim p \wedge q \wedge r)$ **by** $p \wedge 1 \Leftrightarrow p$

$\Leftrightarrow (p \wedge q \wedge r) \vee (p \wedge q \wedge \sim r) \vee (\sim p \wedge q \wedge r)$

So $\sim A \Leftrightarrow (p \wedge \sim q \wedge r) \vee (p \wedge \sim q \wedge \sim r) \vee (\sim p \wedge \sim q \wedge \sim r) \vee (\sim p \wedge \sim q \wedge \sim r) \vee (p \wedge \sim q \vee \sim r)$.

Since $A \Leftrightarrow \sim\sim A$,

$A \Leftrightarrow \sim (p \wedge \sim q \wedge r) \wedge \sim (p \wedge \sim q \wedge \sim r) \wedge \sim (\sim p \wedge \sim q \wedge r) \wedge \sim (\sim p \wedge q \wedge \sim r) \wedge \sim (\sim p \wedge \sim q \wedge \sim r)$

$\Leftrightarrow (\sim p \vee q \vee r) \wedge (\sim p \vee q \vee r) \wedge (p \vee q \vee \sim r) \wedge (p \vee \sim q \vee r) \wedge (p \vee q \vee r)$

Example 3.67 Obtain the principal disjunctive and conjunctive normal forms of the following formulae.

a. $q \wedge (p \vee \sim q)$ b. $p \rightarrow (p \wedge (q \rightarrow p))$

c. $(q \rightarrow p) \wedge (\sim p \wedge q)$ d. $p \vee (\sim p \rightarrow (q \vee (\sim q \rightarrow \text{p})))$

Which of the above formulas are tautologies?

Solution

a. Let $A = q \wedge (p \vee \sim q)$

$\Leftrightarrow (q \vee (p \wedge \sim p)) \wedge (p \vee \sim q)$ since $p \vee 0 \Leftrightarrow p$

$\Leftrightarrow (q \vee p) \wedge (q \vee \sim p) \wedge (p \vee \sim q)$

$\Leftrightarrow (p \vee q) \wedge (\sim p \vee q) \wedge (p \vee \sim q)$

is the principal conjunctive normal form. But $\sim A \Leftrightarrow (\sim p \vee \sim q)$

So, $A \Leftrightarrow \sim\sim A \sim (\sim p \wedge \sim q) \Leftrightarrow (p \wedge q)$

which is the principal disjunctive normal form. Clearly A is not a tautology.

b. Let $A = p \rightarrow (p \wedge (q \rightarrow p))$

$\Leftrightarrow \sim p \vee (p \wedge (\sim q \vee p))$

$\Leftrightarrow \sim (p \vee ((p \wedge \sim q) \vee (p \wedge p))$

$\Leftrightarrow (\sim p \wedge (q \vee \sim q)) \vee (p \wedge \sim q) \vee (p \wedge (q \vee \sim q))$ since $p \wedge T \Leftrightarrow p$

$\Leftrightarrow (\sim p \wedge q) \vee (\sim p \wedge \sim q) \vee (p \wedge \sim q) \vee (p \wedge q) \vee (p \wedge \sim q)$

$\Leftrightarrow (\sim p \wedge q) \vee (\sim p \wedge \sim q) \vee (p \wedge \sim q) \vee (p \wedge q)$

which is in principal disjunctive normal form. Since the formula contains all the minterms, the formula is a tautology. Further the principal conjunctive normal form does not exist.

c. Let $A = (q \rightarrow p) \wedge (\sim p \wedge q)$

$\Leftrightarrow (\sim q \vee p) \wedge (\sim p \wedge q)$

$\Leftrightarrow (\sim q \vee p) \wedge (\sim p \vee (q \wedge \sim q)) \wedge (q \vee (p \wedge \sim p))$ since $p \vee 0 \Leftrightarrow p$

$\Leftrightarrow (\sim q \vee p) \wedge (\sim p \vee q) \wedge (\sim p \wedge \sim q) \wedge (q \vee p) \wedge (q \vee \sim p)$

$\Leftrightarrow (p \vee \sim q) \wedge (\sim p \vee q) \wedge (\sim p \vee \sim q) \wedge (p \vee q)$

which is in principal conjunctive normal form. Since the formula contain all the maxterms, it is a contradiction. Further, the principal disjunctive normal form does not exist.

d. Let $A = p \vee (\sim p \rightarrow (q \vee (\sim q \rightarrow r)))$

$\Leftrightarrow p \vee (p \vee (q \vee (q \vee r))) \Leftrightarrow (p \vee q \vee r)$

which is in principal conjunctive normal form.

Now

$\sim A \Leftrightarrow (p \vee q \vee \sim r) \wedge (p \vee \sim q \vee r) \wedge (p \vee \sim q \vee \sim r)$

$\wedge (\sim p \vee q \vee r) \wedge (\sim p \vee q \vee \sim r) \wedge (\sim p \vee \sim q \vee r) \wedge (\sim p \vee \sim q \vee \sim r)$

So $A \Leftrightarrow \sim\sim A \Leftrightarrow \wedge (p \vee q \vee \sim r) \vee \sim (p \vee \sim q \vee r) \vee \sim (p \vee \sim q \vee \sim r)$

$\vee \sim (\sim p \vee q \vee r) \vee \sim (\sim p \vee q \vee \sim r)$

$\vee \sim (\sim p \vee \sim q \vee r) \vee \sim (\sim p \vee \sim q \vee \sim r)$

$\Leftrightarrow (\sim p \wedge \sim q \wedge r) \vee (\sim p \wedge q \wedge \sim r) \vee (\sim p \wedge q \wedge r)$

$\vee (p \wedge \sim q \wedge \sim r) \vee (p \wedge \sim q \wedge r) \vee (p \wedge q \wedge \sim r) \vee (p \wedge q \wedge r)$

which is in principal disjunctive normal form, further the formula is not a tautology.

e. Let $A = (p \vee q) \vee (\sim p \wedge r)$

$\Leftrightarrow ((p \wedge q) \vee \sim p) \wedge ((p \wedge q) \vee r)$

$\Leftrightarrow (p \vee \sim q) \wedge (q \vee \sim p) \wedge (p \vee r) \wedge (q \vee r)$

$\Leftrightarrow (q \vee \sim p \vee (r \wedge \sim r)) \wedge (p \vee r \vee (q \wedge \sim q))$

$\Leftrightarrow (q \vee \sim p \vee (r \wedge \sim r)) \wedge (p \vee r \vee (q \wedge \sim q))$

$\wedge (q \vee r \vee (p \wedge \sim p))$ since $p \wedge 0 \Leftrightarrow p$

$\Leftrightarrow (q \vee \sim p \vee r) \wedge (q \vee \sim p \vee \sim r) \wedge (p \vee r \vee q)$

$\wedge (p \vee r \vee \sim q) \wedge (q \vee r \vee p) \wedge (q \vee r \vee \sim p)$

$\Leftrightarrow (\sim p \vee \sim q \vee r) \wedge (\sim p \vee q \vee \sim r) \wedge (p \vee q \vee r) \wedge (p \vee \sim q \vee r)$

which is in principal conjuctive normal form. But

$\sim A \Leftrightarrow (p \vee q \vee \sim r) \wedge (p \vee \sim q \vee \sim r) \wedge (\sim p \vee \sim q \vee r) \wedge (\sim p \vee \sim q \vee \sim r)$

$A \Leftrightarrow \sim\sim A \Leftrightarrow \sim (p \vee q \vee \sim r) \vee \sim (p \vee \sim q \vee \sim r) \vee \sim (\sim p \vee \sim q \vee r) \vee \sim (\sim p \vee \sim q \vee \sim r)$

$\Leftrightarrow (\sim p \wedge \sim q \wedge r) \vee (\sim p \wedge q \wedge r) \vee (p \wedge q \wedge \sim r) \vee (p \wedge q \wedge r)$

which is in principal disjunctive normal form. Here also A is not tautology.

EXERCISE III

1. Obtain the principle conjunctive normal form of

   a. $\sim (p \vee q)$ b. $\sim (p \to q)$ c. $\sim (p \leftrightarrow q)$

2. Obtain the principle conjunctive normal forms of

   a. $(p \wedge q \wedge r) \vee (\sim p \wedge r \wedge q) \vee (\sim p \wedge \sim q \wedge \sim r)$ b. $(p \wedge q) \vee (\sim p \wedge q) \vee (p \wedge \sim q)$

   c. $(p \wedge q) \vee (\sim p \wedge q \wedge r)$

3. Obtain the principle disjunctive normal forms of

   a. $(p \to (q \wedge r)) \wedge (\sim p \to (\sim q \wedge \sim r))$ b. $(\sim p \vee \sim q) \to (p \leftrightarrow \sim q)$

   c. $(p \vee (\sim p \to (q \vee (\sim q \to r)))$

4. Obtain the principle disjunctive normal forms of

   a. $p \wedge (p \vee q)$ b. $p \wedge (\sim p \vee q)$ c. $(p \vee q) \wedge (\sim p \vee q) \wedge (q \vee r)$

5. Obtain the principle disjunctive and conjunctive normal forms of the following formulas and hence state which of the formulas are tautologies?

   a. $q \vee (p \wedge \sim q) \vee (\sim p \wedge \sim q)$ b. $q \vee (p \vee \sim q)$

   c. $(q \to p) \wedge (\sim p \wedge q)$ d $p \to (p \wedge (q \to p))$

PREDICATE CALCULUS

Consider the two statements "Rohit is Brilliant" and "Manaswine is Brilliant". As propositions, there is no relation between them, but they have something in common. Instead of writing two statements we can write a single statement like "x is brilliant", because both Rohit and Manaswine share the same nature brilliant. By replacing x by any other name we get many propositions. The common feature expressed by "is brilliant" is called predicate. Predicate calculus deals with sentences involving predicates.

A part of a declaritive sentence describing the properties of an object or relation among objects can be referred as predicate, e.g."is brilliant".

Note: A statement of the form $p(x_1, x_2, ..., x_n)$ is the value of the propositional function P at the nth tuple $(x_1,...,x_n)$ and P is also called a predicate.

Example Statements involving variables such as $x > 5$, $x = y + 6$, and $x + y = z$.

These statement are neither true nor false. When the values of the variables are not specified.

Let us consider the statement x > 5.

Here the variable x is the subject of the statement. The second part predicate is greater than 5 and → refers to a property that the subject of the statement can have. Therefore $x > 5$ can be written in the form $p(x)$. The statement $p(x)$ is also said to be the value of the propositional function P at x. Once a value has been assigned to the variable x, the statement can be written as $p(x)$.

Statement Calculus

In logical reason, a certain number of propositions is assumed to be true and based on that assumption some other propositions are derived.

Hypothesis The propositions that are assumed to be true. It may also be referred to as premises.

Conclusion The proposition derived by using the rules of inference.

Valid argument The process of deriving conclusions based on the assumptions of a premise.

Table 3.13

Rules of Inference	Implication Form
Addition I_1	
$\dfrac{p}{\therefore p \vee \varphi}$	$P \Rightarrow p \vee \varphi$
Conjunction I_2	
$\dfrac{\varphi}{\therefore p \wedge \varphi}$	$P \wedge \varphi \Rightarrow p \wedge \varphi$
Simplification I_3	
$\dfrac{p \wedge \varphi}{\therefore p}$	$(p \wedge \varphi) \Rightarrow p$
Modus tollens I_4	
$\neg\varphi$ $p \Rightarrow \varphi$ $\therefore \neg p$	$[\neg\varphi \wedge (p \Rightarrow \varphi)] \Rightarrow \neg p$
Disjunctive Syllogism I_5	
$\neg p$ $\dfrac{p \vee \varphi}{\therefore \varphi}$	$[\neg p \wedge (p \vee \varphi)] \Rightarrow \varphi$

Modus ponens

$$\frac{\begin{array}{c} p \\ p \Rightarrow \varphi \end{array}}{\therefore \varphi} \qquad [p \wedge (p \Rightarrow \varphi)] \Rightarrow \varphi$$

Hypothetical syllogism I_7

$$\frac{\begin{array}{c} p \Rightarrow \varphi \\ \varphi \Rightarrow R \end{array}}{\therefore p \Rightarrow R} \qquad [(P \Rightarrow \varphi) \wedge (\varphi \Rightarrow R)] \Rightarrow (P \Rightarrow R)$$

Conjunctive Dilemma

$$\frac{\begin{array}{c} [(P \Rightarrow \varphi) \wedge (R \Rightarrow s)] \\ p \vee R \end{array}}{\therefore \varphi \vee s} \qquad (p \Rightarrow \varphi) \wedge (R \Rightarrow S) \wedge (p \vee R) \Rightarrow (\varphi \vee S)$$

Destructive dilemma

$$\frac{\begin{array}{c} (p \Rightarrow \varphi) \wedge (R \Rightarrow S) \\ \neg\varphi \vee \neg s \end{array}}{\therefore p \vee R} \qquad [(p \Rightarrow \varphi) \wedge (R \Rightarrow S)] \wedge [\neg\varphi \vee \neg S] \Rightarrow (\neg p \vee \neg R)$$

The two rules of inference are called rules p and T.

Rule P A premise may be introduced at any point in the derivation.

Rule T A formula S may be introduced in a derivation if S is tautologically implied by any one or more of the preceeding formulae in the derivation of a truth value.

Example 3.68 For the given set of arguments check the validity of conclusion

a. If determinism is true then we have no free will.
b. If Heisenberg's interpretation of quantum physics is correct, then there are events not necessiated by prior events.
c. If there are events not necessitated by prior events, then we have free will.

Conclusion If Heisenberg interpretation of physics is correct then we have free will.

Solution D: Determinism is true.

Q: We have no free will.

R: Heisenberg interpretation of quantum physics is correct.

S: There are events not necessitated by prior events.

Arguments $(D \rightarrow \varphi)(R \rightarrow S)(S \rightarrow 7\varphi)$

Conclusion $R \rightarrow \neg\varphi$

(1) $R \rightarrow S$ Rule P

(2) $R \rightarrow \neg\varphi$ Rule P

From (1) and (2), (3) $R \rightarrow \neg\varphi$ rule T and hypothetical syllogism.

Free and Bound Variables

The variable is said to be bound if it is concerned with either universal ($\forall$) or existential ($\exists$) quantifier and the scope of the variable is the formulae immediately following the quantifier. The variable which is not concerned with any quantifier is called free variable.

Example $\forall x\, [p(x, y)]$ in the statement given above x is said to be bound and the scope of x is upto $p(x, y)$, while y is called free variable.

Quantifiers

When all the variables in a propositional function are assigned values, the resulting statement has a truth value. However, there is another important way to change propositional functions into propositions, called quantification. It has been broadly classified into two types, namely

i. Universal Quantification
ii. Existential Quantification

Universe of Discourse Many mathematical statements assert that a property is true for all values of a variable in a particular domain, called the universe of discourse. Such a statement is expressed using an universal quantification.

Universal Quantification The universal quantification of $p(x)$ is a proposition only when $p(x)$ is true for all values of x in the universe of discourse.

Notation $\forall x\, B(x) \rightarrow$ universal quantification of $p(x)$

It is also expressed as

"for all $x\, p(x)$"

or "for every $x\, p(x)$"

Example 3.69 "Every student in this class has studied Logic"

Solution Let $p(x)$ denote the statement "x has studied logic",

$\forall x\, [s(x) \rightarrow p(x)]$

where $s(x)$ is the statement "x is in this class"

Example 3.70 What is the truth value of $\forall x\, p(x)$, where $p(x)$: $x^2 < 10$ and the universe of discourse consists of the positive integers not exceeding 4!

Solution The statement $\forall x\, p(x)$ is the same as the conjunction. $p(1) \wedge p(2) \wedge p(3) \wedge p(4)$

since the universe of discourse consists of the integers 1, 2, 3,and 4.

$p(4)$, in the statement "$4^2 < 10$" is false, it follows $\forall x\, p(x)$ is also false.

Reason:

$p(1) \wedge p(2) \wedge p(3) \wedge p(4)$

$T \wedge T \wedge T \wedge F =$ false

Existential quantification The existential quantification of $p(x)$ is the proposition "There exists an element x in the universe of discourse such that $p(x)$ is true."

Notation $\exists x\, p(x)$

It is also expressed as

"There is an x such that $p(x)$"

"There is atleast one x such that $p(x)$",

or for some $x\, p(x)$.

Example 3.71 Let $p(x) : x > 3$, what is the truth value of the quantification $\exists x\, p(x)$ where the universe of discourse is the set of real numbers.

Solution Since $x > 3$ is true for instance, when $x = 4$ the existential quantification of $p(x)$ is $\exists x\, p(x)$ is true.

Example 3.72 Write the predicate "x is the father of the mother of y".

Solution Let $p(x) : x$ is a person

$p(x, y) : x$ is the father of y

$m(x, y) : x$ is the mother of y

We assume that the there exists a person Z such that x is the father of Z and Z is the mother of y.

Universal Specification (US)

For a given predicate $(\forall x)\,[p(x)]$ one can conclude $p(b)$

Existential specification (ES) From $(\exists x)\,[p(x)]$ one can conclude the value of $p(b)$ provided that b is not free in any given premise and also not free in any prior step of the derivation. These requirements can easily be met by choosing a new variable each time *ES* is used.

Existential Generalization (EG) From $p(b)$ one can conclude $(\exists x)\,[p(x)]$

Universal Generalization (UG) From $p(b)$ one can conclude $\forall x\,[p(x)]$ provided that b is not free in any of the given premises and provided that if b is free in the prior step which resulted from the use of *ES*, then no variables introduced by that use of *ES* appear free in $p(b)$.

Example 3.73 Some cats are black but all buffaloes are black.

Solution $C(x) : x$ is a cat

$B(y) : y$ is a buffalo

$b(x) : x$ is black

Thus, $(\exists x)\ (\forall y)[C(x) \wedge b(x)] \wedge [B(y) \rightarrow b(y)]$

Example 3.74 Sum of two positive integers is greater than either of the integers.

Solution $I(x) : x$ is a positive integer

$GT(x, y) : x$ is greater than y

$Su(x, y)$: sum of x and y

Thus, $(\forall x)\ (\forall y)\ [(I(x) \wedge I(y))] \rightarrow [GT(Su(x, y), x] \vee GT[Su(x, y) y]$

Example 3.75 Every student in this school is either good at studies or good in sports.

Solution $S(x) : x$ is a student of this school

$ST(x) : x$ is good at study

$SP(x)$: x is good at sports

Thus, $(\forall x)\ [S(x) \rightarrow ST(x) \vee SP(x)]$

1. $(\exists x)\,[A(x) \vee B(x)] \Leftrightarrow (\exists x)\ \{A(x)\} \vee (\exists x)\,\{B(x)\}$
2. $(\forall x)\,[A(x) \wedge B(x)] \Leftrightarrow (\forall x) \wedge [A(x)] \wedge (\forall x)\ [B(x)]$
3. $\neg\,(\exists x)\ A(x) \Leftrightarrow \forall x \neg A(x)$
4. $\neg\,(\forall x)\ A(x) \Leftrightarrow (\exists x)\ \neg A(x)$

Example 3.76 Give an argument which will establish the validity of the following inference.

All integers are rational numbers. Some integers are power of 2. Therefore, some rational numbers are power of 2.

Solution

$I(x)$: x is an integer

$R(x)$: x is a rational number

$P(x)$: x is a power of 2

Inference pattern

$$\begin{array}{l} (\forall x)[I(x) \rightarrow R(x)] \\ (\exists x)[I(x) \wedge P(x)] \\ \hline (\exists x)[R(x) \wedge P(x)] \end{array}$$

Argument

1.	$(\exists x) I(x) \wedge P(x)]$	*premise*
2.	$I(b) \wedge P(x)$	*ES* (1)
3.	$I(b)$	
4.	$P(b)$	
5.	$(\forall x)\,[I(x) \rightarrow R(x)]$	*premise*
6.	$I(b) \rightarrow R(b)$	*US*
7.	$R(b)$	
8.	$R(b) \wedge P(b)$	
9.	$(\exists x)\,[R(x) \wedge P(x)]$	*EG*

Example 3.77 Verify the validity of the following argument:

All men are mortal. Socrates is a man. Therefore Socrates is a mortal.

Solution $H(x)$: x is a man

$M(x)$: x is a mortal

S : Socrates

We have to show $(\forall x)\ [H(x) \rightarrow M(x) \wedge H(s)] \Rightarrow M(s)$

Argument

1.	$(\forall x)[H(x) \rightarrow M(x)$	*premise*
2.	$H(s) \rightarrow M(s)$	*US*
3.	$H(s)$	*premise*
4.	$M(s)$	

Thus the influence is valid.

Example 3.78 Show that from

a. $(\exists x)[F(x) \wedge S(x)] \rightarrow \forall y[m(y) \rightarrow w(y)]$

b. $(\exists y)[m(y) \wedge \neg w(y)]$

The conclusion $(\forall x)[F(x) \rightarrow \neg S(x)]$ follows.

1.	$(\exists y)[m(y) \wedge \neg w(y)]$	*premise*
2.	$m(z) \wedge \neg w(z)$	*ES*
3.	$\neg[M(z) \rightarrow W(z)]$	
4.	$(\exists y)\{\neg[m(y) \rightarrow w(y)]\}$	*EG*
5.	$\neg(\forall y)[m(y) \rightarrow w(y)]$	
6.	$(\exists x)[f(x) \wedge s(x)] \rightarrow \forall y[m(y) \rightarrow w(y)]$	*premise*
7.	$\exists, \forall\ [(f(x) \wedge s(x))]$	
8.	$\forall x \neg(f(x) \wedge s(x))$	
9.	$\neg[f(x) \wedge s(x)]$	*US*
10.	$F(x) \rightarrow \neg s(x)$	
11.	$(\forall x)[F(x) \rightarrow \neg s(x)]$	*UG*

Example 3.79 **Is the following conclusion validly derivable from the premises given?**

If $(\forall x)[p(x) \rightarrow \varphi(x)]$, $(\exists y)\, p(y)$, then $(\exists z)\, \varphi(z)$.

Solution

1.	$\neg(\exists z)\varphi(z)$	*premise (assumed)*
2.	$(\forall z)\neg\varphi(z)$	
3.	$(\exists y)\, p(y)$	*premise*
4.	$p(a)$	*ES* (3)
5.	$\neg\varphi(a)$	*US* (2)
6.	$p(a) \wedge \neg\varphi(a)$	
7.	$\neg[p(a) \rightarrow \varphi(a)]$	
8.	$(\forall x)[p(x) \rightarrow \varphi(x)]$	*premise*
9.	$p(a) \rightarrow \varphi(a)$	*US* (8)
10.	$[p(a) \rightarrow \varphi(a)] \wedge \neg[p(a) \rightarrow \varphi(a)]$	*Contradiction*

Rule CP (conditional Proof)

If we can derive S from R and a set of promises then we may derive $R \rightarrow S$ from set of premises alone.

Example 3.80 Derive $P \rightarrow (Q \rightarrow S)$, using the rule CP if necessary from $P \rightarrow (Q \rightarrow R)$, $Q \rightarrow (R \rightarrow S)$

Step 1

In C.P the conditional part can be taken as an additional premise i.e. $P \rightarrow (Q \rightarrow S)$

$\therefore$ we p include P as an additional premise.

Step 2

	1.	$P \rightarrow (Q \rightarrow R)$	*premise*
	2.	P	*additional premise*
{1,2}	3.	$Q \rightarrow R$	$P \rightarrow Q, P \Rightarrow R$
	4.	$Q \rightarrow (R \rightarrow S)$	*premise*
	5.	$\neg Q \vee R$	$P \rightarrow Q, \neg P \vee R$
	6.	$\neg Q \vee (R \rightarrow S)$	
	7.	$\neg Q \vee (R, R \rightarrow S)$	
	8.	$\neg Q \vee S$	
	9.	$Q \rightarrow S$	
	10.	$P \rightarrow (Q \rightarrow S)$	*CP*

Example 3. 81 Show that $P \rightarrow S$ can be derived from the premises. $\neg P \vee Q, \neg Q \vee R, R \rightarrow S$

Step 1

P is additional premise

Step 2

	1.	$\neg P \vee Q$	*premise*
	2.	P	*additional premise*
{1,2}	3.	Q	
	4.	$\neg Q \vee R$	*premise*
{3,4}	5.	R	
	6.	$R \rightarrow S$	*premise*
{5,6}	7.	S	
	8.	$P \rightarrow S$	*CP*

Consistent and inconsistent

A set of formulae $S_1, S_2, \ldots S_m$ is said to be inconsistent if their conjunction implies a contradiction, that is $S_1 \wedge S_2 \wedge S_3 \wedge \ldots \wedge S_m \Rightarrow R \wedge \neg R \Rightarrow F$ for same formula R.

A set of formula $S_1, S_2, \ldots S_m$ is said to be consistent if it is not inconsistent.

Indirect Method of proof

By using the rule of conditional proof and the notion of an inconsistent set of premises, we introduce a method called proof by contradiction or indirect method of proof.

Example 3.82 **Prove by indirect method $P \to (q \wedge r)$, $(q \vee S) \to t$, and $p \vee S \Rightarrow t$.**

Step 1

7*t* is an additional premise

Step 2

	1.	$P \to (q \wedge r)$	*premise*
	2.	$(q \vee s) \to t$	*premise*
	3.	$\neg t$	*additional premise*
{1}	4.	$P \to q$	*premise*
{1}	5.	$P \to r$	
{2}	6.	$S \to t$	*premise*
{3,6}	7.	$\neg S$	$\neg Q, P \to Q \Rightarrow \neg P$
	8.	$P \vee S$	*premise*
{8}	9.	S	
	10.	$S \times \neg S = F$	

Example 3.83 **Prove by indirect method that**

$$E \to S,\ S \to H,\ A \to 7H \Rightarrow 7(E \wedge A)$$

Solution

Step 1

$\neg\neg(E \wedge A)$

$\Rightarrow E \wedge A$ **is an additional premise**

Step 2

	1.	$E \to S$	*premise*
	2.	$S \to H$	*premise*
{1,2}	3.	$E \to H$	
	4.	$E \to \neg H$	*premise*
	5.	$E \wedge A$	*additional premise*
{5}	6.	E	
{5}	7.	A	
{3,6}	8.	$E, E \to H = H$	
{4,7}	9.	$A, A \to \neg H = \neg H$	
	10.	$H \wedge \neg H = F$	

Example Using indirect proof show that $p \to q,\ q \to r,\ \neg(p \wedge r),\ p \vee r \Rightarrow r$

Solution

Step 1

$\neg r$ is an additional premise

1.	$q \to r$	*premise*
2.	$\neg r$	*premise*
{1,2} 3.	$\neg q$	$p \to q, \neg q \Rightarrow \neg q$
4.	$p \to q$	*premise*
{3,4} 5.	$\neg p$	$p \to q, \neg q \Rightarrow \neg q$
6.	$p \vee r$	*premise*
{5,6} 7.	r	
8.	$r \wedge \neg r = f$	

contradiction hence premise.

Example 3.85 If an integer is divisible by 10 then it is divisible by 2. If an integer is divisible by 2, then it is divisible by 3. Prove that an integer divisible by 10 is divisible by 3.

Solution

$D_{10}(x) : x$ is divisible by 10

$D_2(x) : x$ is divisible by 2

$D_3(x) : x$ is divisible by 3

$$\frac{\begin{array}{l}\forall x[D_{10}(x) \to D_2(x)]\\ \forall x[D_2(x) \to D_3(x)]\end{array}}{\forall x[D_{10}x \to D_3(x)]}$$

1.	$\forall x[D_{10}(x) \to D_2(x)]$	*premise*
2.	$D_{10}(b) \to D_2(b)$	*US*
3.	$\forall x[D_2(x) \to D_3(x)]$	*premise*
4.	$D_2(b) \to D_3(b)$	*US*
{2,4} 5.	$D_{10}(b) \to D_3(b)$	$P \to Q, Q \to R \Rightarrow P \to R$
6.	$\forall x[D_{10}(x) \to D_3(x)]$	*Contusion*

Example 3.86 Prove $(\forall x)[P(x) \to Q(x)],\ \forall x[R(x) \to 7Q(x)] \Rightarrow \forall x[R(x) \to 7P(x)]$

Step 1

Taking negated conclusion

$$\neg(\neg P(x))$$
$$\Rightarrow P(x)$$

Step 2

	1.	$(\forall x)[P(x) \to Q(x)]$	*premise*
	2.	$(\forall x)[R(x) \to \neg Q(x)]$	*premise*
{2}	3.	$R(x) \to \neg Q(x)$	*US*
	4.	$R(x)$	*additional premise*
	5.	$\neg Q(x)$	(3, 4)
	6.	$P(x) \to Q(x)$	*US*
{5, 6}	7.	$\neg P(x)$	using $\neg Q, P \to Q = \neg P$
	8.	$R(x) \to \neg P(x)$	*CP*
	9.	$(\forall x)[R(x) \to \neg P(x)]$	*UG*

CHAPTER

4

Induction, Recursion and Recurrence Relations

INTRODUCTION

In this chapter, first we will examine mathematical induction, a technique for proving results over natural numbers. We will use a variety of examples to illustrate how results are proved using mathematical induction. Next we discuss recursion, an important application of mathematical induction.

Recursion involves the definition of sequences and we discuss how sequences can be defined recursively. Recursive definitions can be used to solve counting problems. The rule for finding terms of a sequence is called a recurrence relation. We will study the theory of linear recurrence relations and their solutions. Recurrence relations occur constantly in practical applications, most often in the analysis of certain discrete time systems, analysis of algorithms, error-correcting code, etc. Finally, we will introduce " generating function", which can be used to solve linear recurrence relations.

MATHEMATICAL INDUCTION

Mathematical induction is the process of inferring the truth form a general statement for particular cases.

A statement may be true with reference to more than hundred cases yet we cannot conclude it to be true in general. It is possible to disprove the statement by a counter-example, a statement need not be accepted to be true. Such a statement inferred from particular cases is called a conjecture. If the conjecture is a statement involving natural numbers, we can use the principle of the mathematical induction to prove the same.

Principle of Mathematical Induction

For a given statement involving a natural number n, if we can show that:

1. The statement is true for $n = n_0$ and
2. The statement is true for $n = k+1$, assuming that the statement is true for $n = k (k \geq n_0)$ then we can conclude that the statement is true for all natural numbers $n \geq n_0$.

(1) is usually referred to as the basis of induction, and (2) is usually referred to as the induction step. Also, the assumption that the statement is true for $n = k$ in (2) is usually referred to as the induction hypothesis.

Now let us see how the principle of mathematical induction is applied to prove statements involving natural numbers.

Example 4.1 Prove that the sum of the first n integers is $\frac{n(n+1)}{2}$ for all n.

(or)

Prove that $1+2+3+...+n=\frac{n(n+1)}{2}$ for all $n \in N$ by induction principle.

Solution

1. Basis of induction: For $n=1$, we have

$$1=\frac{1(1+1)}{2}$$

2. Induction step: Assume that

$$1+2+3+...+k=\frac{k(k+1)}{2} \tag{1}$$

We have

$$1+2+3+...+k+(k+1)=\frac{k(k+1)}{2}+(k+1) \quad \text{by(1)}$$

$$=\frac{k(k+1)+2(k+1)}{2}=\frac{(k+1)(k+2)}{2}$$

Therefore, by the principle of mathematical induction, the result is true for all $n \in N$.

Example 4.2 Show that $1^2+2^2+...+n^2=\frac{n(n+1)(2n+1)}{6}$ $n \geq 1$ by mathematical induction. (*B.E.Oct.* 96)

Solution

1. Basis of induction: For $n = 1$, we have

$$1^2=\frac{1(1+1)(2+1)}{6}$$

2. Induction step: Assume that

$$1^2+2^2+...+k^2=\frac{k(k+1)(2k+1)}{6} \tag{1}$$

We have

$$1^2+2^2+...+k^2+(k+1)^2=\frac{k(k+1)(2k+1)}{6}+(k+1)^2$$

$$=\frac{(k+1)[k(2k+1)+6(k+1)]}{6} \quad \text{by(1)}$$

$$=\frac{(k+1)(2k^2+7k+6)}{6}=\frac{(k+1)(k+2)(2k+3)}{6}$$

$$=\frac{(k+1)[(k+1)+1][2(k+1)+1]}{6}$$

Hence by the principle of mathematical induction, the statement is true for all $n \geq 1$.

Example 4.3 Show that $1^3+2^3+3^3+...+n^3=\frac{n^2(n+1)^2}{4}$; $n \in N$ (*B.E. Apr.* 97)

Solution

1. Basis of induction: For $n = 1$, we have

$$1^3=1=\frac{1^2(1+1)^2}{4}$$

2. Induction step: Assume that

$$1^3+2^3+3^3+...+k^3=\frac{k^2(k+1)^2}{4} \qquad (1)$$

We have

$$1^3+2^3+3^3+\cdots+k^3+(k+1)^3=\frac{k^2(k+1)^2}{4}+(k+1)^3 \qquad \text{by (1)}$$

$$=\frac{k^2(k+1)^2+4(k+1)^3}{4}=\frac{(k+1)^2[k^2+(4k+1)]}{4}$$

$$=\frac{(k+1)^2[k^2+4k+4]}{4}=\frac{(k+1)^2(k+2)^2}{4}$$

$$=\frac{(k+1)^2[(k+1)+1]^2}{4}$$

Hence by the principle of mathematical induction, the statement is true for all $n \in N$.

Example 4.4 Prove that the sum of the first n odd integers is n^2 for all integers n.

(or)

Prove that $\sum_{i=1}^{n}(2i-1)=n^2$ for all $n \in N$.

Solution

1. Basis of induction: For n = 1, we have $1 = 1^2$
2. Induction step: Assume that

$$\sum_{i=1}^{k+1}(2i-1)=k^2 \qquad (1)$$

We have

$$\sum_{i=1}^{k+1}(2i-1)=\sum_{i=1}^{k}(2i-1)+[2(k+1)-1]$$

$$=k^2+[2(k+1)-1]=k^2+2k+1 \qquad \text{by (1)}$$

$$=(k+1)^2$$

Therefore, by mathematical induction, the statement is true for all n.

Example 4.5 Using mathematical induction, prove that $2+5+8+...+(3n-1)=\frac{n(3n+1)}{2}$ (*B.E. Oct.97*)

Solution

1. Basis of induction: For $n = 1$, we have

$$2=\frac{1.4}{2}=\frac{1.(3(1)+1)}{2}$$

2. Induction step: Assume that

$$2+5+8+...+(3k-1)=\frac{k(3k+1)}{2} \qquad (1)$$

Consider,

$$2+5+8+\cdots+(3k-1)=\frac{k(3k+1)}{2}+(3k+2) \qquad \text{by (1)}$$

$$= \frac{k(3k+1)+2(3k+2)}{2} = \frac{3k^2+k+6k+4}{2}$$

$$= \frac{3k^2+7k+4}{2} = \frac{(k+1)(3k+4)}{2} = \frac{(k+1)(3k+3+1)}{2}$$

$$= \frac{(k+1)[3(k+1)+1]}{2}$$

Hence by the principle of mathematical induction the result follows.

Example 4.6 Prove that for $n \geq 1$, $1+2+4+...+2^n = 2^{n+1}-1$.

Solution

1. Basis of induction: If $n=1$, we have $1+2=2^{1+1}-1=4-1$
2. Induction step: Assume that

$$1+2+4+\cdots+2^k = 2^{k+1}-1 \qquad (1)$$

We have

$$1+2+4+...+2^k+2^{k+1} = (2^{k+1}-1)+2^{k+1}$$
$$= 2^{k+1}+2^{k+1}-1 \qquad \text{by (1)}$$
$$= 2(2^{k+1})-1 = 2^{k+2}-1$$
$$= 2^{(k+1)+1}-1$$

Hence by the principle of mathematical induction, the result is true for all n.

Example 4.7 Prove that if $n \geq 1$, then $1(1!)+2(2!)+\cdots+n(n!) = (n+1)!-1$.

Solution

1. Basis of induction: For $n = 1$, we have

$$1(1!) = 1(1) = 1 = (1+1)!-1 = 2-1.$$

2. Induction step: Assume that

$$1(1!)+2(2!)+...+k(k!) = (k+1)!-1. \qquad (1)$$

We have

$$1(1!)+2(2!)+\cdots+k(k!) = (k+1)((k+1)!)$$
$$= [(k+1)!-1]+(k+1)((k+1)!)$$
$$= (k+1)!+(k+1)((k+1)!)-1 \qquad \text{by (1)}$$
$$= (k+1)!(1+k+1)-1 = (k+1)!(k+2)-1$$
$$= (k+2)!-1$$

Hence by mathematical induction, the statement is true for all $n \geq 1$.

Example 4.8 Use mathematical induction to show that $\frac{1}{1.2}+\frac{1}{2.3}+\frac{1}{3.4}+...+\frac{1}{n(n+1)} = \frac{n}{n+1}$ for all $n \geq 1$.

Solution

1. Basis of induction: For $n = 1$, we have

$$\frac{1}{1.2} = \frac{1}{2} = \frac{1}{1+1}$$

2. Induction step: Assume that

$$\frac{1}{1.2} = \frac{1}{2.3} = \frac{1}{3.4}+...+\frac{1}{k(k+1)} = \frac{k}{k+1} \qquad (1)$$

We have

$$\frac{1}{1.2}+\frac{1}{2.3}+\frac{1}{3.4}+...+\frac{1}{k(k+1)}+\frac{1}{(k+1)[(k+1)+1]} \quad \text{by (1)}$$

$$=\frac{k}{k+1}+\frac{1}{(k+1)[(k+1)+1]}$$

$$=\frac{k}{k+1}+\frac{1}{(k+1)(k+2)}=\frac{k(k+2)+1}{(k+1)(k+2)}$$

$$=\frac{k^2+2k+1}{(k+1)(k+2)}=\frac{(k+1)^2}{(k+1)(k+2)}$$

$$=\frac{k+1}{(k+1)+1}$$

Therefore by the principle of mathematical induction, the statement is true for all n.

Example 4.9 Use mathematical induction to prove that $n^3 - n$ is divisible by 3 whenever n is a positive integer.

Solution

1. Basis of induction: For $n = 1$, we have $1^3 - 1 = 1 - 1 = 0$ and it is divisible by 3.
2. Induction step: Assume that $k^3 - k$ is divisible by 3. We must show that $(k + 1)^3 - (k + 1)$ is divisible by 3. Note that

$$(k+1)^3 - (k+1) = (k^3 + 3k^2 + 3k + 1) - (k+1) = (k^3 - k) + 3(k^2 + k)$$

Since both terms in this sum are divisible by 3, it follows that $(k + 1)^3 - (k + 1)$ is also divisible by 3. Thus, by the principle of mathematical induction, $n^3 - n$ is divisible by 3 for all $n \in N$.

Example 4.10 Use mathematical induction to show that 3 divides $n^3 + 2n$ whenever n is a non-negative integer.

(or)

Prove that for all $n \geq 1$, n^3+2n is a multiple of 3. *(B.E. Apr. 98)*

Solution

1. Basis of induction: For $n=1$, we have $1^3 + 2(1) = 3$ and it is divisible by 3.
2. Induction step: Assume that $k^3 + 2k$ is divisible by 3. We show that $(k + 1)^3+2(k + 1)$ is divisible by 3. Note that

$$(k+1)^3 + 2(k+1) = (k^3 + 3k^2 + 3k + 1) + 2k + 2 = (k^3 + 2k) + 3(k^2 + k + 1)$$

Since both terms in this sum are divisible by 3, it follows that $(k + 1)^3 + 2(k + 1)$ is also divisible by 3. Thus, by the principle of mathematical induction, $n^3 + 2n$ is divisible by 3 for all $n \in N$

Example 4.11 Use mathematical induction to show tht 3^n is a multiple of 5. *(B.E. Oct, 98)*

Solution

1. Basis of induction: For $n = 1$, we have $8 - 3 = 5$ and it is a multiple of 5.
2. Induction step: Assume that $8^k - 3^k$ is a multiple of 5. We must show that $8^{k+1} - 3^{k+1}$ is a multiple of 5. Note that

$$8^{k+1} - 3^{k+1} = 8^k(8) - 3^{k+1}$$

$$= 8^k(5+3) - 3^{k+1} = 5(8^k) + 3(8^k) - 3^{k+1}$$

$$= 5(8^k) + 3(8^k - 3^k)$$

Since both forms in this sum are multiples of 5 (the first because it is 5 times an integer, and the second by the assumption of the induction step), it follows that $8^{k+1} - 3^{k+1}$ is also a multiple of 5. Thus, by the principle of mathematical induction, $8^n - 3^n$ is a multiple of 5.

Example 4.12 Use mathematical induction to show that $n^2 - 7n + 12$ is non-negative whenever n is an integer greater than 3.

Solution

1. Basis of induction: For $n = 4$, we have $4^2 - 7(4) + 12 = 16 - 28 + 12 = 0$ and it is non-negative.
2. Induction step: Assume that $k^2 - 7k + 12$ is non-negative and $k > 3$. We must show that $(k+1)^2 - 7(k+1) + 12$ is non-negative. Note that

$$(k+1)^2 - 7(k+1) + 12 = k^2 + 2k + 1 - 7k - 7 + 12$$
$$= (k^2 - 7k + 12) + (2k - 6).$$

By the assumption of the inductive step, the first term is non-negative and for $k > 3$ the second term is also non-negative. Hence $(k+1)^2 - 7(k+1) + 12$ is non-negative. Thus, by the principle of mathematical induction, $n^2 - 7 + 12$ is nonnegative whenever n is an integer greater than 3.

The following examples use the principle of mathematical induction to prove the inequalities.

Example 4.13 Use mathematical induction to prove the inequality, $n < 2^n$ for all positive integers n.

Solution

1. Basis of induction: For $n = 1$, we have $1 < 2 = 2^1$.
2. Induction step: Assume that $k < 2^k$ (1)

We have,

$$k + 1 < 2^k + 1$$
$$< 2^k + 2^k \text{ since } 1 < 2^k \quad \text{by(1)}$$
$$= 2 \times 2^k = 2^{k+1}$$

Hence by the principle of mathematical induction, the result is true for all $n \in N$.

Example 4.14 Prove that $n + 10 \le 2$ for all $n \in N$ and $n \ge 4$.

Solution

1. Basis of induction: For $n = 4$, we have $4 + 10 = 14 < 16 = 2^4$
2. Induction step: Assume that $k + 10 \le 2^k$ (1)

we have

$$(k+1) + 10 = (k + 10) + 1$$
$$\le 2^k + 1 \quad \text{by (1)}$$
$$\le 2^k + 2k \text{ since } 1 < 2^k$$
$$= 2.2^k = 2^{k+1}$$

Therefore by the mathematical induction, the results follows.

Note that the result is false for n = 1, 2 and 3.

Example 4. 15 Prove that $2^n < n!$ for $n \ge 4$ and $n \in N$.

Solution

1. Basis of induction: For n = 4, we have $2^4 < 4!$ (16 < 24)

(Here the result is false for $n = 1,2$ and 3).

2. Induction step: Assume that $2^k < k!$ (1)

we have

$$\begin{aligned} 2^{k+1} &= 2^k \cdot 2 \\ &< 2^k(k+1) \text{ since } 2 < (k+1) \quad \text{by (1)} \\ &< k!(k+1) \text{ by assumption} \\ &= (k+1)! \end{aligned}$$

Here the result is true for all $n \geq 4$ by principle of induction.

Example 4.16 Suppose that there are n people in a room, $(n \geq 1)$ and they all shakes hands with one another. Prove that $\dfrac{(n-1)n}{2}$ handshakes will have occurred.

Solution

1. Basis of induction: If $n = 1$, we have $0 = \dfrac{(1-1)(1)}{2}$

   There is no handshake.

2. Induction step: Assume that the number of handshakes occurred as $\dfrac{(k-1)k}{2}$. If one enters the room he will shake hands with k people, so

$$\begin{aligned} \frac{(k-1)k}{2} + k &= \frac{k^2 - k + 2k}{2} \\ &= \frac{k^2 + k}{2} = \frac{k(k+1)}{2} \\ &= \frac{[(k+1)-1](k+1)}{2} \end{aligned}$$

Hence by the principle of mathematical induction, the result follows.

EXERCISE I

By principle of induction:

1. Show that the sum as first n even integers is n^2+n.
2. Prove that $1^2 + 3^2 + 5^2 + \ldots + (2n-1)^2 = \dfrac{4n^3 - n}{3}$.
3. Show that $2 + 2^2 + 2^3 + \ldots + 2^n = 2^{n+1} - 2$
4. Prove that $1.2 + 2.3 + \ldots + n(n+1) = \dfrac{n(n+1)(n+2)}{3}$.
5. Prove that $1 + 4 + 7 + \ldots + (3n-2) = \dfrac{n(3n-1)}{2}$.
6. Prove that $\dfrac{1}{2} + \dfrac{1}{2^2} + \dfrac{1}{2^3} + \ldots + \dfrac{1}{2^n} = 1 - \dfrac{1}{2^n}$.
7. Show that $\dfrac{1}{1.3} + \dfrac{1}{3.5} + \dfrac{1}{5.7} + \ldots + \dfrac{1}{(2n-1)\,(2n+1)} = \dfrac{n}{2n+1}$.
8. Show that $n^4 - 4n^2$ is divisible by 3 for all $n \in N$ by induction.

9. Prove by induction that the sum of the cubes of three consecutive integers is divisible by 9.
10. Show that $11^n - 4^n$ is divisible by 7 for $n \in N$.
11. Show that $2^{2n} - 1$ is divisible by 3 for all $n \in N$.
12. Prove that $a^n - b^n$ is divisible by $a - b$ for all $n \in N$.
13. Prove by induction that $5^n - 4n - 1$ is exactly divisible by 16 for $n \in N$.
14. Show that $2n < 3^n$ for all $n \in N$.
15. Show that $n + 1 < n^2$ for $n \geq 2$.
16. Show that $n^3 < 2^n$ for $n \geq 10$.
17. Prove that $2n + 1 \leq 2n$ for all $n \geq 3$.

RECURSION

The numbers in the sequence 0, 1, 2, 3, 5, 8, 13, 21,... in which each new term is the sum of the previous two terms are called the Fibonacci numbers. If we denote the $(n+1)$th Fibonacci number by f_n, we have

$$f_n = f_{n-1} + f_{n-2} \text{ for } n \geq 2$$

and $f_0 = 0$ and $f_1 = 1$. This is called a recursive definition, in which each element of the sequence is defined in terms of previous numbers in the sequence.

We can define a function with the set of non-negative integers as its domain by

(a) specifying the value of the function at zero, and

(b) giving a rule for finding its value at an integer from its values at smaller integers.

Such a function is called a recursively defined function or such a definition is called recursive definition.

Example Consider the sequence of powers of 3 given by

$$a_n = 3^n \text{ for } n = 0,1,2...$$

This sequence can also be defined by giving the first term of the sequence, namely $a_0 = 1$ and a rule for finding a term of the sequence from the previous one, namely

$$a_{n+1} = 3a_n \text{ for } n = 0,1,2,....$$

Example 4.17 Find a recursive definition of binomial coefficients.

Solution Denote the binomial coefficient ${}^nC_K$ by $C(n,k)$. Then the recursive definition for $C(n,k)$ where $n \geq 0$, $k \geq 0$ and $n \geq k$ is given by

$$C(n,0) = 1;$$

$$C(n,n) = 1; \text{ and}$$

$$C(n,k) = C(n-1,k) + C(n-1,k-1) \text{ for } n > k > 0.$$

Example 4.18 Suppose that f is defined recursively by

$$f(0) = 2$$

$$f(n+1) = 3f(n)+2$$

Find $f(1), f(2), f(3)$ and $f(4)$.

Solution From the recursive definition, it follows that

$$f(1) = 3f(0) + 2 = 3.2 + 2 = 8$$

$$f(2) = 3f(1) + 2 = 3.8 + 2 = 26$$
$$f(3) = 3f(2) + 2 = 3.26 + 2 = 80$$
$$f(4) = 3f(3) + 2 = 3.80 + 2 = 242$$

Example 4.19 Give a recursive definition of $f(n) = n!$.

Solution Since $0! = 1$ and $(n+1)! = (n+1)n!$ the desired rule is

$$f(0) = 1; \text{ and}$$
$$f(n+1) = (n+1)f(n)$$

Example 4.20 Give a recursive definition of a^n where a is a non zero real number and n is non-negative integer.

Solution It is easy to see that

$$a^0 = 1; \text{ and } \quad a^{n+1} = a.a^n \text{ for } n \geq 0$$

uniquely define a^n for all non-negative integers n.

Example 4.21 Give a recursive definition for polynomial expression.

Solution An expression of the form

$$f(x) = a_0x^n + a_1x^{n-1} + \ldots + a_{n-1}x + a_n$$

where a's are constants ($a_0 \neq 0$) and $n \geq 0$, is called polynomial in x of degree n.

Let S be a set of coefficients. Then recursive definition for polynomial is

(a) a zeroth degree (constant) polynomial is an element of S.

(b) for $n \geq 1$, an nth degree a polynomial expression is an expression of the form $q(x)x+a$ where $q(x)$ is an $(n-1)$th degree polynomial and $a \in S$.

Example 4.22 Using recursive definition for polynomial expression, prove that $f(n) = 3n^3 - 8n^2 + 2n + 4$ is a third-degree polynomial expression.

Solution Given

$$f(n) = 3n^3 - 8n^2 + 2n + 4$$
$$= (3n^2 - 8n + 2)\, n + 4 = ((3n - 8)\, n - 8)\, n + 4$$
$$= (((3)\, n - 8)\, n + 2)n + 4.$$

Now, 3 is a zeroth degree polynomial

$\Rightarrow (3)n - 8$ is a first degree polynomial

$\Rightarrow ((3)n - 8)\, n + 2$ is a second degree polynomial

$\Rightarrow f(n) = (((3)n - 8)\, n + 2)\, n + 4$ is a third degree polynomial.

Note: The final expression for

$$f(n) = (((3)n - 8)\, n + 2)\, n + 4$$

is called its telescoping form. If we use it to calculate $f(6)$, we need only three multiplications and three additions or subtractions. This is called Horner's method for evaluating a polynomial expression.

Example 4.23 Write $p(x) = x^3 - 6x^2 + 11x - 6$ in telescoping form. *(B.E. Apr. 98)*

Solution The telescoping form of

$$p(x) = x^3 - 6x^2 + 11x - 6$$
$$= (((1)x - 6)\, x + 11)\, x - 6$$

Example 4.24 Write $g(x) = -x^5 + 3x^4 + 3x^3 + 2x^2 + x$ in telescoping form. *(B.E. Apr. 98)*

Solution The telescoping form of

$$g(x) = x^5 + 3x^4 + 3x^3 + 2x^2 + x$$
$$= (((((-1)\,x + 3)\,x + 3)\,x + 2)\,x + 1)\,x + 0.$$

RECURSION AND ITERATION

There is another way to evaluate a function from its recursive definition. Instead of successively reducing the computation to evaluate of the function at small integers, we can start with the basis and successively apply the recursive definition to find the values of the function at successive larger integers. Such a procedure is called iterative procedure.

Example To find $n!$ using an iterative procedure, we start with 1, the value of the factorial function at 0, and multiply successively by each positive integer less than or equal to n.

Note: Often an iterative approach for the evaluation of a recursively defined sequence requires much less computation than a procedure using recursion. But it is sometimes preferable to use a recursive procedure even if it is less efficient than the iterative procedure.

CLOSED FORM EXPRESSION

Let $f(x_1, x_2, ..., x_n)$ be an algebraic expression involving variables. $x_1, x_2, ..., x_n$ which are allowed to take values from some predetermined set. f is a closed form expression if there exists a number F such that evaluation of f with any allowed values $x_1, x_2 ..., x_n$ will take no more than F operations.

Example (i) A closed form expression for

$$1 + 2 + ... + n = \frac{n(n+1)}{2}$$

(ii) A closed form expression for

$$1^2 + 2^2 + ... + n^2 = \frac{n(n+1)(2n+1)}{2}.$$

Sequence of integers

A sequence of integers is a function from the natural numbers into integers. That is, if f is a sequence of integers, then $f: N \rightarrow Z$ is a function. We use the notation f_n or $f(n)$ to denote the image of any natural n. We call f_n an nth term of the sequence.

Note: A sequence is often called a discrete function.

Example 4.25 Prove by induction that $f(k) = 3k + 2$, $k \geq 0$ is a closed form expression of the sequence f defined recursively by $f(0) = 2$ and $f(k) = f(k-1) + 3$ for $k \geq 1$.

Solution

1. Basis of induction: If $k = 0$, then $f(0) = 3(0) + 2 = 2$, as defined.
2. Induction step: Now assume that for some $k > 0$,

$$f(k) = 3k + 2 \qquad (1)$$

we have,

$$f(k+1) = f(k) + 3 \text{ by the recursive definition}$$
$$= 3k + 2 + 3 = 3k + 3 + 2$$
$$= 3(k+1) + 2$$

Therefore, by the principle of mathematical induction the result follows. Hence $f(k) = 3k + 2$, $k \geq 0$ is a closed form expression for the sequence f.

Example 4. 26 Define the sequence of numbers f by $f_0 = 100$ and $f_n = 1.08\, f_{n-1}$ for, $n \geq 1$. Prove by induction that $f_n = 100\,(1.08)^n$, $n \geq 0$ is a closed form of expression for the sequence f.

Solution

1. Basis of induction: If $k = 0$, then $f_0 = 100(1.08)^0 = 100$, as defined.
2. Induction step: Now assume that for some $k > 0$

$$f_k = 100(1.08)^k$$

Then

$$f_{(k+1)} = (1.08) f_k \text{ by the recursive definition}$$

$$= (1.08)100^k$$

$$= 100(1.08)^{k+1} \qquad \text{by}(1)$$

Therefore, by the principle of mathematical induction the result follows. Hence $f_n = 100(1.08)^n$, $n \geq 0$ is a closed form expression for the sequence f.

Note:

i The sequence $f: N \to Z$ defined by $f(n) = n^3 - n$, is a sequence of integers.

ii. The sequence $f: N \to Z$ defined by $f(0) = 1$ and $f(n) = f(n-1) + 2$ for $n \geq 1$ is a sequence of integers.

iii. The codomain of a sequence can be any set. We use the notation $\{f_n\}$ to describe the sequence. We describe sequences by listing the terms of the sequence in the order of increasing subscripts.

Example

i. Consider the sequence $f: N \to S$ defined by $f(n) = f_n = \dfrac{1}{n+1}$. The list of the terms of this sequence $f_0, f_1, f_2, \ldots$ begins with

$$1, \frac{1}{2}, \frac{1}{3}, \frac{1}{4}, \ldots$$

ii. Consider the sequence $f: N \to Z$ defined by $f_n = 4^n$. The list of the terms of the sequence $f_0, f_1, f_2, \ldots$ begins with

$$1, 4, 16, 64, \ldots$$

EXERCISE II

1. Find $f(1)$, $f(2)$, $f(3)$ and $f(4)$ if $f(n)$ is defined recursively by $f(0) = 1$ and for $n = 0, 1, 2, \ldots$
   a. $f_{(n+1)} = f_{(n)} + 2$
   b. $f_{(n+1)} = 3f_{(n)}$
   c. $f_{(n+1)} = 2^{f(n)}$
   d. $f_{(n+1)} = [f(n)^2 + f(n) + 1]$
2. Give a recursive definition of the sequence $\{f_n\}$, $n = 1, 2, 3, \ldots$ if
   a. $f_n = 6n$
   b. $f_n = 2n + 1$
   c. $f_n = 10^n$
   d. $f_n = 5$
3. Let f be the function such that $f(n)$ is the sum of the first n positive integers. Give a recursive definition of $f(n)$.
4. Write $x^5 + 3x^4 - 15x^3 + 4x - 10$ in the telescoping form.

ANSWERS

1. a. 3,5,7,9 b. 3,9,27,81
   c. 2,4,16,65536 d. 3,13,183,33673
2. a. $f_{n+1} = f_{n+6}$ for $n \geq 1; f_1 = 6$ b. $f_{n+1} = f_{n+2}$ for $n \geq 1; f_1 = 3$
   c. $f_{n+1} = 10 f_n$ for $n \geq 1; f_1 = 10$ d. $f_{n+1} = f_n$ for $n \geq 1; f_1 = 5$.
3. $f(n) = f_{(n-1)} + n$ for $n \geq 1; f(0) = 0$
4. $(((((1)x+3)x-15)x+0)x+4)x-10$.

Recurrence Relations

The expressions for permutations and combinations are one of the most fundamental tools for counting the elements of finite sets. They often prove to be inadequate and many problems of computer science require a different approach. An important alternate approach uses recurrence relations (often called recurrence equations or difference equation) to define the terms of a sequence. A formal definition of recurrence relations is difficult because of the wide variety of forms in which such relations can be written, but the concept is straight forward. We have already seen an example of a recurrence relation in the definition of the Fibonacci sequence, where for $n \geq 2$, the term f_n is defined by the recurrence relation

$$f_n = f_{n-1} + f_{n-2}$$

The salient characteristic of a recurence relation is the specification of the term f_n as a function of the terms $f_0, f_1, \ldots, f_{n-1}$. By itself, however, a recurrence relation is not sufficient to define the terms of a sequence; we must also specify the values of some initial terms of the sequence. Thus in our definition of the Fibonacci sequence, we set $f_0 = 0$ and $f_1 = 1$. These are called the boundary conditions or initial conditions of the sequence.

Recall that a recursive definition of a sequence specifies one or more initial terms and a rule for determining subsequent terms from those that precede them. Recursive definitions can be used to solve counting problems. When they are, the rule for finding terms from those that precede them is called a recurrence relation.

Recurrence relation A recurrence relation for the sequence $\{f_n\}$ is a formula that express f_n in terms of one or more of the previous terms of the sequence, namely, $f_0, f_1, \ldots, f_{n-1}$, for all integers n with $n \geq n_0$, when n_0 is a non-negative integer. A sequence is called a solution of a recurrence relation if its terms satisfy the recurrence relation.

Example 4.27 Determine whether the sequence $\{f_n\}$ is a solution of the recurrence relation

$$f_n = 2f_{n-1} - f_{n-2}$$

for $n = 2, 3, 4, \ldots$ where $f_n = 3n$ for every non-negative integer n.

Solution Suppose that $f_n = 3n$ for every non-negative integer n, Then, for $n \geq 2$,

$$\begin{aligned} f_n &= 2f_{n-1} - f_{n-2} \\ &= 2[3(n-1)] - 3(n-2) \text{ since } f_n = 3n \\ &= 6n - 6 - 3n + 6 = 3n \end{aligned}$$

Therefore $\{f_n\}$, where $f_n = 3n$, is a solution of the recurrence relation.

Example 4.28 Show that the sequence $\{f_n\}$ is a solution of the recurrence relation $f_n = -3f_{n-1} + 4f_{n-2}$ if $f_n = 2(-4)^n + 3$.

Solution Suppose that $f_n = 2(-4)^n + 3$. Then

$$f_n = -3f_{n-1} + 4f_{n-2}$$

$$
\begin{aligned}
&= -3\,[2(-4)^{n-1}+3]+4\,[2(-4)^{n-2}+3]\\
&= -6(-4)^{n-1}-9+8(-4)^{n-2}+12\\
&= -6(-4)^{n-1}-2(-4)^{n-1}+3\\
&= -8(-4)^{n-1}+3\\
&= 2(-4)^{n}+3
\end{aligned}
$$

Therefore, $\{f_n\}$, where $f_n = 2(-4)^n + 3$, is a solution of the recurrence relation.

In the following, we shall study a class of recurrence relations known as linear recurrence relations with constant coefficients.

Linear recurrence relation A recurrence relation of the form

$$a_0 f_n + a_1 f_{n-1} + a_2 f_{n-2} + \ldots + a_k f_{n-k} = f(n) \quad (1)$$

Where a_1's are constants and is called a linear recurrence relation with constant coefficients. The recurrence relation (1) is known as a kth-order recurrence relation, provided that both a_0 and a_k are non zero.

Note: The phrase "kth-order" means that each term in the sequence depends only on the k previous terms.

Example Consider the Fibonacci sequence defined by the recurrence relation $f_n = f_{n-1} + f_{n-2}$, $n \geq 2$ and the initial condition $f_0 = 0$ and $f_1 = 1$. The recurrence relation is called a second-order relation because f_n depends on the two previous terms of f.

Example Consider the recurrence relation $f(k) - 5f(k-1) + 6f(k-2) = 4k + 10$ defined for $k \geq 2$, together with the initial condition $f(0) = \frac{7}{3}$ and $f(1) = 5$. Clearly, it is a second-order linear recurrence relation.

Homogeneous recurrence A kth-order linear relation is a homogeneous recurrence relation if $f(n) = 0$ for all n. Otherwise, it is called non-homogeneous.

Example Consider the recurrence relation $c(k) - 5c\,(k-1) + 8c\,(k-2) = 0$ together with the initial condition $c(0) = 5$ and $c(1) = 2$. It is a second-order homogeneous recurrence relation.

Example 4.29 Which of the recurrence relations of the following are homogenous and which are non-homogenous?

(i) $f_n = nf_{n-2}$ (ii) $a_n = a_{n-1} + a_{n-3}$ (iii) $b_n = b_{n-1} + 2$ (iv) $s\,(n-2) + s\,(n-4)$

Solution The relation $f_n = nf_{n-2}$, $a_n = a_{n-1} + a_{n-3}$, $s(n) = s(n-2) + s\,(n-4)$ are all homogeneous and the relation $b_n = b_{n-1} + 2$ is non-homogeneous.

Recurrence Relations Obtained from Solutions

Before giving an algorithm for solving a recurrence relations, we will examine a few recurrence relations that arise from certain closed form expressions. The procedure is illustrated by the following examples.

Example 4. 30 Form the recurrence relation given $f_n = 3.5^n, n \geq 0$.

Solution If $n \geq 1$, then

$$
\begin{aligned}
f_n &= 3.5^n = 3.5.5^{n-1}\\
&= 5.3.5^{n-1} = 5.f_{n-1}.
\end{aligned}
$$

So the recurrence relation is $f_n - 5f_{n-1} = 0$ with $f_0 = 3$.

Example 4. 31 Find the recurrence relation, satisfying $y_n = A(3)^n + B(-2)^n$. *(B.E. Oct. 96)*

Solution Given $y_n = A(3)^n + B(-2)^n$

Therefore, $y_{n+1} = A(3)^{n+1} + B(-2)^{n+1} = 3A(3)^n - 2B(-2)^n$

and $y_{n+2} = A(3)^{n+2} + B(-2)^{n+2} = 9A(3)^n + 4B(-2)^n$

Eliminating A and B from these equation, we get

$$\begin{vmatrix} y_n & 1 & 1 \\ y_{n+1} & 3 & -2 \\ y_{n+2} & 9 & 4 \end{vmatrix} = 0$$

(or) $y_{n+2} - y_{n+1} - 6y_n = 0$ which is the required recurrence relation.

Example 4. 32 Find the recurrence relation satifying $y_n = A(3)^n + B(-4)^n$. *(B.E. Oct. 97).*

Solution Given $y_{n+1} = A(3)^n + B(-4)^n$

Therefore, $y_{n+1} = 3A(3)^n - 4B(-4)^n$

and $y_{n+2} = 9A(3)^n + 16B(-4)^n$. Eliminating A and B from these equation, we get

$$\begin{vmatrix} y_n & 1 & 1 \\ y_{n+1} & 3 & -4 \\ y_{n+2} & 9 & 16 \end{vmatrix} = 0$$

(or) $y_{n+2} + y_{n+1} - 12y_n = 0$ which is the required recurrence relation.

Example 4. 33 Find the recurrence relation satisfing $y_n = (A+Bn)4^n$ *(B.E. Apr. 97).*

Solution Given $y_n = (A + Bn)4^n$

$= A4^n + nB4^n$

Therefore, $y_{n+1} = 4A4^n + 4(n+1)B4^n$

and $y_{n+2} = 16A4^n + 16(n+2)B4^n$.

Eliminating A and B from these equation, we get

$$\begin{vmatrix} y_n & 1 & n \\ y_{n+1} & 4 & 4(n+1) \\ y_{n+2} & 16 & 16(n+2) \end{vmatrix} = 0$$

(or) $y_{n+2} - 8y_{n+1} - 16y_n = 0$, which is the required recurrence relation.

Solving Linear Homogeneous Recurrence Relations

Consider a linear homogeneous recurrence relations of degree k with constant coefficients

$$f_n = a_1 f_{n-1} + a_2 f_{n-2} + \ldots + a_k f_{n-k}$$

Where $a_1, a_2, \ldots, a_k$ are real numbers and $ak \neq 0$. The basic approach for solving linear homogeneous recurrence relations is to look for solutions of the form $f_n = r^n$, where r is a constant. Note that $f_n = r^n$ is solution of the recurrence relation $f_n = a_1 f_{n-1} + a_2 f_{n-2} + \ldots + a_k f_{n-k}$ if and only if

$$r^n = C_1 r^{n-1} + C_2 r^{n-2} + \ldots + C_k r^{n-k}$$

when both sides of this equation are divided by r^{n-k} and the right-hand side is substracted from the left, we obtain

$$r^k - C_1 r^{k-1} - C_2 r^{k-2} - \ldots - C_{k-1} T - C_k = 0$$

consequently, the sequence $\{f_n\}$ with $f_n = r^n$ is a solution if and only if r is a solution of this last equation,

Characteristic equation The characteristic equation of the homogeneous k^{th} order linear relation

$$f_n + a_1 f_{n-1} + a_2 f_{n-2} + \ldots + a_k f_{n-k} = 0$$

is the kth degree polynomial equation

$$r^k + C_1 r^{k-1} + C_2 r^{k-2} + \cdots + C_{k-1} r + C_k = 0$$

The solutions of this equation are called the characteristic roots of the recurrence relation.

Example 4. 34 What is the characteristic equation of

$$Q(k) + 2Q(k-1) - 3Q(k-2) - 6Q(k-4) = 0 \qquad (B.E.\ Oct\ 98).$$

Solution The characteristic equation of

$$Q(k) + 2Q(k-1) - 3Q(k-2) - 6Q(k-4) = 0$$

is $r^4 + 2r^3 - 3r^2 - 6 = 0$.

Note that the absence of a $Q(k-3)$ term means that there is no $r^{4-3} = r$ term in the characteristic equation.

Example 4. 35 What is the characteristic equation of $T(k) - 7T(k-2) + 6T(k-3) = 0$

The characteristic equation is $r^3 - 7r + 6 = 0$ and 1,2 and -3 are the characteristic roots.

Algorithm for solving k^{th} order homogeneous linear recurrence relation.

Step 1. If $f_n + a_1 f_{n-1} + a_2 f_{n-2} + \ldots + a_k f_{n-k} = 0$ is a given recurrence relation, then write its characteristic equation. That is, $r^k + C_1 r^{k-1} + C_2 r^{k-2} + \ldots + C_{k-1} r + C_k = 0$

Step 2. Find all the characteristic roots of this equation.

Step 3. *Case* (i) If there are k distinct roots, say $c_1, c_2, \ldots, c_k$ then the general solution of the recurrence relation is

$$f_n = A_1 c_1^{\,k} + A_2 c_r^{\,k} + \ldots + A_k c^k_{\,k}$$

Case (ii) Suppose that c_1 is a root of multiplicity m. Then the corresponding solution is

$$f_n = (A_1 r^{m-1} + A_2 r^{m-2} + \ldots + A_{m-2} r^2 + A_{m-1}{}^T + A_m) c^T{}_1 + \ldots.$$

Step 4. Use the boundary conditions to determine the constants $A_1, A_2, \ldots, A_k$.

Example 4.36 Solve the Fibonacci sequence $\{f_n\}$ defined by

$$f_n = f_{n-1} + f_{n-2} \text{ for } n \geq 2 \text{ with the initial conditions } f_0 = 1 \text{ and } f_1 = 1.$$

Solution The first step is to form the characteristic equation corresponding to the given difference equation. In this case it is

$$r^2 - r - 1 = 0$$

Solving, we get $c_1 = \dfrac{1+\sqrt{5}}{2}$,and $c_2 = \dfrac{1-\sqrt{5}}{2}$ the characteristic roots. It follows that the general solution is

$$f_n = A_1 c_1^{\,m} + A_2 c_2^{\,m}$$

where A_1 and A_2 are constants. Since $f_0 = 1$ and $f_1 = 1$, we get

$$0 = A_1 + A_2$$

and

$$1 = A_1 c_1 + A_2 c_2$$

$$1 = A_1\left(\frac{1+\sqrt{5}}{2}\right) + A_2\left(\frac{1-\sqrt{5}}{2}\right)$$

On solving, we get

$$A_1 = \frac{1}{\sqrt{5}} \text{ and } A_2 = \frac{-1}{\sqrt{5}}$$

Hence the solution is

$$f_n = \frac{1}{\sqrt{5}}\left[\left(\frac{1+\sqrt{5}}{2}\right)^n - \left(\frac{1-\sqrt{5}}{2}\right)^n\right]$$

Example 4.37 If the recurrence relation is $u_{n+1} - 2u_n = 0$, find the closed form expression (solution) for u_n. (*B.E. Oct.* 98)

Solution The characteristic equation of the given recurrence relation is $r-2 = 0$, i.e., $r = 2$. Therefore, the general solution is $u_n = A2^n$. Hence $u_n = u_0 \cdot 2^n$ is the closed form expression, where the value of u_0 is the initial condition.

Example 4. 38 Find $f(n)$ if $f(n) = 7f(n-1) - 10f(n-2)$, given that $f(0) = 4$ and $f(1) = 17$ (*B.E April* 98)

Solution The characteristic equation of the given recurrence relation is $r^2 - 7r + 10 = 0$

It's characteristic roots are $r = 2, 5$. So the general solution of the recurrence relation is $f(n) = A_1(2)^n A_2(5)^n$.

Since $f(0) = 4$, $4 = A_1 + A_2$

Again, $f(1) = 17$ implies $17 = 2A_1 + 5A_2$

Solving, we get $A_1 = 1$ and $A_2 = 3$. Therefore, $f(n) = (2)^n + 3(5)^n$

Example 4. 39 Find $T(k)$ if $T(k) - 7T(k-2) + 6T(k-3) = 0$, where $T(0) = 8$, $T(1) = 6$ and $T(2) = 22$.

Solution The characteristic equation is $r^3 - 7r + 6 = 0$

Its roots are 1,2 and -3. Therefore, the general solution is $T(k) = A_1(1)^k + A_2(2)^k + A_3(-3)^k$

Now

$$T(0) = 8 \Rightarrow A_1 + A_2 + A_3 = 8$$
$$T(1) = 6 \Rightarrow A_1 + 2A_2 - 3A_3 = 6$$
$$T(2) = 22 \Rightarrow A_1 + 4A_2 + 9A_3 = 22$$

Solving, we get $A_1 = 5$, $A_2 = 2$ and $A_3 = 1$

Hence,

$$T(k) = 5 + 2(2)^k + 1(-3)^k$$
$$= 5 + 2^{k+1} + (-3)^k$$

Example 4.40 Solve $f_k - 8f_{k-1} + 16f_{k-2} = 0$ where $f_2 = 16$ and $f_3 = 80$.

Solution The characteristic equation is $r^2 - 8r + 16 = 0$ (or) $(r-4)^2 = 0$

So $r = 4$ is a double characteristic root. Therefore, the general solution is $f_k = (A_1 + A_2 k)4^k$

Now, $\quad f_2 = 16 \Rightarrow (A_1 + 2A_2)16 = 16$

and $\quad f_3 = 80 \Rightarrow (A_1 + 3A_2)64 = 80$

Solving, we get $A_1 = \frac{1}{2}$ and $A_2 = \frac{1}{4}$

Hence the solution is

$$f_k = \left(\frac{1}{2} + \frac{1}{4}k\right)4k$$
$$= (2+k)4^{k-1}$$

Example 4. 41 Find a solution to the recurrence relation $C_n = -3C_{n-1} - 3C_{n-2} - C_{n-3}$ for $n \geq 3$ with initial conditions $C_0 = 1$, $C_1 = -2$ and $C_2 = 1$.

Solution The characteristic equation is $r^3 + 3r^2 + 3r + 1 = 0$ (or) $(r+1)^3 = 0$.

So $r = -1$ is a characteristic root of multiplicity 3. Therefore, the general solution is

$$C_n = (A_1 + A_2 n + A_3 n^2)(-1)^n$$

Now,
$$C_0 = 1 \Rightarrow A_1 = 1$$
$$C_1 = -2 \Rightarrow -(A_1 + A_2 + A_3) = -2$$
$$C_2 = 1 \Rightarrow A_1 + 2A_2 + 4A_3 = 1$$

Solving, we get $A_1 = 1$, $A_2 = 2$, $A_3 = 1$. Hence the solution is $C_n = (1 + 2n - n^2)(-1)^n$.

Solving Linear Non Homogeneous Recurrence Relations

The solution of a linear non homogeneous recurrence relation with constant coefficient is the sum of the two parts, the homogeneous solution, which satisfies the recurrence relation when the right hand side of the equation is set 1to 0, and the particular solution, which satisfies the difference equation with $f(n)$ on the right hand side.

There is no general procedure for determining the particular solution of a difference equation. However, in simple cases, this solution can be obtained by the method of inspection. To determine the particular solution we use the following rules.

***Rule* 1**. When $f(n)$ is of the form of a polynomial of degree m in n

$$k_0 + k_1 n + k_2 n^2 + \ldots + k_{m-1} n^{m-1} + k_m n^m$$

the corresponding particular solution will be of the form

$$Q_0 + Q_1 n + Q_2 n^2 + \ldots + Q_{m-1} n^{m-1} + Q_m n^m$$

***Rule* 2**. When $f(n)$ is of the form

$$(k_0 + k_1 n + k_2 n^2 + \ldots + k_{m+1} n^{m-1} + k_m n^m)a^n$$

the corresponding particular solution is of the form

$$(Q_0 + Q_1 n + Q_2 n^2 + \ldots + Q_{m-1} n^{m-1} + Q_m n^m)a^n$$

if a is not a characteristic root of the recurrence relation.

Example 4.42 Solve $S(k) - S(k-1) - 6S(k-2) = -30$, where $S(0) = 20$, $S(1) = -5$. (*B.E. Oct.* 98)

Solution The associated homogeneous relation is $S(k) - S(k-1) - 6S(k-2) = 0$.

The characteristic equation is $r^2 - r - 6 = 0$.

Its characteristic roots are $r = -2, 3$. So the homogeneous solution is $A_1(-2)^k + A_2(3)^k$.

Since the right hand side of $S(k) - S(k-1) - 6S(k-2) = -30$ (1)

is a constant, by Rule 1, the particular solution will also be a constant, say Q into (1), we obtain

$$Q - Q - 6Q = -30 \Rightarrow -6Q = -30$$
$$\Rightarrow Q = 5$$

Therefore, the general solution is $S(k) = A_1(-2)^k + A_2(3)^k + 5$

Using the initial conditions, $S(0)=20$, $S(1)=-5$

$$20 = A_1 + A_2 + 5$$
$$-5 = -2A_1 + 3A_2 + 5$$

which yield, $A_1 = 11$ and $A_2 = 4$. Hence the complete solution is $S(k) = 11(-2)^k + 4(3)^k + 5$.

***Rule* 3.** If a is a characteristic root of multiplicity r-1, when $f(n)$ is of the form

$$(k_0 + k_1 n + k_2 n^2 + ... + k_{m-1} n^{m-1} + k_m n^m)a^n$$

the corresponding particular solution is of the form

$$n^{r-1}(Q_0 + Q_1 n + Q_2 n^2 + ... + Q_{m-1} n^{m-1} + Q_m n^m)a^n.$$

Note: The general solution of the recurrence relation is the sum of the homogeneous solution and particular solution. If no initial conditions are given, then you have finished. If m initial conditions are given, obtain m linear equations in m unknowns and solve the system, if possible, to get a complete solution.

Example 4.43 Solve the recurrence relation $f_n - 5f_{n-1} + 6f_{n-2} = 1$. (1)

Solution The associated homogeneous relation is $f_n - 5f_{n-1} + 6f_{n-2} = 0$.

The characteristic equation is $r^2 - 5r + 6 = 0$

Its characteristic roots are $r = 2, 3$ so the homogeneous solution is

$$A_1(2)^n + A_2(3)^n \quad (2)$$

Since the right hand side $f(n) = 1$ is a constant, by Rule 1, the particular solution will also be a constant, say Q. Substituting Q into (1), we obtain

$$Q - 5Q + 6Q = 1$$

This implies that $Q = \frac{1}{2}$. Therefore, the complete solution

$$f_n = A_1(2)^n + A_2(3)^n + \frac{1}{2}.$$

Example 4.44 Find the particular solution of the recurrence relation

$$f(n) + 5f(n-1) + 6f(n-2) = 3n^2 - 2n + 1 \quad (1)$$

Solution By Rule 1, the particular solution is of the form

$$Q_0 + Q_1 n + Q_2 n^2 \quad (2)$$

Substituting (2) into (1), we obtain

$$(Q_0 + Q_1 n + Q_2 n^2) + 5(Q_0 + Q_1(n-1) + Q_2(n-1)^2)$$
$$+ 6(Q_0 + Q_1(n-2) + Q_2(n-2)^2) = 3n^2 - 2n + 1$$

which simplifies to

$$(12Q_0 - 17Q_1 + 29Q_2) + (12Q_1 - 34Q_2)n + 12Q_2 n^2 = 3n^2 - 2n + 1 \quad (3)$$

Comparing the two sides of (3), we obtain

$$12Q_2 = 3; \quad 12Q_1 - 34Q_2 = -2; \quad 12Q_0 - 17Q_1 + 29Q_2 = 1$$

which yields $Q_2 = \frac{1}{4}$, $Q_1 = \frac{13}{24}$, $Q_0 = \frac{71}{288}$. Therefore, the particular solution is $\frac{71}{288} + \frac{13}{24}n + \frac{1}{4}n^2$.

Example 4. 45 Solve $a_r + 5a_{r-1} = 9$ with initial condition $a_0 = 6$.

Solution The associated homogeneous recurrence relation is $a_r + 5a_{r-1} = 0$.

It's characteristic equation is $r + 5 = 0$.

Therefore, $r = -5$. So the homogeneous solution is $A_1(-5)^r$. Since the right hand side of the given relation is a constant, the particular solution will also be a constant Q. Substituting in the recurrence relation, we get

$$Q + 5Q = 9$$

So $Q = \frac{3}{2}$. Therefore the general solution is $a^r = A_1(-5)^r + \frac{3}{2}$. Using the initial condition $a_0 = 6$, we get

$$6 = A_1 + \frac{3}{2} \Rightarrow A_1 = \frac{9}{2}$$

Hence the complete solution is

$$a_r = \frac{9}{2}(-5)^r + \frac{3}{2}$$

Example 4. 46 Solve the recurrence relation $f(n) - 7f(n-1) + 10f(n-2) = 6 + 8n$ with $f(0) = 1$ and $f(1) = 2$. (1)

Solution The homogeneous solution is $A_1(2)^n + A(2)^n + A_2(5)^n$. By rule 1, the particular solution is of the form $Q_0 + Q_1{}^n$.

Substituting in (1), we get

$$(Q_0 + Q_1{}^n) - 7(Q_0 + Q_1(n-1)) + 10(Q_0 + Q_1(n-2)) = 6 + 8n$$

Comparing the two sides, we obtain

$$4Q_0 - 13Q_1 = 6 \text{ and } 4Q_1 = 8$$

which yields, $Q_0 = 8$, $Q_1 = 2$. Therefore, the particular solution is $8 + 2n$ and the general solution is

$$f(n) = A_1(2)^n + A_2(5)^n + 8 + 2n$$

Now the initial conditions,

$$f(0) = 1 \Rightarrow A_1 + A_2 + 8 = 1$$

and

$$f(1) = 2 \Rightarrow 2A_1 + 5A_2 + 10 = 2$$

Solving, we get $A_1 = -9$ and $A_2 = 2$. Hence the complete solution is

$$f(n) = -9(2)^n + 2(5)^n + 8 + 2n$$

Example 4.47 Find the particular solution of the recurrence relation $a_n + 5a_{n-1} + 6a_{n-2} = 42(4)^n$. (1)

Solution Now $r^2 + r + 6 = 0$, $r = -2, -3$ are characteristic roots. Since 4 is not a characteristic root, by Rule 2, we assume that the general form of the particular solution is $Q.(4)^n$. Substituting in (i), we obtain

$$Q.(4)^n + 5Q.(4)^{n-1} + 6Q.(4)^{n-2} = 42(4)^n$$
$$\Rightarrow Q.4^{n-2}[16 + 20 + 6] = 42(4)^n \Rightarrow Q.4^{n-2}(42) = 42(4)^n$$
$$\Rightarrow Q = 4^2 = 16$$

Therefore, the particular solution is $16(4)^n = 4^{n+2}$.

Example 4.48 Find the particular solution of the recurrence relation $f_n + f_{n-1} = 3n2^n$ (1)

Solution The characteristic equation is $r + 1 = 0$. Therefore $r = -1$ is a characteristic root. Since 2 is not a characteristic root, by Rule 2, the general form of the particular solution is $(Q_0 + Q_1 n)2^n$. Substituting in (1), we obtain

$$(Q_0 + Q_1 n)2^n + (Q_0 + Q_1(n-1))2^{n-1} = 3n2^n$$

which simplifies to

$$Q_0 2^n + Q_1 n2^n + \frac{1}{2}Q_0 2^n + \frac{1}{2}Q_1 n2^n - \frac{1}{2}Q_1 n2^n = 3n2^n$$

$$\Rightarrow (\frac{3}{2}Q_0 - \frac{1}{2}Q_1)2^n + \frac{3}{2}Q_1 n2^n = 3n2^n$$

Comparing the two sides, we obtain

$$\frac{3}{2}Q_0 - \frac{1}{2}Q_1 = 0$$
$$\frac{3}{2}Q_1 = 3$$

Thus, $Q_0 = \frac{2}{3}$ and $Q_1 = 2$ and the particular solution is $(\frac{2}{3} + 2n)2^n$.

Example 4. 49 Find the particular solution of the recurrence relation $f(n) - 2f(n-1) = 3.2^n$ (1)

Solution The characteristic equation is $r - 2 = 0$. Since $r = 2$ is the characteristic root of multiplicity 1, by Rule (3), the general form of the particular solution is $Q_n 2^n$. Substituting in (1), we obtain

$$Qn2^n - 2Q.(n-1)2^{n-1} = 3.2^n$$
$$\Rightarrow Q \cdot 2^n = 3 \cdot 2^n$$
$$\Rightarrow Q = 3$$

Thus the particular solution is $3n2^n$.

Example 4. 50 Find the general solution of $f(n) - 3f(n-1) - 4f(n-2) = 4^n$ (1)

Solution The associated homogeneous relation is

$$f(n) - 3f(n-1) - 4f(n-2) = 0.$$

Its characteristic equation is $r^2 - 3r - 4 = 0$.

Solving, we get $r = -1, 4$ the characteristic roots. Therefore, the homogeneous solution is $A_1(-1)^n + A_2(4)^n$. Since 4 is a characteristic root, by Rule 3, we assume that the general form of the particular solution is $Qn4^n$.

Substituting in (1), we get

$$Qn4^n - 3Q(n-1)4^{n-1} - 4Q(n-2)4^{n-2} = 4^n$$
$$\Rightarrow Qn4^n - 3Qn4^{n-1} + 3Q4^{n-1} - 4Qn4^{n-2} + 8Q4^{n-2} = 4^n$$
$$\Rightarrow (16Qn - 12Qn + 12Q - 4Qn + 8Q)4^{n-2} = (16)4^{n-2}$$
$$\Rightarrow 20Q = 16 \Rightarrow Q = \frac{4}{5}$$

Therefore, the particular solution is $\frac{4}{5}n4^n$. Hence, the general solution of the recurrence relation of

$$f(n) = A_1(-1)^n + A_2(4)^n + \frac{4}{5}n4^n$$

Note: What if the characteristic equation gives rise to complex roots? Here, our methods are still valid, but the method for expressing the solutions of the recurrence relations is different. Since an understanding of these representations require some background in complex numbers, we suggest that an interested reader refer to a more advanced treatment of recurrence relations.

EXERCISE III

1. Find the general solution of the following recurrence relations.

a. $f_n - 3f_{n-1} - 10f_{n-2} = 0$

b. $f_{n+2} + 6f_{n+1} + 9f_n = 0$

c. $2f_n + 2f_{n-1} - f_{n-2} = 0$

d. $f_{n+2} - 3f_{n-1} - 4f_{n-2} = 0.$

2. Solve the following sets of recurrence relations and initial conditions.

   a. $f(n) - 10f(n-1) + 9f(n-2) = 0; f(0) = 3, f(1) = 11$

   b. $f(n) - 9f(n-1) + 18f(n-2) = 0; f(0) = 1, f(1) = 4$

   c. $f(n+2) - 8f(n+1) + 16f(n) = 0; f(0) = 0, f(1) = 8$

   d. $f(n) - 3f(n-1) + 3f(n-2) - f(n-3) = 0; f(1) = 0, f(2) = 1, f(3) = 0$

   e. $f(n+2) - 2f(n+1) + f(n) = 0; f(0) = 1, f(1) = 2$

   f. $f(n) - 20f(n-1) + 100f(n-2) = 0; f(0) = 2, f(1) = 30$

3. Find the recurrence relation satisfying

   a. $y_n = A(3)^n + B(8)^n$

   b. $y_n = (A + Bn)(-2)^n$

   c. $y_n = (A + Bn)(6)^n$

   d. $y_n = A(3)^n + B(5)^n$ *(B.E. Oct. 98)*

   e. $y_n = 2(3)^n$

4. Solve the following sets of recurrence relations and initial conditions.

   a. $y_n - 2y_{n-1} = 6n; y_1 = 2$

   b. $y_{n+2} + 2y_{n+1} - 15y_n = 6n + 10; y_0 = 1, y_1 = -\frac{1}{2}$

   c. $y_{n+1} + 2y_n = 3 + 4^n; y_0 = 2$

   d. $y_{n+2} - 2y_{n+1} + y_n = 1; y_0 = 1, y_1 = \frac{1}{2}$

   e. $y_{n+1} + y_n = 5; y_0 = 1$

   f. $y_n - 3y_{n-1} + 2y_{n-2} = n^2; y_0 = 0, y_1 = 0$

   g. $y_n - 4y_{n-1} + 4y_{n-2} = 3n + 2^n; y_0 = 1, y_1 = 1$

   h. $y_n - 5y_{n-1} = 5^n; y_0 = 3$

ANSWERS

1. a. $f_n = A_1(5)^n + A_2(-2)^n$

   b. $f_n = A_1(-3)^n + A_2 n(-3)^n = (A_1 + nA_2)(-3)^n$

   c. $f_n = A_1[(-1+\sqrt{3})/2]^n + A_2[(-1-\sqrt{3})/2]^n$

   d. $f_n = A_1(4)^n + A_2(-1)^n$

2. a. $f(n) = 2 + (9)^n$

   b. $f(n) = \left(\frac{2}{3}\right)(3)^n + \left(\frac{1}{3}\right)(6)^n$

   c. $f(n) = 2n(4)^n$

   d. $f(n) = -3 + 4n - n^2$

   e. $f(n) = 1 + n$

   f. $f(n) = (2 + n)(10)^n$

3. a. $y_{n+2} - 11y_{n+1} + 24y_n = 0$

   b. $y_{n+2} + 4y_{n+1} + 4y_n = 0$

   c. $y_{n+2} - 12y_{n+1} + 36y_n = 0$

   d. $y_{n+2} - 8y_{n+1} + 15y_n = 0$

   e. $y_n - 3y_{n-1} = 0$

4. a. $y_n = 10(2)^n - 6n - 12.$

   b. $y_n = \left(\frac{5}{8}\right)(-5)^n + \left(\frac{11}{8}\right)(3)^n - \left(\frac{1}{2}\right)n - 1.$

   c. $y_n = \left(\frac{5}{6}\right)(-2)^n + 1 + \left(\frac{1}{6}\right)(4)^n.$

   d. $y_n = 1 - n + \left(\frac{1}{2}\right)n^2.$

   e. $y_n = \left(\frac{-3}{2}\right)(-1)^n + \frac{5}{2}.$

   f. $y_n = 11(2)^n - 11 - \left(\frac{1}{3}\right)n^3 - \left(\frac{5}{2}\right)n^2 - \left(\frac{49}{6}\right)n.$

   g. $y_n = 12 + 3n + (n^2 + 7n + 22)2^{n\ 1}.$

   h. $y_n = (3 + n)5^n.$

GENERATING FUNCTIONS

In this section we will show how recurrence relations can be solved using the powerful generating function method. Generating functions are an important tool in discrete mathematics, and their use is by no means confined to the solution of recurrence relations.

If $a_0, a_1, a_2, ..., a_n$ is a finite sequence of numbers, the generating function for the a_n's is the polynomial

$$G(z) = \sum_{k=0}^{n} a_k z^k = a_0 + a_1 z + a_2 z^2 + ... + a_k z^k$$

where z is an indeterminate (ie. an abstract symbol). If $a_0, a_1, a_2, ...$ is an infinite sequence of numbers, its generating function is defined to be

$$G(z) = \sum_{k=0}^{n} a_k z^k = a_0 + a_1 z + a_2 z^2 + ...$$

The symbol z is just the name given to a variable, and has no special significance. For any sequence $\{a_n\}$, we write $G(z)$ to denote the generating function of $\{a_n\}$. Clearly, given a sequence, we can easily obtain its generating function and its converse. For example, the generating function of $a_n = \alpha_n$, $n \geq 0$ is

$$\alpha^0 + \alpha z + \alpha^2 z^2 + \alpha^3 z^3 + ... \quad (1)$$

We note that the infinite series in (1) can be written in closed form as $\frac{1}{1-\alpha z}$ which is a rather compact way to represent the sequence $\{a_n\}$ or $(1, \alpha, \alpha^2, ...)$.

Example 4.51 Find the generating function for the sequence 1, 1, 1, 1, 1, 1.

Solution By definition, the generating function of 1, 1, 1, 1, 1, 1 is

$$G(z) = 1 + z + z^2 + z^3 + z^4 + z^5 = \frac{z^6 - 1}{z - 1}$$

Example 4.52 Let n be a positive integer. Let $a_k = C(n, k)$, for $k = 0, 1, 2, ..., n$. Find the generating function for the sequence a_0, a_1, a_n.

Solution The generating function for this sequence is

$$G(z) = C(h, 0) + C(h, 1)z + C(h, 2)z^2 + ... + C(n, n)\, z^n$$
$$= (1+z)^n, \text{ by binomial theorem.}$$

Example 4.53 Find the generating function for the infinite sequence $1, \alpha, \alpha^2, \alpha^3, ...$ where α is a fixed constant.

Solution The generating function for this sequence is

$$G(z) = 1 + \alpha z + \alpha^2 z^2 + \alpha^3 z^3 + ... \quad (1)$$

Now we can derive a closed-form expression for $G(z)$. From (1),

$$G(z) = -1 = \alpha z + \alpha^2 z^2 + \alpha^3 z^3 + ...$$

$$\Rightarrow \frac{G(z) - 1}{\alpha z} = 1 + \alpha z + \alpha^2 z^2 + ... = G(z) \quad \text{by (1)}$$

$$\Rightarrow G(z) - 1 = \alpha z G(z) \Rightarrow G(z) - \alpha z G(z) = 1$$

$$\Rightarrow G(z) = \frac{1}{1 - \alpha z} \text{ (closed form)}$$

Therefore, the required generating function is $\frac{1}{1-\alpha z}$

Example 4.54 Find the generating function for $f_n = 3^n, n \geq 0$ in closed form.

Solution The generating function for this sequence is

$$G(x) = 1 + 3z + (3z)^2 + (3z)^3 + ... \quad (1)$$

$$\Rightarrow G(z) - 1 = 3z + (3z)^2 + (3z)^3 + ...$$

$$\Rightarrow \frac{G(z)-1}{3z} = 1 + 3z + (3z)^2 + ... = G(z) \quad \text{by (1)}$$

$$\Rightarrow G(z) = \frac{1}{1-3z}$$

Therefore, the generating function in closed form is $\frac{1}{1-3z}$

Example 4.55 Find the generating function (in closed form) of the fibonacci sequence $\{f_n\}$ defined by $f_n = f_{n-1} + f_{n-2}; f_0 = 0, f_1 = 1$. *(B.E. April. 98).*

Solution The generating function is

$$G(z) = f_0 + f_1 z + f_2 z^2 + f_3 z^3 + ... = \sum_{n=0}^{\infty} f_n z^n \quad (1)$$

Consider, $f_n = f_{n-1} + f_{n-2}, n \geq 0$

Multiply both sides by z^n, and sum over all $n \geq 2$, we get

$$\sum_{n=2}^{\infty} f_n z^n = \sum_{n=2}^{\infty} f_{n-1} z^n + \sum_{n=2}^{\infty} f_{n-2} z^n \quad (2)$$

Consider the first sum

$$\sum_{n=2}^{\infty} f_n z^n = f_2 z^2 + f_3 z^3 + ...$$

$$= G(z) - f_0 - f_1 z \text{ from (1)}$$

Similarly,

$$\sum_{n=2}^{\infty} f_{n-1} z^n = f_1 z^2 + f_2 z^3 + ...$$

$$= z(f_1 z + f_2 z^2 + ...)$$

$$= z(G(k) - f_0) \text{ from (1)}$$

and

$$\sum_{n=2}^{\infty} f_{n-2} z^n = f_0 z^2 + f_1 z^3 + f_2 z^4 + ...$$

$$= z^2(f_0 + f_1 z + f_2 z^2 + ...)$$

$$= z^2 G(z)$$

Substituting these three expressions in (2), we get

$$G(z) - f_0 - f_1 z = z(G(z) - f_0) + z^2 G(z)$$

Since $f_0 = 0$ and $f_1 = 1$, we get

$$G(z) - z = zG(z) + z^2 G(z) \Rightarrow G(z)[1 - z - z^2] = z$$

$$\Rightarrow G(z) = \frac{z}{1 - z - z^2}$$

Therefore, the required generating function is $G(z) = \frac{z}{1 - z - z^2}$

Example 4.56 Find the generating function of the sequence $y_0, y_1, y_2, \ldots$ defined as follows:

$$y_n + 2y_{n-1} - 15y_{n-2} = 0 \text{ for } n \geq 2 \quad (1)$$

with $y_0, y_1 = 1$.

Solution The generating function is

$$G(z) = y_0 + y_1 z + y_2 z^2 + \ldots + = \sum_{n=2}^{\infty} y_n z^n \quad (2)$$

Multiply (1) by z^n and sum over all $n \geq 0$, we get

$$\sum_{n=2}^{\infty} y_n z^n + 2\sum_{n=2}^{\infty} y_{n-1} z^n - 15\sum_{n=2}^{\infty} y_{n-2} z^n = 0$$

$$\Rightarrow (G(z) - y_0 - y_1{}^2) + 2[z(G(z) - y_0)] - 15z^2 G(z) = 0$$

Since $y_0 = 0, y_1 = 1$, we get

$$(G(z) - z) + 2[zG(z) - 15z^2 G(z)] = 0 \Rightarrow G(z)(1 + 2z - 15z^2) = z$$

$$\Rightarrow G(z) = \frac{z}{1 + 2z - 15z^2} = \frac{z}{(1+3z)(1+5z)}$$

Solution of Recurrence Relation using Generating Function

We can find the solution to a recurrence relation and its initial conditions by finding an explicit formula for the associated generating function. This is illustrated in the following examples.

Example 4. 57 Using generating function solve the recurrence relation $f_n = 3f_{n-1}$ for $n = 1, 2, 3, \ldots$ and initial condition $f_0 = 2$.

Solution Let the generating function for the sequence be

$$G(z) = \sum_{n=0}^{\infty} f_n z^n \quad (1)$$

Multiply $f_n = 3f_{n-1}$ by z^n and sum for all $n \geq 1$, we get

$$\sum_{n=2}^{\infty} f_n z^n = 3\sum_{n=1}^{\infty} f_{n-1} z^n \quad (2)$$

The first sum

$$\sum_{n=1}^{\infty} f_n z^n = f_1 z + f_2 z^2 + \ldots = G(z) - f_0 \quad \text{using (1)}$$

and the second sum

$$\sum_{n=1}^{\infty} f_{n-1} z^n = f_0 z + f_1 z^2 + \ldots = z(f_0 + f_1 z + \ldots)$$

$$= zG(z) \quad \text{using (2)}$$

So (2) becomes,

$$G(z) - f_0 = 3zG(z)$$

Since $f_0 = 2$,

$$G(z) - 2 = 3zG(z)$$

$$\Rightarrow G(z) = \frac{2}{1-3z} = 2(1-3z)^{-1} \quad (3)$$

$$= 2\sum_{n=0}^{\infty} 3^n z^n = \sum_{n=0}^{\infty} 2 \cdot 3^n z^n$$

Comparing (1) and (2), we get

$$f_n = 2.3^n$$

which is the required solution.

Note: The following are some important results.

1. $(1+x)^n = \sum_{r=0}^{\infty} \frac{n(n-1)...(n-r+1)}{r!} x^r$

2. $(1+x)^{-n} = \sum_{r=0}^{\infty} (-1)^r \frac{n(n-1)...(n-r+1)}{r!} x^r$

3. $(1+x)^{-1} = \sum_{r=0}^{\infty} (-1)^r x^r = \sum_{r=0}^{\infty} (-x)^r$

4. $(1+x)^{-2} = \sum_{r=0}^{\infty} (-1)^r (r+1)x^r = \sum_{r=0}^{\infty} (r+1)(-x)^r$

5. $(1-x)^{-1} = \sum_{r=0}^{\infty} (x)^r$

6. $(1-x)^{-2} = \sum_{r=0}^{\infty} (r+1)x^r$

7. $e^x = \sum_{r=0}^{\infty} \frac{x^r}{r!}$

Example 4. 58 Using generating function, solve the recurrence relation

$$y_n = 3y_{n-1} + 2; n \geq 1 \text{ with } y_0 = 1$$

Solution Let the generating function for the required sequence be

$$G(z) = \sum_{n=0}^{\infty} y_n z^n \quad (1)$$

$$\text{given } y_n = 3y_{n-1} + 2, \quad (2)$$

Multiply both sides of (2) by z^n and sum for $n \geq 1$, we get

$$\sum_{n=1}^{\infty} y_n z^n = 3\sum_{n=1}^{\infty} y_{n-1} z^n + 2\sum_{n=1}^{\infty} z^n \quad (3)$$

Consider the first sum,

$$\sum_{n=1}^{\infty} y_n z^n = y_1 z + y_2 z^2 + ... = G(z) - y_0 \quad \text{Using (1)}$$

Second sum

$$\sum_{n=1}^{\infty} y_{n-1} z^n = y_0 z + y_1 z^2 + ... = z(y_0 + y_1 z + ...)$$

$$= zG(z)$$

Third sum

$$\sum_{n=1}^{\infty} z^n = z + z^2 + z^3 + ... = z(1 + z + z^2 + ...)$$

$$= z(1-z)^{-1}$$

$$= \frac{z}{1-z}$$

So (3) becomes,

$$G(z) - y_0 = 3zG(z) + \frac{2z}{1-z}$$

Using $y_0 = 1$,

$$G(z)(1-3z) = 1 + \frac{2z}{1-z} = \frac{1+z}{1-z}$$

Let

$$G(z) = \frac{1+z}{(1-z)(1-3z)}$$

Using partial fraction $\dfrac{1+z}{(1-z)(1-3z)} = \dfrac{A}{1-3z} + \dfrac{B}{1-z}$

$\Rightarrow 1 + z = A(1-z) + B(1-3z) = (A+B) - (A+3B)z$

$\Rightarrow A + B = 1$ and $A + 3B = -1 \Rightarrow A = 2$ and $B = -1$

Therefore,

$$G(z) = \frac{2}{1-3z} - \frac{1}{(1-z)} = 2(1-3z)^{-1} - (1-z)^{-1}$$

$$= 2\sum_{n=0}^{\infty}(3z)^n - \sum_{n=0}^{\infty} z^n \qquad \text{by note (5)}$$

$$= 2\sum_{n=0}^{\infty} 3^n z^n - \sum_{n=0}^{\infty} z^n$$

$$= 2\sum_{n=0}^{\infty}(2\cdot 3^n - 1)z^n \qquad (4)$$

comparing (1) and (4), we get

$$y_n = 2.3^n - 1$$

Example 4.59 **Using generating function, solve the difference equation**

$$y_{n+2} - 4y_{n+1} + 3y_n = 0;\ y_0 = 2, y_1 = 4 \qquad \textit{(B.E. Oct. 96)}$$

Solution Let the generating function for the required sequence be

$$G(z) = \sum_{n=0}^{\infty} y_n z^n \qquad (1)$$

Consider the given equation

$$y_{n+2} - 4y_{n+1} + 3y_n = 0 \qquad (2)$$

Multiply (2) by z^n and sum for $n \geq 2$, we get

$$\sum_{n=2}^{\infty} y_{n+2} z^n - 4\sum_{n=2}^{\infty} y_{n+1} z^n + 3\sum_{n=2}^{\infty} y_n z^n = 0 \qquad (3)$$

Consider the first sum

$$\sum_{n=0}^{\infty} y_{n+2}\, z^n = y_2 + y_3 z + y_4 z^2 + \ldots\ldots = \frac{1}{z^2}(y_2 z^2 + y_3 z^3 + y_4 z^4 + \ldots\ldots)$$

$$= \frac{1}{z^2}(G(z) - y_0 - y_1 z) \qquad \text{Using (1)}$$

the second sum,

$$\sum_{n=0}^{\infty} y_{n+1} z^n = y_1 + y_2 z + y_3 z^2 + \ldots = \frac{1}{z}(y_1 z + y_2 z^2 + y_3 z^3 + \ldots)$$

$$= \frac{1}{z}(G(z) - y_0) \qquad \text{Using (1)}$$

the third sum,

$$\sum_{n=0}^{\infty} y_n z^n = G(z) \qquad \text{Using (1)}$$

Substituting these in (3), we get

$$\frac{1}{z^2}(G(z) - y_0 - y_1 z) - \frac{4}{z}(G(z) - y_0) + 3G(z) = 0$$

put $y_0 = 2$ and $y_1 = 4$,

$$\frac{1}{z^2}(G(z) - 2 - 4z) - \frac{4}{z}(G(z) - 2) + 3G(z) = 0$$

$$\Rightarrow (G(z) - 2 - 4z) - 4z(G(z) - 2) + 3z^2 G(z) = 0$$

$$\Rightarrow G(z)(1 - 4z + 3z^2) = 2 - 4z$$

$$\Rightarrow (G(z) = \frac{2-4z}{1-4z+3z^2} = \frac{2-4z}{(1-z)(1-3z)}$$

Using partial fractions $\dfrac{2-4z}{(1-z)(1-3z)} = \dfrac{A}{1-z} + \dfrac{B}{1-3z}$

$$\Rightarrow 2 - 4z = A(1-3z) + B(1-z)$$

$$\Rightarrow A + B = 2 \text{ and } 3A + B = 4$$

$$\Rightarrow A = 1 \text{ and } B = 1$$

Therefore,

$$G(z) = \frac{1}{1-z} + \frac{1}{1-3z} = (1-z)^{-1} + (1-3z)^{-1}$$

$$= \sum_{n=0}^{\infty} z^n + \sum_{n=0}^{\infty} 3_n z^n \qquad \text{by note (5)}$$

$$= \sum_{n=0}^{\infty} (1+3^n) z^n \qquad (4)$$

Comparing (1) and (4), we get the required solution

$$y_n = 1 + 3^n$$

Example 4.60 Using generating function solve the difference equation

$$y_{n+2} - 6y_{n+1} + 8y_n = 0, y_0 = 1, y_1 = 4 \qquad \textit{(B.E.Oct. 98)}$$

Solution Let the generating function for the required sequence be

$$G(z) = \sum_{n=0}^{\infty} y_n z^n \qquad (1)$$

Consider the given equation

$$y_{n+2} - 6y_{n+1} + 8y_n = 0 \qquad (2)$$

Multiply (2) by z^n and sum for $n \geq 0$, we get

$$\sum_{n=0}^{\infty} y_{n+2} z^n - 6\sum_{n=0}^{\infty} y_{n+1} z^n + 8\sum_{n=0}^{\infty} y_n z^n = 0 \qquad (3)$$

As in the previous example, we obtain

$$\frac{1}{z^2}(G(z)-y_0-y_1)-\frac{6}{z}(G(z)-y_0)+8G(z)=0.$$

Since $y_0 = 1$ and $y_1 = 4$, we get

$$\frac{1}{z^2}(G(z)-1-4z)-\frac{6}{z}(G(z)-1)+8G(z)=0.$$

$$\Rightarrow (G(z)-1-4z)-6z(G(z)-1)+z^2 8G(z)=0$$

$$\Rightarrow G(z)[1-6z+8z^2]=1-2z$$

$$\Rightarrow G(z)=\frac{1-2z}{1-6z+8z^2}$$

$$=\frac{1-2z}{(1-2z)(1-4z)} \qquad (4)$$

$$=\frac{1}{1-4z}=(1-4z)^{-1}=\sum_{n=0}^{\infty}4^n z^n$$

Comparing (1) and (4), we get the required solution.

$$y_n = 4^n$$

Example 4.61 Solve $S(k) - 7S(k-2) + 6S(k-3) = 0$, $S(0) = 8$, $S(1) = 6$ and $S(2) = 22$.

Solution Let the generating function for the sequence be

$$G(z)=\sum_{k=0}^{\infty}S(k)z^k \qquad (1)$$

Consider the given equation $\quad S(k) - 7S(k-2) + 6S(k-3) = 0 \qquad (2)$

Multiply (2) by z^k and sum for all $k \geq 3$, we get

$$\sum_{k=3}^{\infty}S(k)z^k - 7\sum_{k=3}^{\infty}S(k-2)z^k + 6\sum_{k=3}^{\infty}S(k-3)z^k = 0 \qquad (3)$$

The first sum,

$$\sum_{k=3}^{\infty}S(k)z^k = S(3)z^3 + S(4)z^4 + \ldots = G(z)-S(0)-S(1)z-S(z)z^2 \qquad \text{Using (1)}$$

the second sum,

$$\sum_{k=3}^{\infty}S(k-2)z^k = S(1)z^3 + S(2)z^4 + \ldots = z^2(S(1)z + S(2)z^2 + \ldots)$$

$$= z^2(G(z) - S(0)) \qquad \text{Using (1)}$$

the third sum,

$$\sum_{k=3}^{\infty}S(k-3)z^k = S(0)z^3 + S(1)z^4 + \ldots = z^3(S(0) + S(1)z + \ldots)$$

$$= z^3 G(z) \qquad \text{Using (1)}$$

Substituting these values in (3), we get

$$[G(z)-S(0)-S(1)z-S(2)z^2] - 7z^2[G(z)-S(0)] + 6z^3G(z) = 0$$

Since $S(0) = 8$, $S(1) = 6$ and $S(2) = 22$, it reduces to

$$[G(z) - 8 - 6z - 22)z^2] - 7z^2[G(z) - 8] + 6z^3G(z) = 0$$

$$\Rightarrow G(z)[1-7z^2+6z^3]=8+6z-34z^2$$

$$\Rightarrow G(z)=\frac{8+6z-34z^2}{1-7z^2+6z^3}$$

$$=\frac{8+6z-34z^2}{(1-z)(1-2z)(1+3z)}$$

$$\frac{8+6z-34z^2}{(1-z)(1-2z)(1+3z)}=\frac{A}{(1-z)}+\frac{B}{(1-2z)}+\frac{C}{(1+3z)} \quad \text{(Partial fractions)}$$

$$\Rightarrow 8+6z-34z^2=A(1-2z)(1+3z)+B(1-z)(1+3z)+C(1-z)(1-2z).$$

If $z=1$ then $-20=-4A\Rightarrow A=5$

If $z=\frac{1}{2}$ then $\frac{10}{4}=\frac{5}{4}B\Rightarrow B=2$

If $z=-\frac{1}{3}$ then $\frac{20}{9}=\frac{20}{9}C\Rightarrow C=1$

$$\therefore G(z)=\frac{5}{1-z}+\frac{2}{1-2z}+\frac{1}{1+3z}$$

$$=5(1-z)^{-1}+2(1-2z)^{-1}+(1+3z)^{-1}$$

$$=5\sum_{k=0}^{\infty}z^k+2\sum_{k=0}^{\infty}2^kz^k+\sum_{k=0}^{\infty}(-3)^kz^k$$

$$=\sum_{k=0}^{\infty}(5+2^{k+1}+(-3)^k)z^k \tag{4}$$

Comparing (1) and (4), we get

$$S(k)=5+2^{k+1}+(-3)^k$$

which is the required solution.

Example 4.62 Suppose that a valid code word is an n-digit number in decimal notation containing an even number of 0's. Let a_n denote the number of valid code words of length n. The sequence $\{a_n\}$ satisfies the recurrence relation $a_n=8a_{n-1}+10^{n-1}$ and the initial condition $a_1=9$. Use generating functions to find an explicit formula for a_n.

Solution To make our work with generating functions simpler, we extend this sequence by setting $a_0=1$ so that $a_1=8a_0+10^0=9$.

Let the generating function for the sequence $\{a_n\}$ be

$$G(z)=\sum_{n=0}^{\infty}a_nz^n \tag{1}$$

Multiply both sides of $a_n=8a_{n-1}+10_{n-1}$ by z^n and sum for all $n\geq 1$, we get

$$\sum_{n=1}^{\infty}a_nz^n=8\sum_{n=1}^{\infty}a_{n-1}z^n+\sum_{n=1}^{\infty}10_{n-1}z^n \tag{2}$$

$$\Rightarrow (G(z)-a_0)=8zG(z)+z\sum_{n=1}^{\infty}(10z)^{n-1}$$

Since $a_0=1$ and $a_1=9$, it reduces to

$$G(z)-1=8zG(z)+z(1-10z)^{-1}$$

$$\Rightarrow G(z)(1-8z)=1+\frac{z}{1-10z}=\frac{1-9z}{1-10z} \quad (3)$$

$$\Rightarrow G(z)=\frac{1-9z}{(1-8z)(1-10z)}$$

$$\frac{1-9z}{(1-8z)(1-10z)}=\frac{A}{(1-8z)}+\frac{B}{(1-10z)} \quad \text{(Partial fractions)}$$

$$\Rightarrow 1-9z = A(1-10z)+B(1-8z)$$

$$\Rightarrow A+B = 1 \text{and } 10A+8B = 9 \Rightarrow A+B = \frac{1}{2}$$

$$\therefore G(z) = \frac{1}{2}\left\{\frac{1}{1-8z}+\frac{1}{1-10z}\right\} = \frac{1}{2}(1-8z)^{-1}+(1-10z)^{-1}$$

$$= \frac{1}{2}\sum_{n=0}^{\infty}(8z)^n + \sum_{n=0}^{\infty}(10z)^n$$

$$= \sum_{n=0}^{\infty}\left[\frac{1}{2}8^n + 10^n\right]z^n \quad (4)$$

Comparing (1) and (4), we get

$$a_n = \frac{1}{2}(8^n + 10^n)$$

Example 4.63 Solve $S(n)-2S(n-1)-3S(n-2)=0, n\geq 2$ with $S(0)=3$ and $S(1)=1$ using generating function.

(B.E.Oct. 98)

Solution Let the generating function for the required sequence be

$$G(z)= \sum_{n=0}^{\infty} S(n)z^n \quad (1)$$

Consider the given equation.

$$S(n) - 2S(n-1) - 3S(n-2) = 0 \quad (2)$$

Multiply (2) by z^n and sum for $n\geq 2$, we get

$$\sum_{n=2}^{\infty} S(n)z^n - 2\sum_{n=2}^{\infty} S(n-1)z^n - 3\sum_{n=2}^{\infty} S(n-2)z^n = 0 \quad (3)$$

Consider the first sum,

$$\sum_{n=2}^{\infty} S(n)z^n = S(2)z^2 + S(3)z^3 + ... = G(z) - S(0) - S(1)z \quad \text{using (1)}$$

the second sum,

$$\sum_{n=2}^{\infty} S(n-1)z^n = S(1)z^2 + S(2)z^3 + ... = z(S(1)z + S(2)z^2 + ...) \quad \text{using (1)}$$

$$= z(G(z) - S(0)$$

the third sum,

$$\sum_{n=2}^{\infty} S(n-2)z^n = S(0)z^2 + S(1)z^3 + ... = z^2(S(0) + S(1)z + ...)$$

$$= z^2G(z)$$

Substituting in (3), we obtain

$$[G(z)-S(0)-S(1)z]-2z[G(z)-S(0)]-3z^2G(z)=0$$

Put $S(0) = 3$ and $S(1) = 1$,

$$[G(z)-3-z]-2z[G(z)-3]-3z^2G(z)=0$$
$$\Rightarrow G(z)[1-2z-3z^2]=3-5z$$
$$\Rightarrow G(z)=\frac{3-5z}{1-2z-3z^2}$$

Using partial fractions, $\frac{3-5z}{1-2z-3z^2}=\frac{3-5z}{(1-3z)(1+z)}$

$$=\frac{A}{1-3z}+\frac{B}{1+z}$$
$$\Rightarrow 3-5z = A(1+z)+B(1-3z)$$
$$\Rightarrow 3 = A+B$$
$$-5 = A-3B$$
$$\Rightarrow A=1 \text{ and } B=2$$

Therefore,

$$G(z)=\frac{1}{1-3z}+\frac{2}{1+z}=(1-3z)^{-1}+2(1+z)^{-1}$$
$$=\sum_{n=0}^{\infty}(3z)^n+2\sum_{n=0}^{\infty}(-1)^n z^n=\sum_{n=0}^{\infty}3^n z^n+\sum_{n=0}^{\infty}2(-1)^n z^n$$
$$=\sum_{n=0}^{\infty}(3^n+2(-1)^n)z^n \qquad (4)$$

Comparing (1) and (4), we get

$$S(n)=3^n+2(-1)^n$$

EXERCISE IV

1. Find the generating function of the following sequences.
   a. 2, 2, 2, 2, 2, 2
   b. $a_n = (-2)^n$
   c. 0, 0, 1, 1, 1, 1,.....
   d. 1, 0, –1, 0, 1, 0, 0, 0, –1, 0, 1, 0, –1, 0, ...
2. For the following expressions, identify the sequences having the expression as a generating function.
   a. $\frac{5+2z}{1-4z^2}$
   b. $\frac{6-29z}{1-11z+30z^2}$
   c. $\frac{32-22z}{2-3z+z^2}$
   d. $\frac{3+7z}{1+3z-4z^2}$
   e. $\frac{3+5z}{1-2z-3z^2}$

   (Hint: Use partial fractions).
3. Find the generating function for the sequence satisfying the given initial condition.
   a. $y_{n+2}+2y_{n+1}-15y_n=0; y_0=0, y_1=1$
   b. $y_{n+1}-y_{n-1}=0; y_0=0, y_1=1.$
   c. $y_{n+2}-2y_{n+1}-3y_n=0; y_0=1, y_1=1$
   d. $S(n)-2S(n-1)-3S(n-2)=0; S(0)=3, S(1)=1.$
   e. $S(k)+3S(k-1)-4S(k-2)=0; S(0)=3, S(1)=-2.$
4. Using generating function solve the following recurrence relations.
   a. $y_n=7y_{n-1}; y_0=5$
   b. $y_n=3y_{n-1}+4^{n-1}; y_0=1$

c. $y_n 2y_{n-1} - 15y_{n-2} = 0; y_0 = 0, y_1 = 1$

d. $y_{n+2} - 8y_{n+1} + 16y_n = 0; y_0 = 0, y_1 = 8$

e. $y_{n+2} - 2y_{n+1} + y_n = 0; y_0 = 1, y_1 = 2$

f. $y_{n+2} - y_{n+1} - 6y_n = 0; y_0 = 2, y_1 = 1$. (*B.E. Oct. 97*)

g. $y_{n+1} - 5y_{n+1} + 6y_n = 0; y_0 = 1, y_1 = 3$ (*B.E. Apr. 97*)

5. **Use a generating function to find an explicit formula for the Fibonacci numbers.**
(Hint: See Example 4.36).

ANSWERS

1. a. $2\dfrac{(z^6 - 1)}{(2-1)}$ b. $\dfrac{1}{(1+2z)}$

   c. $\dfrac{z^2}{1-z}$ d. $\dfrac{1}{1+z^2}$

2. a. $3(2)^n + 2(-2)^n$ b. $7(6)^n - (5)^n$

   c. $10 + 6(\frac{1}{2})^n$ d. $2 + (-4)^n$

   e. $3^n + 2(-1)^n$

3. a. $G(z) = \dfrac{z}{(1-3z)(1+5z)}$ b. $G(z) = \dfrac{z}{1-z^2}$

   c. $G(z) = \dfrac{1-2z}{(1-3z)(1+z)}$ d. $G(z) = \dfrac{3-5z}{(1-3z)(1+z)}$

   e. $G(z) = \dfrac{3+7z}{(1-z)(1+4z)}$

4. a. $y_n = 5 \cdot (7)^n$ b. $y_n = (4)^n$

   c. $y_n = \frac{1}{8}(3)^n - \frac{1}{8}(-5)^n$ d. $y_n = 2n \cdot 4^n$

   e. $y_n = 1 + n$ f. $y_n = 3^n + (-2)^n$

   g. $y_n = 3^n$

5. $$f_n = \frac{1}{\sqrt{5}}\left[\left(\frac{1+\sqrt{5}}{2}\right)^n - \left(\frac{1-\sqrt{5}}{2}\right)^n\right]$$

CHAPTER

5

Algebraic Systems

INTRODUCTION

Originally algebra was a study of number systems. Later on some properties of numbers were picked out and models were built having these properties as axioms. Algebraic systems deal with non empty sets satisfying some axioms. Group is one of the Algebraic systems. The French mathematician Galois (1811–1832) was the first to use the word "Group" in any technical sense when he considered the group of permutations of roots of polynomial equations.

Today, group theory has embraced all branches of mathematics and wide variety of other fields. The concept of group theory is used in Quantum mechanics, Physics of fundamental particles, Chemistry and even in Biology.

BINARY OPERATIONS AND GENERAL PROPERTIES

Binary operation A binary operation $*$ on a non empty set is a rule which assigns to each ordered pair of elements of the set, a unique element of the set. That is, a binary operation on a non empty set A is a function $*$ from $A \times A$ into A. The image of the ordered pair $(a, b) \in A \times A$ under the function $*$, is denoted by $a * b$, where $a, b \in A$.

The fact that $*$ is a binary operation defined on a set A, is usually denoted by $(A, *)$. Further we say that A is closed under the binary operation $*$.

Algebraic systems A non-empty set A with one or more binary operations on it is called an algebraic system. *(MCA 98)*

Examples

1. $N = \{1, 2, 3, \ldots\}$ (the set of Natural numbers).
   The usual addition '+', is a binary operation on N.

   Since $a + b \in N \ \forall \ a, b \in N$. Similarly + is a binary operation on Z (the set of integers), Q (the set of rational numbers), R (the set of real numbers) and C (the set of complex numbers). $(N, +)$, $(Z, +)$, $(R, +)$, $(C, +)$ are algebraic systems.

2. The usual subtraction '–', is not a binary operation on N.
   For $2, 5 \in N$. But $2 - 5 = -3 \notin N$. Hence $(N, -)$ is not an algebraic system.

General Properties

i. Closure property: A binary operation $* : A \times A \rightarrow A$ is said to be closed if for all $a, b \in A$, $a * b \in A$. [This is the definition of binary operation]

ii. Associativity: $a * (b * c) = (a * b) * c \ \forall \ a, b, c, \in A$

iii. Existence of identity: There exists an element $e \in A$ such that $a * e = e * a = a \ \forall \ a \in A$. [$e$ is called the identity element].

iv. Existence of inverse: For $a \in A$, if there exists an element $a' \in A$ such that $a * a' = a' * a = e$, then the element a' is called inverse of a and a' is denoted by a^{-1}.

v. Commutativity: If $a * b = b * a \ \forall \ a, b \in A$, then $*$ is commutative in A. $(A, *)$ is commutative or abelian.

vi. Cancellation properties: $\forall \ a, b, c \in A, a * b = a * c \Rightarrow b = c$ (left cancellation), $\forall \ a, b, c \in A, b * a = c * a \Rightarrow b = c$ (right cancellation).

Semi group Let S be a non-empty set and $*$ be a binary operation in S. If $*$ associative in S then $(S, *)$ is called a semigroup. *(M.C.A 96 N, 97 Oct, B.E, 92 A, 93A)*

Example $(N, +)$ is a semigroup.

Example $(Z, -)$ is not a semigroup. For, $-$ is closed in Z. But it is not associative in Z.

$$2, 5, 6 \in Z$$

$$(2-5)-6 = -3-6 = -9, \quad 2-(5-6) = 2-(-1) = 2+1 = 3$$

$$(2-5)-6 \neq 2-(5-6)$$

Monoid Let $(M, *)$ be a semigroup. If M has an identity element then $(M, *)$ is called a monoid. *(MCA 96N, 97 Oct, B.E. 92A, 93A)*

Example Let $W = \{0, 1, 2, 3, \ldots\}$. $+$ is a binary operation and associative in W, $0 \in W$ is the identity element. $(W, +)$ is a monoid.

Example Let $N = \{1, 2, 3, \ldots\}$. $(N, +)$ is a semigroup but not a monoid since the identity element $0 \notin N$.

Groups A non-empty set G together with a binary operation $*$ is called a group if the following axioms are satisfied. *(MCA 96 N, B.E. 96 Oct)*

i. $a * b \in G \ \forall \ a, b \in G$ (closure axiom)

ii. $a * (b * c) = (a * b) * c \ \forall \ a, b, c \in G$ (associative axiom)

iii. There exists an identity element $e \in G$ such that $a * e = e * a = a \ \forall \ a \in G$. e is the identity element of G (existence of identity)

iv. $\forall \ a \in G$, there exists an element $a' \in G$ such that $a * a' = a' * a = e$. a' is called inverse of a and is denoted by a^{-1}. (existence of inverse).

Note 1: The set G itself is not a group with respect to operation $*$. $(G, *)$ satisfying the above four axioms is a group.

Note 2: The statement that a binary operation is defined on G implies closure. Still we have given closure axiom in the definition because it should be checked up first before checking up the other three axioms.

Note 3: Let $(G, *)$ be a monoid. $\forall a \in G$, if there exists $a^{-1} \in G$ such that $a * a^{-1} = a^{-1} * a = e$ then $(G, *)$ is a group.

Note 4: A group is a monoid and a semigroup but a semigroup or monoid need not be a group.

Abelian group If $(G, *)$ is a group such that $a * b = b * a \forall a, b \in G$, then $(G, *)$ is an abelian group or commutative group.

Order of Group

The group $(G, *)$ is said to be finite or infinite according as the underlying set is finite or infinite. If $(G, *)$ is a finite group then the number of elements in G is called the order of the group, written as $O(G)$ or $|G|$.

***Example* 5.1** Test whether $(Z, +)$ is a group or not.

Solution $\forall a, b \in Z, a+b \in Z$ (closure).

$+$ is associative in Z (associative). $0 \in Z$ is the identity for $0 + a = a + 0 = a \forall a \in Z$ (existence of identity) $\forall a \in Z$, there exists $-a \in Z$ such that $a+(-a)=(-a)+a=0$ (existence of inverse) $\therefore$ $(Z, +)$ is a group. Further $a+b=b+a \ \forall \ a, b \in Z$ $\therefore$ $(Z, +)$ is an infinite abelian group.

As in example 5.1 $(Q, +)$, $(R, +)$, $(C, +)$ are also infinite abelian groups.

Example 5.2 Test whether the set of all non zero rational number forms a group with respect to usual multiplication. i.e.,test whether $(Q-\{0\}, \cdot)$ is a group.

Solution $\forall a, b \in Q-\{0\}, a \cdot b \in Q-\{0\}$ (closure)

. is associative in $Q-\{0\}$ (associative)

$1 \in Q-\{0\}$ is the identity for $a \cdot 1 = 1 \cdot a = a \ \forall \ a \in Q-\{0\}$ (identity)

$\forall a \in Q-\{0\}$, $\frac{1}{a} \in Q-\{0\}$ exists such that $a \cdot \frac{1}{a} = \frac{1}{a} \cdot a = 1$ (inverse)

$(Q-\{0\}, \cdot)$ is a group.

Further $a \cdot b = b \cdot a \ \forall \ a, b \in Q-\{0\}$

$\therefore$ $(Q-\{0\}, \cdot)$ is an infinite abelian group.

Note:1 $(Q, \cdot)$ is a monoid but not a group for $0 \in Q$ does not have an inverse in Q.

2 As in example (5.2) $(R-\{0\}, \cdot)$, $(C-\{0\}, \cdot)$ are also infinite abelian groups.

Example 5.3 Test whether the set of all non zero real numbers namely $R-\{0\}$ forms an abelian group with respect to $*$ defined by $a * b = \frac{ab}{2} \forall a, b \in R-\{0\}$

Solution $\forall a, b \in R, a \neq 0, b \neq 0, a * b = \frac{ab}{2} \in R$ and $\frac{ab}{2} \neq 0$ (closure)

$\forall a, b, c \in R, a \neq 0, b \neq 0, c \neq 0,$

$$(a*b)*c = \left(\frac{ab}{2}\right)*c = \left(\frac{\frac{ab}{2}\cdot c}{2}\right) = \frac{abc}{4}; \qquad a*(b*c) = a*\left(\frac{bc}{2}\right) = \frac{\left(a\cdot\left(\frac{bc}{2}\right)\right)}{2} = \frac{abc}{4}$$

$(a*b)*c = a*(b*c)$ (associative)

$a*e = a \therefore \frac{ae}{2} = a \ e = 2 \in R-\{0\}$ is the identity (identity)

$a*a' = e = 2$

$\frac{aa'}{2} = 2 \ a' = \frac{4}{a} \in R-\{0\}$ is the inverse of a (inverse) also $a*b=\dfrac{ab}{2}=\dfrac{ba}{2}=b*a\,\forall\ a,b\in R-\{0\}$

$\therefore (R-\{0\},*)$ is an infinite abelian group.

Example 5.4 Prove that $(R-\{1\},*)$ where $*$ is defined by $a*b=a+b-ab$ is an abelian group.

Solution Let $a,b\in R-\{1\}$

$$\therefore\ a,b\in R, a\neq 1, b\neq 1.$$

$$a*b=a+b-ab\in R.$$

Let us prove $a+b-ab\neq 1$

Suppose $a+b-ab=1$

then $a-1+b-ab=0$

$$1(a-1)-b(a-1)=0$$

$$(a-1)(1-b)=0$$

$\therefore\ a=1$ or $b=1$ which is a contradiction to the fact that $a\neq 1, b\neq 1$

$$\therefore\ a+b-ab\neq 1$$

$a*b=a+b-ab\in R-\{1\}$ (closure)

Let $a,b,c\in R-\{1\}$

$$\begin{aligned} a*(b*c) &= a*(b+c-bc)\\ &= a+(b+c-bc)-a(b+c-bc)\\ &= a+b+c-ab-bc-ac+abc \qquad (1)\\ (a*b)*c &= (a+b-ab)*c\\ &= a+b-ab-(a+b-ab)c\\ &= a+b-ab-bc-ac+abc \qquad (2)\end{aligned}$$

(1) = (2); hence $a*(b*c)=(a*b)*c$ (associative).

Suppose $a*e=a$ then $a+e-ae=a$

$$\therefore e(1-a)=0$$

But $a\neq 1$

$\therefore e=0\in R-\{1\}$ is the identity (existence of identity)

Suppose $a*b=e=0$.

Then $a+b-ab=0$

$$(1-a)b=-a$$

$b=\dfrac{-a}{1-a}=\dfrac{a}{a-1}\in R-\{1\}\,(a\neq 1)$ is the inverse of $a\in R-\{1\}$

$$a*b=a+b-ab=b+a-ba=b\neq a\,\forall\, a,b\in R-\{1\}$$

$$\therefore (R-\{1\},*) \text{ is an infinite abelian group.}$$

Example 5.5 Prove that the cube roots of unity namely $\{1,\omega,\omega^2\}$ is abelian under multiplication of complex numbers.

Solution Let $G=\{1,\omega,\omega^2\}$ where $1,\omega,\omega^2$ are cubic roots of unity. $1,\omega,\omega^2$ are complex numbers. is multiplication of complex numbers. $(G,\cdot)$ is an abelian group.

[Note that $1+\omega+\omega^2=0,\ \omega^3=1$]

The composition table known as Cayley's table is given below.

$\cdot$	1	ω	ω^2
1	1	ω	ω^2
ω	ω	ω^2	1
ω^2	ω^2	1	ω

From the table, we see that . is closed.

We know that multiplication of complex numbers is associative and commutative.

$1 \in G$ is the identity

$1^{-1}=1,\ \omega^{-1}=\omega^2,\ (\omega^2)^{-1}=\omega$

Inverse exists in G for every element in G.

$(G,.)$ is a finite abelian group.

***Example* 5.6** Prove that four fourth roots of unity namely $\{1, i, -1, -i\}$ is an abelian group under multiplication of complex numbers.

Solution Let $G=\{1, i, -1, -i\}$. The multiplication table is as follows.

$\cdot$	1	i	-1	$-i$
1	1	i	-1	$-i$
i	i	-1	$-i$	1
-1	-1	$-i$	1	i
$-i$	$-i$	1	i	-1

From the table, we see that . is closed in G. Multiplication of complex numbers is associative and commutative.

$1 \in G$ is the identity.

The inverses of $1, i, -1, -i$ are respectively $1, -i, -1, i$ which are in G.

$\therefore (G, .)$ is a finite abelian group.

***Example* 5.7** Let $G=\{f_1, f_2, f_3, f_4\}$ where $f_1(x)=x, f_2(x)=-x, f_3(x)=\dfrac{1}{x}, f_4(x)=-\dfrac{1}{x}$ and $\circ$ be the composition of functions. Show that $(G, \circ)$ is an abelian group with respect to $\circ$.

The table for composition of functions is given below.

$\circ$	f_1	f_2	f_3	f_4
f_1	f_1	f_2	f_3	f_4
f_2	f_2	f_1	f_4	f_3
f_3	f_3	f_4	f_1	f_2
f_4	f_4	f_3	f_2	f_1

From the table, we see that G is closed with respect to $\circ$.

We know that the composition of functions is associative.

$f_1 \in G$ is the identity.

The inverses of f_1, f_2, f_3, f_4 are respectively f_1, f_2, f_3, f_4 which are in G.

The table is symmetric about the leading diagonal.

Hence $\circ$ is commutative.

$(G, \circ)$ is a finite abelian group.

Note: In general $\circ$ (composition of functions) is not commutative.

Example 5.8 Let $a \in R$ be a fixed real number. Let $G = \{a^n : n \in Z\}$. **Show that** G **is an abelian group under usual multiplication.**

Solution Let $x, y \in G$. $\therefore x = a^m,\ y = a^n$ where $m, n \in Z$

$x \cdot y = a^m \cdot a^n = a^{m+n} \in G$ since $m + n \in Z$ (closure)

We know that the usual multiplication is associative and commutative.

$1 = a^0 \in G\ (\because 0 \in Z)$ is the identity.

$\forall a^m \in G, (m \in Z)$, there exists $a^{-m} \in G$. (since $-m \in Z$) such that $a^m \cdot a^{-m} = a^{-m} \cdot a^m = a^0.a^{-m} \in G$ is the inverse of a^m.

$x \cdot y = a^m \cdot a^n = a^{m+n} = a^{n+m} = a^n \cdot a^m = y \cdot x\ \forall x, y \in G$. $\therefore$ (G,.) is an abelian group.

Example 5.9 Let G be the set of all 2×2 matrices over R and + be matrix addition. Show that $(G, +)$ is an abelian group.

Solution Let $G = \left\{\begin{pmatrix} a & b \\ c & d \end{pmatrix} : a, b, c, d \in R\right\}$

Let $A = \begin{pmatrix} a_1 & b_1 \\ c_1 & d_1 \end{pmatrix}$ $B = \begin{pmatrix} a_2 & b_2 \\ c_2 & d_2 \end{pmatrix}$ where $a_1, b_1, c_1, d_1, a_2, b_2, c_2, d_2 \in R$. Then $A, B \in G$.

$A + B = \begin{pmatrix} a_1 + a_2 & b_1 + b_2 \\ c_1 + c_2 & d_1 + d_2 \end{pmatrix} \in G$

since $a_1 + a_2, b_1 + b_2, c_1 + c_2, d_1 + d_2 \in R$ (closure).

We know that in general, matrix addition is associative and commutative.

$0 = \begin{pmatrix} 0 & 0 \\ 0 & 0 \end{pmatrix} \in G$ is the identity. Since $A + 0 = 0 + A = A\ \forall A \in G$.

$\forall A = \begin{pmatrix} a_1 & b_1 \\ c_1 & d_1 \end{pmatrix} \in G$, there exists $-A = \begin{pmatrix} -a_1 & -b_1 \\ -c_1 & -d_1 \end{pmatrix}$ in G such that $A + (-A) = (-A) + A = 0$.

$\therefore$ $(G, +)$ is an abelian group.

Example 5.10 Let G be the set of all 2×2 non-singular matrices over R and $\cdot$ be the matrix multiplication. Then find whether $(G, \cdot)$ is a group.

Solution Let $G = \left\{A = \begin{pmatrix} a & b \\ c & d \end{pmatrix} : a, b, c, d \in R, |A| \neq 0\right\}$

Let $A, B \in G \therefore |A| \neq 0, |B| \neq 0$.

$A \cdot B$ is also a 2×2 matrix over R, and $|A \cdot B| = |A||B| \neq 0$

$A \cdot B \in G$ (closure)

We know that matrix multiplication is associative.

$I = \begin{pmatrix} 1 & 0 \\ 0 & 1 \end{pmatrix}$ is the identity element and $I \in G$ since $|I| = 1 \neq 0$.

Let $A \in G \therefore |A| \neq 0$ A^{-1} exists and $AA^{-1} = A^{-1}A = I$

$$|A||A^{-1}| = |I| = 1 \neq 0 \; |A^{-1}| \neq 0$$

$\therefore A^{-1} \in G$ is the inverse of G.

But matrix multiplication is non commutative.

$\therefore (G, \cdot)$ is a non-abelian group.

Congruent Modulo *n*

Let n be a positive integer. If a and b are two integers and n divides $a - b$ then we say that "a is congruent to b modulo n" and we write $a \equiv b \pmod{n}$. The integer n is called modulus.

Note: If $a \equiv b \pmod{n}$, then we have $a - b = kn$, k being an integer.

For example, $23 \equiv 3 \pmod 5$ since 5 divides 23–3=20. $15 \equiv 0 \pmod 5$ since 5 divides $15 - 0 = 15$.

Congruence Classes Modulo *n*

Let a be an integer. Let $[a]$ denote the set of all integers congruent to $a \pmod n$

$$\text{i.e. } [a] = \{x : x \in Z, x \equiv a \bmod (n)\}$$
$$= \{x : x \in Z, x = a + kn \text{ for some integer } k\}$$

$[a]$ is known as the equivalent class, modulo n, represented by a. Then set of all congruence classes modulo n is denoted by Z_n.

$$\therefore Z_n = \{[0], [1], \ldots [n-1]\}$$

Z_n is also called residue classes modulo n. For example consider the congruence classes modulo 5. $Z = \{0, \pm 1, \pm 2, \ldots\}$

$$[0] = \{x : x \in Z, x = 5k \text{ where } k \in Z\}$$
$$= \{\ldots -10, -5, 0, 5, 10 \ldots\}$$
$$[1] = \{x : x \in Z, x = 5k + 1 \text{ where } k \in Z\}$$
$$= \{\ldots -9, -4, 1, 6, 11 \ldots\}$$
$$[2] = \{x : x \in Z, x = 5k + 2 \text{ where } k \in Z\}$$
$$= \{\ldots -8, -3, 2, 7, 12 \ldots\}$$
$$[3] = \{x : x \in Z, x = 5k + 3 \text{ where } k \in Z\}$$
$$= \{\ldots -7, -2, 3, 8, 13 \ldots\}$$
$$[4] = \{x : x \in Z, x = 5k + 4 \text{ where } k \in Z\}$$
$$= \{\ldots -6, -1, 4, 9, 14 \ldots\}$$

Note:

1. $[5] = \{x : x \in Z, x = 5k + 5 = 5(k+1), k + 1 \in Z\} = [0]$
2. $Z_5 = \{[0], [1], [2], [3], [4]\}$
3. The union of all classes in Z_5 is Z.
4. The intersection of any two classes in Z_5 is a null set.

Addition of Residue Classes

Let $[a], [b] \in Z_n$. Then their sum is denoted by $+_n$ and is defined as follows:

$$[a] +_n [b] = \begin{cases} [a+b] & \text{if } a + b < n \\ [r] & \text{if } a + b \geq n \end{cases}$$

where r is the least non-negative remainder when $a + b$ is divided by n. Hence $0 \le r < n$. For example, let us consider Z_5

$$[1]+_5[2]=[1+2]=[3] \text{ for } 1+2=3<5$$
$$[3]+_5[4]=[2] \text{ for } 3+4=7\ge 5 \text{ and } 3+4=7=1\times 5+2$$
$$[3]+_5[2]=[0] \text{ for } 3+2=5=1\times 5+0$$

Multiplication of Residue Classes

Let $[a], [b] \in Z_n$. Then their product is denoted by X_n and is defined as follows:

$$[a]\times_n[b]=\begin{cases}[ab] & \text{if } ab<n\\ [r] & \text{if } ab\ge n\end{cases}$$

where r is the least non-negative integer when ab is divided by n. Hence $0 \le r < n$. For example, let us consider Z_5

$$[2]\times_5[2]=[4] \text{ since } 2.2=4<5$$
$$[2]\times_5[4]=[3] \text{ since } 2\times 4=8>5 \text{ and } 8=1.5+3.$$

Example 5.11 Show that $(Z_n, +_n)$ is an abelian group, where $Z_n = \{[0],[1],...,[n-1]\}$.

Solution Let $[a], [b], [c] \in Z_n$.

$$[a]+_n[b]=[a+b]\in Z_n \textit{ if } a+b<n$$
$$[a]+_n[b]=[r]\in Z_n \textit{ if } a+b\ge n\ (0\le r<n)$$
$$\therefore [a]+_n[b]\in Z_n \text{ (closure)}$$

$$\begin{aligned}([a]+_n[b])+[c] &= [a+b]+_n[c]\\ &=[(a+b)+c]\\ &=[a+(b+c)]\\ &=[a]+_n[b+c]\\ &=[a]+_n([b]+_n[c]) \text{ (associative)}\end{aligned}$$

$\forall\ [a]\in Z_n$, there exists $[0] \in Z_n$ such that

$$[a]+_n[0]=[a+0]=[a] \text{ and}$$
$$[0]+_n[a]=[0+a]=[a]$$

$\therefore [0] \in Z_n$ is the identity (existence of indentity).

$\forall\ [a]\in Z_n$, there exists $[n-a] \in Z_n$ such that

$$[a]+_n[n-a]=[n]=[0] \text{ and}$$
$$[n-a]+_n[a]=[n]=0.$$

$\therefore [n-a] \in Z_n$ is the inverse of $[a]$. (existence of inverse).

$$\begin{aligned}\text{Also } [a]+_n[b] &= [a+b]\\ &=[b+a]\\ &=[b]+_n[a] \text{ (commutative)}\end{aligned}$$

$\therefore (Z_n, +_n)$ is an abelian group.

Properties of Groups

***Theorem* 1** Let $(G, *)$ be a group. Then

i. The identity element of G is unique.

ii. For any $a \in G$, the inverse of is unique.

Proof

i. If possible, let e_1 and e_2 be two identities of G.

$$e_1 = e_2 * e_1 \text{[since } e_2 \text{ is the identity]}$$
$$= e_2 \text{[since } e_1 \text{ is the identity]}$$
$$\therefore e_1 = e_2$$

$\therefore$ The identity element of G is unique.

ii. If possible, let a' and a'' be two inverses of a.

$$\therefore a * a' = a' * a = e \qquad (1)$$

[$\because a'$ is to inverse of a]

$$\therefore a * a'' = a'' * a = e \qquad (2)$$

[$\because a''$ is to inverse of a]

Now $a' = a' * e$[since e is the identity]

$$= a' * (a * a'') \text{ by (2)}$$
$$= (a' * a) * a'' \text{ [from associativity]}$$
$$= e * a'' \text{ by (1)}$$
$$= a'' \text{ [since } e \text{ is the identity]}$$
$$\therefore a' = a''$$

Hence the inverse of a is unique.

Theorem 2 (Cancellation laws)

If G is a group, then

i. $a * b = a * c \Rightarrow b = c$ [left cancellation law]

ii. $b * a = c * a \Rightarrow b = c$ [right cancellation law] for all $a, b, c \in G$

Proof

i. Let $a * b = a * c$ (1)

Since $a \in G, a^{-1} \in G$ exists such that $a * a^{-1} = a^{-1} * a = e$ (2)

Premultiplying (1) by a^{-1}

$$a^{-1} * (a * b) = a^{-1} * (a * c)$$
$$(a^{-1} * a) * b = (a^{-1} * a) * c \text{ [by associativity]}$$
$$e * b = e * c \text{ by (2)}$$
$$b = c \text{ [since } e \text{ is the identity]}$$

$\therefore a * b = a * c \Rightarrow b = c$ [left cancellation law]

ii. Let $b * a = c * a$

Post multiplying by a^{-1}

$$(b * a) * a^{-1} = (c * a) * a^{-1}$$
$$b * (a * a^{-1}) = c * (a * a^{-1}) \text{ [by associativity]}$$
$$b * e = c * e \text{ [Using equation 2]}$$
$$b = c \text{ [since } e \text{ is the identity]}$$

$\therefore b * a = c * a \Rightarrow b = c$. [Right Cancellation law]

Theorem 3 Let G be a group and $a, b, \in G$. Then the equations $a * x = b$ and $y * a = b$ have unique solutions in G.

Proof Let $a * x = b$

Since $a, b \in G, a^{-1} * b \in G$.

$$\text{Then } a * (a^{-1} * b) = (a * a^{-1}) * b \text{ (by associativity)}$$
$$= e * b (\because a * a' = e)$$
$$= b (\because e \text{ is the identity})$$

$\therefore x = a^{-1} * b \in G$ is the soultion of $a * x = b$.

To prove uniqueness, let x_1 and x_2 be two solutions of

$$a * x = b \quad \therefore a * x_1 = b,$$
$$a * x_2 = b$$
$$\therefore a * x_1 = a * x_2$$

By left cancellation law, $x_1 = x_2$

$\therefore$ $a * x = b$ has unique solution $x = a^{-1} * b$ in G.

Let $y * a = b$. Since $a, b \in G$, $b * a^{-1} \in G$.

Then $(b * a^{-1}) * a = b * (a^{-1} * a)$ [by associativity]

$= b * e$ [$\because a^{-1} * a = e$]

$= b$ [$\because$ e is theidentity]

$\therefore$ $y * a = b$ has the solution $y = b * a^{-1} \in G$.

To prove uniqueness, let y_1, y_2 be two solutions of the equation $y * a = b \therefore y_1 * a = b, y * a = b$.

$$\therefore y_1 * a = y_2 * a$$
$$\therefore y_1 = y_2 \text{ [by right cancellation law]}$$

$\therefore$ $y = b * a^{-1}$ is the unique solution for $y * a = b$.

Theorem 4 Let G be a group and $a, b \in G$. Then (i) $(a^{-1})^{-1} = a$ (ii) $(a * b)^{-1} = b^{-1} * a^{-1}$.

Proof i. a^{-1} is the inverse of a.

$$\therefore a * a^{-1} = a^{-1} * a = e$$
$$\therefore a \text{ is the inverse of } a^{-1}$$
$$\therefore (a^{-1})^{-1} = a$$

ii. Since $a, b \in G, b^{-1} * a^{-1} \in G$

Let us consider $(a * b) * (b^{-1} * a^{-1}) = a * [b * (b^{-1} * a^{-1})]$ [by associativity]

$= a * [(b * b^{-1}) * a^{-1}]$ [by associativity]

$= a * (e * a^{-1})$ [$\because b * b^{-1} = e$]

$= a * a^{-1} = e$ [$a * a^{-1} = e$] [$\because$ e is the identity]

Similarly, we can prove

$$(b^{-1} * a^{-1}) * (a * b) = e$$
$$\therefore (a * b) * (b^{-1} * a^{-1}) = (b^{-1} * a^{-1}) * (a * b) = e$$
$$\therefore b^{-1} * a^{-1} \text{ is the inverse of } a * b$$
$$\therefore (a * b)^{-1} = b^{-1} * a^{-1}$$

Corollary If $a_1, a_2, ..., a_n \in G$, then $(a_1 * a_2 * ... * a_n)^{-1} = a_n^{-1} * a_{n-1}^{-1} * ... * a_2^{-1} * a_1^{-1}$

This is known as inverse reversal law.

Theorem 5 If a has inverse b and b has inverse c, then $a = c$.

Proof Since b is the inverse of a, $a * b = b * a = e$. (1)

Since c is the inverse of b, $b * c = c * b = e$. (2)

By (1) and (2) $a * b = c * b$, and hence

$a = c$ [by right cancellation]

Theorem 6 **A group cannot have any idempotent element except the identity element.**

Proof If possible, let $a \in G$, $a \neq e$ be the idempotent element,

i.e., $a * a = a$ (1)

But $e * a = a$ (2)

$\therefore a * a = e * a$ by (1) and (2)

By right cancellation,

$a = e$ which is a contradiction to the hypothesis that $a \neq e$.

$\therefore$ e is the only idempotent element of G.

Theorem 7 For any group $(G, *)$ if $a^2 = e$ with $a \neq e$ then G is abelian.

(or)

If every element of a group $(G, *)$ is its own inverse, then G is abelian. *(MCA 96N, B.E. 96 Oct.)*

Proof Let $a^2 = e$

$$\therefore (a * a) * a^{-1} = e * a^{-1}$$
$$\therefore a * (a * a^{-1}) = a^{-1}$$
$$\therefore a * e = a^{-1}$$
$$\therefore a = a^{-1}$$

If $a^2 = e$, a^{-1} is the inverse of a.

Let $a, b \in G$, then $a^{-1} = a$, $b^{-1} = b$

$a * b \in G \therefore (a * b)^{-1} = a * b$

$\therefore b^{-1} * a^{-1} = a * b$ [by (1)]

$b * a = a * b$

$a * b = b * a$ for every $a, b \in G$

$\therefore (G, *)$ is abelian

Note: The converse need not be true.

For example $(Z, +)$ is an abelian group.

Except 0, there is no element in Z which is its own inverse.

Properties of Exponentiation

i. $a^n = a * a * ... * a(n \text{ times})$

ii. $a^m * a^n = a^{m+n} \, (m, n \in Z)$

iii. $(a^m)^n = a^{mn} \, (m, n \in Z)$

iv. $(a^n)^{-1} = a^{-n} \quad (n \in Z)$

***Theorem* 8** A group $(G, *)$ is abelian iff $(a * b)^2 = a^2 * b^2$.

Proof Let us assume that G is abelian.

For $a, b \in G, a * b = b * a$ (1)

$$(a * b)^2 = (a * b) * (a * b)$$
$$= a * (b * a) * b \text{ (by associativity)}$$
$$= a * a * b * b \quad \text{[Using (1)]}$$
$$= a^2 * b^2$$

Conversely let us assume $(a * b)^2 = a^2 * b^2$

$$\therefore (a * b) * (a * b) = (a * a) * (b * b)$$
$$a * (b * a) * b = a * (a * b) * b$$

By let and right cancellation

$$b * a = a * b$$
$$\therefore (G, *) \text{ is an abelian)}$$

***Theorem* 9** Let $(G, *)$ be an abelian group and $a, b \in G$. Then for all $n \in Z$, $(a * b)^n = a^n * b^n$. *(MCA 98 N)*

Proof *Case 1* Let $n = 0$.

$$a^0 = e, \, b^0 = e, \, (a * b)^0 = e = e * e = a^0 * b^0$$
$$\therefore (a * b)^0 = a^0 * b^0$$

The theorem is true for $n = 0$.

Case 2 Let n be a positive integer.

$$(a * b)^1 = a * b = a^1 * b^1$$

So the result is true for $n = 1$.

Let it be true for $n = k$ so that $(a * b)^k = a^k * b^k$ (1)

Now $(a * b)^{k+1} = (a * b)^k * (a * b)$

$$= (a^k * b^k) * (a * b) \quad \text{by (1)}$$
$$= a^k * (b^k * a) * b \text{ (by associativity)}$$
$$= a^k * (a * b^k) * b \; [\because G \text{ is an abelian}]$$
$$= (a^k * a) * (b^k * b) \text{ [by associativity]}$$
$$= a^{k+1} * b^{k+1}$$

The result is true for $n = k+1$. By principle of induction the theorem is true for $n = 1, 2, 3\ldots$

Case 3 Let n be a negative integer.

Let $n = -m$, where m is a positive integer.

$$(a*b)^n = (a*b)^{-m}$$
$$= [(a*b)^m]^{-1}$$
$$= [a^m * b^m]^{-1} \text{ [by case 2]}$$
$$= [b^m * a^m]^{-1} \text{ [}\because G \text{ is an abelian]}$$
$$= (a^m)^{-1} * (b^m)^{-1}$$
$$= a^{-m} * b^{-m} = a^n * b^n$$

Hence $(a * b)^n = a^n * b^n \ \forall\ n \in Z$.

***Theorem* 10** [Alternate set of axioms for a group]

Let $(G, *)$ be a semigroup. $(G, *)$ is a group iff $a * x = b$ and $y * a = b$ have unique solutions in G for all $a, b \in G$.

Proof Let $(G, *)$ be a group. We had already proved the equations $a * x = b$, $y * a = b$ have unique solutions in G for all $a, b \in G$ [by theorem 3].

Conversely, let $(G, *)$ be a semigroup.

Assume that $a * x = b$ and $y * a = b$ have unique solutions in $G\ \forall\ a, b \in G$.

Let $a \in G$. Then $a,\ a \in G$

The equations $a * x = a$, $y * a = a$ have solutions in G. Let e and e' be the solutions respectively.

$$a * e = a \qquad (1)$$
$$e' * a = a \qquad (2)$$

Now let $b \in G$. There exists unique solutions x and y in G such that

$$a * x = b \qquad (3)$$
$$y * a = b \qquad (4)$$

Now $\quad b * e = (y * a) * e \qquad$ by (4)

$= y * (a * e)$ (by associativity)

$= y * a \qquad$ (1)

$= b \qquad$ by (4)

$$b * e = b \qquad (5)$$

$e' * b = e' * (a * x) \qquad$ by (3)

$= (e' * a) * x$ (by associativity)

$= a * x \qquad$ by (2)

$= b \qquad$ by (3)

$$e' * b = b \qquad (6)$$

Taking $b = e'$ and e in (5) and (6) respectively,

$$e' * e = e' \text{ and } e' * e = e$$
$$e' = e$$

Hence $a * e = e * a \; \forall \; a \in G$.

$\therefore \; e \in G$ is the identity element in G. (existence of identity).

Now $a, e \in G$. The equations $a * x = e, y * a = e$ have unique solutions p and q in G.

$$a * p = e, \; q * a = e$$

Now $a * p = e \; \therefore q * (a * p) = q * e$

$$(q * a) * p = q * e$$
$$e * p = q * e$$
$$\therefore \; p = q$$

i.e., $a * p = p * a = e \; \forall a \in G$.

Inverse exists $\forall \; a \in G$.

$\therefore (G, *)$ is a group.

***Theorem* 11** A finite semigroup satisfying cancellation laws is a group.

Proof Let $(G, *)$ be a finite semigroup where $G = \{a_1, a_2 \ldots a_n\}$. Let cancellation laws hold good in G. Let $a, b \in G$. Consider $\{a * a_1, a * a_2 \ldots a * a_n\}$. The elements in this set are elements of G itself. Also these are distinct since for $r \neq s$, $aa_r = aa_s$. By left cancellation, $a_r = a_s$, which is not true. The elements $a * a_1, a * a_2 \ldots a * a_n$ are distinct elements of G taken in some order. Now, since $b \in G$, we have $a * a_k = b$ for some unique $a_k \in G$. $\therefore$ The equation $a * x = b$ has unique solution in G.

Similarly considering $a_1 * a, a_2 * a \ldots, a_n * a$ we can prove the equation $y * a = b$ has unique solution in G. $(G, *)$ is a group [by theorem (10)].

Note: The above theorem need not be true for infinite semigroup. For example, Cancellation laws hold good in the infinite semigroup $(N, +)$ but it is not a group.

***Theorem* 12** Every group of order 4 is abelian. *(B.E. 98 Nov.)*

Proof Let $(G, *)$ be a group of order 4 where $G = \{e, a, b, c\}$. Since G is of even order, there exists atleast one element (say) a such that $a^{-1} = a$. Then two cases arise (i) $b^{-1} = b, c^{-1} = c$ (ii) $b^{-1} = c, c^{-1} = b$.

Case 1 $e^{-1} = e, a^{-1} = e, a^{-1} = a, b^{-1} = b, c^{-1} = c$.

Every element is its own inverse. By a known result $(G, *)$ is abelian.

Case 2 $a^{-1} = a, b^{-1} = c, c^{-1} = b$.

$$\therefore \; a^2 = e, \; b * c = e, \; c * b = e$$

*	e	a	b	c
e	e	a	b	c
a	a	e	<u>c</u>	<u>b</u>
b	b	<u>c</u>	<u>a</u>	e
c	c	<u>b</u>	e	<u>a</u>

Since $(G, *)$ is a group, its elements will appear in a row (or column) only once. Since a, e appears in the 2nd row and b appears in the 3rd column, c will appear as $(2, 3)^{th}$ element.

$\therefore \; (2, 4)^{th}$ element is b

$(3, 3)^{th}$ element is a

$(3, 2)^{th}$ element is c

$(4, 2)^{th}$ element is b

$(4, 4)^{th}$ element is a

The table is symmetric about leading diagonal $(G, *)$ is abelian.

Hence a group of order 4 is abelian.

Order of an Element

Let $(G, *)$ be a group and $a \in G$. The least positive integer m, if it exists such that $a^m = e$ (identity of G) is called order of a and is written as $O(a) = m$.

In case, such m does not exist we say that a is of infinite order or zeroth order.

Note: In the case of additive group, m is the least positive integer such that $ma = e$ and $O(a) = m$.

Example Let $G = \{1, \omega, \omega^2\}$. $(G, \cdot)$ is an abelian group with identity 1.

Since $1' = 1, O(1) = 1;\ \omega^3 = \omega^6 = \omega^9 = \cdots = 1,\ O(\omega) = 3;\ (\omega^2)^3 = (\omega^2)^6 = \cdots = 1, O(\omega^2) = 3.$

Example Let $G = \{1, i, -1, -i\}$. $(G, \cdot)$ is an abelian group with identity 1.

$$O(1) = 1, O(i) = 4, O(-1) = 2, O(-i) = 4$$

Example $(Z, +)$ is a group with identity 0. Since $1.0 = 2.0 = ... = 0$, $O(0) = 1$.

But for $2 \in Z$, there is no positive integer m such that $m. 2 = 0$.

2 is said to be of infinite order or zeroth order.

***Theorem* 1** Let $(G, *)$ be a group. Then the identity of G is the only element in G of order 1.

Proof Let $a \neq e$ in G be of order 1. Then by definition, 1 is the least positive integer such that $a^1 = e$, i.e., $a = e$ which is a contradiction to the hypothesis that $a \neq e$. $\therefore$ e is the only element of order 1, in G.

***Theorem* 2** If n is a positive integer such that $a^n = e$ then $O(a) \leq n$.

Proof If n is the least positive integer such that $a^n = e$ then $O(a) = n$. Otherwise, if there exists a positive integer $m < n$ such that $a^m = e$ then $O(a) < n$.

Note:

1. The order of every element of a finite group is finite.
2. The order of any element of a finite group can never exceed the order of the group.

***Theorem* 3** The order of an element of a group is the same as that of its inverse.

Proof Let $(G, *)$ be a group and $a \in G$. Let $O(a) = m$ and $O(a^{-1}) = n$.

Since $O(a) = m$, m is the least positive integer such that

$$a^m = e \therefore (a^m)^{-1} = e^{-1} \therefore (a^{-1})^m = e \therefore O(a^{-1}) \leq m \quad \therefore n \leq m \tag{1}$$

Since $O(a^{-1}) = n$, n is the least positive integer such that

$$(a^{-1})^n = e \therefore (a^n)^{-1} = e \therefore a^n = e^{-1} = e \therefore O(a) \leq n \text{ i.e. } m \leq n \tag{2}$$

By (1) and (2) $m = n$, i.e., $O(a) = O(a^{-1})$

***Theorem* 4** If $(G, *)$ is a group of even order, it has an element a different from the identity element e such that $a^2 = e$. *(MCA 94M, N)*

Proof Every element in the group G possesses a unique inverse and the identity is its own inverse. Let $O(G) = 2n$. Omitting e, there are $2n - 1$ elements in G. If we pair off these $2n - 1$ elements and their inverses, we will be left with only one element, say, $a \neq e$ which is its own inverse.

$$\therefore a = a^{-1} \quad \therefore a * a = a^{-1} * a = e \quad \therefore a^2 = e$$

Definition Let $(G_1, *)$ and (G_2, Δ) be two groups with identities e_1 and e_2. The direct product of G_1 and G_2 is defined by $G_1 \times G_2 = \{(a_1, b_1) : a_1 \in G_1, b_1 \in G_2\}$.

Note: The direct product $(G_1 \times G_2, \cdot)$ is an algebraic structure in which the binary operation $\cdot$ is defined by

$$(a_1,b_1)\cdot(a_2,b_2) = (a_1 \times a_2, b_1 \Delta b_2)$$

$$\text{for any } (a_1,b_1),(a_2,b_2) \in G_1 \times G_2$$

Theorem 1 The direct product of the groups $(G_1, *)$ and (G_2, Δ) is again a group.

Proof Let $(G_1, *)$ and (G_2, Δ) be two groups with identities e_1 and e_2 respectively. Let us now define binary operation $\odot$ on $G_1 \times G_2$ by $(a_1, b_1) \odot (a_2, b_2) = (a_1 * a_2, b_1 \Delta b_2)$ for any $(a_1, b_1), (a_2, b_2) \in G_1 \times G_2$.

By definition of $\odot$ on $G_1 \times G_2$, $(a_1, b_1) \odot (a_2, b_2) \in G_1 \times G_2$ (closure).

Let $(a_1, b_1), (a_2, b_2), (a_3, b_3) \in G_1 \times G_2$

$$\begin{aligned}\text{Let } (a_1,b_1)\odot[(a_2,b_2)\odot(a_3,b_3)] &= (a_1,b_1)\odot(a_2 * a_3, b_2 \Delta b_3)\\ &= (a_1 * (a_2 * a_3), b_1 \Delta (b_1 \Delta b_3))\\ &= ((a_1 * a_2) * a_3, (b_1 \Delta b_2) \Delta b_3)\\ &= (a_1 * a_2, b_1 \Delta b_2)\odot(a_3,b_3)\\ &= [(a_1,b_1)\odot(a_2,b_2)]\odot(a_3,b_3) \text{ (associativity)}\end{aligned}$$

Let $e = (e_1, e_2) \in G_1 \times G_2$

$$\begin{aligned}(a_1,b_1)\odot e &= (a_1,b_1)\odot(e_1,e_2)\\ &= (a_1 * e_1, b_1 \Delta e_2)\\ &= (a_1,b_1)\end{aligned}$$

Similarly $e \odot (a_1, b_1) = (a_1, b_1)$ $e \in G_1 \times G_2$ is the identity. (existence of identity).

Let $(a_1,b_1)^{-1} = (a_1^{-1}, b_1^{-1})$

$$\begin{aligned}(a_1,b_1)\odot(a_1,b_1)^{-1} &= (a_1,b_1)\odot(a_1^{-1},b_1^{-1})\\ &= (a_1 * a_1^{-1}, b_1 \Delta b_1^{-1})\\ &= (e_1,e_2) = e.\end{aligned}$$

Similarly $(a_1, b_1)^{-1} \cdot (a_1, b_1) = e(a_1, b_1^{-1} = (a_1^{-1}, b_1^{-1}) \in G_1 \times G_2$ is the inverse of $(a_1, b_1) \in G_1 \times G_2$.

Note: 1 If $(G_1, *)$ and (G_2, Δ) are both abelian then $(G_1 \times G_2, \cdot)$ is also abelian.

2 If $G = G_1 \times G_2 \times \ldots \times G_n$ is a direct product of n groups, then

i. $e = (e_1, e_2 \ldots e_n)$ is the identity of G.

ii. $(a_1, a_2 \ldots a_n)^{-1} = (a_1^{-1}, a_2^{-1} \cdots a_n^{-1})$ is the inverse of $(a_1, a_2 \ldots a_n)$

iii. $(a_1, a_2, \ldots a_n)^m = (a_1^m, a_2^m \cdots, a_m^m)$

iv. G is abelian if each of $G_1, G_2, \ldots, G_n$ is abelian.

SUBGROUPS AND CYCLIC GROUPS

Subsemigroup Let $(S, *)$ be a semigroup. Let $S_1 \subset S$ be a non empty set. If $(S_1, *)$ is a semigroup then it is called a sub semigroup of $(S, *)$ *(MCA 98 A, 96 M, N, B.E. 94 A)*

Example $(N, +)$ is a semi group.

Let $S_1 = \{3, 6, 9...\}$. $S_1 \subset N$ is non empty and $(S_1, +)$ is a semigroup. $\therefore$ $(S_1, +)$ is a subsemigroup of $(N, +)$

Submonoid Let $(M, *)$ be a monoid. Let $M_1 \subset M$ be a non empty set. If $(M_1, *)$ itself is a monoid, then $(M_1, *)$ is a submonoid of $(M, *)$.

Example Let $W = \{0, 1, 2, 3,...\}$. $(W, +)$ is a monoid. Let $W_1 = \{0, 2, 4, 6, 8...\}$ $W_1 \subset W$ is non empty and $(W_1, +)$ itself is a monoid. $\therefore$ $(W_1, +)$ is a submonoid of $(W, +)$.

Theorem For any commutative monoid $(M, *)$, the set of idempotent elements of M forms a submonoid.

(MCA 98 A, 98 N, 96 N, 96 M, N, B.E. 94A)

Proof Let $(M, *)$ be a monoid with identity element e. Let $M_1 = \{a_1 : a \in M, a * a = a\}$.

Since $e * e = e$, $e \in M_1$ and M_1 is non empty. Since $M_1 \subset M$, associative property is true.

Let $a \in M_1 \therefore a \in M$, and $a * e = e * a = a$ and $e \in M_1$ is the identity of M_1.

Let us now prove closure property.

Let $a, b \in M_1$ $\quad \therefore\ a * a = a$ (1),

$b * b = b$ (2)

Also $a * b \in M$.

Now

$$\begin{aligned}(a*b)*(a*b) &= (a*b)*(b*a) \text{ [since } M \text{ is abelian]}\\ &= a*(b*b)*a\\ &= a*b*a\\ &= a*(a*b) \text{ [since } M \text{ is abelian]}\\ &= (a*a)*b\\ &= a*b \quad \text{by (1)}\end{aligned}$$

$a*b \in M_1$. (closure property)

$\therefore$ $(M_1, *)$ is a monoid and hence it is a submonoid of $(M, *)$.

Subgroups

A non-empty subset H of G is said to be a subgroup of G, if H itself is a group under the same operation defined on G.

Example Let $3Z = \{0, \pm 3, \pm 6, \pm 9 \cdots\}$. $(3Z, +)$ is a subgroup of $(Z, +)$.

Example $(Z, +)$ is a group of $(R, +)$

Example $G = \{1, i, -1, -i\}$ is a subgroup of $C - \{0\}$ with respect to multiplication of complex numbers.

Note: Let $(G, *)$ be a group. Then the two subgroups $(\{e\}, *)$ and $(G, *)$ are called trival subgroups or improper subgroups of $(G, *)$. All other subgroups of $(G, *)$ are called nontrivial subgroups or proper subgroups of $(G, *)$.

***Theorem* 1** A non-empty subset H of a group G is a subgroup of $(G, *)$

(i) $\forall$ iff $a, b \in H, a * b \in H$ (ii) $a \in H \Rightarrow a^{-1} \in H$.

Proof Let $(H, *)$ be subgroup of $(G, *)$. Then $(H, *)$ is a group. Hence $a, b \in H \Rightarrow a * b \in H, a \in H \Rightarrow a^{-1} \in H$. Hence the conditions are true.

Conversely, let us assume $H \subset G$ and $a,b \in H \Rightarrow a*b \in H \cdots$ (i)

$a \in H \Rightarrow a^{-1} \in H \ldots$ (ii)

By (i) closure property is true.

Since $H \subset G$, associative property is true.

Let $a \in H$ by (ii), $a^{-1} \in H$ by (i) $a*a^{-1} \in H$

i.e. $e \in H$

Let $x \in H$ $\therefore x \in G$ and $x*e = e*x = x$

$\therefore e \in H$ is the identity element.

[Note that the identity element of subgroup is the same as the identity element of G].

$$\forall a \in H, a^{-1} \in H \text{ and } a*a^{-1} = a^{-1}*a = e.$$

$\therefore$ (*H*, *) is a group. Since $H \subset G$, (*H*, *) is a subgroup of (*G*, *)

***Theorem* 2** A non-empty subset *H* of a group *G* is a subgroup of (*G*, *) iff $a,b \in H \Rightarrow a*b^{-1} \in H$.

(MCA 94 M, B.E. 97 Oct., 96 Oct., 96 Oct.).

Proof Let (*H*, *) be a subgroup of (*G*, *). Then (*H*, *) itself is a group. Hence $a,b \in H, b^{-1} \in H$. $\therefore a*b^{-1} \in H$. Conversely let $H \subset G$ and $a*b^{-1} \in H \forall a,b \in H$.

Let $a \in H$. $\therefore a*a^{-1} \in H$ i.e., $e \in H$. $\forall x \in H$, $x \in G$,

$e*x = x*e = x \therefore e \in H$ $\forall x \in H, x \in G, e*x = x*e = x \therefore e \in H$ is the identity.

Since $H \subset G$, * is associative in *G*.

Now let $a \in H, a, e \in H$, $e*a^{-1} \in H$ $a^{-1} \in H$ and $a*a^{-1} = a^{-1}*a = e$. $\therefore a^{-1} \in H$ is the inverse of $a \in H$.

Let $a,b \in H \because b \in H, b^{-1} \in H$ and $a,b^{-1} \in H \therefore a*(b^{-1})^{-1} \in H \therefore$ $a*b \in H$

$\therefore$ * is closed in *H*. $\therefore$ (*H*, *) is a group. Since $H \subset G$, (H, *) is a subgroup of (*G*, *).

***Theorem* 3** Let *H* be a non empty finite subset of a group (*G*, *). Then (*H*, *) is a subgroup of (*G*, *) if $a,b \in H \Rightarrow a*b \in H$.

Proof Suppose (*H*, *) is a subgroup of (*G*, *). (*H*, *) itself is a group. $\forall a,b \in H, a*b \in H$.

Conversely, let $H \subset G$ be non empty, finite and $a*b \in H$ $\forall a,b \in H$. $\therefore$ * is closed in *H*. Since $H \subset G$, * is associative in *H*. Suppose $a \in H$. Since *H* is closed under *, $a^2, a^3 \ldots, \in H$. Thus an infinite collection of elements must be in *H*. But *H* is finite. $\therefore$ There must be a repetition of elements in this collection. i.e. for some integers *r*, *s* and $r > s$, $a^r = a^s \therefore a^{r-s} = a^0 = e \in H$. $\therefore e \in H$ is the identity.

Since $r-s-1 \geq 0, a^{r-s-1} \in H$ and $a*a^{r-s-1} = a^{r-s} = e$.

$\therefore a^{r-s-1} \in H$ is the inverse of a. $\therefore (H,*)$ is a group.

Since $H \subset G$, $(H,*)$ is a subgroup of $(G,*)$

***Example* 5.12** If *H* and *K* are two subgroups of *G*, then $H \cap K$ is a subgroup of G. (93A)

Solution $e \in H, e \in K \therefore e \in H \cap K \therefore H \cap K$ is non empty.

Let *a*, *b* $\in$ $H \cap K$. $\therefore a,b \in H, a,b \in K$. Since *a*, *b* $\in K$ and *K* is a subgroup,

$$a*b^{-1} \in K. \therefore a*b^{-1} \in H \cap K.$$

$a*b^{-1} \in H \cap K$ $\forall a,b \in H \cap K$ and $H \cap K \subset G$.

$\therefore (H \cap K,*)$ is a subgroup of $(G,*)$.

Example 5.13 If H and K are subgroups of G $H \cup K$ need not be a subgroup of G.

Solution To prove this, let us give a counter example $(Z, +)$ is a group.

Let $H = 3Z = \{0, \pm 3, \pm 6 \ldots\}$

$K = 2Z = \{0, \pm 2, \pm 4 \ldots\}$

H and K are subgroups of $(Z, +)$.

$$H \cup K = \{0, \pm 3, \pm 6, \ldots \pm 2, \pm 4 \ldots\}$$

Now $3, 2 \in H \cup K$ but $3 + 2 = 5 \notin H \cup K$.

$\therefore$ $H \cup K$ is not a subgroup of $(Z, +)$.

Example 5.14 The union of two subgroups of $(G, *)$ is a subgroup of $(G, *)$ iff one is contained in the other.
(MCA 98)

Solution Suppose H and K be two subgroups of G and $H \subset K$ or $K \subset H$.

If $H \subset K$, $H \cup K = K$ is a subgroup of G.

If $K \subset H$, $H \cup K = H$ is a subgroup of G.

Conversely, $H, K, H \cup K$ are subgroups of G. We have to prove that $H \subset K$ or $K \subset H$.

Suppose $H \not\subset K$ and $K \not\subset H$

$\because$ $H \not\subset K$, there exists $a \in H$ and $a \notin K$.

$\because$ $K \not\subset H$, there exists $b \in K$ and $b \notin H$.

Now $a, b \in H \cup K$. Since $H \cup K$ is a subgroup $a * b \in H \cup K$. $\therefore a * b \in H$ or $a * b \in K$

If $a * b \in H$, we have $(a^{-1} * a) * b \in H$, i.e., $e * b \in H$, i.e., $b \in H$.

This is a contradiction to the assumption that $b \notin H$. *If* $a * b \in K$, $a * (b * b^{-1}) \in K$

i.e., $a * e \in K$, *i.e.,* $a \in K$. This is a contradiction to the assumption that $a \notin k$

$\therefore$ We have either $H \subset K$ or $K \subset H$.

Example 5.15 The set of all elements 'a' of a group $(G, *)$ such that $a * x = x * a \ \forall \ x \in G$ is a subgroup of G. [This subgroup is known as centre of the group $(G, *)$] *(MCA 96 M,N, 97 M,N).*

Solution Let $H = \{a \in G : a * x = x * a \ \forall x \in G\}$

Now $e \in G$ and $e * x = x * e \ \forall x \in G$.

$$\therefore e \in H \ \therefore H \text{ is not empty.}$$

Let $a, b \in H$. $\therefore a * x = x * a \ \forall \ x \in G$ (1)

and $$b * x = x * b \quad \forall x \in G$$

$$\therefore b^{-1} * b * x * b^{-1} = b^{-1} * x * b * b^{-1}$$

$$e * x * b^{-1} = b^{-1} * x * e$$

$$x * b^{-1} = b^{-1} * x \quad \forall x \in G \qquad (2)$$

Now $$(a * b^{-1}) * x = a * (b^{-1} * x)$$

$$= a * (x * b^{-1}) \text{ by (2)}$$

$$= (a * x) * b^{-1}$$

$$= (x * a) * b^{-1} \text{ by (1)}$$

$$= x * (a * b^{-1}) \ \forall \ x \in G.$$

$\therefore a * b^{-1} \in H$

$(H, *)$ is a subgroup of $(G, *)$.

Cyclic Groups *(92A, 91N, 91A)*

A group $(G, *)$ is said to be a cyclic group if there exists an element $a \in G$ such that every element of G can be expressed as some integral power of a. a is called generator of G. We write $G = (a)$.

Note: An additive group $(G, +)$ is a cyclic group generated by an element $a \in G$, iff each element of G is of the form na for some integer n. We write $G = (a)$.

Example Let $G = \{1, i, -1, -i\}$ $(G, \cdot)$ is a group.

It is cyclic for $1 = i^4, i = i^1, -1 = i^2, -i = i^3$. $\therefore$ i is a generator and $G = (i)$.

Similarly $-i$ is also a generator and $G = (-i)$.

Example Let $G = \{1, \omega, \omega^2\}$. $(G, \cdot)$ is a cyclic group generated by ω. ω^2 is also a generator.

Example $(Z, +)$ is a cyclic group generated by 1. For $0 = 0.1, 1 = 1.1, 2 = 2.1$. Similarly $-1 = (-1).1, -2 = (-2).1$, etc. Another generator is -1.

Example $(Q, +)$ is not a cyclic group. Since $\frac{1}{2} \in Q$ but $\frac{1}{2}$ cannot be written as an integral multiple of 1. i.e. there is no $a \in Q$ such that $x = n \cdot a \ (n \in Z, x \in Q)$.

***Theorem* 1** Every cyclic group is abelian. *(B.E. 98A, 95 Oct)*

Proof Let $(G, *)$ be a cyclic group with generator a. Let $x, y \in G$. Then $x = a^m, y = a^n$ for some $m, n \in Z$.

$$x * y = a^m * a^n$$

$$= a^{m+n}$$

$$= a^{n+m} = a^n * a^m$$

$$x * y = y * x \ \forall \ x, y \in G.$$

$\therefore$ $(G, *)$ is abelian.

Note: The converse of the above theorem need not be true i.e. an abelian group need not be cyclic. *(B.E. 98A, 95 Oct.)*

For example, let us consider Klein's 4-group, $G = \{e, a, b, c\}$. The operation is $a^2 = b^2 = c^2 = e$ and $a * b = b * a = c, b * c = c * b = a, a * c = c * a = b$.

The Cayley's table is as follows:

*	e	a	b	c
e	e	a	b	c
a	a	e	c	b
b	b	c	e	a
c	c	b	a	e

$(G, *)$ is an abelian group, but it is not cyclic since no element of G can generate G.

***Theorem* 2** Let $(G, *)$ be a cyclic group generated by a. Then a^{-1} is also a generator.

Proof Let $G = (a)$ be a cyclic group generated by a.

If $x \in G$ then $x = a^n$ for some $n \in Z$. $\therefore$ $x = a^n = (a^{-1})^n (-n \in Z)$. $\therefore$ a^{-1} is also a generator of G.

***Theorem* 3** Any subgroup of a cyclic group is itself a cyclic group. *(MCA 96N, 95N)*

Proof Let $G = (a)$ be a cyclic group and H be a subgroup of G. Elements of H are of the form $a^0 \cdot a^k (k \in Z)$. If $a^k \in H, a^{-k}$ is also an element of H. $\therefore$ Either k or $-k$ is positive.

Let m be the least positive integer such that $a^m \in H$.

We shall prove that $H = (a^m)$. Let $c \in H \because H \subset G, c \in G \therefore c = a^n$ for some $n \in Z$. Now $m, n \in Z$. By division algorithm, there exists integers q and r such that $n = mq + r$ where $0 \leq r < m$.

$$\text{Now} \quad c = a^n$$
$$= a^{mq+r}$$
$$= a^{mq} * a^r$$

$\therefore a^r = a^{-mq} * c = (a^m)^{-q} * c \in H$. Since $c \in H$, $(a^m)^{-q} \in H$ and H is a subgroup. But $0 \leq r < m$ and m is the least positive integer such that $a^m \in H. \therefore r = 0$.

$$\therefore c = a^{mq} = (a^m)^q$$

$\therefore$ Every element of H can be written as an integer power of a^m. $\therefore H = (a^m)$ is a cyclic group.

Note: The converse of the above theorem is not true. i.e. If every proper subgroup of a group is cyclic the group need not be cyclic. For example let us consider Klein's 4-group $G = \{e, a, b, c\}$ units respect to $a^2 = b^2 = c^2 = e$, $a * b = b * a = c, b * c = c * b = a, c * a = a * c = b$. The proper subgroups $\{e, a\}$, $\{e, b\}$, $\{e, c\}$ are cyclic with generators a, b, c respectively. The group is not cyclic.

***Theorem* 4** The order of a cyclic group is the same as the order of its generator.

Proof Let $G = (a)$ be a cyclic group.

Case 1 Let $O(a) = n$ (finite). Then $a^n = e = a^0$. Let us consider $\{a, a^2 \ldots a^n = e\}$.

By closure property these are elements of G. Also they are distinct for otherwise if $1 \leq r < S \leq n$, $a^S = a^r$ then $a^{s-r} = a^0 = e$, and $0 < s - r < n$. $\therefore O(a) < n$ which is a contradiction.

All elements $a, a^2 \ldots a^n = c$ are distinct elements of G itself.

Now let $a^m \in G, m \in Z$ $m, n \in Z$. By division algorithm, there exists integers q and r such that $m = nq + r, 0 \leq r < n$. $a^m = a^{nq+r} = (a^n)^q * a^r = e * a^r = a^r$. Since $0 \leq r < n, a^m$ is one of the elements of $\{a, a^2 ..., a^n\} = e$.

$$\therefore G = \{a, a^2 \cdots, a^n = e\} \therefore O(G) = n = O(a).$$

Case 2 Let a be of infinite order.

We shall prove G is of infinite order. Suppose G is of finite order then for some integers, $r, s, r > s$, $a^r = a^s$ $\therefore a^{r-s} = a^0 = e$. $\therefore$ $O(a)$ is finite which is a contradiction, $\therefore$ $O(G)$ is infinite and $O(G) = O(a)$.

In both cases, the order of cyclic group is the same as the order of its generator.

***Theorem* 5** A finite group of order n containing an element a of order n is cyclic.

Proof Let $O(G) = n$. Let $a \in G$ such that $O(a) = n$. Let $H = (a)$ be a cyclic group generated by a.

$\therefore H = \{a, a^2 \cdots a^n = e\}$. Every element of H is an element of G (By closure property).

$$\therefore H \subset G \tag{1}$$

$$\text{Now} \quad O(H) = n = O(G) \tag{2}$$

By (1) and (2) $H = G$ $\therefore$ G is cyclic.

Note: Let $G=(a)$ be a cyclic group of order n. Then a^m is a generator of G iff $(m, n)=1$.

i.e., the greatest common divisor of m and n is 1.

Example 5.16 Find the number of generators of a cyclic group of order 5.

Solution Let $G=(a)$ be a cyclic group of order 5. Then $G=\{a,a^2,a^3,a^4,a^5=e\}$. Since $(1, 5)=1$, $(2, 5)=1$, $(3, 5)=1$, $(4, 5)=1$. The generators are a,a^2,a^3,a^4. $\therefore$ The number of generators is 4.

Example 5.17 Find the generators and the number of generators of a cyclic group of order 8.

Solution Let $G=(a)$ be a cyclic group of order 8.

$$G=\{a,a^2,a^3,a^4,a^5,a^6,a^7,a^8=e\}$$

Since $(1, 8)=1$, $(3, 8)=1$, $(5, 8)=1$, $(7, 8)=1$, the generators are a,a^3,a^5,a^7.

The number of generators is 4.

PERMUTATION GROUPS

Permutation

Suppose $S=\{x_1,x_2\cdots x_n\}$ is a finite set of n elements. A bijective mapping ϕ of S onto S itself is defined as a permutation of S. We can write ϕ as

$$\begin{pmatrix} x_1 & x_2 & \cdots & x_n \\ x_{i_1} & x_{i_2} & \cdots & x_{i_n} \end{pmatrix} \text{ where } x_{i_j} \in S \text{ itself.}$$

Leaving x's we can write as

$$\begin{pmatrix} 1 & 2 & 3 & \cdots & n \\ i_1 & i_2 & i_3 & \cdots & i_n \end{pmatrix}$$

Example Suppose $S=\{1, 2, 3\}$, the possible permutations are $\begin{pmatrix} 1 & 2 & 3 \\ 1 & 2 & 3 \end{pmatrix}, \begin{pmatrix} 1 & 2 & 3 \\ 1 & 3 & 2 \end{pmatrix}, \begin{pmatrix} 1 & 2 & 3 \\ 3 & 2 & 1 \end{pmatrix}$.

$\begin{pmatrix} 1 & 2 & 3 \\ 2 & 1 & 3 \end{pmatrix}, \begin{pmatrix} 1 & 2 & 3 \\ 2 & 3 & 1 \end{pmatrix}, \begin{pmatrix} 1 & 2 & 3 \\ 3 & 1 & 2 \end{pmatrix}$.

Let $f=\begin{pmatrix} 1 & 2 & 3 \\ 1 & 3 & 2 \end{pmatrix}$, $g=\begin{pmatrix} 1 & 2 & 3 \\ 2 & 1 & 3 \end{pmatrix}$. Then the product of permutations f and g is denoted by $f \circ g$.

$$f \circ g=\begin{pmatrix} 1 & 2 & 3 \\ 1 & 3 & 2 \end{pmatrix} \circ \begin{pmatrix} 1 & 2 & 3 \\ 2 & 1 & 3 \end{pmatrix} = \begin{pmatrix} 1 & 2 & 3 \\ 3 & 1 & 2 \end{pmatrix}.$$

Note: The composition of permutations is explained below.

$$\begin{matrix} 1 & 2 & 3 \\ \downarrow & \downarrow & \downarrow \\ 2 & 1 & 3 \\ \downarrow & \downarrow & \downarrow \\ 3 & 1 & 2 \end{matrix}$$

Similarly $g \circ f=\begin{pmatrix} 1 & 2 & 3 \\ 2 & 3 & 1 \end{pmatrix}$

$\therefore f \circ g \neq g \circ f$. This is due to the fact that composition of mappings need not be commutative.

$i = \begin{pmatrix} 1 & 2 & 3 \\ 1 & 2 & 3 \end{pmatrix}$ is identity permutation.

If $f = \begin{pmatrix} 1 & 2 & 3 \\ 1 & 3 & 2 \end{pmatrix}$ then $f^{-1} = \begin{pmatrix} 1 & 2 & 3 \\ 1 & 3 & 2 \end{pmatrix}$ is called inverse of f.

Also we can verify that $f \circ f^{-1} = f^{-1} \circ f = i$ (identity permutation).

Symmetric Group

Theorem 1 The set G of all permutations on S forms a group with respect to composition of functions as operation. This group is called symmetric group of degree n on S.

Proof Let $f,\ g \in G$. Then f and g are bijective mappings of S onto S itself. $f \circ g$ is also a bijective mapping of S into S itself. $\therefore f \circ g \in G$ (closure).

We know that composition of mappings is associative (associativity). $i : S \to S$ defined as $i(x) = x$. $\therefore\ i \in G$ is the identity (existence of identity).

If $f \in G,\ f$ is bijective. Then f^{-1} also exists and f^{-1} is also bijective. Also $f^{-1} \circ f = f \circ f^{-1} = i$. $\therefore\ f^{-1} \in G$ is the inverse of f (existence of inverse).

$\therefore\ (G, \circ)$ is a group.

Note: If S is a finite set having n elements then G is denoted by $A(S)$ or S_n. It is called the symmetric group on symbols or symmetric group of degree n. Also order of $S_n = \circ\ (S_n) = n!$

For example $O(S_3) = 3! = 6,\ O(S_4) = 4! = 24$

Symmetries of an Equilateral Triangle (92A, 91A, N)

Let 1, 2, 3 denote the vertices of an equilateral triangle with O as centroid. The symmetries are the three rotation in its plane about O through 120°, 240°, 360° in the clockwise sense and the three reflections about their medians. i.e., an equilateral triangle has three rotational symmetries and three axial symmetries.

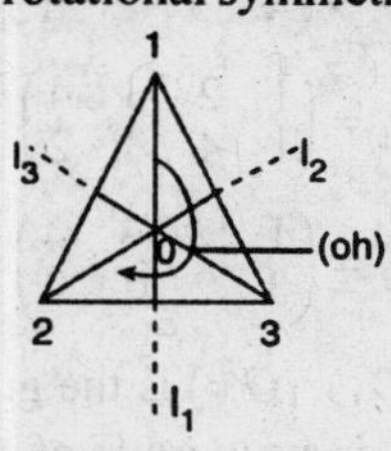

Rotation through 120° is denoted by R_{120} and $R_{120} = \begin{pmatrix} 1 & 2 & 3 \\ 3 & 1 & 2 \end{pmatrix}$

Similarly $R_{240} = \begin{pmatrix} 1 & 2 & 3 \\ 2 & 3 & 1 \end{pmatrix}$ and $R_{360} = R_0 = \begin{pmatrix} 1 & 2 & 3 \\ 1 & 2 & 3 \end{pmatrix}$.

Let L_1, L_2, L_3 be the reflections about the lines l_1, l_2, l_3 respectively.

$$L_1 = \begin{pmatrix} 1 & 2 & 3 \\ 1 & 3 & 2 \end{pmatrix},\ L_2 = \begin{pmatrix} 1 & 2 & 3 \\ 3 & 2 & 1 \end{pmatrix},\ L_3 = \begin{pmatrix} 1 & 2 & 3 \\ 2 & 1 & 3 \end{pmatrix}$$

Now $G = \{R_0, R_{120}, R_{20}, L_1, L_2, L_3\}$ is a group with respect to composition of function. Here (G,O) is called dihedral group.

Actually $G = S_3$.

Symmetrices of a Square

Let 1, 2, 3, 4 denote the vertices of a square and O be its centre. The square has four rotational symmetries obtained by rotating the square in its plane about O through 90°, 180°, 270°, 360° in the clockwise sense. These are denoted by R_{90}, R_{180}, R_{270}, $R_{360} = R_0$ respectively. There are four axial symmetries. The reflections about the lines l_1, l_2 bisecting opposite sides are denoted by L_1 and L_2 respectively. The reflection about the diagonals are denoted by D_1 and D_2 respectively.

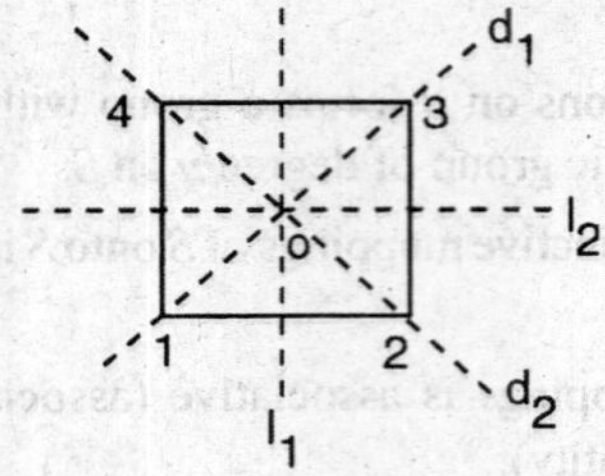

$$R_{90} = \begin{pmatrix} 1 & 2 & 3 & 4 \\ 4 & 1 & 2 & 3 \end{pmatrix}$$

$$R_{180} = \begin{pmatrix} 1 & 2 & 3 & 4 \\ 3 & 4 & 1 & 2 \end{pmatrix}$$

$$R_{270} = \begin{pmatrix} 1 & 2 & 3 & 4 \\ 2 & 3 & 4 & 1 \end{pmatrix}$$

$$R_{360} = R_0 = \begin{pmatrix} 1 & 2 & 3 & 4 \\ 1 & 2 & 3 & 4 \end{pmatrix}$$

$$L_1 = \begin{pmatrix} 1 & 2 & 3 & 4 \\ 2 & 1 & 4 & 3 \end{pmatrix}$$

$$L_2 = \begin{pmatrix} 1 & 2 & 3 & 4 \\ 4 & 3 & 2 & 1 \end{pmatrix}$$

$$D_1 = \begin{pmatrix} 1 & 2 & 3 & 4 \\ 1 & 4 & 3 & 2 \end{pmatrix}$$

$$D_2 = \begin{pmatrix} 1 & 2 & 3 & 4 \\ 3 & 2 & 1 & 4 \end{pmatrix}$$

Let $G = (R_0, R_{90}, R_{270}, R_{360}, L_1, L_2, D_1, D_2)$. $(G, \circ)$ is the group of symmetries of a square and $O(G) = 8$. $(G, \circ)$ is called dihedral group. In fact $(G, \circ)$ is a subgroup of $(S_4, \circ)$.

Orbit Let S be a finite set and $s \in S$. Let Q be a permutation. If there exists a smallest positive integer l such that $Q^l s = s$, then the set $\{s, Qs, Q^2 s, \ldots Q^{l-1} s\}$ is called orbit of S under Q. The ordered set $\{s, Qs, Q^2 s, \ldots Q^{l-1} s\}$ is called a cycle of Q.

Example Let $s = \{1, 2, 3, 4, 5, 6\}$ and

$$Q = \begin{pmatrix} 1 & 2 & 3 & 4 & 5 & 6 \\ 2 & 1 & 3 & 5 & 6 & 4 \end{pmatrix}$$

$Q1 = 2$, $Q^2 1 = Q.Q_1 = Q2 = 1$. The orbit or cycle of 1 is (1 2). Similarly the orbit of 3 is (3) and the orbit of 4 is (4 5 6). Q can be written as $Q = (1\ 2)\ (3)\ (4\ 5\ 6)$. Hence every permutation is a product of distinct cycles.

Length of a cycle The number of elements in a cycle is called the length of the cycle.

Transposition A cycle of length 2 is called a transposition. Every permutation is product of transpositions.

Example Let us consider an m-cycle (1 2 3...m). This can be written as (1 2 3...m) = (1 2) (1 3)..(1m).

Note 1: A permutation is a product of distinct cycles. A cycle is a product of transpositions. Hence a permutation is a product of transpositions. The way of expression of a permutation into a transposition is not unique.

Definition Let S_n be the symmetric group of degree n on a set S. A permutation $Q \in S_n$ is said to be an even permutation if it can be expressed as a product of an even numbers of transpositions. Otherwise, it is said to be odd.

Note:

1. The product of two even permutations is even.
2. The product of one odd permutation and an even permutation is odd.
3. The product of two odd permutations is even.

Theorem 2 The set of all even permutations on a set S forms a group with respect of composition of permutations.

Proof Let $S = \{1, 2..., n\}$. Let S_n be the symmetric group of degree n. $O(S_n) = n!$ Let A_n denote the set of all even permutations. $A_n \subset S_n$ and A_n is finite. Let $f, g \in A_n$. $\therefore f$ and g are even. $f \circ g$ is also even. $\therefore f \circ g \in A_n$. Since A_n is finite, $(A_n, \circ)$ is a subgroup of $(S_n, \circ)$. $\therefore (A_n, \circ)$ itself is a group. This group is called alternating group $O(A_n) = \dfrac{n!}{2}$.

Note: 1. The identity permutation $i = \begin{pmatrix} 1 & 2 & \cdots & n \\ 1 & 2 & \cdots & n \end{pmatrix}$ is even and hence $i \in A_n$. Also i = (1 2) (21).

2. If $f \in A_n, f^{-1} \in A_n$ is also even.

3. Since the product of two odd permutations is even, the set of all odd permutations does not form a group.

Theorem 3 Every row (or column) in the composition table of a group (G, *) is a permutation of the elements of G. Further each row (or column) is a distinct permutation. (MCA 97 M)

Proof If possible, let the row corresponding to $a \in G$ be such that $a * b_1 = k$ and $a * b_2 = k$ for some $b_1, b_2 \in G$ and $b_1 \neq b_2$. $\therefore a * b_1 = a * b_2$.

By cancellation law, $b_1 = b_2$ which is a contradiction. Hence a row cannot contain an element of G more than once. A similar result holds good for the columns of G.

Let us now prove that every element of G appears in each row of the table.

Let us consider the row corresponding to $a \in G$ and $b \in G$. Since $b = a * (a^{-1} * b)$, b must appear in that row. This is true for column also. Hence every row (or column) of the composition table is obtained by a permutation of elements of G and each row (or column) is a distinct permutation of G.

COSETS AND LAGRANGE'S THEOREM

Let $(G, *)$ be a group and (H, *) be a subgroup (G, *). For $a, b \in G$, we say that a is congruent to mod H, written as $a \equiv b \bmod H$ if $a * b^{-1} \in H$.

Note: In additive group, $a \equiv b \bmod H$ if $a - b \in H$.

Theorem 1 The relation $a \equiv b \bmod H$ is an equivalence relation.

Proof Since H is a subgroup of G, $e \in H$.

$\forall a \in G,\ a * a^{-1} = e \in H\ \therefore\ a \equiv a \bmod H$ (Reflexive)

If $a \equiv b \bmod H$ then $a * b^{-1} \in H. \therefore (a * b^{-1})^{-1} \in H. \therefore b * a^{-1} \in H. \therefore b \equiv c \bmod H.$

$\therefore a \equiv b \bmod H \Rightarrow b \equiv a \bmod H$. (symmetry).

Suppose $a \equiv b \bmod H$, $b \equiv c \bmod H$, $a, b, c \in G$.

$a * b^{-1} \in H, b * c^{-1} \in H.$

Since H is a subgroup, $(a * b^{-1}) * (b * c^{-1}) \in H$. i.e., $a * c^{-1} \in H. \therefore a \equiv c \bmod H$. (Transitivity).

Hence $\equiv$ is an equivalence relation.

Cosets

(MCA 97 M, 94 N)

Let $(H, *)$ be a subgroup of $(G, *)$ and $a \in G$. Then $H * a = \{h * a : h \in H\}$ is called right coset of H in G. $H * a$ is usually denoted as Ha.

Similarly $a * H = \{a * h : h \in H\}$ is called left coset of H in G. $a * H$ is usually denoted as $a . H$.

Example Let $G = \{1, -1, i, -i\}$ and $\cdot$ be usual multiplication. $H = \{1, -1\}$ is a subgroup of G.

Right cosets:

$$H1 = \{1.1, 1.(-1)\} = \{1, -1\} = H.$$
$$H(-1) = \{-1, 1\} = \{1, -1\} = H.$$
$$Hi = \{i, -i\}$$
$$H(-i) = \{-i, i\} = \{i, -i\} = Hi$$

Note:

1. H and Hi are the two distinct right cosets of H in G.
2. $H \cup Hi = G$, i.e., G is partitioned into two distinct right cosets of H in G.
3. $H \cap Hi = \{\}$. Any two right cosets of H in G are either identical or disjoint.
4. The right cosets of H in G are also subsets of H in G.
5. H itself is a right coset and the number of elements in each right coset is the same as the number of elements in H.
6. If $a \in H, Ha = H$. If $a \in G, a \notin H, Ha \neq H$.
7. In a similar way we can form the left cosets of H in G and the above six results are true.

Example Let us consider $(Z_4, +_4)$ and $H = \{[0], [2]\}$ $(H, +_4)$ is a subgroup of $(Z_4, +_4)$. The right cosets of H in G are as follows

$$H + [0] = \{[0], [2]\} = H$$
$$H + [1] = \{[1], [3]\}$$
$$H + [2] = \{[2], [0]\} = \{[0], [2]\} = H$$
$$H + [3] = \{[3], [1]\} = \{[1], [3]\} = H + [1]$$

Theorem 2 Let $(G, *)$ be a group and $(H, *)$ be a subgroup. Any two right cosets of H in G are either identical or disjoint.

(95 Oct. 91 N, 96 N)

Proof Let $a, b \in G$ and Ha, Hb be two right cosets of H in G. Let $Ha \cap Hb \neq \phi$. Then there exists an element $c \in Ha \cap Hb$.

$\therefore c \in Ha$ and $c \in Hb$. $\therefore$ There exists $h_1, h_2 \in H$ such that $c = h_1 * a$, $c = h_2 * b$.

$$\therefore h_1 * a = h_2 * b \therefore h_2^{-1} * h_1 = b * a^{-1}, \text{ but } h_2^{-1} * h_1 \in H.$$

$\therefore b * a^{-1} \in H. \therefore H = H(b * a^{-1}) \therefore Ha = Hb$. This proves the theorem.

Theorem 3 Let $(G, *)$ be a group and $(H, *)$ be a subgroup. There is one-to-one correspondence between any two right cosets of H in G.

Proof Let $a, b \in G$. Ha and Hb are two right cosets of H in G.

Let $f : Ha \to Hb$ be a function defined by

$$f(h * a) = h * b \ \forall \ h * a \in Ha.$$

$\forall \ k * b \in Hb,\ k \in H,$

there exists $k * a \in Ha$ such that

$$f(k * a) = k * b \therefore f \text{ is onto} \tag{1}$$

Suppose $f(h_1 * a) = f(h_2 * a)$ for $h_1, h_2 \in H$.

We have $h_1 * b = h_2 * b$ and hence $h_1 = h_2$

$\therefore f$ is one-one (2)

By (1) and (2) the theorem follows.

Corollary If G is finite and $a \in G$ then Ha and H have the same number of elements.

Theorem 4 **(Lagrange's theorem)**

If $(G, *)$ is a finite group and $(H, *)$ is a subgroup then the order of H is a divisor of order of G.

(MCA 94 N).

Proof Since G is finite, $H \subset G$, H is also finite, Let $O(G) = n$ and $O(H) = M$. Now G is partitioned into finite number of distinct right cosets of H in G, say k right cosets namely $H, Ha, Ha_2, \ldots, Ha_{k-1}$. Also order of each coset is the order of H.

$$\therefore O(Ha_1) = O(Ha_2) = \ldots = O(Ha_{k-1}) = O(H) = m$$

Also
$$O(G) = O(H) + O(Ha_1) + \ldots + O(Ha_{k-1})$$
$$n = m + m + \ldots + m \,(k \text{ times})$$
$$= km$$

$\therefore m$ divides n. i.e., the order of H divides the order of G.

Definition Let $(H, *)$ be a subgroup of $(G, *)$. The index of H in G is the number of distinct right cosets of H in G. It is denoted by $I_G(H)$. If G is finite

$$I_G(H) = \frac{O(G)}{O(H)}$$

Note: The converse of Lagrange's theorem need not be true. i.e. If m divides $O(G)$ then G need not have a subgroup of order M. For example (A_4, O) (alternating group on 4 symbols) is a group and $O(A_4) = \frac{4!}{2} = 12$. Also 6 divides 12. But (A_4, O) has no subgroup of order 6.

Corollary 1 If $(G, *)$ is a finite group and $a \in G$ then $O(a)$ divides $O(G)$.

Proof Let us consider the cyclic subgroup generated by a.

Let $O(a) = m \therefore (a) = \{a, a^2, \ldots, a^m = e\} \therefore O((a)) = m$ divides $O(G)$. (i.e.) $O(a)$ divides order of G.

Corallary 2 If $(G, *)$ is a finite group and $a \in G$ then $a^{O(G)} = e$.

Proof Let $O(a) = m$ and $O(G) = n$.

m is the least positive integer such that $a^m = e$

Also by corollary (1) m divides n.

Therefore $n = mq$ for some integer q.

$$a^{O(G)} = a^n = a^{mq} = (a^m)^q = e^q = e$$

Corollary 3 If $(G, *)$ is a finite group whose order is a prime number P then $(G, *)$ is a cyclic group.

Proof Let $(H, *)$ be a subgroup of $(G, *)$. Since $O(G) = p$, $O(H)$ divides p. But p is a prime number. We have either $O(H) = 1$ or $O(H) = p$.

If $O(H) = 1$ then $H = \{e\}$

If $O(H) = p$ then $H = G$.

Therefore, G has no nontrivial subgroups.

Let $a \in G, a \neq e$ and $H = (a)$. Now H is a subgroup of G and $H \neq \{e\}$

Therefore $H = G$,

Therefore $G = (a)$

Therefore $(G, *)$ is cyclic.

Therefore group of prime order is cyclic.

Note 1: Since a cyclic group is abelian, a group of prime order is abelian. For example, since 13 is a prime number, a group of order 13 will be cyclic and hence abelian.

Note 2: Every group of order ≤ 5 is abelian.

NORMAL SUBGROUPS AND FACTOR GROUPS

Normal Subgroups

Let $(N, *)$ be a subgroup of $(G, *)$ $\forall g \in G$ and $n \in N, g * n * g^{-1} \in N$ then $(N, *)$ is said to be a normal subgroup of $(G, *)$.

Also, suppose $gNg^{-1} = \{g * n * g^{-1} : n \in N\}$. Then $(N, *)$ is normal in G if $gNg^{-1} \subset N \; \forall g \in G$.

Theorem 1 $(N, *)$ is a normal subgroup of $(G, *)$ iff and only $gNg^{-1} = N \; \forall g \in G$. *(B.E 98 Oct.)*

Proof Suppose $gNg^{-1} = N \; \forall g \in G$. Then $gNg^{-1} \subset N \; \forall g \in G$. Hence N is normal in G. Conversely suppose N is normal in G. We have

$$gNg^{-1} \subset N \; \forall \; g \in G \qquad (1)$$

Also $g^{-1}Ng = (g^{-1})N(g^{-1})^{-1} \subset N. \therefore g^{-1}Ng \subset N$

Now $N = g(g^{-1}Ng)g^{-1} \subset gNg^{-1} \forall \; g \in G$

$\therefore N \subset gNg^{-1} \; \forall \; g \in G$ (2)

By (1) and (2) $gNg^{-1} = N$

Theorem 2 The subgroup $(N, *)$ of $(G, *)$ is a normal subgroup of $(G, *)$ if and only if every left coset of N in G is a right coset of N in G. *(B. W. 98 Oct.)*

Proof Suppose N is normal in G.

Then $\forall g \in G$, $gNg^{-1} = N$ i.e. $(gNg^{-1})g = Ng$.

i.e. $gN = Ng$.

$\therefore$ Every left coset is a right coset.

Conversely, suppose that every coset of N in G is a right coset of N in G.

$$\therefore \text{ for } g \in G, gN = Ng, g_1 \in G, g_1 \neq g.$$

Now $e \in N$ and $g = g * e \in gN = Ng_1$

$$\therefore g \in Ng, \text{ Also } g \in Ng \text{ Therefore } g \in Ng \cap Ng_1$$

This is possible only if $Ng = Ng_1$.

$$\therefore gN = Ng_1 = Ng.$$

$$\therefore gN = Ng$$

$$gNg^{-1} = Ng * g^{-1} = Ne = N$$

$$\therefore gNg^{-1} = N$$

Therefore N is normal in G.

Theorem 3 If $(G, *)$ is a group and $(H, *)$ is a subgroup of index 2 then $(H, *)$ is a normal subgroup of $(G, *)$.

Proof Since $(H, *)$ is a subgroup of index 2 in G, the number of distinct right cosets of H in G is 2. Let g be any element of G.

If $g \in H$ then $gH = Hg = H$.

If $g \notin H$ then Hg and H are two distinct right cosets of H. Similarly gH and H are two distinct left cosets of H in G.

Therefore $H \cup Hg = H \cup gH = G$ and $H \cap Hg = \phi$, $H \cap gH = \phi$

This is possible only if $gH = Hg \; \forall g \in G$.

$\therefore$ H is normal in G.

Theorem 4 The intersection of two normal subgroups of a group of $(G, *)$ is a normal subgroup of $(G, *)$ *(MCA 98M)*.

Proof Let $(H, *)$, $(K, *)$ be two normal subgroups of $(G, *)$.

Now $e \in H, e \in K$ therefore $e \in H \cap K$. Therefore $H \cap K$ is nonempty (1)

Let $a, b \in H \cap K$. Therefore $a, b \in H, a, b \in K$. Now $a, b \in H$ and H is subgroup, $a * b^{-1} \in H$. $a, b \in K$ and K is a subgroup $a * b^{-1} \in K$.

$$\therefore a * b^{-1} \in H \cap K \; \forall a, b \in H \cap K.$$

$\therefore$ $H \cap K$ is a subgroup of G (2)

Let $a \in H \cap K$ $\therefore$ $a \in H$, $a \in K$.

Since H is normal in G, $g * a * g^{-1} \in H$, $\forall g \in G$ and $a \in H$

Since K is normal in G, $g * a * g^{-1} \in K \; \forall g \in G$ and $a \in K$.

$\therefore$ $g * a * g^{-1} \in H \cap K \; \forall g \in G$ and $a \in H \cap K$ (3)

By (1), (2), (3), $(H \cap K, *)$ is a normal subgroup of $(G, *)$

Theorem 5 Every subgroup of an abelian group is a normal subgroup.

Proof Let $(G, *)$ be an abelian group and $(H, *)$ be a subgroup. $\forall\, g \in G,\ h \in H$, we have $h \in G$ and

$$\begin{aligned} g*h*g^{-1} & \\ &= g*g^{-1}*h \;[\because G \text{ is abelian}] \\ &= e*h \\ &= h \in H \end{aligned}$$

$\therefore g*h*g^{-1} \in H\ \forall g \in G$ and $h \in H$.

$\therefore (H, *)$ is a normal subgroup of $(G, *)$

Theorem 6 Let $(G, *)$ be a group and $H = \{h : h \in G, h*x = x*h\ \forall x \in G\}$. Then $(H, *)$ is a normal subgroup of $(G, *)$. [$(H, *)$ is called the centre of G]

Proof Now $e \in G$ and $e*x = x*e\ \forall\ x \in G$.

$$\therefore\ e \in H\ \therefore\ H \text{ is nonempty} \tag{1}$$

Let $h_1, h_2 \in H$. Then $h_1 * x = x * h_1\ \forall x \in G$,

$$h_2 * x = x * h_2\ \ \forall x \in G \tag{2}$$

$$\begin{aligned} \therefore (h_2^{-1} * h_2) * x * h_2^{-1} &= h_2^{-1} * x * (h_2 * h_2^{-1}) \\ e * x * h_2^{-1} &= h_2^{-1} * x * e \\ x * h_2^{-1} &= h_2^{-1} * x \forall x \in G \end{aligned} \tag{3}$$

Now

$$\begin{aligned} (h_1 * h_2^{-1}) * x &= h_1 * (h_2^{-1} * x) \\ &= h_1 * (x * h_2^{-1}) \qquad \text{(by 3)} \\ &= (h_1 * x) * h_2^{-1} \\ &= (x * h_1) * h_2^{-1} \\ &= x * (h_1 * h_2^{-1}) \end{aligned}$$

$$(h_1 * h_2^{-1}) * x = x * (h_1 * h_2^{-1})\ \forall x \in G \tag{4}$$

$$\therefore (H, *) \text{ is a subgroup of } (G, *) \text{ by (1) and (4)} \tag{5}$$

$\forall\ g \in G,\ h \in H,$ we have

$$\begin{aligned} g*h*g^{-1} &= g*(h*g^{-1}) \\ &= g*(g^{-1}*h) \qquad \text{(by 3)} \\ &= (g*g^{-1})*h \\ &= e*h \\ &= h \in H. \end{aligned}$$

$$g*h*g^{-1} = h \in H\ \forall\ g \in G \tag{6}$$

By (5) and (6) $(H, *)$ is a normal subgroup of $(G, *)$

Factor Group or Quotient Group

Let $(G, *)$ be a group and $(N, *)$ be a normal subgroup of $(G, *)$. Let $G/N = \{Na : a \in G\}$ i.e. G/N is the set of all right cosets of N in G. Then $(G/N, \odot)$ is a group with respect to multiplication of cosets defined by

$$(Na) \odot (Nb) = Na * b \forall\ a, b \in G$$

$(G/N, \odot)$ is called a factor group or quotient group of G by N.

Note: 1. $O\left(\frac{G}{N}\right) = \frac{O(G)}{O(N)}$

2. $N = Ne$ is the identity element of G/N.

3. $(Na)^{-1} = Na^{-1}$ is the inverse of Na, for $a \in G$.

Theorem 7 Every quotient group of an abelian group is abelian.

Proof Let $(G, *)$ be an abelian group and $(N, *)$ be a subgroup. Since $(G, *)$ is abelian, N is normal in G. $(G/N, \odot)$ is a group.

Let Na, $Nb \in G/N$, $a, b \in G$.

$$\begin{aligned}(Na)\odot(Nb) &= N(a*b) \\ &= N(b*a) \ (\because G \text{ is abelian}) \\ &= (Nb)\odot(Na)\end{aligned}$$

$$\therefore \left(\frac{G}{N}, \odot\right) \text{ is abelian.}$$

Note: The converse of the above theorem is not true. i.e., if the quotient group is abelian, the group need not be abelian. For example, $(S_3, \circ)$ is the symmetric group of degree 3. $(A_3, \circ)$ is a normal subgroup $\therefore (S_3/A_3, \odot)$ is a quotient group and $O(S_3/A_3) = \frac{n!}{n!/2} = 2$. S_3/A_3 is abelian. But $(S_3, \circ)$ is not abelian.

Theorem 8 The quotient group of a cyclic group is cyclic.

Proof Let $G = (a)$ be cyclic and N be a subgroup of G. Therefore N is cyclic and hence N is abelian therefore N is normal in G.

$\therefore (G/N, \bullet)$ is the quotient group.

Let $a^n \in G. \therefore Na^n \in G/N$

$$\begin{aligned}Na^n &= Na * a * \ldots * a (n \text{ times}). \\ &= (Na) \odot (Na) \odot \ldots \odot Na \ (n \text{ times}) \\ &= (Na)^n\end{aligned}$$

$\therefore$ Every element of G/N is an integral power of Na.

$\therefore$ $G/N = (Na)$ is cyclic.

Note: The converse of the theorem is not true.

For example S_3/A_3 is of order 2 and cyclic. But (S_3, O) is not cyclic.

HOMOMORPHISMS AND ISOMORPHISMS

Morphisms of Semigroups

Let $(S, *)$, $(\bar{S}, \Delta)$ be any two semigroups. If $\phi : S \to \bar{S}$ is a mapping such that $\phi(a*b) = \phi(a)\Delta\phi(b) \forall a, b \in S$, then ϕ is said to be a homomorphism of S into $\bar{S}$.

Further, if ϕ is surjective, it is said to be a semigroup epimorphism.

If ϕ is injective, it is said to be a semigroup monomorphism. If ϕ is bijective, it is said to be a semigroup isomorphism and in this case, the two semigroups are said to be isomorphic. A homomorphism of a semigroup (S, *) into (S, *) is called a semigroup endomorphism.

An isomorphism of a semigroup (S, *) into (S, *) is called a semigroup automorphism.

Example We know that $(N, +)$, $(Z_m, +_m)$ are two semigroups. Define $\phi : N \to Z_m$ by $\phi(a) = [a]$ $\forall a \in N \cdot \forall\ a, b \in N$, $\phi(a+b) = [a+b] = [a] +_m [b] = \phi(a) +_m \phi(b)$.

$\therefore \phi$ is a semigroup homomorphism

Example Let $\overline{N} = \{2, 3, 4 \ldots\}$. Now $(N, +)$ and $(\overline{N}, +)$ are two semigroups. Define $\phi : N \to \overline{N}$ by $\phi(a) = a + 1 \forall\ a \in N, \forall\ a, b \in N$

$$\phi(a+b) = a+b+1$$
$$\phi(a) + \phi(b) = a+1+b+1 = a+b+2$$
$$\phi(a+b) \neq \phi(a) + \phi(b)$$

$\therefore \phi$ is not a homomorphism.

Example Let $(S, *)$, $(\overline{S}, \Delta)$ be two semigroups and $\phi : S \to \overline{S}$ be a homomorphism. If $a \in S$ is idempotent $\phi(a) \in \overline{S}$ is also for idempotent.

For
$$a * a = a$$
$$\phi(a * a) = \phi(a)\ \Delta \phi(a)$$

and hence $\phi(a) = [\phi(a)]^2 . \therefore \phi(a)$ is idempotent.

Theorem 1 Let $(S, *), (T, \Delta)$, and $(V, \oplus)$ be three semigroups and $f : S \to T$ and $g : T \to V$ be semigroup homomorphism. Then $h = g \circ f$ is a semigroup homomorphism from $(S, *)$ to $(V, \oplus)$.

Proof Let $a, b \in S$. Then we have

$$\begin{aligned} h(a*b) &= (g \circ f)(a*b) \\ &= g(f(a*b)) \\ &= g(f(a) \Delta f(b)) \\ &= g(f(a)) \oplus g(f(b)) \\ &= (g \circ f)(a) \oplus (g \circ f)(b) \\ &= h(a) \oplus h(b). \end{aligned}$$

$\therefore h = g \circ f$ is a semigroup homomorphism from $(S, *)$ to $(V, \oplus)$.

Theorem 2 Let (S, *) be a semigroup and $(S^S, \circ)$ be a semigroup where S^S is the set of all functions from S to S. Then there exists a homomorphism. $\phi : S \to S^S$. *(MCA 97 M, B.E. 93 N)*

Proof Let $a \in S$ and define $\phi(a) = f_a$ where $f_a \in S^S$ and $f_a(x) = a * x \forall\ x \in S$.

Let $a, b \in S$

$$\begin{aligned} f_{a*b}(c) = (a*b)*c = a*(b*c) &= f_a(f_b(c)) \\ &= (f_a \circ f_b)(c) \end{aligned}$$

$\therefore f_{a*b} = f_a \circ f_b$.

Now $\phi(a*b) = f_{a*b} = f_a \circ f_b = \phi(a) \circ \phi(b)$. $\therefore \phi$ is a homomorphism from $(S, *)$ to $(S^S, \circ)$.

Theorem 3 Let X be a set containing n elements and X^* be the free semigroup generated by X. Let $(S,\oplus)$ be any other semigroup generated by any n generators. Then there exists a homomorphism $g: X^* \rightarrow S$ *(MCA 96N)*

Proof Let y be the set of n generators of S.

Let $g: X \rightarrow Y$ be a one-to-one mapping given by $g(x_i) = y_i$ for $i = 1,2,\ldots,n$, $x_i \in X, y_i \in Y$.

Let $\alpha = x_1, x_2, \ldots, x_m \in X^*$. Let us define $g: X^* \rightarrow S$ by $g(\alpha) = g(x_1) \oplus g(x_2) \oplus \ldots \oplus g(x_m)$.

Now let $\alpha, \beta \in X^*$. Then we have $\alpha = x_{i_1}, x_{i_2}, \ldots, x_{ik}$, $\beta = x_{j_1}, x_{j_2}, \ldots, x_{j_e}$ where x_i's and x_j's are in X.

$\alpha\beta = x_{i_1}, x_{i_2} \ldots x_{i_k}\, x_{j_i} x_{j_2} \ldots x_{j_i}$

$$\begin{aligned} g(\alpha\beta) &= g(x_{i_1}) \oplus \cdots \oplus g(x_{i_k}) \oplus g(x_{j_1}) \oplus \cdots \oplus g(x_{j_e}) \\ &= (g(x_{i_1}) \oplus \cdots \oplus g(x_{i_k}) \oplus g(x_{j_1}) \oplus \cdots \oplus g(x_{j_e}) \\ &= g(\alpha) \oplus g(\beta). \end{aligned}$$

$\therefore\ g: X^* \rightarrow S$ is a homomorphism.

Morphisms of Monoid

Let $(M, *)$ and $(\overline{M}, \Delta)$ be two monoids with identities e, $\bar{e}$ respectively. A mapping $\phi: M \rightarrow \overline{M}$ is said to be a homomorphism from M to $\overline{M}$ if $\phi(a * b) = \phi(a) \Delta \phi(b) \forall a, b \in M$ and $\phi(e) = \bar{e}$.

Further, if ϕ is surjective, it is said to be epimorphism; if ϕ is injective, it is said to be monomorphism; if ϕ is bijective, it is said to be isomorphism.

Example Let $W = \{0,1,2,\ldots\}$. Then $(W,+)$ is a monoid. Also $(Z_m, +_m)$ is also a monoid.

Let $\phi: W \rightarrow Z_m$ be a function defined by $\phi(a) = [a]\ \forall\ a \in W$

Now,
$$\begin{aligned} \phi(a+b) &= [a+b]\ \forall a, b \in W \\ &= [a] +_m [b] = \phi(a) +_m \phi(b) \end{aligned}$$

Also $\phi(0) = [0]$ where 0 is the identity of M and $[0]$ is the identity of Z_m.

$\therefore\ \phi$ is a homomorphism from $(W,+)$ to $(Z_m, +_m)$.

Example Let $W = \{0, 1, 2, \ldots\}$. Then $(W, +)$ is a monoid with identity 0. Let $S = \{e, 0, 1\}$ and $*$ be the operation as in table

*	e	0	1
e	e	0	1
0	0	0	0
1	1	0	1

Then $(S, *)$ is a monoid with identity e.

Let $\phi: W \rightarrow S$ be defined as $\phi(0) = 1$ and $\phi(i) = 0$ for $i \neq 0$.

We can see that $\phi(a+b) = \phi(a) * \phi(b)$ but $\phi(0) \neq e$.

$\therefore\ \phi$ is not a monoid homomorphism.

Morphisms of Groups

Let $(G, *)$ and $(\overline{G}, \Delta)$ be two groups.

A mapping $\phi: G \rightarrow \overline{G}$ is said to be a homomorphism.

If $\phi(a*b)=\phi(a)\Delta\phi(b)\ \forall a,b\in G$

Further,

If ϕ is surjective, it is said to be epimorphism.

If ϕ is injective, it is said to be monomorphism.

If ϕ is bijective, it is said to be isomorphism and in this case we say that (G, $*$) is isomorphic to $(\overline{G},\Delta)$ and we write $G\cong\overline{G}$.

Definition: Let (G, *) be a group. If $\phi:G\to G$ is an isomorphism, then ϕ is said to be an automorphism of G.

Example Let (G, *) be a group and $\phi:G\to G$ be defined by $\phi(x)=e\,\forall\, x\in G$.

Now $\phi(x*y)=e$

$= e*e$

$= \phi(x)*\phi(y)\,\forall\, x_1 y\in G$

ϕ is called a trivial homomorphism.

Example Let (G, +) be the additive group of real numbers and $(\overline{G},\circ)$ be the multiplicative group of nonzero real numbers.

Let $\phi:G\to\overline{G}$ be defined by $\phi(a)=2a\,\forall\, a\in G$.

$\forall\, a,b\in G,\ \phi(a+b)=2^{a+b}$

$= 2^a.2^b$

$= \phi(a).\phi(b)$

$\therefore\ \phi$ is a homomorphism from G to $\overline{G}$.

Example Let (G, +) be the additive group of integers.

Let $\phi:G\to\overline{G}$ be defined by $\phi(a)=a+1\ \forall\ a\in G$,

Now $\forall\, a,b\in G,\ \phi(a+b)=a+b+1$

But $\phi(a)+\phi(b)=a+1+b+1=a+b+2$

$\therefore\ \phi(a+b)\neq\phi(a)+\phi(b)$

$\therefore\ \phi$ is not homomorphism.

Example Let (G, *) be a group and (N, *) be a normal subgroup. $(G/N,\odot)$ is also a group.

Let $\phi:G\to G/N$ be defined by $\phi(x)=Nx\,\forall\, x\in G$

$\forall\ x,y\in G,\ \phi(x*y)=Nx*y$

$= (Nx)\odot(Ny)$

$= \phi(x)\odot\phi(y)$

$\therefore\ \phi$ is a homomorphism from G to G/N.

Also $\forall\ Nx\in G/N$, there exists $x\in G$ such that $\phi(x)=Nx$.

$\therefore\ \phi$ is an epimorphism.

Kernal Let ϕ be a homomorphism from (G, *) onto $(\overline{G}\ \Delta)$. Let e and $\overline{e}$ be the identities, then the kernal of ϕ, K is defined as $K=\{x\in G:\phi(x)=\overline{e}\}$

Theorem 4 If ϕ is a homomorphism from $(G, *)$ onto $(\overline{G}, \Delta)$ with identities e and $\bar{e}$ respectively then

i. $\phi(e) = \bar{e}$

ii. $\phi(x^{-1}) = [\phi(x)]^{-1} \forall\ x \in G$

Proof

$$\begin{aligned} \forall x \in G,\ &\phi(x) \Delta \bar{e} \\ &= \phi(x) \\ &= \phi(x * e) \\ &= \phi(x) \Delta \phi(e) \end{aligned}$$

$\therefore \phi(x) \Delta \bar{e} = \phi(x) \Delta \phi(e)$. Since $(\overline{G}, \Delta)$ is a group, by left cancellation law, $\bar{e} = \phi(e)\ \therefore\ \phi(e) = \bar{e}$. Hence (i) is true.

Now,
$$\begin{aligned} \bar{e} &= \phi(e) \\ &= \phi(x * x^{-1}) \\ &= \phi(x) \Delta \phi(x^{-1}) \end{aligned}$$

Similarly $\phi(x^{-1}) \Delta \phi(x) = \bar{e}$

$\therefore\ \phi(x)\ \Delta \phi(x^{-1}) = \phi(x^{-1}) \Delta \phi(x) = \bar{e}$

$\therefore\ \phi(x^{-1})$ is the inverse of $\phi(x)$

$\therefore\ \phi(x^{-1}) = [\phi(x)]^{-1}$.

Hence (ii) is true.

Theorem 5 If ϕ is a homomorphism of $(G, *)$ into $(\overline{G}, \Delta)$ with identities e, $\bar{e}$ respectively and k is kernal of ϕ, then $(K, *)$ is a normal subgroup of $(G, *)$. *(MCA 97M, 96N, 96N, 93N)*

Proof Since $e \in G$ and $\phi(e) = \bar{e}, e \in K$. Hence K is non-empty.

Let $x, y \in K$. $\therefore\ \phi(x) = \bar{e},\ \phi(y) = \bar{e}\ \ \therefore\ [\phi(y)]^{-1} = \bar{e}^{-1} = \bar{e}$ (1)

$\therefore\ \phi(y^{-1}) = \bar{e}$.

Now $\phi(x * y^{-1}) = \phi(x) \Delta \phi(y^{-1}) = \bar{e} \Delta \bar{e} = \bar{e}$

$\therefore\ x * y^{-1} \in K.\ \ \therefore\ (K, *)$ is a subgroup (2)

$\forall\ g \in G,\ \ k \in K,\ \ \phi(k) = \bar{e}$ and

$$\begin{aligned} \phi(g * k * g^{-1}) &= \phi(g) \Delta \phi(k) \Delta \phi(g^{-1}) \\ &= \phi(g) \Delta \bar{e} \Delta [\phi(g)]^{-1} \\ &= \phi(g) \Delta [\phi(g)]^{-1} \\ &= \bar{e} \end{aligned}$$

$$\therefore\ g * k * g^{-1} \in K \tag{3}$$

By (1), (2) and (3) $(K, *)$ is a normal subgroup of $(G, *)$.

Theorem 6 A homomorphism ϕ from $(G, *)$ onto $(\overline{G}, \Delta)$ with kernal K is an isomorphism if and only if $K = \{e\}$.

Proof Suppose $K = \{e\}$.

For $x, y \in G$, let $\phi(x) = \phi(y)$

$$\therefore\ \phi(x)\Delta[\phi(y)]^{-1} = \overline{e}$$
$$\therefore\ \phi(x)\Delta\phi(y^{-1}) = \overline{e}$$
$$\therefore\ \phi(x * y^{-1}) = \overline{e}$$
$$\therefore\ x * y^{-1} \in K \text{ But } K = \{e\}$$
$$\therefore\ x * y^{-1} = e$$
$$\therefore\ x = y$$

$\therefore \phi$ is one-to-one and hence it is an isomorphism.

Conversely, suppose ϕ is an isomorphism, then ϕ is one-to-one. Also $\phi(e) = \overline{e}$.

For no other element $x \in G,\ x \neq e,\ \phi(x) = \overline{e}$

$$\therefore\ K = \{e\}$$

Theorem 7 **(Fundamental theorem of homomorphism)**

Let ϕ be a homomorphism of (G, *) onto $(\overline{G}, \Delta)$ with kernal *K*. Then $(G/K, \odot)$ is isomorphic to $(\overline{G}, \Delta)$

Proof

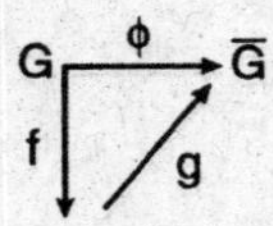

Let $f : G \to G/K$ be defined by $f(x) = Kx\ \forall\, x \in G$.

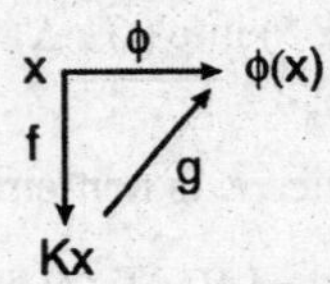

Let $g : G/K \to G$ be defined by $g(Kx) = \phi(x) \forall\ Kx \in G/K$

Let $\overline{x} \in \overline{G}$. Since ϕ is onto, there exists $x \in G$ such that $\phi(x) = \overline{x}$.

Now, $g(Kx) = \phi(x) = \overline{x}$

$\therefore\ \forall\, \overline{x} \in \overline{G}$, there exists $Kx \in G/K$ such that $g(Kx) = \overline{x}$

$$\therefore\ g \text{ is onto} \tag{2}$$

Let $Kx, Ky \in G/K,\ x, y \in G$.

$$\begin{aligned} g(Kx \odot Ky) & \\ &= g(Kx * y) \\ &= \phi(x * y) \\ &= \phi(x)\Delta\phi(y) \\ &= g(Kx) \odot g(Ky) \end{aligned}$$

$\therefore g$ is a homomorphism (2)

Suppose for some $x \in G$, $g(Kx) = \overline{e}$ $\therefore \phi(x) = \overline{e}$

$\therefore x$ is in the kernal of ϕ. $\therefore x \in K$

In this case, $Kx = K$.

$\therefore$ Kernal of g is $\{k\}$ (identity element of G/K). (3)

By (1), (2), (3) $(G/K, \odot)$ is isomorphic to $(\overline{G}, \Delta)$.

Theorem 8 Any infinite cyclic group is isomorphic to the additive group of integers.

Proof Let $G = (a)$ be an infinite cyclic group generated by a and $(Z, +)$ be the additive group of integers.

Let $f: Z \to G$ be defined by $f(x) = a^x \; \forall x \in Z$.

For $x, y \in Z$, $x \neq y$, $a^x \neq a^y$ and hence $f(x) \neq f(y)$

$\therefore f$ is one-one. (1)

Also $\forall a^x \in G$, there exists $x \in Z$ such that $f(x) = a^x$

$\therefore f$ is into. (2)

Further for $x, y \in Z$, $x + y \in Z$ and we have

$$f(x+y) = a^{x+y} = a^x \cdot a^y = f(x) \cdot f(y)$$

$\therefore f$ is a homomorphism (3)

By (1), (2) and (3), $(Z, +)$ is isomorphic to $(G, \cdot)$

By symmetry, $(G, \cdot)$ is isomorphic to $(Z, +)$.

Theorem 9 Any cyclic group of order n is isomorphic to the additive group of residue classes of integers modulo n.

Proof Let $G = \{a, a^2, \ldots, a^n = e\}$ be a cyclic group of order n generated by a. We know that $(Z_n, +_n)$ is the additive group of residue classes modulo n.

$\therefore Z_n = \{[1], [2] \ldots, [n] = [0]\}$.

Isomorphism $f: G \to Z_n$ by $f(a^r) = [r] \; \forall \; a^r \in G$.

$\forall [r] \in Z_n$, there exists $a^r \in G$ such that $f(a^r) = [r]$

$\therefore f$ is onto (1)

For $r \neq s$, $[r] \neq [s]$ and hence $f(a^r) \neq f(a^s)$

$\therefore f$ is one-one (2)

$$\forall a^r, a^s \in G, \; f(a^r \cdot a^s) = f(a^{r+s}) = [r+s] = [r] +_n [s] = f(a^r) +_n f(a^s)$$

$\therefore f$ is a homomorphism (3)

By (1), (2) and (3) $(G, \cdot)$ is isomorphic to $(Z_n, +n)$.

Theorem 10 **(Cayley's theorem or Cayley's representation theorem)**

Every finite group is isomorphic to a permutation group. *(B.E. 96, MCA 96 N, 95 Oct, 95 N)*

Proof Let $(G, *)$ be a finite group of order n and $a \in G$.

Consider a mapping $f_a: G \to G$ defined by $f_a(x) = a * x \; \forall x \in G$.

Suppose $x, y \in G$ and $f_a(x) = f_a(y)$ then $a * n = a * y$

By left cancellation, we have $x = y$. $\therefore f$ is one-one. Since f_a is a one-one mapping of a finite set onto itself, f_a is onto. $\therefore fa$ is a permutation of G.

Let $\overline{G} = \{f_a : a \in G\}$. Define a mapping $\phi : G \to \overline{G}$ by $\phi(a) = f_a \ \forall a \in G$.

Suppose $\phi(a) = \phi(b)$. We have

$$f_a = f_b \text{ and } f_a(x) = f_b(x) \ \forall x \in G$$

$\therefore a * x = b * x.$

By right cancellation $a = b$

$\therefore \phi(a) = \phi(b) \Rightarrow a = b$

$\therefore \phi$ is one – one. (1)

Since G and $\overline{G}$ are finite containing the same number of elements and

ϕ is one-one, we see that ϕ is onto (2)

Moreover $\forall a, b \in G, \phi(a * b) = f_{a*b}$

But

$$\begin{aligned} f_{a*b}(x) &= (a*b)*x \ \forall x \in G \\ &= a*(b*x) \\ &= (f_a(f_b(c)) \\ &= (f_a \circ f_b)(x) \\ \therefore f_{a*b} &= f_a \circ f_b \\ \therefore \phi(a*b) &= f_{a*b} = f_a \circ f_b = \phi(a) \circ \phi(b) \end{aligned}$$

$\therefore \phi$ is a homomorphism. (3)

By (1), (2) and (3), $(G, *)$ is isomorphic to $(\overline{G}, \circ)$ which is a subgroup of $(S_n, \circ)$.

Definition The permutation group which is isomorphic to a given finite group is called the regular permutation group of the given group.

Example 5.18 Find the regular permutation group isomorphic to the multiplication group $G = \{1, i, -1, -i\}$.

Solution Let $G = \{1, i, -1, -i\}$ and $S = \{1, 2, 3, 4\}$

Let $\overline{G} = \{p_1, p_2, p_3, p_4\}$ where $p_1 = \begin{pmatrix} 1 & 2 & 3 & 4 \\ 1 & 2 & 3 & 4 \end{pmatrix}$

$$p_2 = \begin{pmatrix} 1 & 2 & 3 & 4 \\ 2 & 3 & 4 & 1 \end{pmatrix}, \ p_3 = \begin{pmatrix} 1 & 2 & 3 & 4 \\ 3 & 4 & 1 & 2 \end{pmatrix}, \ p_4 = \begin{pmatrix} 1 & 2 & 3 & 4 \\ 4 & 1 & 2 & 3 \end{pmatrix}$$

The composition table is

$\circ$	p_1	p_2	p_3	p_4
p_1	p_1	p_2	p_3	p_4
p_2	p_2	p_3	p_4	p_1
p_3	p_3	p_4	p_1	p_2
p_4	p_4	p_1	p_2	p_3

$(\overline{G}, \circ)$ is a group.

Let $f : G \to \overline{G}$ be defined as follows.

$$f : 1 \to p_1, i \to p_2, -1 \to p_3, -i \to p_4$$

Now f is one-one and onto.

Now $f(i \cdot (-1)) = f(-i) = p_4 = p_2 \circ p_3 = f(i) \circ f(-1)$

Similarly we can verify for all other possible pairs of elements of G. $\therefore \phi$ is a homomorphism and (G,.) is isomorphic to $(\overline{G},\circ).(\overline{G},\circ)$ is a regular permutation group of $(G,\cdot)$ and it is a subgroup of $(S_4,\dot{\circ})$

Additional Examples

Example 5.19 Show that the set of all natural numbers, N is a semigroup under the operation $x * y = \max(x, y)$. Is it a monoid? *(MCA 98, B.E. 93N)*

Solution Let $N = \{1, 2, 3,...\}$

Let $x * y = \max(x, y)$, $\forall\ x,\ y \in N$.

$$\begin{aligned} \forall\ x, y \in N, \text{ if } x > y, x * y &= \max(x, y) = x \in N \\ \text{if } x < y, x * y &= \max(x, y) = y \in N \\ \text{if } x = y,\ x * y &= \max(x, y) = x \in N. \end{aligned}$$

$\therefore *$ is closed in N.

$\forall\ x, y, z \in N$, we have

$$\begin{aligned} (x * y) * z &= \max\ (x, y) * z \\ &= \max\ ((x, y), z) \\ &= \max(x, (y, z)) \\ &= x * \max(y, z) \\ &= x * (y * z) \end{aligned}$$

$\therefore *$ is associative in N.

$\therefore (N, *)$ is a semigroup.

Further $1 \in N$, we have $1 * 1 = \max(1, 1) = 1$

Also $\forall x \in N$, $x > 1$, $1 * x = \max\ (1, x) = x$.

Similarly $x * 1 = x$.

$\therefore 1 \in N$ is the identity.

$\therefore (N, *)$ is a monoid.

Example 5.20 Show that any semigroup S can be extended to a monoid by adjoining an identity element.

Solution Let $(S, *)$ be a semigroup.

$*$ is closed and associative in S.

Let us adjoin e (identity) to S. Let $M = \{e\} \cup S$. Let us now prove that $(M, *)$ is a monoid.

$\forall\ x, y \in M$, either $x, y \in S$ or $x = e,\ y \in S$ or $x \in S,\ y = e$ or $x = e,\ y = e$

If $x, y \in S, x * y \in S$ and hence $x * y \in M$.

If $x = e,\ y \in S$ then $x * y = e * y = y \in M$.

Similary $x \in S, y = e,\ x * y \in M$.

If $x = e,\ y = e, x * y = e * e = e \in M$.

$\therefore *$ is closed in M.

Example 5.21 Prove that every finite semigroup has an idempotent element.

(MCA 97 Oct., 96 Nov., B.E 93 Apr., 92 Apr.)

Solution Let S be a finite semigroup with respect to the binary operation $*$. Let $a \in S$. Let us consider (a), cyclic subsemigroup generated by a.

Now $(a) = \{a, a^2, a^3, \ldots, a^n, \ldots\}$

Now $(a) \subset S$ and S is finite. $\therefore$ (a) is also finite.

There exists integers n, k with $k < n$ such that $a^n = a^k$, $1 \le k \le n-1$ (1)

Case 1 Let $2k = n$

Then by (1) $a^{2k} = a^k$

i.e., $a^k * a^k = a^k$

$\therefore$ a^k is an idempotent element in S.

Case 2 Let $2k < n$

We have

$$\begin{aligned} a^{n-k} * a^{n-k} &= a^{n-k+n-k} \\ &= a^n * a^{n-2k} \\ &= a^k * a^{n-2k} \quad \text{by (1)} \\ &= a^{n-k} \end{aligned}$$

$\therefore$ $a^{n-k} \in S$ is an idempotent element.

Case 3 Let $n < 2k$.

$$a^n * a^{n-k} = a^k * a^{n-k} = a^n = a^k \quad \text{by (1)}$$

Also

$$\begin{aligned} a^{n+(n-k)} * a^{n-k} &= a^k * a^{n-k} \quad \text{by (} \\ &= a^n \\ &= a^k \quad \text{by (} \end{aligned}$$

$$\therefore a^{n+2(n-k)} = a^k$$

Similarly

$$a^{n+2k(n-k)} = a^k \quad (4)$$

Put $n' = n + 2k(n-k)$

Now $n' > 2k$ $\therefore$ $2k < n'$

Hence by case 2 $a^{n'-k} \in S$ is an idempotent.

$\therefore$ $(S, *)$ has an idempotent element.

Also $\forall\, x, y, z \in M$, $(x * y) * z = x * (y * z)$

$\therefore$ $*$ is associative in M.

$\forall\, x \in M$, we have $x * e = e * x = x$

$\therefore$ $e \in M$ is not an identity.

$\therefore$ $(M, *)$ is a monoid.

Example 5.22 Let Z_n denote the set of all integers $\{0, 1, 2, ... n-1\}$. Let $\odot$ be a binary operation on Z_n such that $a \odot b$ = the remainder of ab divided by n.

a. Construct the table for the operation $\odot$ for $n = 4$.

b. Show that $(Z_n, \odot)$ is a semigroup for any n. *(MCA 91N)*

Solution (a) The composition table for $n = 4$ is as follows.

$\cdot$	0	1	2	3
0	0	0	0	0
1	0	1	2	3
2	0	2	0	2
3	0	3	2	1

$(Z_4, \odot)$ is semigroup.

(b) Let $a, b \in Z_n, 0 \le a < n, 0 \le b < n$ and r be the remainder when ab is divided by n. $\therefore 0 \le r < n$ $a \cdot b = r \in Z_n$ (closure)

$\forall\ a, b, c \in Z_n$, we have

$$\begin{aligned}(a \odot b) \odot c &= (ab)_n \odot c \\ &= ((ab)c)_n \\ &= (a(bc))_n \\ &= a \odot (bc)_n \\ &= a \odot (b \odot c) \text{ (associativity)}\end{aligned}$$

$\therefore (Z_n, \odot)$ is a semigroup.

Example 5.23 Show that a semigroup with more than one idempotent element cannot be a group. Give an example which is not a group.

Solution Let $S = \{e, 0, 1\}$. Let $*$ be the binary operation defined as follows.

$*$	e	0	1
e	e	0	1
0	0	0	0
1	1	0	1

Now $(S, *)$ is a semigroup with three idempotent elements e, 0, 1 since $e * e = e$. $0 * 0 = 1$, $1 * 1 = 1$. But $(S, *)$ is not a group. Suppose it is a group. Since $e * 1 = 1, 1 * 1 = 1 \therefore e * 1 = 1 * 1$. By cancellation $e = 1$ which is not true $\therefore (S, *)$ is not a group.

Now $(N, +)$ is a semigroup. Since the additive identity $0 \notin N$ and the additive inverse for $a \in N$ does not exist in N.

$\therefore (N, +)$ is not a group

Example 5.24 Show that the set of all invertible elements of a monoid forms a group under the same operation as that of a monoid. *(MCA 96A,, 97M)*

Solution Let $(M, *)$ be a monoid. Let M' be the set of all invertible elements of M.

Let $a_1, a_2 \in M'$ $\therefore a_1^{-1}, a_2^{-1} \in M'$.

Now
$$\begin{aligned}(a_1 * a_2) * (a_2^{-1} * a_1^{-1}) &= a_1 * (a_2 * a_2^{-1}) * a_1^{-1} \\ &= a_1 * e * a_1^{-1} \\ &= e \in M'\end{aligned}$$

Similarly,
$$(a_2^{-1} * a_1^{-1}) * (a_1 * a_2) = e \in M'$$

$\therefore a_1 * a_2$ is invertible and $a_1 * a_2 \in M'$ (closure). Since $M' \subset M$ associativity is true.

Now $e * e = e \therefore e$ is invertible and $e \in M'$ is the identity of M' since $a * e = e * a = a \forall a \in M'$.

$\forall\, a_1 \in M'$, we have $a_1 * a_1^{-1} = a_1^{-1} * a_1 = e$

$\therefore (a_1^{-1})^{-1} = a_1 \in M' \therefore a_1^{-1} \in M'$ is the inverse of $a_1 \in M' \therefore (M', *)$ is a group.

Example 5.25 Prove that every proper subgroup of an infinite cyclic group is infinite. *(95 Oct.)*

Solution Let $G = (a)$ be an infinite cyclic group generated by a and H be a proper subgroup of G. Let m be the least positive integer such that $a^m \in H$. Then H is a cyclic group generated by a^m.

Now, if possible let H be finite and $O(H) = n$.

$$O(a^m) = O(H) = n$$

$$\therefore (a^m)^n = e, \; a^{mn} = e \;\therefore O(a) \le mn$$

i.e., (a) is finite.

$\therefore G$ is of finite order which is a contradiction.

$\therefore (H, *)$ is infinite.

Example 5.26 Let $(G, *)$ be a group of order 2, where $G = \{e, a\}$. Find $(G \times G, \circ)$ the direct product of $(G, *)$ with itself. *(MCA 97 Oct.)*

Solution $G = \{e, a\}$

$*$	e	a
e	e	a
a	a	e

The direct product $(G \times G, \circ)$ is as follows:

$\circ$	(e,e)	(e,a)	(a,e)	(a,a)
(e,e)	(e,e)	(e,a)	(a,e)	(a,a)
(e,a)	(e,a)	(e,e)	(a,a)	(a,e)
(a,e)	(a,e)	(a,a)	(e,e)	(e,a)
(a,a)	(a,a)	(a,e)	(e,a)	(e,e)

Example 5.27 Show that the cosets determined by a subgroup S in a group $(G, *)$, only one of the cosets is a subgroup. *(MCA 96 Mar.)*

Solution Let S be a subgroup of G. If possible let Sa be a subgroup of G where $a \in G$.

Now $e \in S$ and $e \in Sa$ ($\because Sa$ is a subgroup).

$\therefore e \in S \cap Sa$ and $\therefore S \cap Sa \neq \phi$

$\therefore Sa = S$ is the only subgroup of G.

Similar result holds goods for left cosets also.

Example 5.28 Let $(H, *)$ be a subgroup of $(G, *)$. Prove that the right cosets of H in G form a partition of G and every element of G belongs to one and only one right coset of H in G *(88 Nov.)*

Solution For $a, b \in G$, $a \equiv b \bmod H$ if $a * b^{-1} \in H$.

The relation $\equiv$ is an equivalence relation and it partitions G into mutually disjoint equivalence classes. Let us now prove that $[a] = Ha$.

Let $x \in [a]$. $\therefore x * a^{-1} \in H$ $\therefore x \in Ha$

$$[a] \subset Ha \tag{1}$$

Let $y \in Ha$. $y = h * a$ for some $h \in H$

$\therefore y * a^{-1} = h \in H$ $\therefore y \in [a]$

$$\therefore Ha \subset [a] \tag{2}$$

By (1) and (2), $Ha = [a]$.

$\therefore G$ can be expressed as the union of mutually disjoint and distinct right cosets of H and G.

Suppose $a \in G, a = e * a \in Ha$. $\therefore$ Every element of G appears in atleast one right coset. Also the right cosets of H in G are either disjoint or identical. $\therefore$ Each element of G appears in only one right coset of H in G. Similar result holds good for left coset also.

Example 5.29 If $f : G \to H$ is any morphism of groups, prove that

a. the image of any subgroup of G is a subgroup of H.
b. the inverse image $f^{-1}(T)$ of any subgroup $T \subseteq H$ is a subgroup of G. *(91 Apr.)*

Solution Let $(G, *)$ and (H, Δ) be two groups.

Let $f : G \to H$ be a homomorphism.

a. Let $(K, *)$ be any subgroup of G. $f(K) \subset H \ \forall\ y_1, y_2 \in f(K)$,

there exist $x_1, x_2, \in K$ such that $y_1 = f(x_1), y_2 = f(x_2)$.

$$\begin{aligned} \text{Now } y_1 \,\Delta\, y_2^{-1} &= f(x_1) \Delta (f(x_2))^{-1} \\ &= f(x_1) \Delta f(x_2^{-1}) \\ &= f(x_1 * x_2^{-1}) \in f(K) \text{ since } x_1, x_2 \in K \end{aligned}$$

and K is a subgroup, $x_1 * x_2^{-1} \in K$.

$\therefore f(K)$ is a subgroup of H.

b. Let $x_1, x_2 \in f^{-1}(T)$. $\therefore$ There exists $y_1, y_2 \in T$ such that $x_1 = f^{-1}(y_1)$, $x_2 = f^{-1}(y_2)$

$\therefore y_1 = f(x_1)$, $y_2 = f(x_2)$.

$$\begin{aligned} \text{Now } \ y_1 \,\Delta\, y_2^{-1} &= f(x_1) \Delta [f(x_2)]^{-1} \\ &= f(x_1) \Delta f(x_2^{-1}) \\ &= f(x_1 * x_2^{-1}) \end{aligned}$$

Since $y_1, y_2 \in T$ and (T, Δ) is a subgroup of (H, Δ), $y_1 \,\Delta\, y_2^{-1} \in T$.

$\therefore f(x_1 * x_2^{-1}) \in T$

$\therefore x_1 * x_2^{-1} \in f^{-1}(T)$

$\therefore$ For $x_1, x_2 \in f^{-1}(T)$, we have $x_1 * x_2^{-1} \in f^{-1}(T)$

$\therefore f^{-1}(T)$ is a subgroup of G.

Example 5.30 If $(G, *)$ is a finite group then show that for any $a \in G$, $a^n = e$, e is the identity element of G *(MCA96Nov.)*

Solution Let $(G, *)$ be a finite group and $a \in G$. Let us consider a cyclic subgroup H generated by a. $\therefore H = (a)$.

Since $H \subset G$ and G is finite, H is also finite. Let $O(H) = n$.

$\therefore O(a) = O(H) = n$, $\therefore a^n = e$ for some $a \in G$.

EXERCISE I

1. Say 'True' or 'False'
   (a) The set of integers is a group under usual multiplication. *(B.E. Oct. 96)*
   (b) Abelian group is cyclic. *(B.E. Oct. 97)*
   (c) Every cyclic group has more than one generator. *(B.E. Apr. 97)*
2. Fill in the blanks.
   (a) A group of prime order is always ________ *(B.E. Oct. 97)*
   (b) In the symmetric group S_4, the number of elements is ________ *(B.E. Apr. 98)*
3. Give an example for a non abelian finite group. *(B.E. Oct. 97)*
4. Find the identity element of a group of integers with binary operation defined by $a * b = a + b + 2$. *(B.E. Apr. 98)*
5. State any two properties of a group. *(B.E. Apr. 97)*
6. Define a natural homomorphism.
7. Show that every finite semigroup has an idempotent. *(MCA 97 Oct, 96N, B.E. 93A, 97A).*
8. Let $(M, *)$ be a monoid. Prove that there exists a subset $T \leq M^M$ such that $(M, *)$ is isomorphic to the monoid $(T, *)$. *(MCA Nov. 96).*
9. Let $(\{a, b\}, *)$ be a semigroup, where $a * a = b$ Show that (i) $a * b = b * a$ (ii) $b * b = a$. *(MCA Nov. 89)*
10. $S = \{a, b\}$. Show that the semigroup $(S^S, 0)$ is not commutative.
11. Define a semigroup and a monoid. Let the alphabet be $V = \{a, b\}$ and A be the set including the empty string λ of all sequences of V, beginning with a. Show that $(A, \oplus, \lambda)$ is a monoid. *(MCA 95 Nov., 90 Apr.)*
12. Give any one application of a group in Modular arithmetic. *(MCA 98Apr.)*
13. Show that the direct product of groups $(G, *)$ and (H, Δ) contains two normal subgroups which are isomorphic to $(G, *)$ adn (H, Δ) respectively. *(MCA 95 Mar.)*
14. State and prove fundamental theorem of homomorphism.
15. State and prove Cayley's theorem.

GROUP CODES

Introduction

Today communication plays an important role. It may take place in a variety of ways. The three essential parts in an ideal communication system are

1. Transmitter 2. Channel 3. Receiver.

In practice, the communication channel is subject to a variety of disturbances called noise. It may be due to sun spots, lightning, folds in a magnetic tape or many other things. The important task of a communication system is to minimize loss due to noise.

A device used to improve the efficiency of communication channel is an encoder. The device used to transform the encoded message into original form is a decoder.

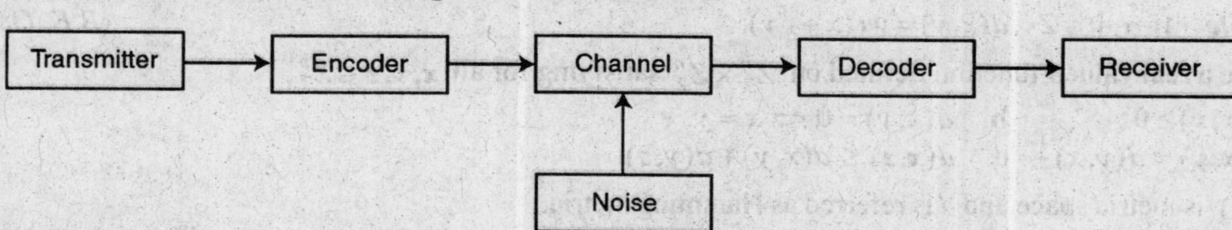

Basic Assumptions

In many cases, the information is sent by a sequence of 0's and 1's. 0 or 1 is called a digit. We denote $Z_2 = \{0, 1\}$. The element of Z_2 is a digit. A word is a sequence of digits. The number of digits in the word is called the length of the word.

For example 1100 is of length 4. 00101 is of length 5.

A word is transmitted by sending its digits, one after the other across a binary channel. $Z_2^2 = \{(0,0), (0,1), (1,0), (1,1)\}$. But we write this as $Z_2^2 = \{00, 01, 10, 11\}$. This set is binary code of words of length 2. Let us consider $(Z_2^2, +_2)$

$+_2$	00	01	10	11
00	00	01	10	11
01	01	00	11	10
10	10	11	00	01
11	11	10	01	00

Note that $1 +_2 1 = 0$. $(Z_2^2, +_2)$ is a group. In a similar manner, we can extend and prove that $(Z_2^2, +_2)$ is a group. 0 = 00... 0 is the identity and every element is its own inverse.

A block code is a code having all its words of the same length and this number is the length of the code. The words that belong to a given code C are called code words. $|C|$ denotes the number of code words in the code C.

The Hamming Metric

For any element $x = x_1, x_2, \ldots x_n \in Z_2^n$ where $n \in N$, the weight of x denoted by $wt(x)$ is the number of 1's in x. For example, if $x = 10101 \in Z_2^s$, $wt(x) = 3$. If $x = 000 \in Z_2^3, wt(x) = 0$.

Hamming distance Let $x = x_1, x_2, \ldots, x_n$ and $y = y_1, y_2, \ldots, y_n \in Z_2^n$, the Hamming distance between x and y, denoted by $d(x, y)$, is the number of components such that $x_i \neq y_i$ for $1 \leq i \leq n$.

For example, let $x = 110010, y = 010110 \in Z_2^6$ then $d(x, y) = 2$ since x and y differ in 2 places.

Also $x +_2 y = 100100$ and $wt\ (x +_2 y) = 2$.

$\therefore d(x, y) = wt\ (x +_2 y)$

Example 5.31 Find the hamming distance between x and y

(a) $x = 110110, y = 000101$ (b) $x = 001100, y = 010110$

Solution

a. $x +_2 y = 110011, d(x, y) = wt\ (x +_2 y) = 4$

b. $x +_2 y = 001100 +_2 010110 = 011010$
$d(x, y) = wt\ (x +_2 y) = 3$

Example If $x, y \in Z_2, d(x, y) = wt\ (x +_2 y)$. *(B.E. Oct. 98)*

Let d be a real valued function defined on $Z_2^n \times Z_2^n$ satisfying for all $x, y, z \in Z_2^n$,

a. $d(x, y) \geq 0$ b. $d(x, y) = 0 \Leftrightarrow x = y$.

c. $d(x, y) = d(y, x)$ d. $d(x, z) \leq d(x, y) + d(y, z)$.

(Z_2^n, d) is metric space and d is referred as Hamming metric.

Let $C \subset Z_2^n$ be a code. The minimum distance between the code words of C is min. $\{d(x, y) : x, y \in C\}$

For example, let $C = \{1001, 0100, 1000\}$

Let $x = 1001,\ y = 0100,\ z = 1000$

$x +_2 y = 1101$ $\quad d(x, y) = 3$

$x +_2 z = 0001$ $\quad d(x, z) = 1$

$y +_2 z = 1100$ $\quad d(y, z) = 2$

Minimum distance = 1

Encoding Function

Let n and m be positive integers such that $n > m$. A one-to-one function $E : Z_2^m \to Z_2^n$ is called an (m, n) encoding function which represents every word in Z_2^m as a word in Z_2^n. If $x \in Z_2^n$, then $E(x)$ is called code word representing x.

Due to noise, the receiver can make an uncorrectable and undetectable mistake like the interchange of 0 or 1 in a particular code. For example, the code 000 may be changed to 100 or 010 or 001 due to noise. Hamming developed a code by introducing redundant (extra) digits. This method is still popular in detecting and correcting errors. In a message of n digits long, m digits $(m < n)$ are used to represent the information part of the message and the remaining $k = n - m$ digits are used to detect and correct errors. The later digits are called parity checks.

Even parity checks Let $E = Z_2^m \to Z_2^{m+1}$ be an encoding function such that for $x = x_1\ x_2 \ldots x_m \in Z_2^n$, we have $E(x) = x_1, x_2, \ldots x_m, x_{m+1}$, where

$$x_{m+1} = \begin{cases} 0 & \text{if } wt\ (x) \text{ is even} \\ 1 & \text{if } wt\ (x) \text{ is odd} \end{cases}$$

This encoding procedure is called even parity check. For example, let us consider $E : Z_2^3 \to Z_2^4$ and define $E(x_1\, x_2\, x_3) = x_1\, x_2\, x_3\, x_4$ where

$$x_4 = \begin{cases} 0 & \text{if } wt(x) \text{ is even} \\ 1 & \text{if } wt(x) \text{ is odd} \end{cases}$$

Original code	Code with redundant digit
000	0000
001	0011
010	0101
011	0110
100	1001
101	1010
110	1100
111	1111

Let us consider the word 000, $E(000) = 0000$. If a single error occurs in 0000, it must be 1000 or 0100 or 0010 or 0001. But this combination is not available for any of the other codes.

Hence by supplying redundant digit, we are able to detect and correct a single error appearing in one of the codes.

Odd parity check Let $E : Z_2^m \to Z_2^{m+1}$ be an encoding function such that for $x = x_1\, x_2, \ldots, x_m \in Z_2^m$, we have $E(x) = x_1, x_2, \ldots, x_m, x_{m+1}$ where

$$x_{m+1} = \begin{cases} 1 & \text{if } wt(x) \text{ is even} \\ 0 & \text{if } wt(x) \text{ is odd.} \end{cases}$$

This encoding procedure is called Odd Parity Check. For example, $E : Z^3{}_2 \to Z^4{}_2$ and define $E(x_1\, x_2\, x_3) = x_1\ x_2\ x_3\ x_4$

where $x_4 = \begin{cases} 1 & \text{if } wt(x) \text{ is even} \\ 0 & \text{if } wt(x) \text{ is odd} \end{cases}$

Original code	Code with redundant digit
000	0001
001	0010
010	0100
011	0111
100	1000
101	1011
110	1101
111	1110

Note: If there is more than one error in the code then we have to supply more than one extra digit.

Decoding and Error Correction

A onto function $D: Z_2^n \rightarrow Z_2^m$ is called an (*n, m*) decoding function associated with E, if $D(y) = x' \in Z_2^m$ and is such that when the transmission channel has no noise then $x' = x$ i.e., $D \circ E = I$, where I is the identity function on Z_2^m. The decoding function decodes properly received words correctly but the decoding of improperly received words may or may not be correct.

For example, let us consider even parity check given by $E: Z_2^3 \rightarrow Z_2^4$. Let $D: Z_2^4 \rightarrow Z_2^3$ be the decoding function.

Code(x)	$y = E(x)$	$D(y) = (D \circ E)(x)$
000	0000	000
001	0011	001
010	0101	010
011	0110	011
100	1001	100
101	1010	101
110	1100	110
111	1110	111

Let C denote a code. Let the word be transmitted into a word v. If $v \in C$, we cannot detect error, even if there is an error.

A code can detect all combination of k or fewer errors if and only if the minimum distance of $x, y \in C$ is atleast $k+1$. It can correct all combinations of k or fewer errors if and only if the minimum is atleast $2k+1$.

For example, if the minimum distance of a code is 5 it can detect 4 or less errors and correct 2 or less errors.

The Parity and Generator Matrices

The encoding and decoding functions can be given by matrices over Z_2. One of these matrices will help us to locate the nearest code word for a received word.

Let $G = \begin{pmatrix} 1 & 0 & 0 & 1 & 1 & 0 \\ 0 & 1 & 0 & 0 & 1 & 1 \\ 0 & 0 & 1 & 1 & 0 & 1 \end{pmatrix}$ be a 3×6 matrix over Z_2. Then $G = [I_3 / A]$ where

$$I_3 = \begin{pmatrix} 1 & 0 & 0 \\ 0 & 1 & 0 \\ 0 & 0 & 1 \end{pmatrix} \text{ and } A = \begin{pmatrix} 1 & 1 & 0 \\ 0 & 1 & 1 \\ 1 & 0 & 1 \end{pmatrix}$$

The partitioned matrix G is called a generator matrix.

Let $E: Z_2^3 \rightarrow Z_2^6$ be an encoding function defined by $E(W) = W\,G \in Z_2^6$

Now $\quad Z_2^3 = \{000, \; 100, \; 010, \; 001, \; 110, \; 101, \; 011, \; 111\}$

$$E(000) = (000) \begin{pmatrix} 1 & 0 & 0 & 1 & 1 & 0 \\ 0 & 1 & 0 & 0 & 1 & 1 \\ 0 & 0 & 1 & 1 & 0 & 1 \end{pmatrix} = (0\,0\,0\,0\,0\,0)$$

$$E(100) = (100)\begin{pmatrix} 1 & 0 & 0 & 1 & 1 & 0 \\ 0 & 1 & 0 & 0 & 1 & 1 \\ 0 & 0 & 1 & 1 & 0 & 1 \end{pmatrix} = (1\,0\,0\,1\,1\,0)$$

In a similar way, we can calculate for all other elements of Z_2^3 and write the set of code words as $C = \{000000, 100110, 010011, 001101, 110101, 101011, 011110, 111000\} \subset Z_2^6$.

We can recapture the message by simply dropping the last three digits of the code word. If the word u is transformed to the word υ and $\upsilon \in C$, we cannot detect error though there may be an error. We can calculate the distances of words in C and find the minimum distance as 3. Hence we can detect 2 or less errors and correct single error.

Let $w = w_1 w_2 w_3 \in Z_2^3$ and $E(w) = w_1 w_2 w_3 w_4 w_5 w_6 \in Z_2^6$. We have

$$E(w) = (w_1 w_2 w_3)\begin{pmatrix} 1 & 0 & 0 & 1 & 1 & 0 \\ 0 & 1 & 0 & 0 & 1 & 1 \\ 0 & 0 & 1 & 1 & 0 & 1 \end{pmatrix}$$

$$= (w_1 w_2 w_3 (w_1 +_2 w_3)(w_1 +_2 w_2)(w_2 +_2 w_3))$$

$$w_1 w_2 w_3 w_4 w_5 w_6 = w_1 w_2 w_3 (w_1 +_2 w_3)(w_1 +_2 w_2)(w_2 +_2 w_3)$$

we have $w_4 = w_1 +_2 w_3$, $w_5 = w_1 +_2 w_2$, $w_6 = w_2 +_2 w_3$.

These equations are called parity check equations. Since w_4, w_5, w_6 are either 0's or 1's and $0 = -0 \bmod 2$, $1 = -1 \bmod 2$ we can write these equations as follows:

$$w_1 +_2 w_3 +_2 w_4 = 0, \quad w_1 +_2 w_2 +_2 w_5 = 0, \quad w_2 +_2 w_3 +_2 w_6 = 0.$$

We can easily see that

$$\begin{pmatrix} 1 & 0 & 1 & 1 & 0 & 0 \\ 1 & 1 & 0 & 0 & 1 & 0 \\ 0 & 1 & 1 & 0 & 0 & 1 \end{pmatrix}\begin{pmatrix} w_1 \\ w_2 \\ w_3 \\ w_4 \\ w_5 \\ w_6 \end{pmatrix} = \begin{pmatrix} w_1 +_2 w_3 +_2 w_4 \\ w_1 +_2 w_2 + w_5 \\ w_2 +_2 w_3 +_2 w_6 \end{pmatrix} = \begin{pmatrix} 0 \\ 0 \\ 0 \end{pmatrix}$$

$$H \cdot (E(w))^{tr} = \begin{pmatrix} 0 \\ 0 \\ 0 \end{pmatrix}$$

where $H = [A^{tr} / I_3]$ and $(E(w))^{tr}$ is the transpose of $E(w)$. H is called parity check matrix associated with the generator matrix G.

Now $r = r_1 r_2 \ldots r_6 \in Z_2^6$ is a code word if and only if

$$H \cdot r^{tr} = \begin{pmatrix} 0 \\ 0 \\ 0 \end{pmatrix}$$

Suppose we receive a word $r = 110110$. We see that $r \notin C$. i.e. r is not a code word.

$$H \cdot r^{tr} = \begin{pmatrix} 1 & 0 & 1 & 1 & 0 & 0 \\ 1 & 1 & 0 & 0 & 1 & 0 \\ 0 & 1 & 1 & 0 & 0 & 1 \end{pmatrix} \begin{pmatrix} 1 \\ 1 \\ 0 \\ 1 \\ 1 \\ 0 \end{pmatrix} = \begin{pmatrix} 0 \\ 1 \\ 1 \end{pmatrix}$$

which is the same as second column of H. Hence we can say that an error occurred in second component of r. By changing the second component, we get 100110 and dropping the last three digits, we can conclude that the message sent is 100. Now $r = 110110$, $c = 100110$ (code word), $e = 010000$ is the transmission error of weight 1 then we see that $r = C +_2 e$ or $C = r +_2 e$.

Suppose if $H \cdot r^{tr}$ is not one of the columns of H but can be obtained as the sum of the two columns of H, we cannot correct multiple errors.

Hence, if the parity check matrix H has no column of 0's and no two columns of H are the same then it is helpful in correcting single error. In this case, we get the following decoding algorithm. For any $r \in Z_2^n$, if $T(C) = r$ then:

1. If $H \cdot r^{tr} = 0$, the transmission is correct and r is the code word that was transmitted. The decoded message consists of the first m components of r.
2. If $H \cdot r^{tr}$ equals the ith column of H, there is a single error in the ith component of r and changing it, we get correct code word C. The first m components of C yield the original message.
3. If neither (1) nor (2) occurs, there is more than one error and it does not provide a reliable way to decode it.

GROUP CODES - DECODING WITH COSET LEADERS

Definition Let $E : Z_2^m \rightarrow Z_2^n$ be an encoding function. The code $C = E(Z_2^m)$ is called a group code if C is a subgroup of Z_2^n.

For example, $Z_2^2 = \{00, 10, 01, 11\}$. Let $E : Z_2^2 \rightarrow Z_2^6$ be an encoding function where $E(00) = 000000$, $E(10) = 101010$, $E(01) = 010101$, $E(11) = 111111$.

Now $C = \{000000, 101010, 010101, 111111\} \subset Z_2^6$ is a subgroup of Z_2^6. Since C is finite and it is closed with respect to $+_2$, it is a subgroup of Z_2^6. C is a group code.

In a group code, the minimum distance between distinct code words is the minimum of the weights of the nonzero elements of the code. In the above example, the minimum distance of C is 3.

Also if $E : Z_2^m \rightarrow Z_2^n$ is an encoding function given by a generator matrix G or the associated parity check matrix H then $C = E(Z_2^m)$ will be a group code.

Now we shall explain how the group structure of C together with its cosets in Z_2^n helps for decoding by giving an example step by step.

Let $E : Z_2^3 \rightarrow Z_2^6$ be an encoding function given by $E(000) = 000000$, $E(100) = 100110$, $E(010) = 010011$ $E(001) = 001101$, $E(110) = 110101$, $E(101) = 101011$, $E(011) = 011110$, $E(11) = 111000$.

Let $C = \{000000, 100110, 010011, 001101, 110101, 101011, 011110, 111000\} \subset Z_2^6$ is a subgroup with respect to $+_2$.

Step 1 Write the elements of C in a row starting with identity.

000000 100110 010011 001101 110101 101011 011110 111000

Step 2 Let $x \in Z_2^6$, which does not appear in the above row and $wt(x)$ be the least. For example $x = 100000 \in Z_2^6$ and it does not appear in the above row and $wt(x) = 1$ is the least. Put it in the left hand corner. Find $x +_2 \upsilon \, \forall \upsilon \in C$ and write the two rows as below.

000000	100110	010011	001101	110101	101011	011110	111000
100000	000110	110011	101101	010101	001011	111110	011000

Step 3 Let $x \in Z_2^6$, which does not appear in the above two rows and $wt(x)$ be the least. For example $x = 010000 \in Z_2^6$ and it is not appearing in the above two rows and $wt(x) = 1$ is the least. Put it in the left hand corner. Find $x +_2 \upsilon \, \forall \upsilon \in C$ and write three rows as below.

000000	100110	010011	001101	110101	101011	011110	111000
100000	000110	110011	101101	010101	001011	111110	011000
010000	110110	000011	011101	100101	111011	001110	101000

Step 4 Repeat the above procedure until the cosets provide a partition of Z_2^6 and construct the decoding table.

000000	100110	010011	001101	110101	101011	011110	111000
100000	000110	110011	101101	010101	001011	111110	011000
010000	110110	000011	011101	100101	111011	001110	101000
001000	101110	011011	000101	111101	100011	010110	110000
000100	100010	010111	001001	110001	101111	011010	111100
000010	100100	010001	001111	110111	101001	011100	111010
000001	100111	010010	001100	110100	101010	011111	111001
010100	110010	000111	011001	100001	111111	001010	101100

The entries in the first column of this table are called coset leaders. For the last row, the weights of 010100, 100001, 001010 are equal to 2. Instead of taking 010100, we can take even 100001 or 001010 and construct the last row. Hence the table need not be unique. As a result, not all double errors can be corrected because there may not be a unique code word at a minimum distance for each received word in the last row.

Suppose the received words are $r_1 = 101001, r_2 = 111010, r_3 = 001001, r_4 = 111011$. Let the corresponding code words be C_1, C_2, C_3, C_4. Let the respective message be w_1, w_2, w_3, w_4.

$r_1 = 101001$ appears in the 6th row and 6th column. Hence the element in the 1st row and 6th column gives the code word $c_1 = 101011$. $\therefore \; w_1 = 101$.

Similarly for $4C_2 = 111010, c_2 = 111000, w_2 = 111$

$$r_3 = 001001, C_3 = 001101, w_3 = 001$$

$$r_4 = 111011, c_4 = 101011, w_4 = 101$$

Hence the messages sent are given by

$$w_1 = 101, \; w_2 = 111, w_3 = 001, w_4 = 101$$

EXERCISE II

1. Find the length and weights of the following words:

   a. 00110   b. 1010

   c. 000   d. 101001

   e. 0001

2. List all the words of length 3.
3. How many words are there of length n ?
4. Find the distance of each of the following codes.

   a. $C = \{101, 111, 011\}$ b. $C = \{0000, 1001, 0110, 1111\}$.

5. Let $C = (001, 101, 110\}$. Determine whether C will detect the error pattern (a) 011 (b) 001 (c) 000.
6. If the minimum distance of C is 3, how many errors can it detect? How many errors can it correct?
7. Let $C \subset Z_2^7$ be a set of code words. In each of the following two of e (error pattern), r (received word) and C (code word) are given with $r = C +_2 e$. Determine the third term.

   a. $c = 1010110, r = 101111$ b. $c = 1010110, e = 0101101$

   c. $e = 01011111, r = 0000111$

8. Let $E : Z_2^2 \to Z_2^5$ be an encoding function defined by $E(00) = 00000$, $E(01) = 01110$ $E(10) = 10101$, $E(11) = 11011$. Let $C = \{00000, 01110, 10101, 11011\}$. Examine whether C is a group code.

9. Explain the concept of coset leader in coding theory.

CHAPTER

6

Lattices and Boolean Algebra

INTRODUCTION

Boolean algebra, introduced by George Boole in 1854, is very useful in the theory and design of computers. Boolean algebra is an algebraic system in which ordering relations play a very important role. This chapter is devoted to the study of various properties of Boolean algebra and their application to switching theory. The chapter begins with an introduction to partially ordered sets and then proceeds to define a lattice. Boolean algebra is introduced as a special type of lattice.

Partially Ordered Sets

A binary relation R in a set P is called a partial order relation or a partial ordering in P if and only if R is reflexive, antisymmetric and transitive. It is denoted by the symbol $\leq$. If $\leq$ is a partial ordering on P, then the ordered pair $(P, \leq)$ is called a partially ordered set or a poset. If for any $x, y \in p$, if $x \leq y$ or $y \leq x$, then $(p, \leq)$called a totally ordered set.

Example Let R be the set of real numbers. The relation "less than or equal to" or "greater than or equal to" is a partial order on R.

Note: The symbol $\leq$ should not be confused with the standard "less than or equal to" relation on real numbers.

Example 6.1 Prove that "being a subset of is a partial order on the power set of a non-empty set of A. (M.E.,M.U.99).

Solution Let $P(A) = 2^A = X$ be the power set of A i.e X is the set of all subsets of A. For any u, v, w in x, set $U \leq V \Leftrightarrow U \subseteq V$.

i. Reflexive: Since $U \subseteq U, U \leq U$.

ii. Antisymmetric: Suppose $U \leq V$ and $V \leq U$. Then $U \subseteq V$ and $V \subseteq U$. This implies that $U = V$.

iii. Transistive: Suppose $U \leq V$ and $V \leq W$. Then $U \subseteq V$ and $V \subseteq W$. This implies that $U \leq W$. Hence $U \leq W$.

Thus $(x, \leq)$ is a partial order.

Example 6.2 Let X be the set of positive integers. Prove that the relation "divides" and "intergral multiples of" are partial orderings on X.

Solution If a and b are positive integers, then we say, " a divides b" written a/b if and only if there is an integer c such that $ac = b$ (or b is an integral multiple of a). For any a,b $\in$ x, set $a \leq b \Leftrightarrow a/b$ (=) $ac = b$.

i. Reflexive: Since $a.1 = a$, a/a and hence $a \leq a$.

ii. Antisymmetric: Suppose $a \leq b$ and $b \leq a$. Then there exist $c,d \in x$ such that $ac = b$ and $bd = a$. Since $a = bd = acd \Rightarrow cd = 1, c = 1$ and $d = 1$. Therefore $a = b$.

iii. Transitive: Suppose for any $a,b,c \in x$, $a \leq b$ and $b \leq c$. Then there exist x,y $\in$ x such that ax = b and $by = c$. Now $a\,xy = by = c$.
and since $xy \in X$, a/c. Therefore $a \leq c$. Thus $(X, \leq)$ is a partial order.

Example 6.3 Consider the set Z of integers. Define $a \leq b$ if there is a positive integer r such that $b = a^r$. Prove $(z, \leq)$ is a partial ordered set.

Solution Let $a,b,c \in Z$.

i. Reflexive: Since $a = a^1$, $a \leq a$.

ii. Antisymmetric: Suppose $a \leq b$ and $b \leq a$. Then there exist positive integers r, s such that $b = a^r$ and $a = b^s$. Since $a = b^s = (a^r)^s = a^{rs}$ implies $rs = 1$. But $rs \in z \Rightarrow r = 1$ and $s = 1$. Hence $a = b$.

iii. Transitive: Suppose $a \leq b$ and $b \leq c$. Then there exist positive integers r,s such that $b = a^r$ and $c = b^s$. Now $c = b^s = (a^r)^s = a^{rs}$ and $rs \in Z$ implies that $a \leq c$. Thus $(Z, \leq)$ is a partially ordered set.

Example 6.4 If R is a partially ordered relation on a set x and $A \leq x$, show that $R \cap (A \times A)$ is a partial ordering relation on A. (*M.E. MU.*2000)

Solution Denote $R \cap (A \times A)$ by R^1.

i. Reflexive: Let $x \in A$. Then $(x, x) \in A \times A$.
Since R is reflexive, $(x, x) \in R$. (Note that $x, x) \in R$ means xRx). Therefore $(x, x) \in R \cap (A \times A) = R^1$.

ii. Antisymmetric: Suppose $(x, y) \in R^1$ and $(y, x) \in R^1$. This implies that $(x, y) \in R \cap (A \times A)$ and $(y, x) \in R \cap (A \times A)$. Since R is antisymmetric, $(x, y) \in R$ and $(y, x) \in R \Rightarrow x = y$.

iii. Transitive: Suppose $(x, y) \in R^1 = R \cap (A \times A)$ and
$(y, z) \in R^1 = R \cap (A \times A)$. Such R is transitive
$(x, y) \in R$ and $(y,z) \in R \Rightarrow (x, z) \in R$ clearly.
$(x, z) \in A \times A$ and hence $(x, z) \in R \cap (A \times A) = R^1$
Thus $R^1 = R \cap (A \times A)$ is a partial ordering relation on A.

Hasse Diagram of Partially Ordered Sets

The Hasse diagram of a poset P is a picture of P. So it is very useful in describing types of elements of P. Some times we define a partially ordered set by simply presenting its Hasse diagram.

Define a relation $<$ on P by $x < y$ (=) $x \leq y$ but $x \neq y$. Let $(p, \leq)$ be a partially ordered set. An element $y \in p$ is said to cover an element $x \in p$ if $x < y$ and if there does not exist an element $z \in p$ such that $x \leq z$ and $z \leq y$; that is

$$y \text{ cover } x \Leftrightarrow (x < y \text{ and } x \leq z \leq y \Rightarrow x = z \text{ or } z = y)$$

Here we say that y is an immediate predecessor of x or x is an immediate successor of y.

A partial ordering $\leq$ on a set P can be represented by means of a diagram known as a Hasse diagram or a partially ordered set diagram of $(p, \leq)$.

Procedure

1. Represent each element of p by a small circle or a dot.
2. The circle for $x \in p$ is drawn below the circle for $y \in p$ if $x < y$, and a line is drawn between x and y if y covers x.
3. If $x < y$ but y does not cover x, then x and y are not connected directly by a single line. However, they are connected through one or more elements of p.

Note:

1. The Harsse diagram of a part p need not be connected. Also, there can be no directed cycles in the diagram of p since the partial order relation is antisymmetric.
2. For a totally ordered set $(p, \leq)$, the Hasse diagram consists of circles one below the other.
3. It is possible to obtain the set of ordered pairs in $\leq$ from such a Hasse diagram.

Example 6.5 Let $p = \{\phi, \{a\}, \{a,b\}, \{a,b,c\}\}$ and the relation $\leq$ be such that $A \leq B$ if $A \subseteq B$, the inclusion relation on p. Draw Hasse diagram of $(p, \leq)$.

Solution

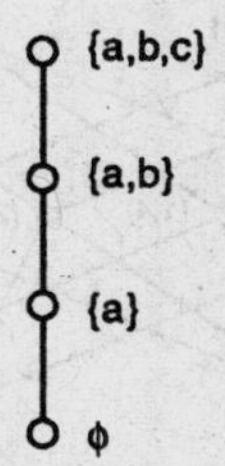

Example 6.6 Let $x = \{2,3,6,12,24,36\}$ and the relation $\leq$ be such that $x \leq y$ if x divides y. Draw the Hasse diagram of $(x, \leq)$.

Solution Here

$\leq = \{(2,2), (2,6), (2,12), (2,24), (2,36), (3,3), (3,6),$
$(3,12), (3,24), (3,36), (6,6), (6,12), (6,24),$
$(6,36), (12,12), (12,24), (12,36), (24,24), (36,36)\}$

So the corresponding Hasse diagram is

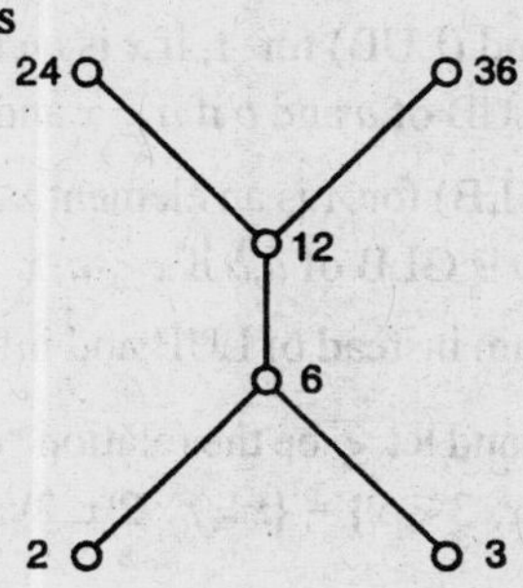

Example 6.7 Let $A = \{a,b,c\}$ and $p(A)$ be its power set. Let $\subseteq$ be the inclusion relation on $p(A)$. Draw Hasse diagram.

Solution Here $p(A) = \{\phi, \{a\}, \{b\}, \{c\}, \{a,b\}, \{a,c\}, \{b,c\}, \{a,b,c\}\}$ and the corresponding Hasse diagram is

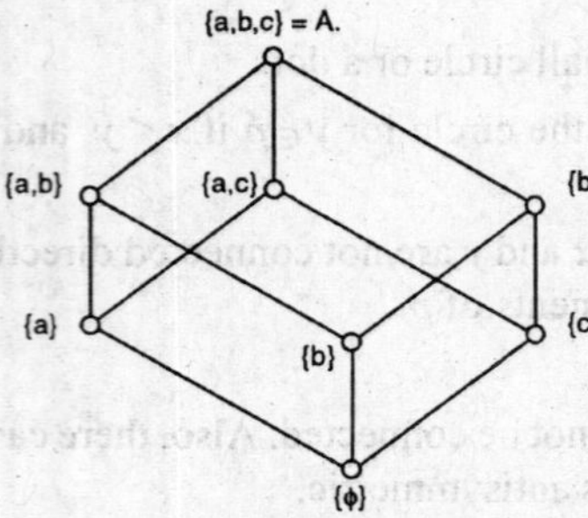

Example 6.8 Let $A = \{1,2,3,4\}$ and $p(A)$ be its powerset. Let $\subseteq$ be the inclusion relation on $p(A)$. Draw the Hasse diagram. *(M.E. M.U 99)*

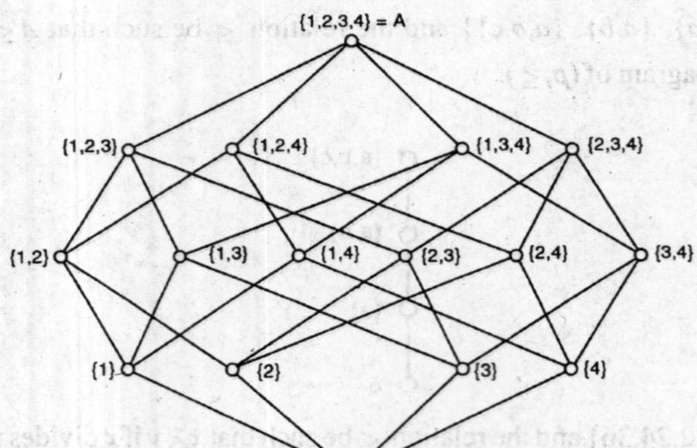

Solution Let $(p, \leq)$ be a poset and let $A \subseteq P$. An element $x \in p$ is an *upper bound* for A if for all $a \in A$, $a \leq x$. Similarly, any element $x \in p$ is a lower bound for A if for all $a \in A$, $x \leq a$.

In other words, $x \in p$ is an upper bound of a and b if $a \leq x$ and $b \leq x$. Similarly, $x \in p$ is called a lower bound of a and b if $x \leq a$ and $x \leq b$.

An element $x \in p$ is a least upper bound (LUB) for A, if x is an upper bound for A and $x \leq y$, where y is any upperbound for A. In otherwords $x \in p$ is LUB of a and b if $a \leq x$ and $b \leq x$ and if for $y \in p$, $a \leq y, b \leq y \Rightarrow x \leq y$.

Similarly the greatest lower bound (GLB) for A is an element $x \in p$ such that x is a lower bound and $y \leq x$ for all lower bounds for y. In otherwords $x \in p$ is GLB of a,b if $x \leq a$, $x \leq b$ and if $y \in p, y \leq a, y \leq b \Rightarrow y \leq x$.

Note: Some authors use the term supremum instead of LUB and infimum instead of GLB.

Example Let $p = \{1, 2, 3, 5, 6, 10, 15, 30\}$ and let $\leq$ be the relation "divides" on p. Then $(p, \leq)$ is a poset. The set of upper bounds of 2 and 3 is $\{x \in p : 2 \leq x,\ 3 \leq x\} = \{x \in p : 2 \backslash x, 3 \backslash x\} = \{6, 30\}$.

The least element of this set is 6. So LUB of 2 and 3 = 6. The set of lower bounds of 6 and 15 is {1, 3}. The greatest element of this set is 3. Hence GLB of 6 and 15 = 3.

The Hasse diagram for this example is

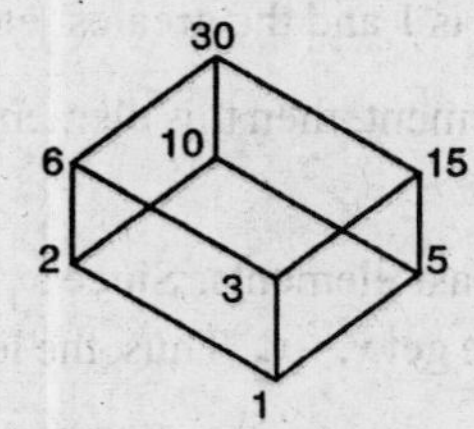

Example 6.9 Draw the Hasse diagram for $(N, \leq)$ where $A = \{1,2,3,4,12\}$ and $x \leq y$ if and only if $x \backslash y$.

(*M.E. M.U.99*)

Solution

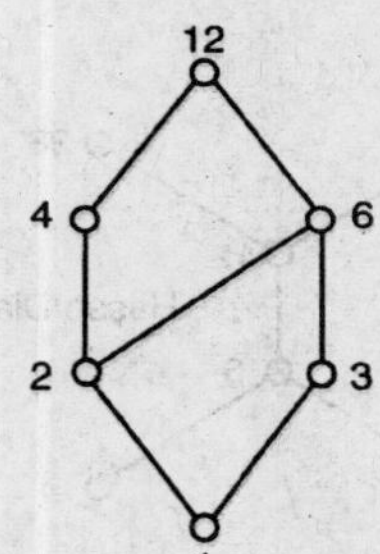

Example 6.10 Let A be the set of factors of a particular positive integer m and e, $\leq$ be the relation "divides". i.e. $\leq = \{(x, y) | x \in A$ and $y \in A$ and $(x$ divides $y)\}$. Draw Hasse diagrams for (a) $m = 12$, (b) $m = 30$, (c) $m = 45$.

Solution

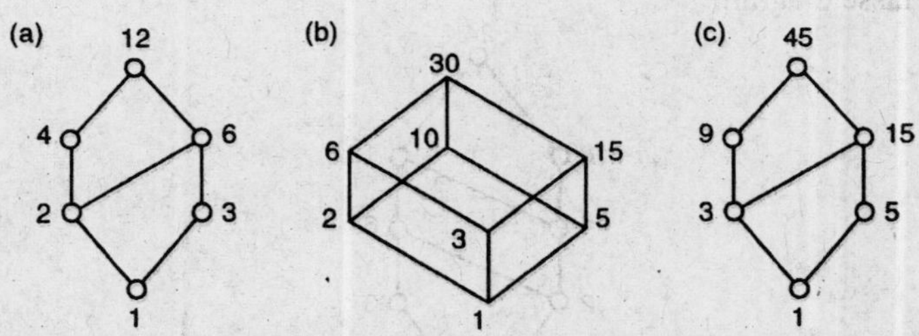

Example 6.11 Show that there are only five distinct Hasse diagram for partially ordered sets that contain three elements.

Solution Let us assume that $a \leq b$ if $a \backslash b$ and consider the sets $A = \{2,3,5\}$, $B = \{2,3,4\}$, $C = \{2,3,6\}$, $D = \{2,4,6\}$, and $E = \{2,4,8\}$. Then its Hasse diagrams are

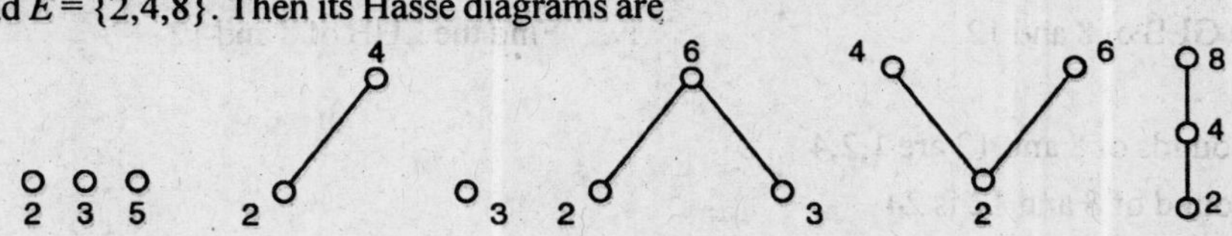

(A, ≤) (B, ≤) (C, ≤) (D, ≤) (E, ≤)

These are the only distinct Hasse diagrams for poset that contain three elements.

Greatest and Smallest elements An element $x \in p$ is called the greater element if for all $a \in p$, $a \leq x$. An element $x \in p$ is called the smallest element or least element if for all $a \in p$, $x \leq a$. The greatest element will be denoted by 1 and the smallest element by 0.

In example 6.10 (b), the least element is 1 and the greatest element is 30.

Theorem If a poset $(p, \leq)$ has a least element, then this element is unique. Similarly, if a poset $(p, \leq)$ has the greatest element, then it is unique.

Proof Suppose x_1 and x_2 are any two least elements. Since x_1 is a least element, $x_1 \leq x_2$. Since x_2 is a least element, $x_2 \leq x_1$. But $\leq$ is antisymmetric we get $x_1 = x_2$. Thus, the least element, if it exists, is unique. Dually, the other result follows.

Example Draw the Hasse diagram for $(X, \leq)$ where $X = \{2,3,6,12,24,36\}$ and $x \leq y$ if $x \backslash y$. Find

i. the LUB and the GLB of $A = \{2,3,6\}$
ii. the LUB and the GLB of $B = \{2,3\}$
iii. the LUB and the GLB of $C = \{6,12\}$

Solution

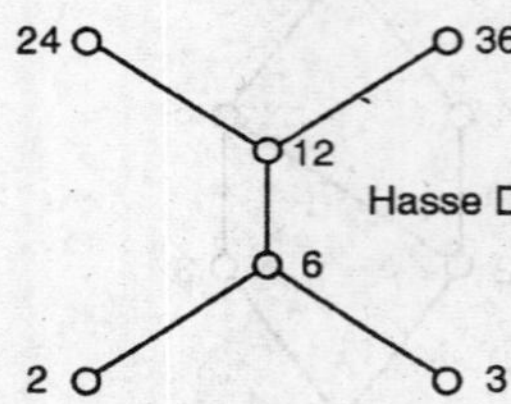

Hasse Diagram

i. LUB of $A = 6$; GLB of A does not exist
ii. LUB of $B = 6$; but there is no GLB
iii. LUB of $C = 12$; GLB of $C = 6$

Example 6.13 For the Hasse diagram

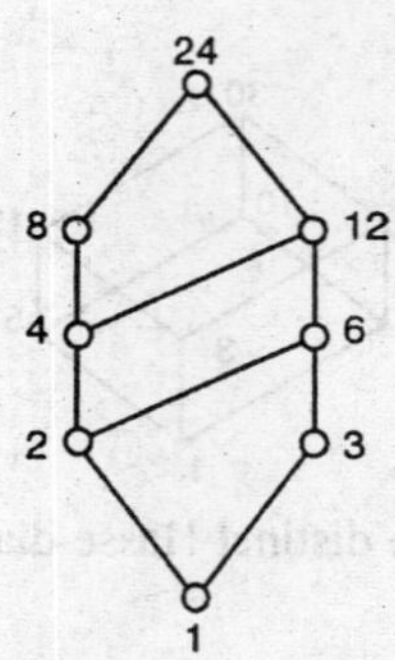

i. Find all lower bounds 8 and 12
ii. Find all upper bounds 8 and 12
iii. Find the GLB of 8 and 12
iv. Find the LUB of 8 and 12

Solution

i. Lower bounds of 8 and 12 are 1,2,4
ii. Upper bound of 8 and 12 is 24
iii. GLB of 8 and 12 is 4
iv. LUB of 8 and 12 is 24.

EXERCISE I

1. Draw the Hasse diagrams of the following sets under the partial ordering relation "divides" and indicate those which are totally ordered.

   {2,6,24}, {3,5,15}, {12,3,6,12}, {2,4,8,16}, {3,9,27,54}
2. Give an example of set X such that $(p(X), \subseteq)$ is a totally ordered set.
3. Give a relation which is both a partial ordering relation and an equivalence relation on a set.
4. Let $X = \{1,2,3,5,6,10,15,30\}$ and the relation "divides" be a partial ordering on x.
   a. Fnd all lower bounds of 10 and 15
   b. Find the GLB of 10 and 15
   c. Find all upper bounds of 10 and 15
   d. Determine the LUB of 10 and 15
   e. Draw the Hasse diagram for $(X, \leq)$
5. Consider the relation of divisibility on the set Z of integers. Is this relation a partial ordering of Z? (Hint: No, for example $3/-3$ and $-3/3$ but $3 \neq -3$, that is, the relation is not antisymmetric).
6. Let $x = \{1,2,3,4,6,8,9,12,18,24\}$ be ordered by divisibility. Draw the Hasse diagram.
7. Let R be a binary relation on the set of all positive integers such that

   $R = \{(x, y)/x - y$ is an odd positive integer$\}$. Is R a partial ordering relation?
8. For a given set A, consider the relation, $R = \{(C, D)\backslash C \in P(A), D \in P(B)$, and $C \subseteq D\}$. Show that R is a partial ordering relation.

LATTICES

A lattice is a partially ordered set $(L, \leq)$ in which every pair of elements $a, b \in L$ has a greatest lower bound (GLB) and a least upper bound (LUB).

The greatest lower bound (GLB) of a subset $\{a, b\} \subseteq L$ will be denoted by $a \wedge b$ and the least upper bound (LUB) by $a \vee b$. So GLB $\{a, b\} = a \wedge b$, called the meet of a and b and LUB $\{a, b\} = a \vee b$, called the join of a and b.

Note that $\wedge$ and $\vee$ are binary operations and we denote the lattice by $(A, \wedge, \vee)$. From the definition of $\wedge$ and $\vee$, it is clear that

i. $a \leq a \vee b$; $b \leq a \vee b$ (i.e., $a \vee b$ is an UB of a and b)
ii. $a \wedge b \leq a$; $a \wedge b \leq b$ (i.e., $a \wedge b$ is a LB of a and b)
iii. If $a \leq c$ and $b \leq c$ then $a \vee b \leq c$ (i.e., $a \vee b$ is the LUB of a and b)
(iv) If $c \leq a$ and $c \leq b$ then $c \leq a \wedge b$. (i.e., $a \wedge b$ is the GLB of a and b).

Example (lattices)

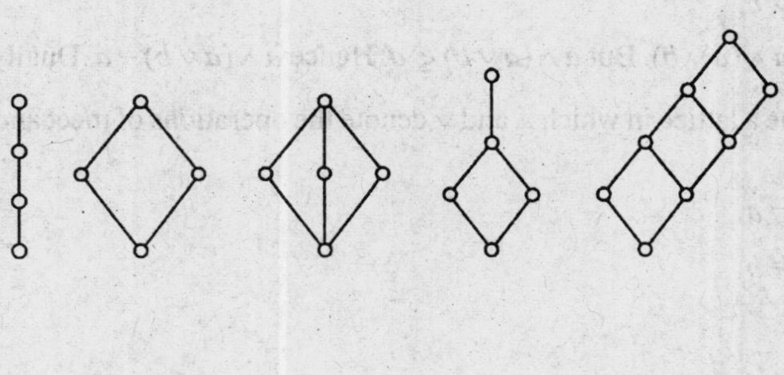

Example (posets but not lattices)

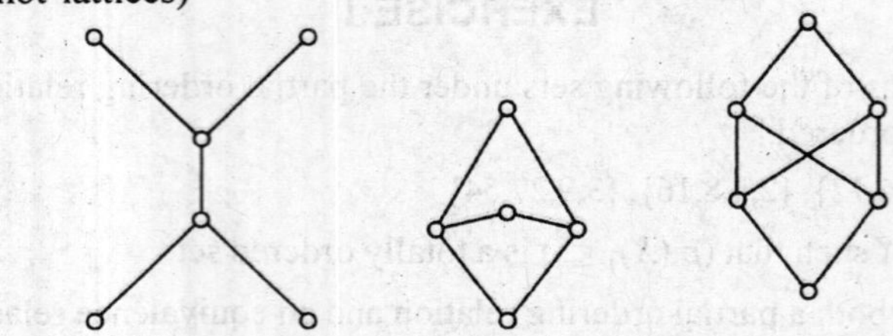

Example Let A be any set and $L = P(A)$ be its power set. The poset $(L, \subseteq)$ is a lattice in which for any $x, y, \in L$, $x \wedge y = x \cap y$ and $x \vee y = x \cup y$

Example Let I be the set of positive integers. For any $x, y \in I$, $x \leq y$ if x/y. Define $x \vee y =$ LCM (x, y) and $x \wedge y =$ GCD$\in (x, y)$. Then $(I, \wedge, \vee)$ is a lattice.

Theorem Let $(L, \leq)$ be a lattice in which $\wedge$ and V denote the operations of meet and join respectively. For any $a, b, c \in L$, we have

i. $a \wedge a = a$; $a \vee a = a$ (idempotent)
ii. $a \wedge b = b \wedge a$; $a \vee b = b \vee a$ (commutative)
iii. $(a \wedge b) \wedge c = a \wedge (b \cap c)$; $(a \vee b) \vee c = a \vee (b \vee c)$ (associative)
iv. $a \wedge (a \vee b) = a$; $a \vee (a \wedge b) = a$ (absorption).

Proof

i. Since $a \leq a$, we know that a is a lower bound of $\{a, a\} = \{a\}$. If b is also a lower bound and if $a \leq b$ then we have $b \leq a$ and $a \leq b$. By antisymmetry, $a = b$. So a is the GLB of $\{a\}$. Therefore, $a \wedge a = a$. Dually, $a \vee a = a$ follows.

ii. Let $x = a \wedge b =$ GLB $\{a, b\}$. Since $\{a, b\} = \{b, a\}$, GLB $\{b, a\} = x$. So $b \wedge a = x$. Hence $x = a \wedge b = b \wedge a$. Dually, $a \vee b = b \vee a$ follows.

iii. Let $x = a \wedge (b \wedge c)$ and $y = (a \wedge b) \wedge c$.

Now $x = a \wedge (b \wedge c) \Rightarrow x \leq a,\ x \leq b \wedge c$

$$\Rightarrow x \leq a, x \leq b, x \leq c$$

$$\Rightarrow x \leq a \wedge b, x \leq c$$

$$\Rightarrow x \leq (a \wedge b) \wedge c = y.$$

Similarly, $y \leq x$ follows. By antisymmetry, $x = y$ and hence $a \wedge (b \wedge c) = (a \wedge b) \wedge c$.

Dually, $a \vee (b \vee c) = (a \vee b) \vee c$ follows.

iv. By definition, for any $a \in L$,

$a \leq a$ and $a \leq a \vee b$

Therefore, $a \leq a \wedge (a \vee b)$. But $a \wedge (a \vee b) \leq a$. Hence $a \wedge (a \vee b) = a$. Dually, $a \vee (a \wedge b) = a$ follows.

Theorem Let $(L, \leq)$ be a lattice in which $\wedge$ and $\vee$ denote the operations of meet and join respectively. For any $a, b \in L$,

$$a \leq b \Leftrightarrow a \wedge b = a$$

$$\Leftrightarrow a \vee b = b$$

Proof Assume that $a \leq a$. Since $a \leq b$, it follows that $a \leq a \wedge b$. But by definition of $\wedge$, $a \wedge b \leq a$. Therefore $a \wedge b = a$. Conversely, suppose $a \wedge b = a$. Then $a \leq b$. Hence $a \leq b \Leftrightarrow a \wedge b = a$. Similarly $a \leq b \Leftrightarrow a \vee b = b$ follows.

Isotonicity law Let $(L, \leq)$ be a lattice in which $\wedge$ and $\vee$ denote the operations of meet and join respectively. For any $a, b, c, \in L$,

$$b \leq c \Rightarrow \begin{cases} a \wedge b \leq a \wedge c \\ a \vee b \leq a \vee c \end{cases}$$

Proof Assume that $b \leq c$. Since $a \wedge b \leq b$, by transitivity, $a \wedge b \leq c$. Since $a \wedge b \leq a$, it follows that,

$$a \wedge b \leq a \wedge c$$

Now, $b \leq c$ and $c \leq a \vee c$ implies $b \leq a \vee c$. But $a \leq a \vee c$. Hence $a \vee b \leq a \vee c$.

Note: For any $a, b, c \in L$, by isotonicity law,

$a \leq b \wedge a \leq c \Rightarrow a \leq b \vee c$

$a \leq b \wedge a \leq c \Rightarrow a \leq b \vee c$

$c \leq b \wedge a \leq a \Rightarrow b \Lambda c \leq a$

$c \leq b \wedge a \leq a \Rightarrow b \vee c \leq a.$

Distributive inequality Let $(L, \leq)$ be a lattice. For any $a, b, c, \in L$, the following inequalities are hold.

i. $a \vee (b \wedge c) \leq (a \vee b) \wedge (a \vee c)$

ii. $(a \wedge b) \vee (a \wedge c) \leq a \wedge (b \vee c)$

Since $a \leq a \vee b$ and $a \leq a \vee c$, we have

$$a \leq (a \vee b) \wedge (a \vee c) \quad (1).$$

Since $b \wedge c \leq b \leq a \vee b$ and $b \wedge c \leq c \leq a \vee c$,

$$b \wedge c \leq (a \vee b) \wedge (a \vee c) \quad (2)$$

From (1) and (2)

$a \vee (b \wedge c) \leq (a \vee b) \wedge (a \vee c)$.

Similarly, (ii) follows.

Modular inequality Let $(L, \leq)$ be a lattice. For any $a, b, c \in L$,

$$a \leq c \Leftrightarrow a \vee (b \wedge c) \leq (a \vee b) \wedge c$$

Proof Suppose $a \leq c$. Then $a \vee c = c$.

By distributive inequality, $a \vee (b \wedge c) \leq (a \vee b) \wedge (a \vee c)$

Since $a \vee c = c$,

$$a \vee (b \wedge c) \leq (a \vee b) \wedge c.$$

Conversely, let us assume that $a \vee (b \wedge c) \leq (a \vee b) \wedge c$.

Since,

$$\begin{aligned} a &\leq a \vee (b \wedge c) \\ &\leq (a \vee b) \wedge c \\ &\leq c \end{aligned}$$

we get $a \leq c$. Hence the proof.

Example 6.14 Prove that in a lattice $(L, \leq)$, for any $a, b, c \in L$, if $a \leq b \leq c \Rightarrow a \vee b = bc$, and $(a \wedge b) \vee (b \wedge c) = b = (a \vee b) \wedge (a \vee c)$.

Solution Since $a \leq b$ and $a \leq c$, $a \leq b \wedge c$. Again $b \leq b$ and $b \leq c$ implies $b \leq b \wedge c$.

Now $a \leq b \wedge c$ and $b \leq b \wedge c \Rightarrow a \vee b \leq b \wedge c$ (3)

Again, $b \wedge c \leq b \leq a \vee b$ (4)

From (3) and (4), $a \vee b = b \wedge c$.

Hence $a \wedge b \leq b$ and $b \wedge c \leq b$, we get $(a \wedge b) \vee (b \wedge c) \leq b$. Since $b \leq c$ and $b \leq b$ implies $b \leq b \wedge c$. Again this implies $b \leq (b \wedge c) \vee (a \wedge b)$. Hence $(a \wedge b) \vee (b \wedge c) = b$. Similarly, $(a \vee b) \wedge (a \vee c) = b$ follows

Example 6.15 Prove that in a lattice $(L, \leq)$, for any $a, b, c, d, \in L$, if $a \leq b$ and $c \leq d$ then $a \wedge c \leq b \wedge d$.

Solution Since $a \wedge c \leq a \leq b$ and $a \wedge c \leq c \leq d$, $a \wedge c \leq b \wedge d$.

Distributive Lattice A Lattice $(L, \leq)$ is said to be distributive lattice if for any $a, b, c, \in L$,

$$a \wedge (b \vee c) = (a \wedge b) \vee (a \wedge c)$$
$$a \vee (b \wedge c) = (a \vee b) \wedge (a \vee c).$$

Theorem Let $a, b, c \in L$, where $(L, \leq)$ is a distributive lattice. Then $a \vee b = a \vee c$ and $a \wedge b = a \wedge c \Rightarrow b = c$.

Proof We know that

$$\begin{aligned} b &= b \vee (b \wedge a) \text{ (absorbtion)} \\ &= b \vee (a \wedge b) \text{ (commutative)} \\ &= b \vee (a \wedge c) \ (\because a \wedge b = a \wedge c) \\ &= (b \vee a) \wedge (b \vee c) \text{ (distributive)} \\ &= (a \vee b) \wedge (c \vee b) \text{ (commutative)} \\ &= (a \vee c) \wedge (c \vee b) \ (\because a \vee b = a \vee c) \\ &= (c \vee a) \wedge (c \vee b) \text{ (commutative)} \\ &= c \vee (a \wedge b) \text{ (distributive)} \\ &= c \vee (a \wedge c) \ (\because a \wedge b = a \wedge c) \\ &= c \vee (c \wedge a) \text{ (commutative)} \\ &= c \text{ (absorption)} \end{aligned}$$

Hence the proof.

Modular lattice A lattice $(L, \leq)$ is said to be modular lattice if $a \leq c \Rightarrow a \vee (b \wedge c) = (a \vee b) \wedge c$.

Bounded lattice A lattice $(L, \leq)$ which has both, a least element denoted by 0, and the greatest element denoted by 1 is called a bounded lattice.

Note: If $L = \{a_1, a_2, ..., a_n\}$ with $\bigwedge_{i=1}^{n} a_i = 0$ and $\bigvee_{i=1}^{n} a_i = 1$. It satisfies $a \vee 0 = a$, $a \vee 1 = 1$, $a \wedge 1 = a$ and $a \wedge 0 = 0$.

Complement of an element In a bounded lattices $(L, \leq)$, an element $b \in L$ is called a complement of an element $a \in L$ if $a \wedge b = 0$ and $a \vee b = 1$, we denote b by a.

Complement lattice A lattice $(L, \leq)$ is said to be complemented lattice if every element of L has at least one complement.

Example 6.16 Show that De Morgan's laws holds in a complemented distributive lattice. (*M.E, MU*98A)

Solution To process that $(a \wedge b)' = a' \vee b'$ and

$(a \vee b)' = (a' \wedge b')$, consider

$$\begin{aligned}(a \wedge b)' \wedge (a' \vee b') &= ((a \wedge b) \wedge a') \vee ((a \wedge b) \wedge b') && \text{(distributive)}\\ &= ((b \wedge a) \wedge a') \vee ((a \wedge b \wedge \wedge b') && \text{(commutative)}\\ &= (b \wedge (a \wedge a')) \vee (a \wedge (b \wedge b')) && \text{(associative)}\\ &= (b \wedge 0) \vee (a \wedge 0)\\ &= 0 \vee 0 = 0\end{aligned}$$

Again,

$$\begin{aligned}(a \wedge b)\ \wedge (a' \vee b') &= (a \vee (a' \vee b') \wedge (b \vee (a' \vee b')) && \text{(distributive)}\\ &= (a \vee (a' \vee b')) \wedge (b \vee (b' \vee a')) && \text{(commutative)}\\ &= ((a \vee a') \vee b') \wedge ((b \vee b') \vee a') && \text{(associative)}\\ &= (1 \vee b') \wedge (1 \vee a')\\ &= 1 \wedge 1 = 1\end{aligned}$$

Hence, $a' \vee b'$ is the complement of $(a \wedge b)$. So $a' \vee b' = (a \wedge b)'$. Similarly, $(a \vee b)' = a' \wedge b'$ follows.

Example 6.17 Show that in a complemented lattice $(L, \leq)$, $a \leq b \Leftrightarrow a' \vee b = 1 \Leftrightarrow a \wedge b' = 0 \Leftrightarrow b' \leq a'$

Solution Consider $a \leq b \Leftrightarrow a \wedge b = a$

$$\begin{aligned}&\Leftrightarrow a' \vee a = a' \vee (a \wedge b) = 1\\ &\Leftrightarrow (a' \vee a) \wedge (a' \vee b) = 1\\ &\Leftrightarrow 1 \wedge (a' \vee b) = 1\\ &\Leftrightarrow a' \vee b = 1\end{aligned}$$

Again, $a \leq b \Leftrightarrow a \vee b = b$

$$\begin{aligned}&\Leftrightarrow b \wedge b' = (a \vee b) \wedge b' = 0\\ &\Leftrightarrow (a \wedge b') \vee (b \wedge b') = 1\\ &\Leftrightarrow (a \wedge b') \vee 0 = 0\\ &\Leftrightarrow a \wedge b' = 0.\end{aligned}$$

To prove the last one,

$$\begin{aligned}a \leq b &\Leftrightarrow a \vee b = b\\ &\Leftrightarrow (a \vee b') = b'\\ &\Leftrightarrow a' \wedge b' = b'\\ &\Leftrightarrow a' \wedge a' = b' && \text{(commutative)}\\ &\Leftrightarrow b' \leq a'.\end{aligned}$$

Example 6.18 Consider the lattice $L = \{1, 2, 3, 4, 6, 12\}$, the divisions of 12 ordered by divisibility. Find

i. the lower bound and upper bound of L

ii. the complement of 4

iii. Is L a complemented lattice?

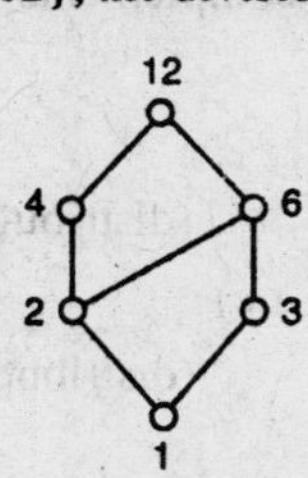

Solution

i. The lower bound of L is 1 and the upper bound is 12.

ii. Since $4 \wedge 3 = \gcd(4,3) = 1$ and
$4 \vee 3 = \text{lcm}(4,3) = 12$,
the complement of 4 is 3.

iii. Since $6 \wedge x = \gcd(6, x)$
$\neq 1$ for $x \neq 1$ and
$6 \vee 1 = \text{lcm}(6,1)$
$\neq 12$,
6 has no complement and hence L is not a complemented lattice.

Sub lattice Let M be a non-empty subset of a Lattice $(L, \leq)$. We say that M is a sublattice of L if M itself is a lattice with respect to the operations of L.

Note: So M is a sublattice of L if and only if M is closed under the operations $\wedge$ and $\vee$ of L.

Example 6.19 Consider the following lattice L.

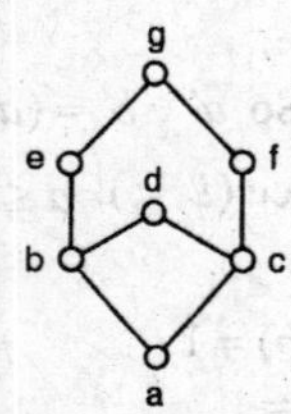

Determine whether each of the following is a sublattice of L.

$M = \{a, b, c, g\}$
$N = \{a, b, f, g\}$
$O = \{b, d, e, g\}$
$P = \{a, d, e, g\}$

Solution Since $b \vee c = d$, and $d \notin M$, M is not a sublattice. Since $d \wedge e = b$ and $b \notin p$, p is not a sublattice. But N and O are sublattices.

Example 6.20 Suppose M is a sublattice of a distributive lattice L. Show that M is a distributive lattice.

Solution For a distributive lattice L, $a \wedge (b \vee c) = (a \wedge b) \vee (a \wedge c)$ and $a \vee (b \wedge c) = (a \vee b) \wedge (a \vee c)$ for all $a, b, c \in L$. Since M is closed, each element of M is also in L, the distributive laws hold for all elements in M. Hence M is a distributive lattice.

Example 6.21 Prove that in a distributive lattice $(L, \leq)$, if an element has a complement then this complement is unique.

Solution Suppose for any $a \in L$ has two complements say b and c in L. Then $a \vee b = 1$; $a \wedge b = 0$ and $a \vee c = 1$; $a \wedge c = 0$.

Consider $b = b \wedge 1 = b \wedge (a \vee c)$
$= (b \wedge a) \vee (b \wedge c)$ (distributive)
$= 0 \vee (b \wedge c) = (a \wedge c) \vee (b \wedge c)$
$= (a \vee b) \wedge c$ (distributive)
$= 1 \wedge c = c$

EXERCISE II

1. Let $(L, \leq)$ be a lattice. Define relation R on L such that for $a, b, \in L$,
   $aRb \Leftrightarrow a \wedge b = a$
   $\Leftrightarrow a \vee b = b$. Prove that R is a partial order on L.
2. In a Lattice show that
   $(a \wedge b) \vee (c \wedge d) \leq (a \vee c) \wedge (b \vee d)$
   $(a \wedge b) \vee (b \wedge c) \vee (c \wedge a) \leq (a \vee b) \wedge (b \vee c) \wedge (c \vee a)$.
3. Show that a lattice is distributive if and only if $(a \wedge b) \vee (b \wedge c) \vee (c \wedge a) = (a \vee b) \wedge (b \vee c) \wedge (c \vee a)$
4. Prove that every finite subset of a lattice has an LUB and a GLB.
5. Show that in a direct product of any two distributive lattices is a distributive lattice.
6. Show that in a lattice with two or more elements, no element is its own complement.
7. Show that in a bounded distributive lattice, the elements which have complements form a sublattice.
8. A lattice is said to the modular if $a \leq c \Rightarrow a \vee (b \wedge c) = (a \vee b) \wedge c$. Show that every distributive lattice is modular but not conversely.
9. Show that a lattice is modular if and only if $a \vee (b \wedge (a \vee c)) = (a \vee b) \wedge (a \vee c)$.
10. Show that for any elements a, b, c in a modular lattice, $(a \vee b) \wedge c = b \wedge c \Rightarrow (c \vee b) \wedge a = b \vee a$.
11. Is the cartesian product of two lattices always a lattice? Prove you claim.

BOOLEAN ALGEBRA

Let B be a non-empty set with two binary opertions + (or $\vee$) and, (or $\wedge$), a unary operation, and two distinct elements 0 and 1. Then B is called a Boolean Algebra if the following axioms hold where a, b, c are any elements in B.

i. $a + b = b + a$; $a . b = b . a$ (Commutative laws)
ii. $a + (b . c) = (a + b).(a + c)$; $a.(b + c) = (a . b)(a . c)$ (Distributive laws)
iii. $a + 0 = a$; $a . 1 = a$ (Identity laws)
iv. $a + a' = 1$; $a . a' = 0$ (Complement laws)

Note:

1. Boolean algebra is a lattice which contains a least element and a greatest element and which is both complemented and distributive.
2. We denote the Boolean algebra B by $(B, +, ., 1, 0, 1)$. Here we call 0 as the zero element, 1 as the unit element and a' is complement of a, + and . are called sum and product.

Example Let $B = \{0,1\}$, the set of binary digits with the binary operations of + and . and the unary operation defined by

+	1	0
1	1	1
0	1	0

.	1	0
1	1	0
0	0	0

′	1	0
	0	1

Then B is a Boolean algebra.

Example Let B = {1, 2, 5, 7, 10, 14, 35, 70}the divisions of 70. Define +,. and on *B* by $a + b = \text{lcm}(a, b)$, $a.b = \gcd(a, b)$, $a' \frac{70}{a}$.

Then *B* is a Boolean algebra with 1, the zero element and 70 the unit element.

Example Let *B* be a collection of set closed under the set operation of union, intersection, and complement. Then *B* is a Boolean algebra with the empty set ϕ as the zero element and the universal set *U* as the unit element.

Subalgebra Suppose *C* is a non-empty subset of a Boolean algebra B. We say C is a subalgebra of B if C itself is a Boolean algebra with respect to the operations of *B*.

Note: Here *C* is a subalgebra of *B* if and only if *C* is closed under the three operations. +,. and' of *B*.

Example Let *C* = {1, 2, 3,35,70}. Then *C* is a subalgebra of *B* = {1,2,5,7,10,14,35,70}, the divisions of 70.

Isomorphic Two Boolean algebras *B* and $\bar{B}$ are said to be isomorphic if f: $B \to \bar{B}$ preserves

i. $f(a+b)=f(a)+f(b)$
ii. $f(a.b)=f(a).f(b)$
iii. $f(a')=f(a))'$ for any $a,b, \in B$.

Duality The dual of any statement in a Boolean algebra B is the statement obtained by interchanging the operations + and . , and interchanging their identity elements 0 and 1 in the original statement.

Example 6.22 Write the dual of the following statements.

i. $(0.a)+(b.1)=b$
ii. $(a+0)+(1+a')=1$
iii. $a.(a'+b)=a.b$

Solution

i. $(1+a).(b+0)=b$
ii. $(a+1)+(0+a')=0$
iii. $a+(a'.b)=a+b$

Basic Boolean Algebra Laws

Let *a*, *b*, *c* be any elements in a Boolean algebra *B*. Then

i. $a+b=b+a$; $a.b=b.a$ (commutative law)
ii. $a+(b+c)=(a+b)+c$; $a.(b.c)=(a.b).c$ (associative law)
iii. $a.(b+c)=(a.b)+(a.c)$; $a+(b.c)=(a+b).(a+c)$ (distributive law)
iv. $a+0=0+a=a$; $a.1=1.a=a$ (Identity law)
v. $a+\bar{a}=1$; $\bar{a}.\bar{a}=0$ (Complement law)
vi. $a+a=a$; $a.a=a$ (Idempotent law)
vii. $a+1=1$; $a.0=0$ (Null law)
viii. $a+(a.b)=a$; $a.(a+b)=a$ (absorbtion law)

ix. $(a+b)' = a'.b'$; $(a+b)' = a'+b'$ (Demorgan's law)

x. $(a')' = a$ (Involution law)

Note: If $a+x=1$ and $a.x=0$, therefore $x=a'$. Therefore, $0'=1$ and $1'=0$.

Atom A non zero element 'a' in a Boolean algebra $(B,+,.,')$ is called an atom if for every $x \in B$, $x \wedge a = a$ or $x \wedge a = 0$.

Note: Here the condition $x \wedge a =$ a means that x is a successor of a and $x \wedge a = 0$ is true only when x and a are "not connected". So in any Boolean algebra, the immediate successors of the 0-element are called atoms.

Example Let A be any non-empty set and $P(A)$ the power set of A. In Boolean algebra $(p(A), \cup, \cap, ')$ over $\subseteq$, the singleton sets are the atoms since each element $p(A)$ can be described completely and uniquely as the union of singleton sets.

Example Let $B = \{1,2,3,5,6,10,15,30\}$ and let the relation $\leq$ be divides. The operation $\wedge$ is GCD and $\vee$ is LCM. The 0-element is 1. Then the set of atoms of the Boolean algebra is $\{2,3,5\}$.

Note:

1. Let $(B,+,.,')$ be any finite Boolean algebra and let A be the set of all atoms. Then $(B,+,.,')$ is isomorphic to $(p(A),\cup,\cap,')$.
2. Every finite Boolean algebra $(B,+,.,')$ has 2^n elements for some positive integer n.
3. All Boolean algebras of order 2^n are isomorphic to each other. Finite Boolean algebras as n- tuples of 0's and 1's.

The simplest nontrivial Boolean algebra is the Boolean algebra $B = \{0,1\}$, the set of binary digits with the binary operations of + and . and the unary operation $'$ by

+	1	0
1	1	1
0	1	0

.	1	0
1	1	0
0	0	0

'	1
1	0
0	1

If we form $B^2 = B \times B$, we obtain the set $B^2 = \{(0,0), (0,1), (1,0), (1,1)\}$. Define +,. and$'$ by

$(0,1)+(1,1)=(0+1, 1+1)=(1,1)$,

$(0,1).(1,1)=(0.1, 1.1)=(0,1)$ and

$(0,1)' = (0',1') = (1,0)$.

Then B^2 is a Boolean algebra.

Note: Here B^2 is a Boolean algebra of order 4 under component wise opertions. Since all Boolean algebra of order 4 one isomorphic to each others, this is a simple way of describing all Boolean algerbras of order 4. In general, any Boolean algebra of order 2^n are isomorphic to B^n.

Example 6.23 Find the atoms of the Boolean algebra (i) B^2 (ii) B^4 (iii) B^n for $n \geq 1$.

Solution

i. (0,1) and (1,0)

ii. (1,0,0,0), (0,1,0,0), (0,0,1,0) and (0,0,0,1)

iii. The n-tuples with exactly one 1.

Boolean Expressions

In this section we will develop a procedure for simplifying Boolean expressions. Let $(B,+,.,')$ be any Boolean algebra. Let $x_1,x_2...,x_n$ be the variables in B. A Boolean expression E in these variables use the Boolean operations t_1.and ^{1}.

Example The following are the Boolean expresions in x,y and z.

i. $E_1 = (x + y'z)' + (xyz' + x'y)'$

ii. $E_2 = (x'z + (y + xy'z'))'$

Literal A literal is a variable or complemented variable fundamental product. A fundamental product is a literal or a product of two or more, literals in which no two literals involve the same variable.

Example The following are the fundamental products.

$x, y', xyz', x'yz, xz'$

and the following are not.

$xyx'z, xyzy, xyzx'$.

Note: Any product of literals can be reduced to either 0 or a fundamental product. For example, $xyx'z = 0 = xyzx'$. Since $xx' = 0$ and $xyzy = xyz$ since $yy = y$.

Contain A fundamental product p_1 is said to contain in another fundamental product p_2 if the literals of p_1 are also literals of p_2.

Sum of products A Boolean expression $E = E(x_1,x_2,...,x_n)$ is said to be in a sum of products form a minterm form if E is a fundamental product or the sum of the two or more fundamental products none of which is in included in another.

Algorithm for minterm form (Transforms E into sum of products)

Step 1. Use Demorgan's laws and involution to move the complement operation into any parenthesis.

Step 2. Use the distributive law to transform E into a sum of products.

Step 3. Use the commutative, idempotent, and complement laws to transform each product in E into 0 or a fundamental product.

Step 4. Use the absorption and identity laws to put E in to a sum of products form.

Example 6.24 Transform $((a' + c)(b_1 + c'))'$ into a sum of products form.

Solution $((a' + c)(b' + c'))' = (a' + c)' + (b' + c') = ac' + bc$

A Boolean expression $E = E(x_1,x_2,...,x_n)$ is called the complete sum-of-products form (or) a minterm normal form (or) full disjunctive normal form or minterm form. If E is a fundamental product or the sum of two or more fundamental products none of which is included in another and each product involves all the n-variables.

Note: We usually drop the symbol and juxtapose instead. For example, $a.(b + c) = a\ (b + c)$ and $a + (b * c) = a + bc$ etc.

Algorithm for finding complete sum of products form

Step 1. Use Demorgan's laws and involution to move the complement operation into any parenthesis until the complement operation applies to variables only. Then E will consist of only form and product of literals.

Step 2. Use the distributive operation next to transform E into a sum of products.

Step 3. Use the commutative, idempotent and complement laws to transform each product in E into 0 or a fundamental product.

Step 4. Use the absorption and identity laws to transform E into a sum of products expression.

Step 5. Find a product p in E which does not involve the variable x_i and then multiply P by $x_i + x_i'$, deleting any repeated products.

Example 6.25 Express $E = x(y'z)'$ in its full disjunctive normal form.

Solution $E = x(y'z)'$

$= x(y+z')$ by DeMorgan's law

$= xy + xz'$ by distributive law

$= xy(z+z') + xz'(y+y')$ by introducing missing variable

$= xyz + xyz' + xyz' + xy'z'$ by distributive and commutative

$= xyz + xyz' + xy'z$ by identity laws.

Example 6.26 Express E in its complete sum-of-products of form.

a. $E = x(xy' + x'y + y'z)$

b. $E = z(x'+y) + y'$

c. $E = (x'+y)' + x'y$

d. $E = y(x+yz)'$

e. $E = x(xy + y' + x'yx)$.

Solution

a. $E = x\,(xy' + x'y + y'z)$

$= xxy' + xx'y + xy'z \qquad \because xx' = 0$

$= xy' + xy'z$

$= xy'(z + z') + xy'z$

$= xy'z + xy'z' + xy'z$

$= xy'z + xy'z'$.

b. $E = z\,(x'+y) + y'$

$= x'z + yz + y'$

$= x'z\,(y + y') + yz\,(x + x') + y'\,(x + x')(z + z')$

$= x'yz + x'y'z + xyz + x'yz + xy'z + xy'z' + x'y'z + x'y'z$

$= xyz + xy'z + xy'z' + x'yz + x'y'z + x'y'z'$.

c. $E = (x'+y)' + x'y$

$= xy' + x'y$ DeMorgan's law

$= xy'(z+z') + x'y\,(z+z')$

$= xy'z + xy'z' + x'yz + x'yz'$

d. $E = y\,(x+yz)'$

$= y\,(x'(yz)')$ DeMorgan's law

$= yx'(y'+z')$ DeMorgan's law

$= yx'y' + yx'z'$

$= x'yz'$ (since $yyl = 0$)

e. $E = x(xy + y' + x'y)$

$= xxy + xy' + xx'y$

$= xy + xy'$

$= xy\,(z+z') + xy'(z+z')$

$= xyz + xyz' + xy'z + xy'z'$.

Example 6.27 Express the following Boolean expressions E in terms of its minterm canonical form:

a. $E = x(xy' + x'y + y'z)$ b. $E = (x + y'z)(y + z')$

c. $E = (x' + y)' + y'z$ d. $E = (x'y)'(x' + xyz')$

e. $E = (x + y')(xy')'$ f. $E = y(x + yz)'$

Solution

a. $E = x(xy + x'y + y'z)$
$= xy' + xy'z$
$= xy'(z + z') + xy'z$
$= xy'z + xy'z'.$

b. $E = (x + y'z)(y + z')$
$= xy + xz'$
$= xy(z + z') + xz'(y + y')$
$= xyz + xyz' + xy'z'.$

c. $E = (x' + y)' + y'z$
$= xy' + y'z$
$= xy'(z + z') + y'z(x + x')$
$= xy'z + xy'z' + x'y'z.$

d. $E = (x'y)'(x' + xyz')$
$= (x + y')(x' + xyz')$
$= xyz' + x'y'$
$= xyz' + x'y'(z + z')$
$= xyz' + x'y'z + x'y'z'.$

e. $E = (x + y)'(xy')'$
$= x'y'(x' + y)$
$= x'y'$
$= x'y'(z + z')$
$= x'y'z + x'y'z'.$

f. $E = y(x + y + z)'$
$= y(x'(yz)')$
$= y(x'(y' + z'))$
$= y(x'y' + x'z')$
$= x'yz'.$

EXERCISE III

1. Write the dual of

a. $a + (b \cdot a) = a$ b. $a + ((b + a).b)' = 1$

c. $(a.b)'b = a + b$

2. Determine the minterm normal form of

a. $x_1 + x_2$ b. $x_1' + x_2'$

c. 0 d. 1

3. Write the Boolean expression $f = (x_1, x_2, x_3) = (x_3'.x_2) + (x_1'.x_3) + x_2x_3$ in minterm normal form.

4. Find the minterm normal form of $f = (x_1, x_2, x_3) = ((x_1 + x_2)x_3)'(x_1 + x_3))$

5. Write the following Boolean expression in conjunctive normal form.

a. $E = ((x_1 + x_2)' + (x_1' + x_3))'$ b. $E = (x_1 x_2 x_3') + (x_1 x_2' x_4) + (x_2 x_3' x_4')$

c. $E = (x_1 x_2) + (x_1 x_3) + (x_2' x_3)$ d. $E = (x_1 x_2) + (x_1 x_2 x_3) + (x_2 x_3)$

e. $E = ((x_1 x_2') + x_3)((x_1 + x_2') x_3)$ f. $E = (x_1 x_2) + (x_1' x_2 x_3') + (x_2 x_3)$

g. $E = (x_1 x_2) + (x_1 x_2' x_3) + (x_2 x_3)$

symbol $\leq$ should not be confused with the standard less than

6. Let $(B, ., +1, 0, 1)$ be a Boolean algebra.
Define the operations $+'$ and $.'$ on B by $+b = (a.b') + (a'.b)\, a.b = ..$ Show that $(B, +, ., 1)$ is with identity 1.

7. Show that a mapping from one Boolean algebra to another preserves the operations + and$'$.

KARNAUGH MAP REPRESENTATION OF LOGICAL FUNCTIONS

Introduction

Karnaugh map is one of the techniques that employs Boolean graphical representation of logical functions. This technique proves to be very useful in simplifying Boolean expressions. In this method, the information contained in a truth table is available in POS or SOP form and represented on Karnaugh map (K-map). The following figure illustrates K - map in two, three & four variables.

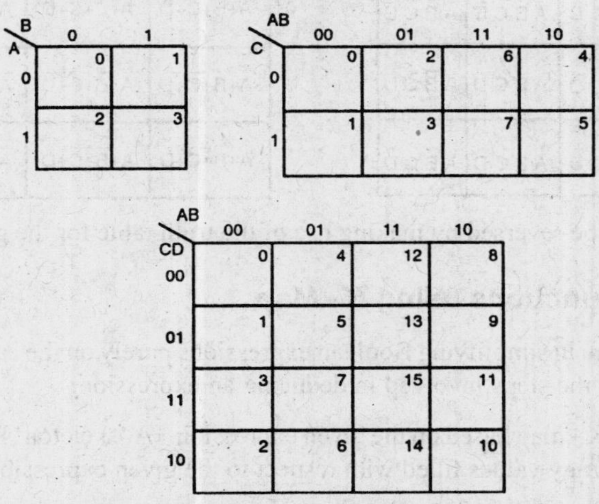

In a K - map each block is referred to as a cell.

In a n - variable K - map there are 2^n cells. Each cell corresponds to one of the combinations of 'n' variables. As each variable can take 2 possible values, for 'n' variables there are 2^n combinations. Hence there is a cell designated for each of the minterms or maxterms of the truth table.

In the above illustration, the variables have been named as 'A', 'B', 'C', &'D' and the binary numbers formed by them are taken as AB, ABC & $ABCD$ for 2, 3, & 4 variables respectively.

In each map, each of the cells have been assigned values taking into account all posssible values. It has been assumed that the first bit corresponds to the first variable and the second bit corresponds to the second variable.

A \ B	0	1
0	$\overline{A}\,\overline{B}$	$\overline{A}\,B$
1	$A\,\overline{B}$	$A\,B$

A \ B	0	1
0	$A+B$	$A+\overline{B}$
1	$\overline{A}+B$	$\overline{A}+\overline{B}$

C \ AB	00	01	11	10
0	$\overline{A}\,\overline{B}\,\overline{C}$	$\overline{A}\,B\,\overline{C}$	$A\,B\,\overline{C}$	$A\,\overline{B}\,\overline{C}$
1	$\overline{A}\,\overline{B}\,C$	$\overline{A}\,B\,C$	$A\,B\,C$	$A\,\overline{B}\,C$

C \ AB	00	01	11	10
0	$A+B+C$	$A+\overline{B}+C$	$\overline{A}+\overline{B}+C$	$\overline{A}+B+C$
1	$A+B+\overline{C}$	$A+\overline{B}+\overline{C}$	$\overline{A}+\overline{B}+\overline{C}$	$\overline{A}+B+\overline{C}$

CD \ AB	00	01	11	10
00	$\overline{A}\,\overline{B}\,\overline{C}\,\overline{D}$	$\overline{A}\,B\,\overline{C}\,\overline{D}$	$A\,B\,\overline{C}\,\overline{D}$	$A\,\overline{B}\,\overline{C}\,\overline{D}$
01	$\overline{A}\,\overline{B}\,\overline{C}\,D$	$\overline{A}\,B\,\overline{C}\,D$	$A\,B\,\overline{C}\,D$	$A\,\overline{B}\,\overline{C}\,D$
11	$\overline{A}\,\overline{B}\,C\,D$	$\overline{A}\,B\,C\,D$	$A\,B\,C\,D$	$A\,\overline{B}\,C\,D$
10	$\overline{A}\,\overline{B}\,C\,\overline{D}$	$\overline{A}\,B\,C\,\overline{D}$	$A\,B\,C\,\overline{D}$	$A\,\overline{B}\,C\,\overline{D}$

CD \ AB	00	01	11	10
00	$A+B+C+D$	$A+\overline{B}+C+D$	$\overline{A}+\overline{B}+C+D$	$\overline{A}+B+C+D$
01	$A+B+C+\overline{D}$	$A+\overline{B}+C+\overline{D}$	$\overline{A}+\overline{B}+C+\overline{D}$	$\overline{A}+B+C+\overline{D}$
11	$A+B+\overline{C}+\overline{D}$	$A+\overline{B}+\overline{C}+\overline{D}$	$\overline{A}+\overline{B}+\overline{C}+\overline{D}$	$\overline{A}+B+\overline{C}+\overline{D}$
10	$A+B+\overline{C}+D$	$A+\overline{B}+\overline{C}+D$	$\overline{A}+\overline{B}+\overline{C}+D$	$\overline{A}+B+C+\overline{D}$

The above process can be reversed by making use of the truth table for the given K - map.

Simplifying Logical Functions using *K* - Map

K - map finds extensive usage in simplifying Boolean expressions purely on the ease of computation involved. The following list describes the steps involved in deducing an expression.

1. Draw the skeleton K - map based on the given expression. [A 'skeleton' K - map refers to a K - map in *n* variables without any values filled with respect to the given expression.]
2. Prepare the truth table of the given Boolean expression.
3. Fill in the cells of the K - map based on the values obtained from the truth table.
4. Group the adjacent ones in the K - map and cancel out variables that undergo a change of state.

While grouping the adjacent ones it should be noted that they can be done in only horizontal or vertical directions and not diagonally.

The grouping of adjacent ones are tabulated for ease of understanding.

Example 6.28: Decipher the truth table for the given K - Map.

CD \ AB	00	01	11	10
00	0	4	12 1	8
01	1	5 1	13	9 1
11	3 1	7	15 1	11
10	2	6	14	10

Solution

***Table A** Grouping of Adjacent Ones*

	A	B	C	D	Y
0	0	0	0	0	0
1	0	0	0	1	0
2	0	0	1	0	0
3	0	0	1	1	1
4	0	1	0	0	0
5	0	1	0	1	1
6	0	1	1	0	0
7	0	1	1	1	0
8	1	0	0	0	0
9	1	0	0	1	1
10	1	0	1	0	0
11	1	0	1	1	0
12	1	1	0	0	1
13	1	1	0	1	0
14	1	1	1	0	0
15	1	1	1	1	1

Cell with decimal No.	Decimal number of adjacent cells		
	2 – Variable	3 – Variable	4 – Variable
0	1,2	1,2,4	1,2,4,8
1	0,3	0,3,5	0,3,5,9
2	0,3	0,3,6	0,3,6,10
3	1,2	1,2,7	1,2,7,11
4		0,5,6	0,5,6,12
5		1,4,7	1,4,7,13
6		2,4,7	2,4,7,14
7		3,5,6	3,5,6,15
8			0,9,10,12

Cell with decimal No.	Decimal number of adjacent cells		
	2 – Variable	3 – Variable	4 – Variable
9			1,8,11,13
10			2,8,11,14
11			3,9,10,15
12			4,8,13,14
13			5,9,12,15
14			6,10,12,15
15			7,11,13,14

Note: Grouping of 'ones' result in SOP form whereas grouping of 'zeros' result in POS form.

Grouping of four adjacent ones In case of a 4 variable K - map, there are 6 possible groupings of 4 variables involving any cell.

Cell with decimal no.	Decimal number of cells forming groups of adjacent fours		
0	(0,2,6,4)	(0,1,2,3)	(0,1,4,5)
1	(1,0,2,3)	(1,3,7,5)	(1,0,4,5)
2	(2,0,6,4)	(2,3,1,0)	(2,3,6,7)
3	(3,1,7,5)	(3,2,1,0)	(3,2,6,7)
4	(4,6,2,0)	(4,5,6,7)	(4,5,0,1)
5	(5,1,3,7)	(5,4,6,7)	(5,4,0,1)
6	(6,0,2,4)	(6,7,4,5)	(6,7,2,3)
7	(7,1,3,5)	(7,6,4,5)	(7,6,2,3)

Grouping of Eight Adjacent Ones

Decimal no. of cells forming groups of adjacent eights in a 4-var. K - map.
0, 4, 12, 8, 1, 5, 13, 9
0, 4, 12, 8, 2, 6, 14, 10
0, 1, 3, 2, 4, 5, 7, 6
0, 1, 3, 2, 8, 9, 11, 10
1, 5, 13, 9, 3, 7, 5, 11
4, 5, 7, 6, 12, 13, 15, 14
12, 13, 15, 14, 8 9 11 10
3, 7, 15, 11, 2, 6, 14, 10

Grouping of adjacent zeros As against the earlier procedure of grouping ones, the zeros of a K - map can also be grouped. In such cases,Group of two adjacent zeros result in the literal which is same in the two maxterm to get eliminated.Similar behaviour can be extended to grouping of 4, 8 or 2^n zeros in an n - variable K - Map.

Example 6.29 Minimize the logical function in POS form.

Solution

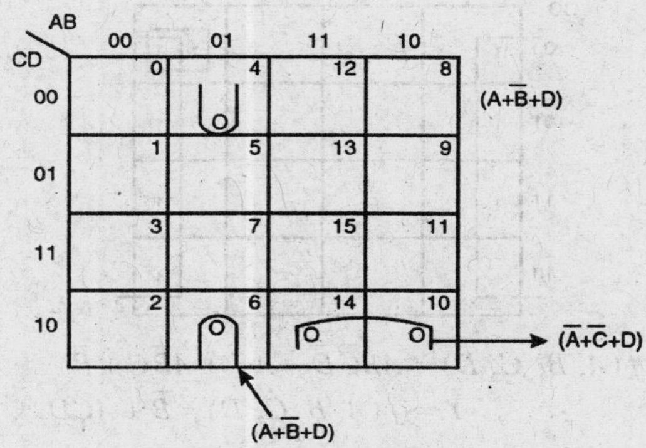

$Y = (A+\overline{B}+D)\ \ \overline{(A+\overline{C}+D)}$

f (A,B,C,D) : åm (2,8,9,10)

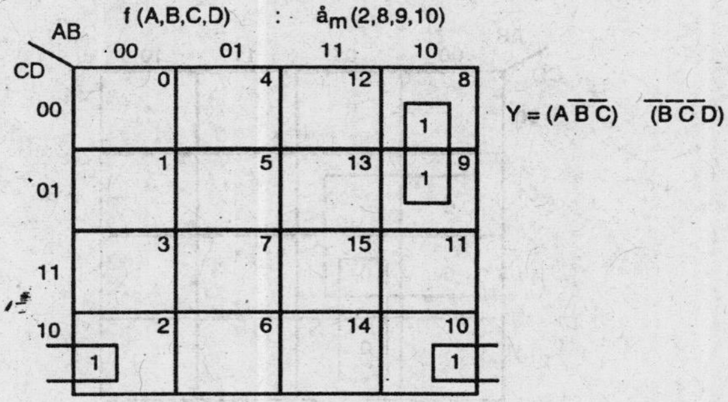

Example 6.30 Minimize the logical expression in *SOP* (minterm) form ($f(A, B, C, D) = \sum m\,(0, 3, 5, 7, 8, 11, 15)$

Solution

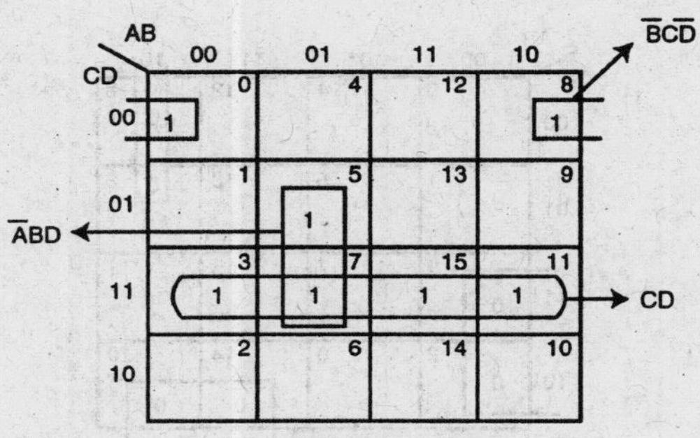

$Y = (\overline{B}\,\overline{C}\,\overline{D}) + CD + \overline{A}BD$

Minimization of logical functions not specified in minterms/maxterms In cases where the given logical expression is not there in standard *SOP* or *POS* form, we can proceed to resolve using K - map either by converting the given expression into standard form and then proceeding further or by directy plotting the K - map. Once the values have been entered, then it can be easily solved using the earlier mentioned technique.

Example

1. $f(A,B,C,D) = AB\overline{C}\,\overline{D} + A\overline{C} + A\overline{B}C + \overline{B}$

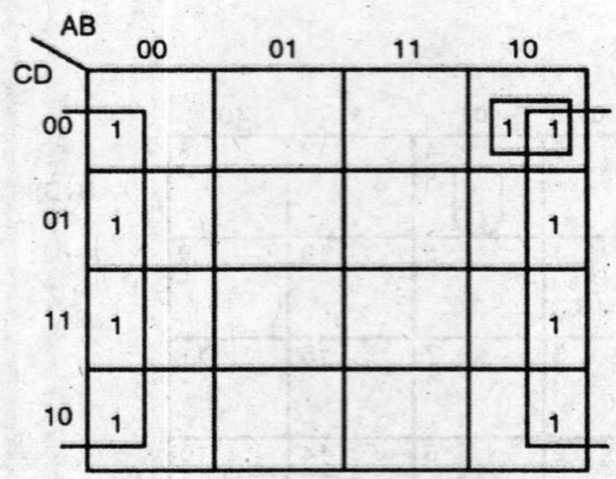

$$f(A, B, C, D) = ABC\overline{D} + A\overline{C} + A\overline{B}C + \overline{B}$$

$$Y = f(A, B, C, D) = \overline{B} + ACD$$

2. $f(A, B, C, D) = (A + \overline{B} + C + \overline{D}).(A + \overline{D}).(A + \overline{B} + \overline{C})$

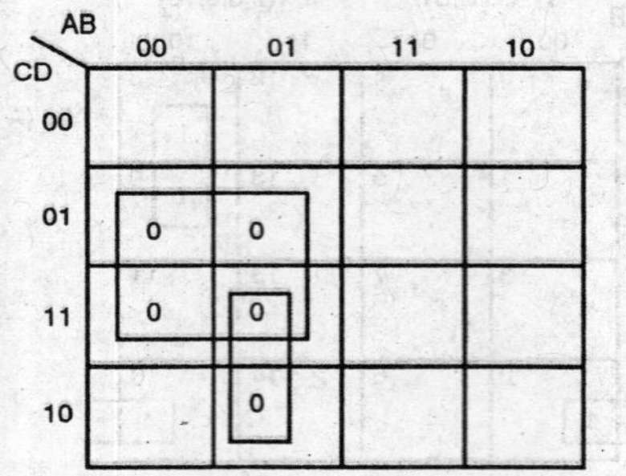

$$Y = (A + \overline{D})\ (A + \overline{B} + C)$$

3. $f(A,B,C,D) = A\ \pi\ m\,(2, 3, 8, 9, 10, 14)$

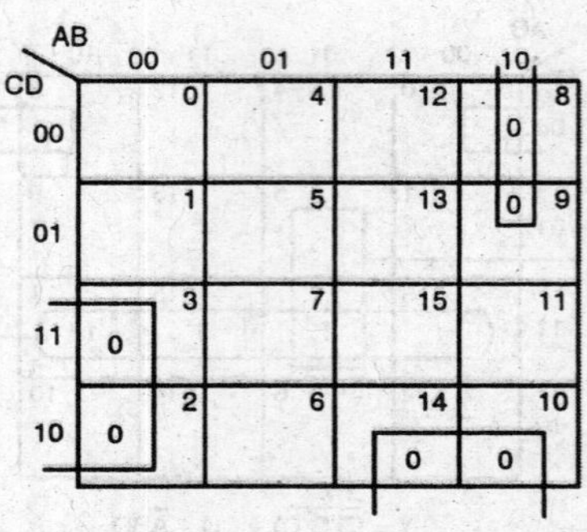

$$Y = (\overline{A} + B + C)\ (\overline{A} + \overline{C} + D)\ (A + B + \overline{C})$$

4. $f(A, B, C, D) = \overline{A} + (C + \overline{B} + D).(\overline{C} + D)$

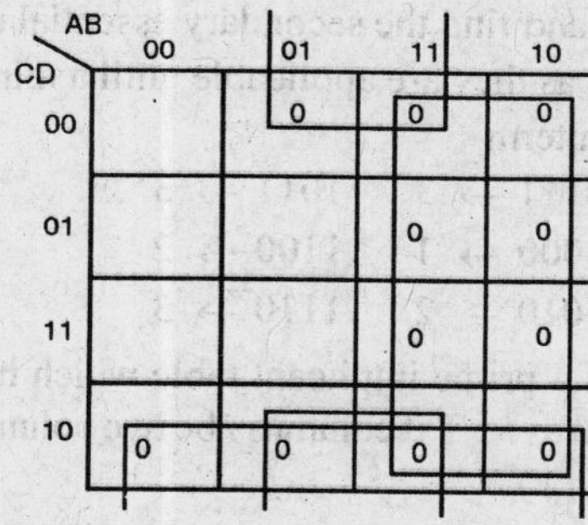

$$Y = (\overline{C} + D).(\overline{B} + D).(\overline{A}) = \overline{A}(\overline{B} + D)(\overline{C} + D)$$

Note: The term implicant may denote the set of adjacent minterms or the simplified product term obtained from the minterms of the set.

Prime implicant An implicant is called a prime implicant if it is not a subset of another implicant of the function.

Essential implicant A prime implicant which includes a '1' cell that is not included in any other prime implicant is called an essential prime implicant.

Don't care conditions Generally, there are certain cases wherein the outputs of function for certain combination of inputs does not matter.

In such cases, the designer is free to assume a '0' state or '1' state for that particular combination of input. This is generally represented by an *X* and is referred to as a don't care condition.

This condition comes in handy in grouping adjacent zeros and adjacent ones in a K-map. These may assume the respective values while grouping in both *SOP* as well as POS forms, while completing a 2, 4, or 8 block.

Drawback of K-map The much easier looking K - map becomes complicated & tedious while extending the variables size to more than 6. Hence other methods are preferred over this method when the number of variables are large.

Quine - Mcclusky Algorithm

The Karnaugh map method is applicable to solve a given Boolean function of upto 4 variables and it is highly difficult to solve functions having more variables using this method. The reason for this is, as the number of variables are increased, the number of cells in K - map are also increased by the power of 2.

Quine - Mcclusky method is used to solve a boolean of *n* - variables. In this method every minterm and a set of 2*k* minterms ($k < n$), forming a set of adjacent minterms or the simplified product term, is obtained by minterms of the set.

Quine - Mcclusky Minimization Procedure

1. Find the prime implicant of the function.
2. Construct the prime implicant table and find the essential prime implicant (i.e. essential rows) of the function.
3. Include the essential prime implicant in the minimal sum.

4. After all the prime implicants are deleted from the prime implicant table, determine the dominated rows and columns in the table. Delete them and find the secondary essential prime applicants.
5. Repeat step 3 and 4 as many times as they are applicable until a minimal cover of the function is found.

Index number representation of minterm

0000 → 0	0011 → 2	0111 → 3	1011 → 3
0001 → 1	0100 → 1	1000 → 1	1100 → 2
0010 → 1	0101 → 2	1010 → 2	1110 → 3

Two rows (or columns) i and j of a prime implicant table which have x's in exactly the same rows (or columns) are said to be equal and written as i = T if columns i be two columns of a prime implicant table. Column i is said to dominate the column j i.e. $i > j$)

If column i has x's in all the rows in which the column j has x's then column i and j are called dominating and dominated columns respectively. Similarly if rows i and j are of the prime implicant table, row i is said to dominate row j. If row i has x's in all the columns in which row j has x then i and j are called dominating and dominated rows respectively.

All the dominating rows and columns of a prime implicant can be removed to obtain a minimal sum without affecting the table.

In minimal sum, we include dominating rows and dominated columns and exclude dominated rows and dominating columns.

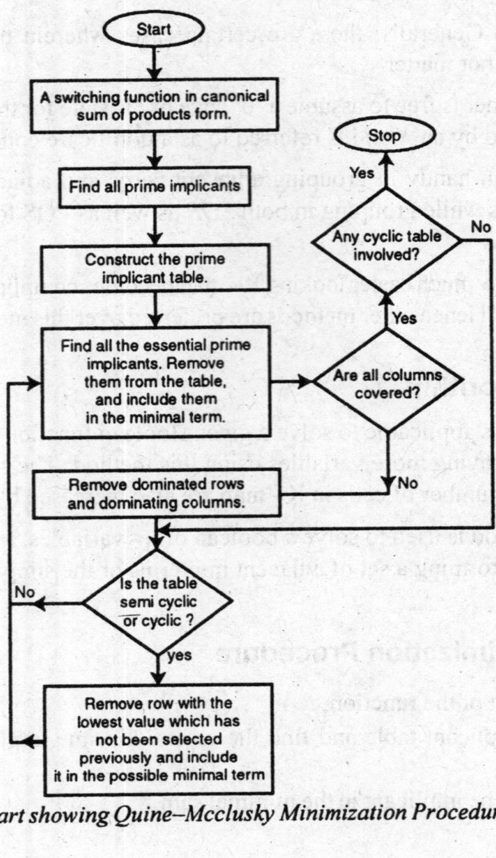

Flowchart showing Quine–Mcclusky Minimization Procedure

Example 6.31 Consider a function with five variables whose canonical sum of products form is

$$f(x_1, x_2, \ldots, x_5) = \Sigma(0, 1, 2, 8, 9, 15, 17, 21, 24, 25, 27, 31)$$

Solution We have to find out prime and essential prime implicants. We obtain 11 prime implicants, which are designated $A, B, \ldots, K$. From the prime implicant table, columns 2, 15, and 4 are covered by rows J, E, and H; thus they are essential prime implicants (indicated by *) which must be selected for inclusion in the minimal sum. After removing the three essential prime implicants the relations are

Column dominance

$9 > 8$
$25 > 24$

Row dominance

$A > G$
$F > D$
$B > I$
$B > K$

The columns 9 and 25 can be removed. Since all the dominated rows G, D, I and K have a cost that is not lower than their respective dominating row, they can also be removed from the table. Therefore columns 24 and 27 are only covered by rows A and F respectively. Thus they must be included in the minimal sum.

Rows A and F are sometimes referred to as secondary essential prime implicants (indicated by *). After rows A and F are removed all that minterm 1 has not been covered. However it can be covered by row B or row C. Suppose that row B is selected, the minimal sum of the function is $E + H + J + A + F + B$

To obtain all the prime implicants

Index	Decimal number	Binary representation of each mintern	Decimal numbers	1st reduction	Decimal numbers	2nd reduction
0	0	00000 ✓	0,1	0000 – K	0,8,1,9	0 – 00 – C
			0,2	000 – 0 J		
	1	00001 ✓	0,8	0 – 000 ✓	1,9,17,25	– 001 - B
1	2	00010 ✓	1,9	0 – 001 ✓	8,9,24,25	– 100 – A
	8	01000 ✓				
2			1,17	– 0001 ✓		
				0100 – ✓		
	9	01001 ✓	8,9			
				– 1000 ✓		
	17	10001 ✓	8,24			
3	24	11000 ✓	9,25	– 1001 ✓		
			17,21	10 – 01 H		
	21	10101 ✓	17,25	1–001 L		
4	25	11001 ✓	24,25	11 00 – G		
	15	01111 ✓	25,27	110 – 1 F		
5	27	11011 ✓	15,31	- 1111 E		
	31	11111 ✓	27,31	11 –11 D		

Prime implicant table (prime implicants E, H and J indicated by *)

	0	1	2	8	9	15	17	21	24	25	27	31
A				×	×				×	×		
B		×			×		×			×		
C	×	×		×	×						×	×
D											*	*
⊛ E	———	———	———	———	———	▶⊛						×
F										*	*	
G								*	*			
⊛ H							*	⊛				
I		×					*					
⊛ J	×		⊛									
K	×	×										

EXERCISE IV

1. Simplify the following functions using Karnaugh diagrams.

   a. $f(x_1, x_2, x_3, x_4) = x_1, x_3, x_1^1, x_3 x_4 + x_2 x_3^1 x_4 + x_2^1 x_3 x_4$

   b. $f = x_1^1 x_2^1 + x_1 x_3 x_4 + x_1 x_2 x_4 + x_1 x_2 x_4^1 + x_2^1 x_3$

2. Simplify.

   a. $f(a, b, c, d) = \Sigma(0, 1, 3, 8, 9, 13, 14, 15, 16, 17, 19, 24, 25, 27, 31)$

   b. $f(a, b, c, d) = \Sigma(0, 2, 6, 7, 8, 9, 13, 15)$

   c. $f(a, b, c, d) = \Sigma(0, 2, 3, 4, 8, 9, 10, 13)$

   d. $f(x_1, x_2, x_3, x_4) = \Sigma(0, 1, 4, 5, 9, 11)$

3. Simplify the Boolean functions represented by the following K - maps.

   a.

1	1		
1			1
1	1		
1			

   b.

1	1		1
1	1		1
1			1

ANSWERS

1. a. $x_1 x_3 + x_3 x_4 + x_2 x_4$ b. $x_1' x_2' + x_1 x_3 + x_1 x_2 x_4'$

2. a. $\bar{a}bcd + \bar{b}ce + \bar{c}d + \bar{a}b\bar{d}e + abde$ b. $\overline{abd} + \overline{abc} + \text{abd} + \bar{a}bc$

   c. $\overline{bd} + \overline{acd} + a\bar{c}d + \bar{a}\,\bar{b}c$ d. $\overline{ac} + a\bar{b}d$

3. a. $\overline{acd} + \bar{b}cd + \bar{a}cd + \bar{b}c\bar{d}$ b. $\overline{ac} + \overline{bc} + \overline{bd}$

CHAPTER

7

Graph Theory

BASIC CONCEPTS

Graph

A graph G, a triplet $(V(G), E(G), \theta_G)$ consisting of a non-empty set $V(G)$ of vertices, a set $E(G)$ of edges, and a function θ_G assigns to each edge, a subset $\{u, v\}$ of $V(G)$ (u, v needn't be distinct). If e is an edge and u, v are vertices such that $\theta_G(e) = uv$, then e is a line (edge) between u and v; the vertices u and v are the end points of the edge e.

Examples (i) $G = (V(G), E(G), \theta_G)$

where $V(G) = \{v_1, v_2, v_3, v_4\}$

$E(G) = \{e_1, e_2, e_3, e_4, e_5, e_6\}$

$\theta_G(e_1) = \{v_1v_2\}, \theta_G(e_2) = \{v_2v_2\}, \theta_G(e_3) = \{v_2v_3\}$

$\theta_G(e_4) = \{v_1v_3\}; \theta_G(e_5) = \{v_4v_5\}; \theta_G(e_6) = \{v_1v_4\}$

(ii) $G = (V(G), E(G), \theta_G)$

where $V(G) = \{v_1, v_2, v_3\}; E(G) = \{e_1, e_2, e_3\}$

$\theta_G(e_1) = \{v_1v_2\}; \theta_G(e_2) = \{v_2v_3\}; \theta_G(e_3) = \{v_3v_1\};$

Every graph has a diagram associated with it. These diagrams are useful for understanding problems involving such a graph. In the pictorial representation, we represent the vertices by small circles and the edges by lines whenever the corresponding pair of vertices forms an edge.

Now the pictorial representation of examples (i) and (ii) are

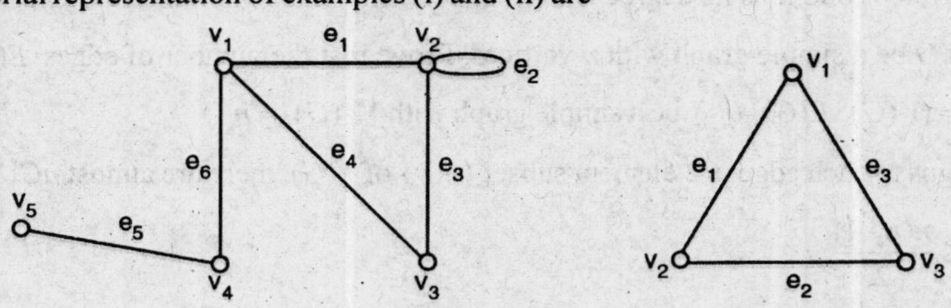

(i) (ii)

Note:

1 In example (i), e_2 joins the vertex v_2 to itself. Such an edge is called self loop (loop).

2 Suppose there is more than one edge between a pair of vertices in a graph, these edges are called parallel edges.

3 Hereafter we denote the graph $G = (V, E)$ for simplicity.

4 A graph which consists of parallel edges is called a multigraph.

Simple graph A graph with no self loops and parallel edges is called a simple graph.

Pseudograph A graph with self loops and parallel edges is called a pseudograph.

Note: Every simple graph and every multigraph is a pseudograph, but the converse is not true.

Example

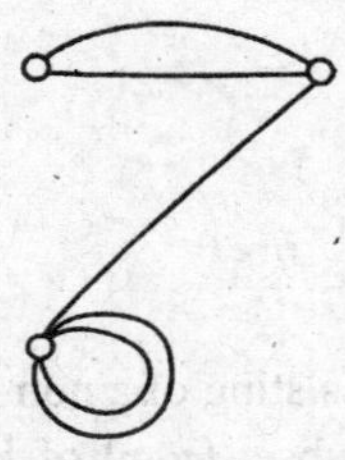

The above graph G is neither a simple graph nor a multigraph.

Degree of a vertex The degree of a vertex v is the number of edges incident with that vertex. In other words, the degree of a vertex is the number of edges having that vertex as an end point, and is denoted by $d(v)$.

Example

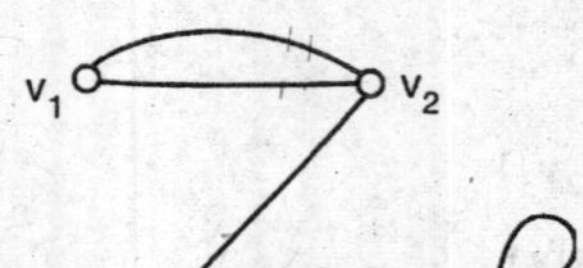

Here $d(v_1) = 2;$
$d(v_2) = 3;$
$d(v_3) = 2;$
$d(v_4) = 3;$

(A loop contributes 2 to the degree of vertex)

Isolated vertex A vertex with degree zero is called an isolated vertex.

Pendant vertex A vertex with degree one is called a pendant vertex.

Adjacent vertices A pair of vertices that determine an edge are called adjacent vertices.

Note: A vertex is even or odd if as its degree is even or odd.

Example 7.1 Let G be a simple graph with n vertices. Prove that the number of edges $E(G)$ is atmost nC_2.

Solution Let $G = (V(G), E(G), \theta_G)$ be a simple graph with $|V(G)| = n$.

Since θ_G assigns to each edge, a 2 element subset $\{u, v\}$ of $V(G)$, there are atmost nC_2 2-element subsets.

Hence $E(G) \leq \dfrac{n(n-1)}{2}$

Theorem Let G be a graph with n vertices and e edges. Then $\sum_{i=1}^{n} d(v_i) = 2.$

Proof Let G be a graph with n vertices and e edges.

Since every edge contributes degree 2 to this sum, $\sum_{i=1}^{n} d(v_i) = 2.e.$

Theorem In a graph G, the number of odd vertices is an even number.

Proof Let G be a graph with n vertices and e edges.

By theorem 1, we have

$$\sum_{i=1}^{n} d(v_i) = 2e = \text{even number} \tag{1}$$

Among n vertices, some are even vertices & some are odd vertices. Let V_e & V_o be the even & odd vertices respectively.

Now (1) can be written as

$$\sum_{v \in V_e}^{n} d(v) + \sum_{v \in V_0} d(v) = \text{even number}$$

$$\therefore \quad \sum_{v \in V_0} d(v) = \text{even number} - \sum_{v \in V_e} d(v) \tag{2}$$

Since every term in the right side of (2) is even, the sum on the left side must contain an even number of terms. i.e. the number of odd vertices in G is even.

Minimum and Maximum degrees Let G be a graph. The minimum and maximum degrees of G are respectively $\delta(G)$ and $\Delta(G)$ and given as

$$\delta(G) = \min\{d(v);\ v \in V(G)\}$$

$$\text{and} \quad \Delta(G) = \max\{d(v);\ v \in V(G)\}$$

k-regular A graph G is k-regular or regular of degree k, if every vertex of G has degree k.

Complete graph A simple graph in which each pair of distinct vertices is joined by an edge is called a complete graph. A complete graph on n vertices is denoted by k_n.

Examples

k_2

k_4

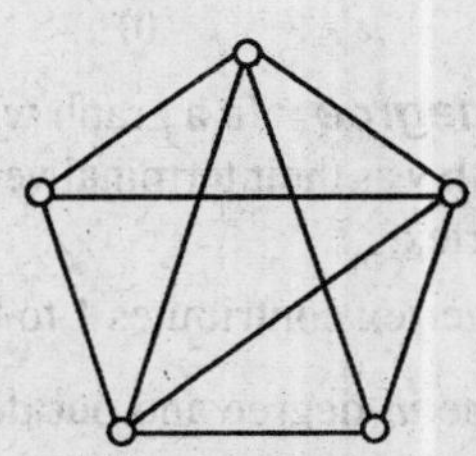

4 - regular graph

Complete graphs on 2 and 4 vertices respectively.

Note:

1 Every complete graph k_n is a $(n-1)$ regular graph.

2 There is no 1-regular or 3-regular graphs with 5 vertices. (since no graph has an odd number of vertices).

Complement of a graph The complement $\overline{G}$ of a graph G is that graph with $V(G) = V(\overline{G})$ and such that uv is an edge of $\overline{G}$ if and only if uv is not an edge of G.

Examples

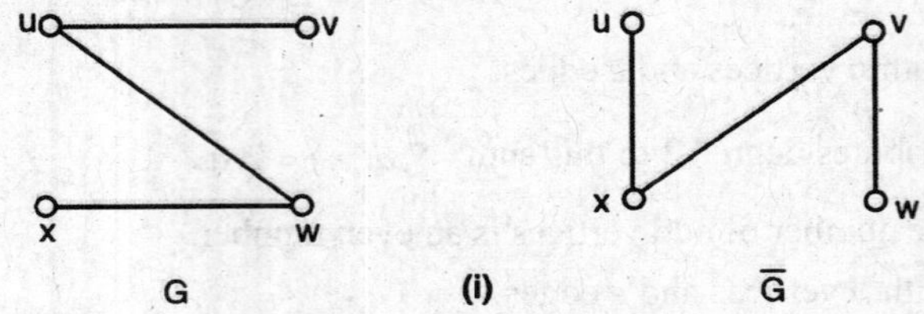

(i)

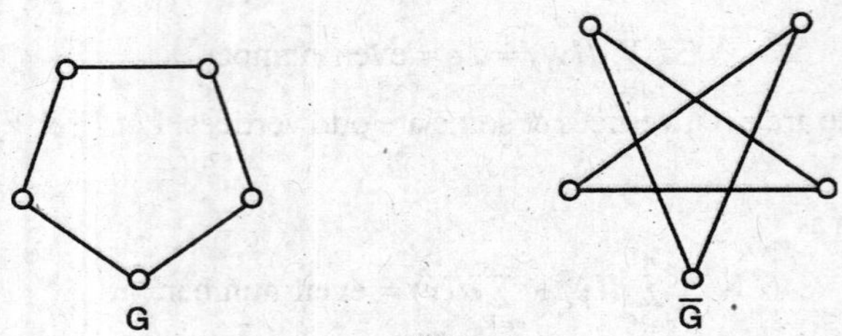

(ii)

There are also some useful terminology for graphs with directed edges.

Graphs with directed edges When (u, v) is an edge of the graph G with directed edges, u is said to be adjacent to v and v is said to be adjacent from u. The vertex u is called the initial vertex of (u, v) and v is called the terminal or end vertex of the edge (u, v).

Examples

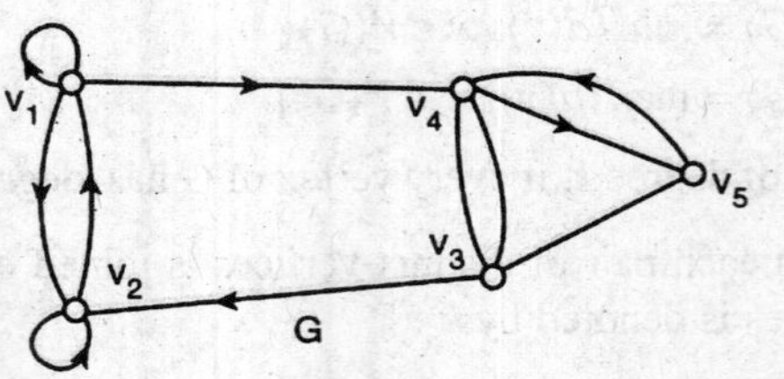

(i)

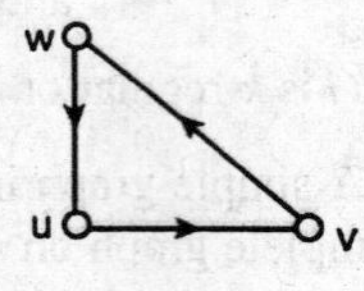

(ii)

In-degree and Out-degree In a graph with directed edges the in-degree of a vertex v denoted by $d^-(v)$ is the number of edges with v as their terminal vertex. The out-degree of v denoted by $d^+(v)$ is the number of edges with v as their initial vertex.

Note: Self-loop at a vertex contributes 1 to both in-degree and out-degree of this vertex.

Examples 7.2 Find the in-degree and out-degree of the following graphs.

Solution

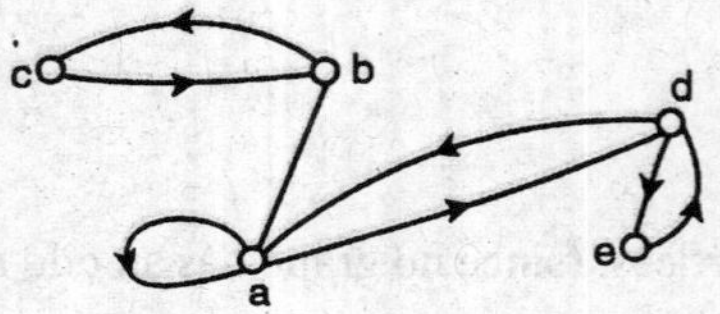

(i)

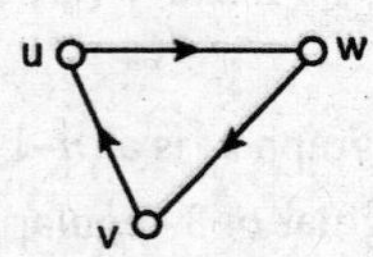

(ii)

i. $d^-(a) = 3$; $d^-(b) = 1$; $d^-(c) = 1$; $d^-(d) = 2$; and $d^-(e) = 1$;

$d^+(a) = 3$; $d^+(b) = 1$; $d^+(c) = 1$; $d^+(d) = 2$; $d^+(e) = 1$

ii. $d^-(u) = 1$; $d^-(v) = 1$; $d^-(w) = 1$ and

$d^+(u) = 1$; $d^+(v) = 1$; $d^+(w) = 1$

Note:

1 "Let $G = (V, E)$ be a graph with directed edges. Then $\sum_{v \in V} d^-(v) = \sum_{v \in V} d^+(v) = e$

2 By ignoring directions of edges in a graph with directed edges, we will get an undirected graph. Such graphs are called underlying undirected graphs.

BIPARTITE GRAPHS

A simple graph G is called bipartite if its vertex set V can be partitioned into two disjoint non-empty sets V_1 and V_2 such that every edge in the graph connects a vertex in V_1 and a vertex in V_2. Note that no edge in G connects either two vertices in V_1 or two vertices in V_2.

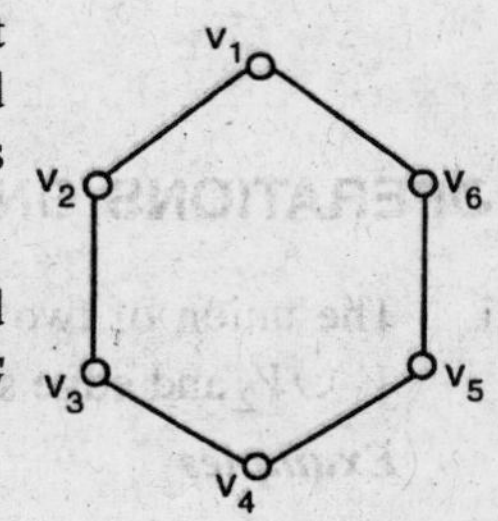

G

Example G is bipartite, because its vertex set $\nabla = \{v_1, v_2, v_3, v_4, v_5, v_6\}$ is partitioned into two non-empty sets $V_1 = \{v_1, v_3, v_5\}$ and $V_2 = \{v_2, v_4, v_6\}$. Also every edge in G connects a vertex in V_1 and a vertex in V_2.

Complete Bipartite Graph

The complete bipartite graph $k_{m,n}$ is the graph that has its vertex set partitioned into two non-empty subsets of m and n vertices, respectively. There is an edge between two vertices, if one vertex is in the first subset then the other vertex is in the second subset.

Examples

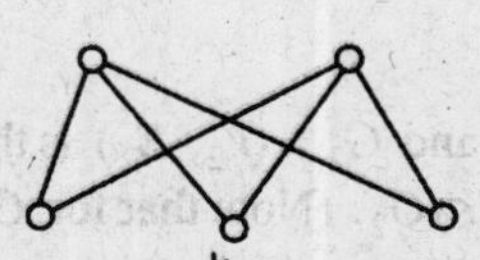
$k_{2,3}$

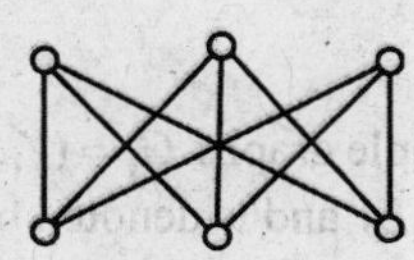
$k_{3,3}$

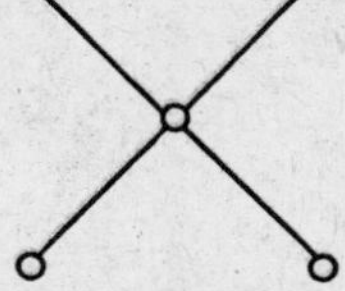
$k_{1,4}$ (star graph)

New Graphs from Old

(i) A graph $H = (V(H), E(H))$ is called a subgraph of a graph $G = (V(G), E(G))$ if

a. $V(H) \leq V(G)$ and b. $E(H) \leq E(G)$.

(ii) A subgraph H of a graph G is called a spanning subgraph if $V(H) = V(G)$.

Examples

G:

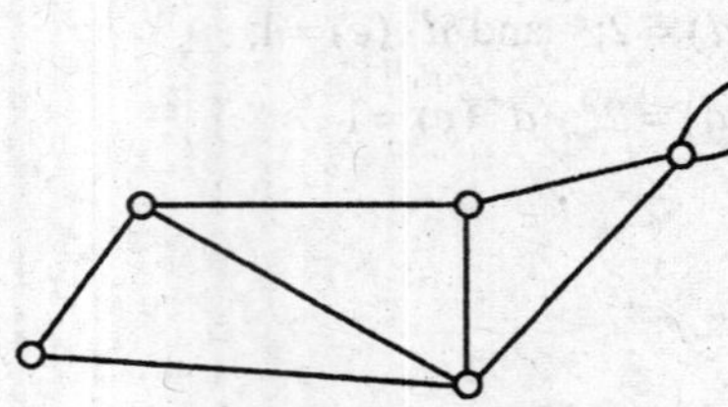

H:

Subgraph

H_1:

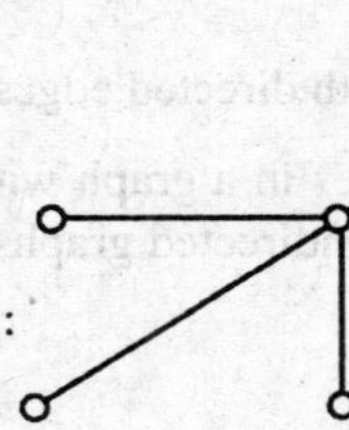

Spanning subgraph

OPERATIONS ON GRAPHS

i. The union of two simple graphs $G_1 = (V_1, E_1)$ and $G_2 = (V_2, E_2)$ is the simple graph with vertex set $V_1 \cup V_2$ and edge set $E_1 \cup E_2$ and is denoted by $G_1 \cup G_2$.

Examples

G_1:

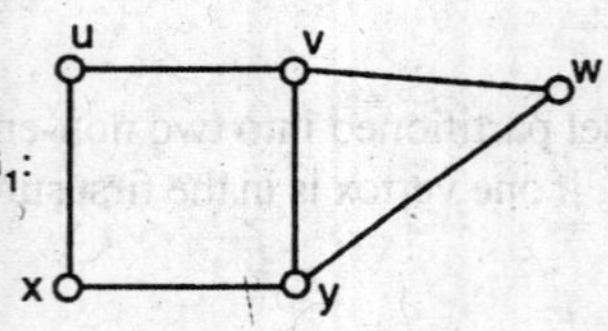

G_2:

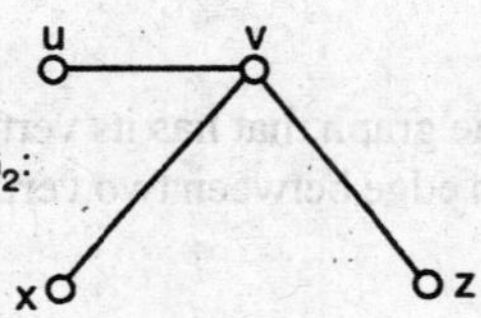

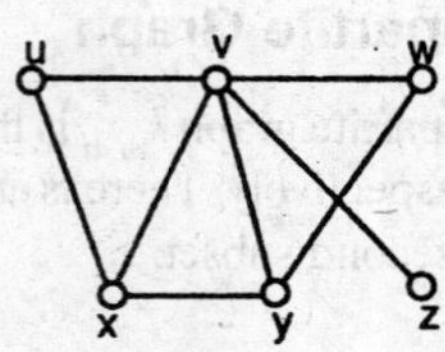

$G_1 \cup G_2$

ii. The intersection of two simple graphs $G_1 = (V_1, E_1)$ and $G_2 = (V_2, E_2)$ is the simple graph with vertex set $V_1 \cap V_2$ and edge set $E_1 \cap E_2$ and is denoted by $G_1 \cap G_2$. (Note that for $G_1 \cap G_2$, $V_1 \cap V_2$ is non-empty always.)

Examples

G_1:

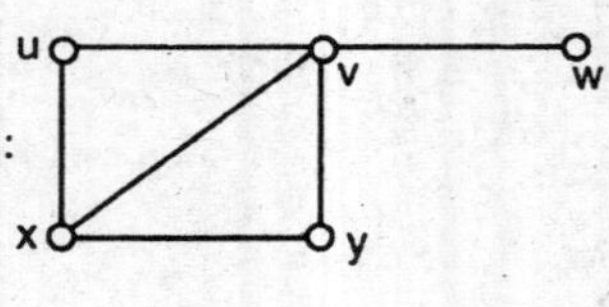

G_2:

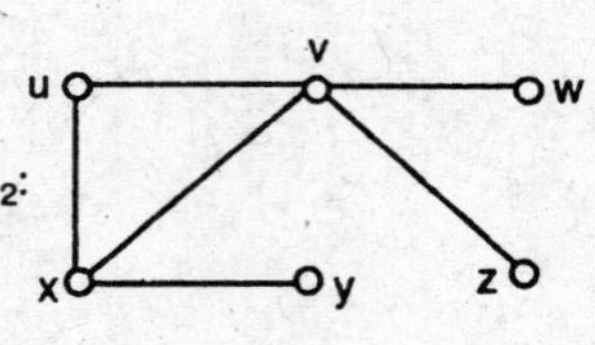

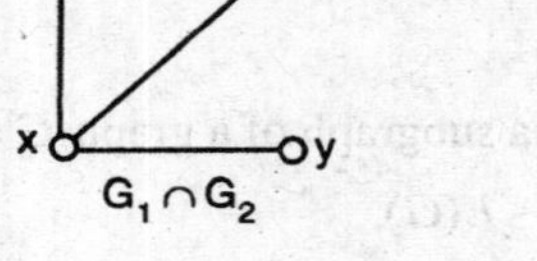

$G_1 \cap G_2$

iii. The ring sum of two graphs G_1 and G_2 is a graph consisting of the vertex set $V_1 \cup V_2$ and of edges that are either in G_1 or in G_2, but not in both and is denoted by $G_1 \oplus G_2$.

$$\text{i.e. } G_1 = (V_1, E_1);\ G_2 = (V_2, E_2)$$

$$\text{then } G_1 \oplus G_2 = (V_1 \cup V_2,\ E_1 \,\Delta\, E_2),$$

where Δ is the symmetric difference.

ISOMORPHISM

Two graphs G and H are said to be isomorphic if there exists bijections $\psi : V(G) \to V(H)$ and $\phi : E(G) \to E(H)$ such that iff $\theta_G(e) = uv$ iff $\theta_H(\phi(e)) = \psi(u)\,\psi(v)$

Such a pair of mappings (ψ, ϕ) is called an isomorphism between G and H and is written as $G \cong H$.

In other words, two simple graphs G and H are isomorphic iff there is a bijection $\psi : V(G) \to V(H)$ such that $uv \in E(G)$ iff $\psi(u)\psi(v) \in E(H)$.

Example

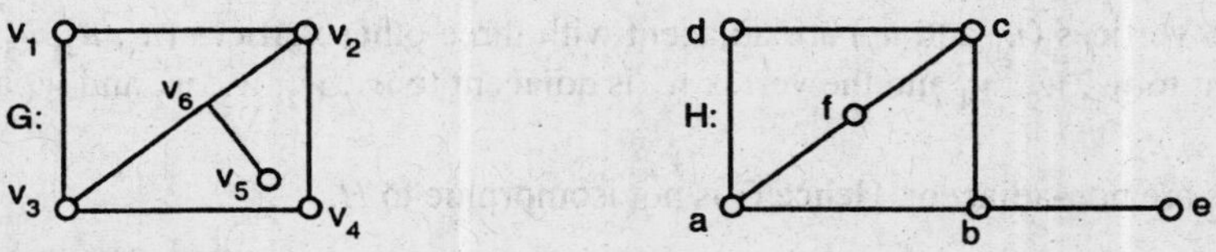

Here G and H are isomorphic

The correspondence which gives isomorphism between G and H is as follows :

$$\begin{aligned}
v_1 v_2 \in E(G) &\iff dc = \psi(v_1)\psi(v_2) \in E(H) \\
v_1 v_3 \in E(G) &\iff da = \psi(v_1)\psi(v_3) \in E(H) \\
v_3 v_6 \in E(G) &\iff ab = \psi(v_3)\psi(v_6) \in E(H) \\
v_6 v_5 \in E(G) &\iff be = \psi(v_6)\psi(v_5) \in E(H) \\
v_3 v_4 \in E(G) &\iff af = \psi(v_3)\psi(v_4) \in E(H) \\
v_6 v_2 \in E(G) &\iff bc = \psi(v_6)\psi(v_2) \in E(H) \\
v_4 v_2 \in E(G) &\iff fc = \psi(v_4)\psi(v_2) \in E(H) \\
&\therefore\ G \cong H
\end{aligned}$$

Note:

i. Two graphs $G_1 = (V_1, E_1)$ and $G_2 = (V_2, E_2)$ are said to be isomorphic if a one-to-one correspondence ϕ exists from V_1 to V_2 such that u and v are adjacent in G_1 iff $\phi(u)$ and $\phi(v)$ are adjacent in G_2.

ii. If $G \cong H$, then degrees of corresponding vertices are equal.

Example 7.3 Prove that the graphs G and H are non-isomorphic

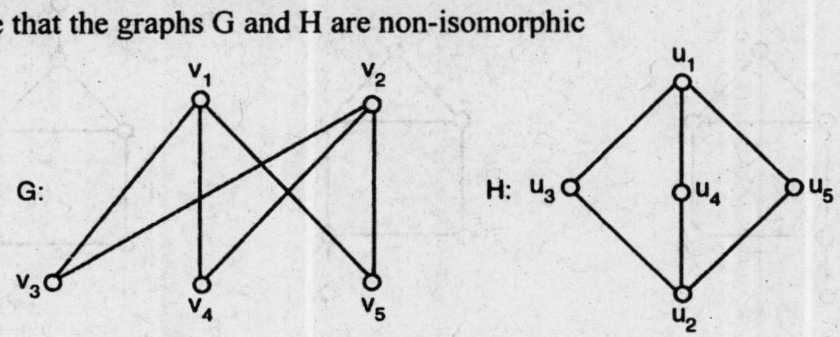

Solution Clearly G and H are isomorphic.

In G, V_1 is adjacent to the vertices V_3, V_4, V_5 ; V_2 is adjacent to the vertices V_3, V_4, V_5.

In H, u_1 is adjacent to u_3, u_4, u_5 and u_2 is adjacent to u_3, u_4, u_5.

Here the function defined by $\phi(v_i) = u_i$, $1 \le i \le 5$ gives the isomorphism.

Example 7.4 Prove that the graphs G and H are non-isomorphic.

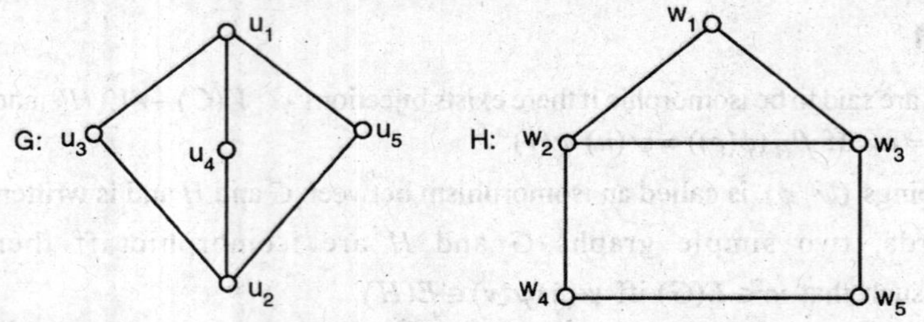

Solution Clearly G and H are non-isomorphic graphs.

In G, these two vertices (u_1 and u_2) are adjacent with three other vertices (u_3, u_4, u_5) whereas in H, the vertex w_2 is adjacent to w_1, w_2, w_4 and the vertex w_3 is adjacent to w_1, w_2, w_5. w_2 and w_3 are adjacent to each other.

In G, u_1, and u_2 are non-adjacent. Hence G is not isomorphic to H.

Note: From the above example, it is clear that two graphs are isomorphic if they have same number of vertices and same number of edges and the degrees of the corresponding vertices are equal, but the converse is not true.

EXERCISE I

1. Prove that every graph with n vertices ($n \ge 2$) has atleast two vertices with the same degree.
2. Mr. X invited 5 friends to a party at his residence. After they all arrive, he asks each friend how many people at the party does he know. Each of the five gives a different answer. Is this possible?
3. a) Construct an r-regular graph on 8 vertices for each r, $0 \le r < 8$
   b) Find the complement of each graph constructed in (a).
   c) Prove that if G is a regular graph, $\overline{G}$ is regular.
4. Draw all eleven non-isomorphic graphs on 4 vertices.
5. Prove that two graphs G_1 and G_2 are isomorphic iff their complements are isomorphic.
6. Determine which pair of graphs G_1, G_2 and G_3 shown below is isomorphic.

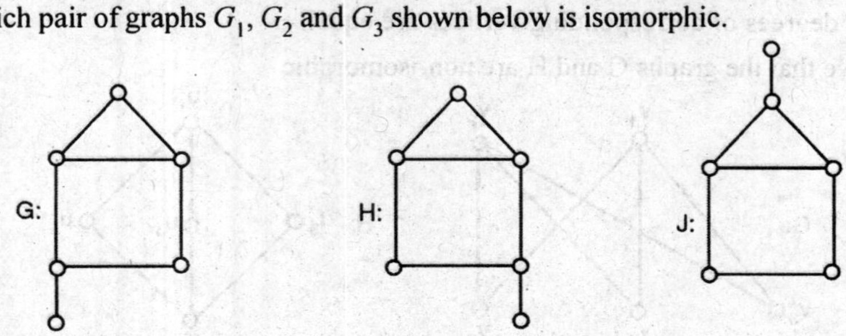

7. Find $G_1 \cup G_2$ and $G_1 \oplus G_2$, where G_1 and G_2 are shown below.

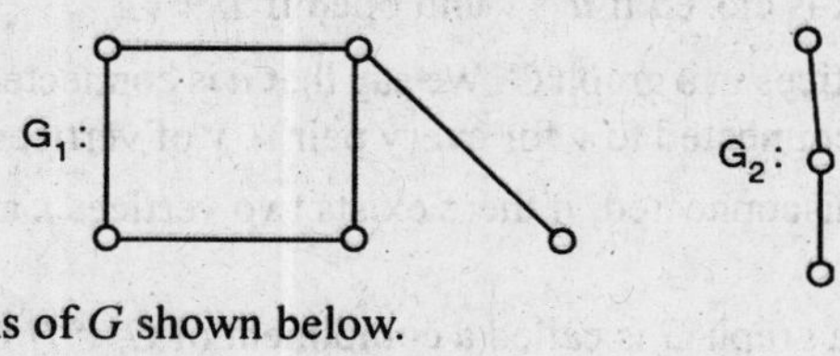

8. Draw all spanning subgraphs of G shown below.

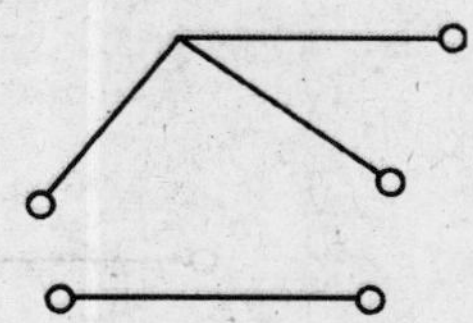

G :

9. A simple graph G is self complementary if $G \cong \overline{G}$. Prove that if G is self complementary then $V(G) \equiv 0,1 \pmod 4$).

10. Draw a self complementary graph on 5 vertices.

CONNECTED GRAPHS

In this section, we study the structure of graphs. A walk in a graph G is an alternating sequence.

$W: v_0,\ e_1,\ v_1,\ e_2, \ldots,\ v_{n-1},\ e_n,\ v_n\ (n \geq 0)$ of vertices and edges, beginning and ending with vertices, such that $e_i = v_{i-1}\ v_i,\ i = 1, 2, \ldots, n$. It is denoted by $(v_0 - v_n)$ walk. The number of edges (not necessarily distinct) is called the length of walk. In graph $G,\ u,\ e_1,\ v,\ e_2,\ w,\ e_6,\ x,\ e_4,\ u$ is a walk of length 4.

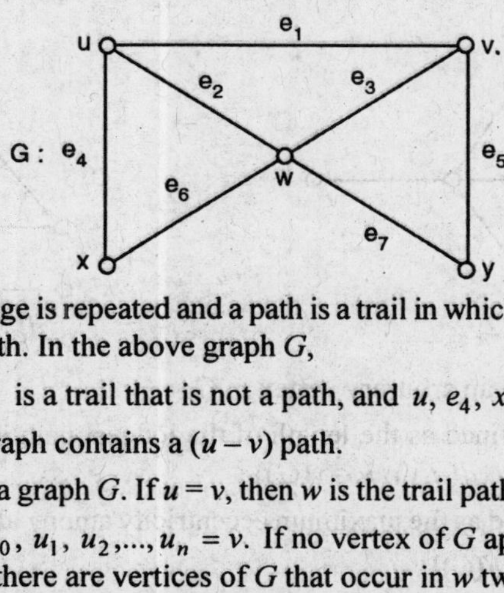

A trail is a walk in which no edge is repeated and a path is a trail in which no vertex is repeated. Thus, a *path* is a trail, but not every trail is a path. In the above graph G,

$x,\ e_6,\ w,\ e_3,\ v,\ e_1,\ u,\ e_2,\ w,\ e_7,\ y$ is a trail that is not a path, and $u,\ e_4,\ x,\ e_6,\ w,\ e_3, v,$ is a path.

Result Every $(u - v)$ walk in a graph contains a $(u - v)$ path.

Proof Let W be a $(u - v)$ walk in a graph G. If $u = v$, then w is the trail path i.e. walk of length zero.

Suppose $u \neq v$ and $W: u = u_0,\ u_1,\ u_2, \ldots,\ u_n = v$. If no vertex of G appears in W more then once, then w itself is a $(u - v)$ path. Otherwise, there are vertices of G that occur in w twice or more. Let i and j be distinct positive integers such that $i < j$ with $ui = uj$. Then say $u_i,\ u_{i+1}, \ldots, u_{j-2}, u_{j-1}$ are removed from w, and the resulting sequence is $(u - v)$ walk w_1 whose length is less than that of w. (By induction hypothesis, this w_1 contains a $(u - v)$ path and hence w has a $(u - v)$ path). If no vertex of G appears more than once in $w1$, then w_1 is a $(u - v)$ path. If not, apply the above procedure, until we get a $(u - v)$ path.

Cycle A cycle is a walk. $v_0, v_1, \ldots, v_n$ is a walk in which $n \geq 3$, $v_0 = v_n$ and the 'n'-vertices $v_1, v_2, \ldots, v_n$ are distinct. We say that a $(u - v)$ walk is closed if $u = v$ and open if $u \neq v$.

Connection Let u and v be vertices in a graph G. We say that u is connected to v if G contains a $(u - v)$ path. The graph G is connected, if u is connected to v for every pair u, v of vertices of G.

Disconnection A graph G is disconnected, if there exists two vertices u and v for which there is no $(u - v)$ path.

Component A subgraph H of a graph G is called a component of G, if H is a maximal connected subgraph of G and component is denoted by $\omega(G)$.

Note: If $\omega(G) > 1$, then G is disconnected.

Examples

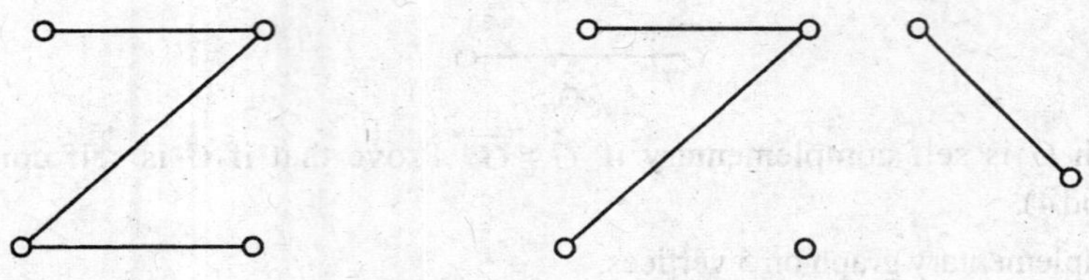

(a) (b)

Graph (a) is connected and (b) is disconnected.

Note that graph (b) has 3 components.

DISTANCE IN A GRAPH

For a non-trial graph G and a pair u, v of vertices of G, the distance $d_G\,(u - v)$ is defined as the length of a shortest $(u - v)$ path in G (if such path exists). If G contains no $(u - v)$ path, then we define $d_G\,(u - v) = \infty$.

Examples

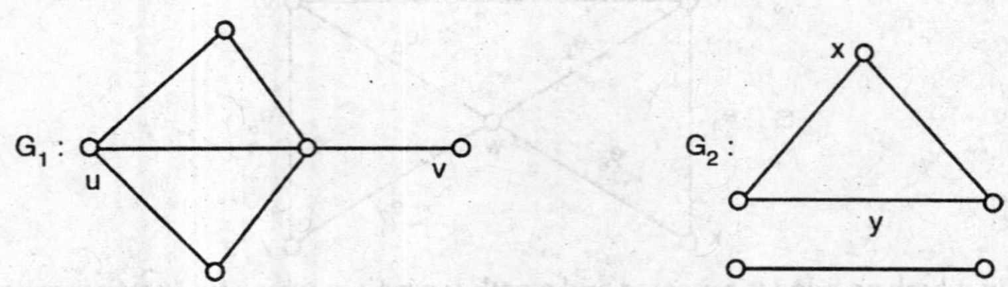

$d_{G_1}(u, v) = 2$; $d_{G_2}(x, y) = \infty$

Let G be a connected graph and v be an arbitrary vertex in G.

i. The eccentricity of v is defined as the length of the longest path in G starting from vertex u and is denoted by $e(v)$ $e(v) = \max\{d(u, v) / u \in v(G)\}$

ii. The diameter of G is defined as the maximum eccentricity among all the vertices of G.
i.e. diam $(G) = \max\{e(v)/v \in V(G)\}$

iii. The radius of G is defined as the minimum eccentricity among all the vertices of G.
i.e. rad $(G) = \min\ \{e(v) \mid v \in V(G)\}$

iv. The centre of G is defined as the set of vertices having minimum eccentricity among all the vertices of G. i.e. cent $(G) = \{v \in V(G)/e(v) = \text{rad}\ (G)\}$

Note:

1 rad(G) $\le$ diam (G) $\le$ 2 rad (G), G is a graph.

2 The median of a connected graph G is defined as the set of vertices having minimum distance.

CUT-VERTICES AND CUT-EDGES

A vertex v in a graph G is said to be a cut-vertex if $\omega(G - v) > \omega(G)$, [where $\omega(G)$ is the component of G and component is a maximal connected subgraph of G]

i.e. a vertex v of a connected graph is a cut-vertex, iff $(G - v)$ is disconnected.

Examples

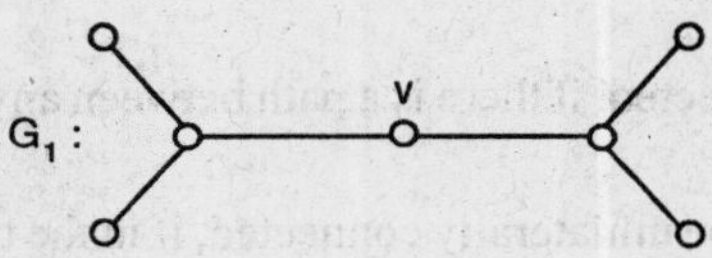

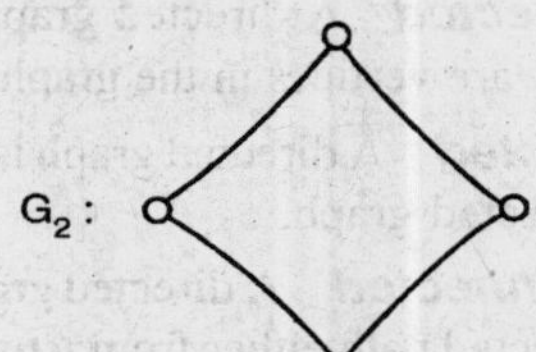

G_1 contains one cut-vertex v and G_2 contains no cut-vertices.

Theorem A vertex v in a connected graph G is a cut-vertex iff there exists vertices u and w (both are different from v) such that every path connecting u and w contain v.

Proof Let G be a connected graph and v be a cut-vertex.

Claim There exists vertices u and w such that every path between u and w contains v.

Since v is a cut-vertex, $(G - v)$ is disconnected and $(G - v)$ contains two components say G_1 and G_2. Let u and w be the vertices of G_1 and G_2 respectively. Clearly there is no $(u - w)$ path in $(G - v)$. Hence every path connecting u and w must contain v.

Conversely, let us assume that there exist vertices u and w such that every $(u - w)$ path contains v.

Claim v is a cut-vertex.

Suppose v is not a cut-vertex. Then $(G - v)$ is connected. Since u, w are vertices in $G - v$, there is a path between u and w in $G - v$, which does not contain the vertex v. This is a contradiction. Hence v is a cut-vertex.

Cut-edge An edge e in a graph G is said to be a cut-edge, if $(G - e)$ is disconnected.

Example

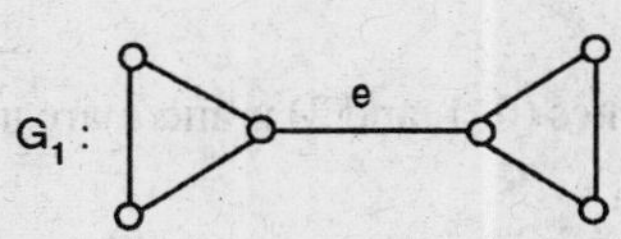

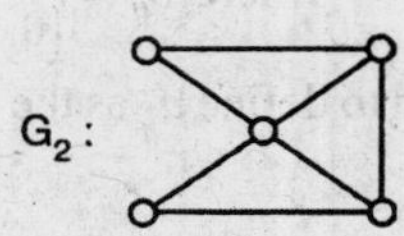

G_1 contains one cut-edge & G_2 has no cut-edge

As in cut-vertex, a similar result can be furnished.

Theorem An edge e in a connected graph G is a cut-edge iff there exists vertices u and w such that every path connecting u and w must contain the edge 'e'.

Proof Let G be a connected graph and e be a cut-edge.

Claim There exist vertices u and w such that every $(u - w)$ path must contain the edge e, since e is a cut-edge in G, $(G - e)$ is disconnected and $(G - e)$ contains atleast two components say G_1 and G_2.

Let u and w be the vertices respectively in G_1 and G_2. Thus there is no path between u and w ln $(G-2)$. Hence every path connecting u and w must contain the edge e.

Conversely, suppose that there exist vertices u and w such that every path connecting u and w must contain the edge e.

Claim e is a cut-edge

Suppose e is not a cut-edge. Then $(G-e)$ is connected and hence e is some circuit in G. Therefore there exists a path connecting u and w, which not contain the edge e. This is a contradiction. Hence e is a cut-edge.

CONNECTEDNESS IN DIRECTED GRAPH

Strongly connected A directed graph is strongly connected if there is a path from u to v and v to u, whenever u and v are vertices in the graph.

Weakly connected A directed graph is weakly connected, if there is a path between any two vertices in the underlying undirected graph.

Unilaterally connected A directed graph is said to be unilaterally connected, if in the two vertices u and v, there exists a directed path either from u to v or from v to u.

Examples

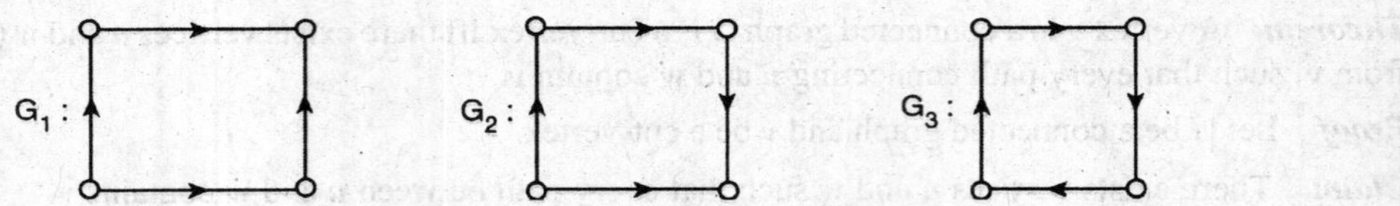

G_1 is weakly connected; G_2 unilaterally connected and G_3 is strongly connected.

INCIDENCE AND ADJACENCY MATRICES

To any graph G, there corresponds a $V \times E$ matrix called the incidence matrix of G and is denoted by $I(G) = [a_{ij}]_{V \times E}$, where

$$a_{ij} = \begin{cases} 1, \text{ if } j\text{th edge is incident with } i\text{th vertex.} \\ 0, \text{ otherwise.} \end{cases}$$

One more matrix associated with graph G is the adjacency matrix, e is denoted by $A(G) = [b_{ij}]_{V \times V}$,

$$\text{where } b_{ij} = \begin{cases} 1, \text{ if } i\text{th vertex and } j\text{th vertex are adjacent.} \\ 0, \text{ otherwise.} \end{cases}$$

[Some authors used to define a_{ij} as the number of times (0, 1, and 2) v_i and e_j are incident ; b_{ij} is the number of edges v_i and v_j].

Examples

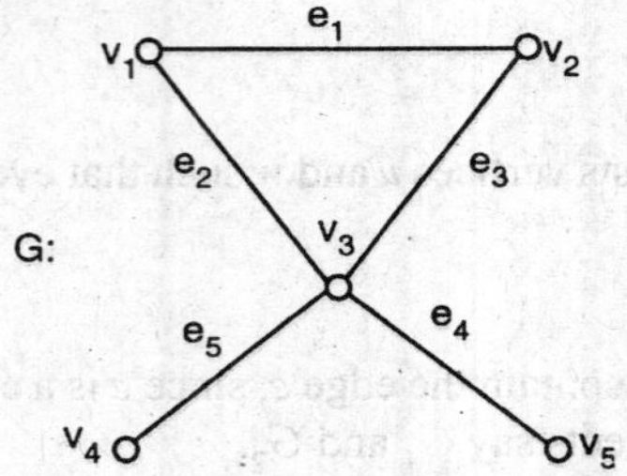

	e_1	e_2	e_3	e_4	e_5
v_1	1	1	0	0	0
v_2	1	0	1	0	0
v_3	0	1	1	1	1
v_4	0	0	0	0	1
v_5	0	0	0	1	0

I(G), incidence matrix of G

	v_1	v_2	v_3	v_4	v_5
v_1	0	1	1	0	0
v_2	1	0	1	0	0
v_3	1	1	0	1	1
v_4	0	0	0	1	0
v_5	0	0	0	1	0

$A(G)$, adjacency matrix of G

The adjacency matrix $A(G) = [b_{ij}]$ of a directed graph is also a $V \times V$ matrix,

$$\text{where } b_{ij} = \begin{cases} 1, & \text{if there is a directed edge from } v_i \text{ to } v_j \\ 0, & \text{otherwise.} \end{cases}$$

(Similarly one can define the incidence matrix of a directed graph)

Example

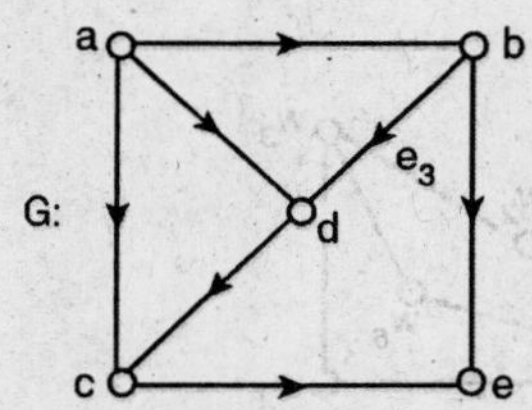

$$A(G) = \begin{bmatrix} 0 & 1 & 1 & 1 & 0 \\ 0 & 0 & 0 & 1 & 1 \\ 0 & 0 & 0 & 0 & 1 \\ 0 & 0 & 0 & 1 & 1 \\ 0 & 0 & 0 & 0 & 0 \end{bmatrix}$$

Example 7.5

Solution Write the adjacency matrix of the following graphs

i.

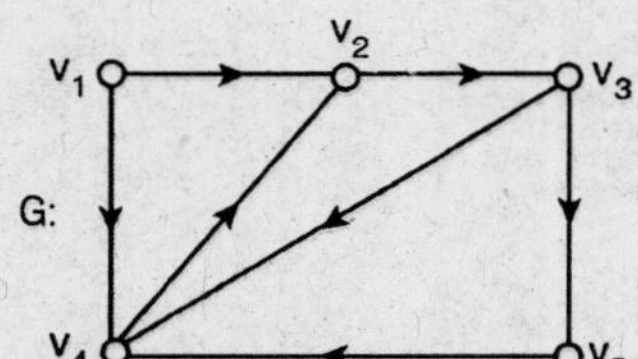

$$A(G) = \begin{bmatrix} 0 & 1 & 0 & 0 & 0 \\ 0 & 0 & 1 & 0 & 0 \\ 0 & 0 & 0 & 1 & 1 \\ 1 & 1 & 0 & 0 & 0 \\ 0 & 0 & 0 & 1 & 0 \end{bmatrix}$$

ii.

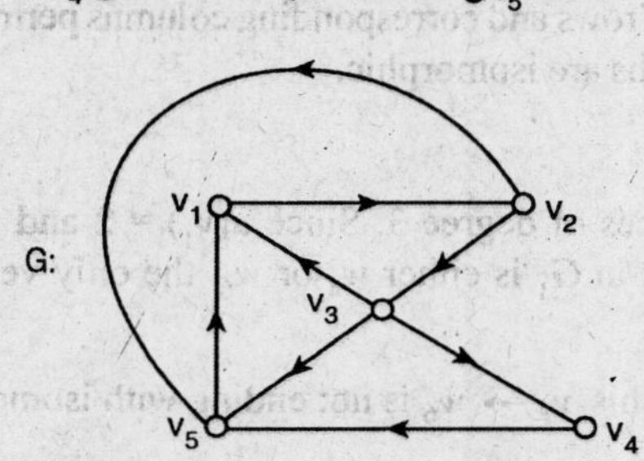

$$A(G) = \begin{bmatrix} 0 & 1 & 0 & 0 & 0 \\ 0 & 0 & 1 & 0 & 1 \\ 1 & 0 & 0 & 1 & 1 \\ 0 & 0 & 0 & 0 & 1 \\ 1 & 0 & 0 & 0 & 0 \end{bmatrix}$$

iii.

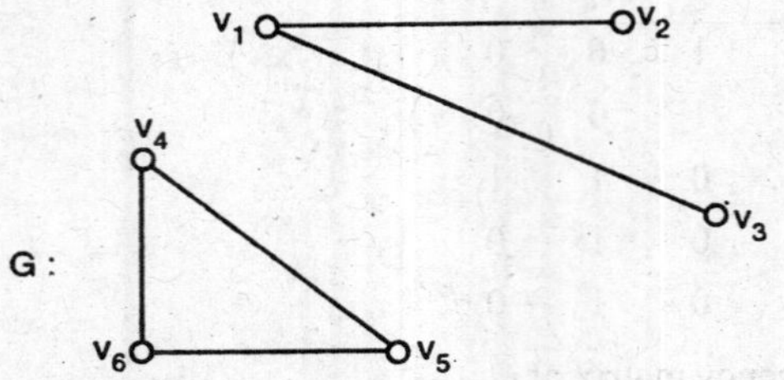

$$A(G) = \begin{bmatrix} 0 & 1 & 1 & 0 & 0 & 0 \\ 1 & 0 & 0 & 0 & 0 & 0 \\ 1 & 0 & 0 & 0 & 0 & 0 \\ 0 & 0 & 0 & 0 & 1 & 1 \\ 0 & 0 & 0 & 1 & 0 & 1 \\ 0 & 0 & 0 & 1 & 1 & 0 \end{bmatrix}$$

Note: From example 7.5 one can conclude that

1 The diagonal entries of an adjacency matrix are all zero, iff the graph is a graph with no self-loops.

2 If G is disconnected and it has two components, then its adjacency matrix $A(G)$ can be written as

$$A(G) = \begin{bmatrix} A(G_1) & 0 \\ 0 & A(G_2) \end{bmatrix}, G_1 \text{ and } G_2 \text{ are components.}$$

With the help of these matrices, one can verify whether the given graphs are isomorphic or not.

Example 7.6 Verify if G and G, are isomorphic

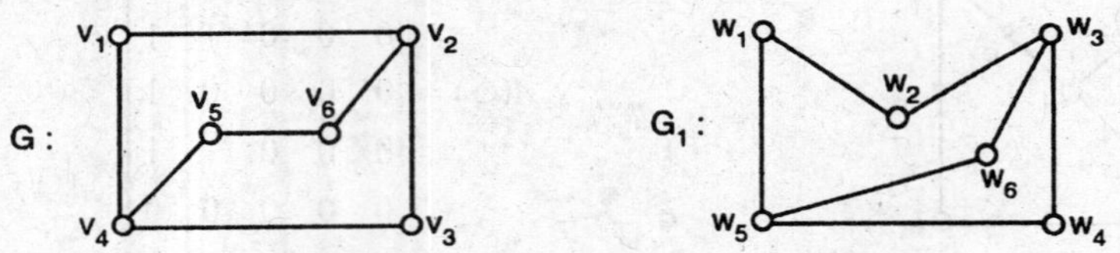

Solution First we shall write the adjacency matrices of G and G_1.

$$A(G) = \begin{bmatrix} 0 & 1 & 0 & 1 & 0 & 0 \\ 1 & 0 & 1 & 0 & 0 & 1 \\ 0 & 1 & 0 & 1 & 0 & 0 \\ 1 & 0 & 1 & 0 & 1 & 0 \\ 0 & 0 & 0 & 1 & 0 & 1 \\ 0 & 1 & 0 & 0 & 1 & 0 \end{bmatrix} \qquad A(G_1) = \begin{bmatrix} 0 & 1 & 0 & 0 & 1 & 0 \\ 1 & 0 & 1 & 0 & 0 & 0 \\ 0 & 1 & 0 & 1 & 0 & 1 \\ 0 & 0 & 1 & 0 & 1 & 0 \\ 1 & 0 & 0 & 1 & 0 & 1 \\ 0 & 0 & 1 & 0 & 1 & 0 \end{bmatrix}$$

By keeping one matrix fixed, and by applying permutation of rows and corresponding columns permutations on the unfixed matrix, yields the fixed one. Then the given graphs are isomorphic.

Here keep $A(G)$ fixed.

Also G and G_1 have 4 vertices of degree 2 and two vertices of degree 3. Since $d(v_1) = 2$ and v_1 is not adjacent to any other vertex of degree 2, corresponding vertex in G_1 is either w_4 or w_6, the only vertices of degree 2 in G_1 not adjacent to a vertex of degree 2.

Without loss of generality, let us take $v_1 \to w_6$. (Suppose this $v_1 \to w_6$ is not ending with isomorphism, we have to take $v_1 \to w_4$).

Similarly, for other vertices of G, we can set $v_2 \to w_3$; $v_3 \to w_4$; $v_4 \to w_5$; $v_5 \to v_1$; $v_6 \to v_2$.

Thus we can modify $A(G_1)$ as

$$A(G_1) = \begin{array}{c} \\ w_6 \\ w_3 \\ w_4 \\ w_5 \\ w_1 \\ w_2 \end{array} \begin{array}{c} \begin{array}{cccccc} w_6 & w_3 & w_4 & w_5 & w_2 & w_1 \end{array} \\ \begin{bmatrix} 0 & 1 & 0 & 1 & 0 & 0 \\ 1 & 0 & 1 & 0 & 0 & 1 \\ 0 & 1 & 0 & 1 & 0 & 0 \\ 1 & 0 & 1 & 0 & 1 & 0 \\ 0 & 0 & 0 & 1 & 0 & 1 \\ 0 & 1 & 0 & 0 & 1 & 0 \end{bmatrix} \end{array}$$

$\therefore$ $A(G) = A(G_1)$ and hence $G \cong G_1$.

EXERCISE II

1. Give an example of a graph G with $w(G) = \delta(G) = \Delta(G) = 2$
2. Let G_1 & G_2 be isomorphic graphs. Prove that if G_1 is connected, then G_2 is connected.
3. If G is a n-vertex graph with $\Delta(G) \geq n/2$, then prove that G need not be connected.
4. P.T a strongly connected graph on n vertices $(n \geq 2)$ contains a cycle of length k for some $k \geq 2$.
5. P.T a directed graph is strongly connected iff D has a closed (directed) spanning walk.
6. Find all cut-vertices and cut-edges in the graph G shown below.

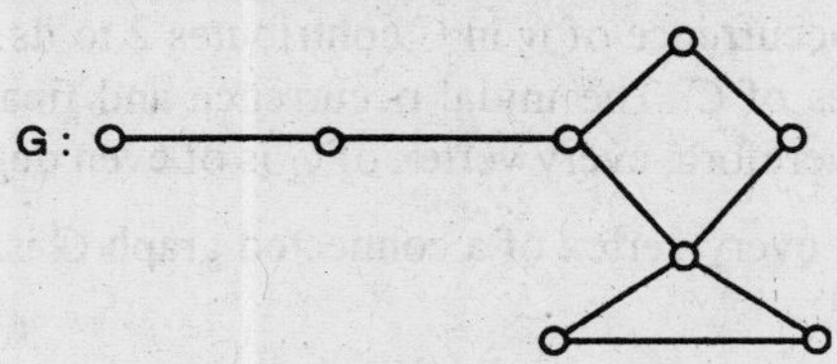

7. Prove or disprove:
   i. If G is a connected graph, then $G + k_1$ contains no cut-vertices.
   ii. If G is bipartite graph, then every edge of G lies on an even cycle.
8. What is the sum of the entries in a row of the incidence matrix?
9. What is the sum of the entries in a column of the incidence matrix?
10. Is every zero-one square matrix that is symmetric and has zero on the diagonal of the adjacency matrix of a simple graph?

EULERIAN AND HAMILTONIAN GRAPHS

In this section, we study some special graphs and also the origin of the graph theory.

The Koingsberg town in Russian Republic was separated into four lands by the river Pregel. These islands were connected by seven bridges. The problem was could people walk from one island, travel across all the seven bridges and return to the island where they had started without using any bridge more than once? For almost two centuries, nobody was in a position to state whether such a walk was possible or not.

In 1736, the great mathematician Lenhered Euler concluded that such a walk was impossible. He used multigraph to study and solve this problem. Euler today is considered as the father of the graph theory.

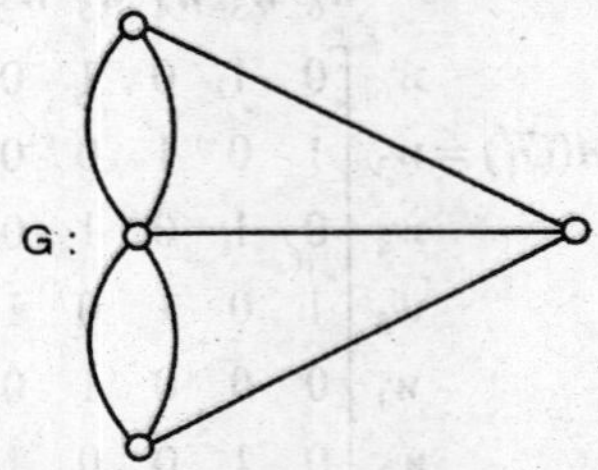

The four lands and the seven bridges are represented by vertices and edges respectively in G. This problem is called as Koingsberg bridge problem.

EULER CIRCUITS (TOURS)

A trail that traverses every edge of G is called an Euler trail of G. A circuit (tour) of G is a closed walk that traverses each edge of G atleast once. An Euler tour is a tour which traverses each edge exactly once. A graph is Eulerian if it contains an Euler tour.

Theorem A connected graph is Eulerian iff it has no vertices of odd degree.

Proof Let G be Eulerian and let C be an Euler tour of G, which begins and ends at some vertex u.

Claim G has no vertices of odd degree i.e. to prove that every vertex of G is even. Consider a vertex $w \neq u$. Since w is neither the first nor the last vertex of C, each time w is encountered, it is reached by some edge and left by another edge. Hence each occurrence of w in C contributes 2 to its degree. Thus w is of even degree. This is true for all internal vertices of C. The initial occurrence and final occurrence of the vertex u in C contributes 1 to the degree of u. Therefore, every vertex of G is of even degree.

Conversely, let us assume that every vertex of a connected graph G is even.

Claim G is Eulerian.

Suppose G be a connected non-Eulerian graph with no vertices of odd degree.

Among such graphs, choose one, say G having least number of edges

Since each vertex of G has atleast two edges, G contains a trail (Exercise problem No. 2) Let C be a closed trail of maximum possible length in G. By assumption, C is not a Euler circuit of G and hence $G - E(C)$ has edges.

Therefore $G - E(C)$ has some component G' with edges. Since C itself is Eulerian, degree of every vertex in C is even. Hence degree of every vertex in $G - E(C)$ is also even. Therefore degree of every vertex in G' is even. Since $E(G') < E(G)$, [by the choice of G in (1)]

G' is Eulerian and hence G' has an Euler circuit (tour) say C'. Since G is connected, there is a vertex v in $V(C) \cap V(C')$ and we may assume without loss of generality that v is the initial and the terminal vertex of both circuits C and C'. Now $(C \cup C')$ is a closed trail of G with $E(C \cup C') > E(C)$. This contradicts the choice of C. Hence every non-empty connected graph with no vertices of odd degree is Eulerian.

Examples

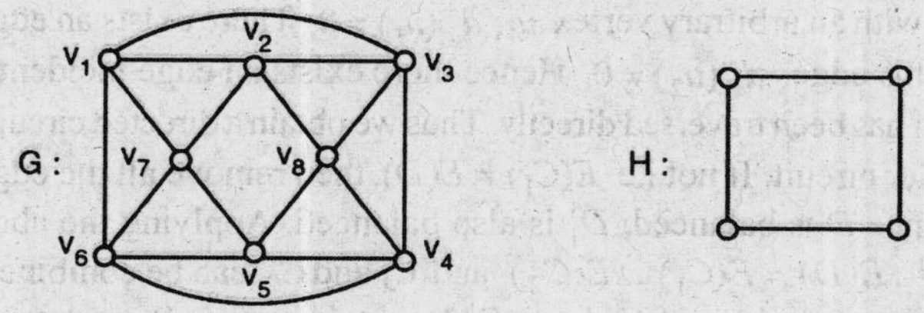

G and *H* are Eulerian graphs

Theorem A connected graph *G* has an Eulerian trail iff *G* has exactly two odd vertices.

Proof Let *G* be a connected graph with an Eulerian $(u - v)$ trail. By the similar argument in the previous theorem, we conclude that all the vertices on the trail except *u* and *v*, have even degree. Conversely, let *G* be connected graph with two odd vertices *u* and *v*. Let G' be the graph obtain from *G* by adding a new edge $e = uv$ between *u* and *v*. By applying the previous theorem to G', we can obtained an Eulerian tour in which the edge *e* is the first edge. Hence this Eulerian trail of *G* can be obtained that starts at *v* and ends at *u*. Therefore *G* has an Eulerian trail.

Eulerian Digraphs

An Eulerian trail of a connected directed graph *D* is a trial of *D* that contains all the edges of *D*; while an Eulerian circuit of *D* is a circuit which contains every edges of *D*. A directed graph that contains an Eulerian circuit is called Eulerian digraph.

Example

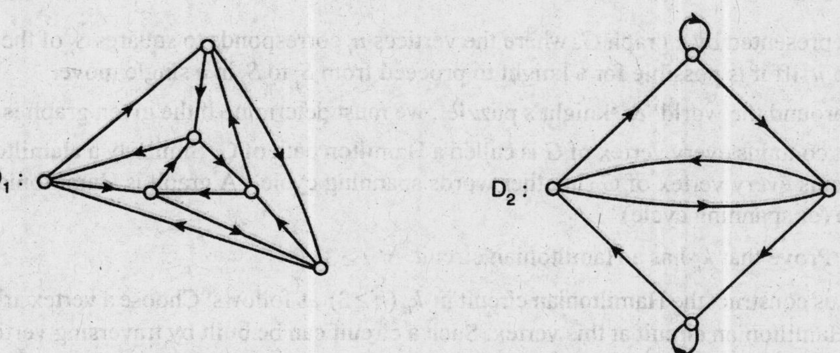

Theorem Let *D* be a connected directed graph. *D* is Eulerian iff $d^+(v) = d^-(v)$, $\forall v \in G$, (*G* is called balanced digraph).

Proof Let *D* be an Euler directed graph. Then *D* contains an Euler circuit *C* with common initial and terminal vertex *v*. Let b_u be the number of occurrence of an internal vertex *u* in *C*.

Whenever *C* enters *u* through some edge incident into *u*, there is another edge incident out of *u* through which *C* leaves *u*. Thus each occurrence of *u* contributes one in-degree and one out-degree. Moreover *C* contains all the edges of *D*. Thus,

$$d^+(u) = d^-(u) = bu. \text{ Similarly } d^+(v) = d^-(v)$$

$$\therefore \text{Hence } d^+(v) = d^-(v), \forall v \in V(D)$$

Similarly, conversely, suppose the connected digraph D is balanced. Then for each vertex u, $d^+(u) = d^-(u) \neq 0$. Start with an arbitrary vertex u_1, $d^+(u_1) \neq 0$. There exists an edge incident out of u_1. Let u_2 be the terminal vertex of this edge, $d^+(u_2) \neq 0$. Hence there exists an edge incident out of u_2. Continuing like this we reach a vertex which has been traversed directly. Thus we obtain a directed circuit C_1 in D. If $E(C_1) = E(D)$, then C_1 is the required Euler circuit. If not i.e. $E(C_1) \neq E(D)$, then remove all the edge of C_1 from D to obtain a spanning subgraph D_1. Since D is balanced, D_1 is also balanced. Applying the above process to D_1, we will obtain a circuit C_2 in D_1. If $E(D) = E(C_1) \cup E(C_2)$ and C_1 and C_2 can be combined to obtain an Euler circuit in D_1. Otherwise, we remove the edges of C_2 from D_1 to obtain a spanning subgraph D_2 of D. We repeat the above process in D_2 and after a finite number of steps, we obtain edge disjoint circuits $C_1, C_2, \ldots, C_k$ such that $E(D) = E(C_1) \cup E(C_2) \cup \ldots \cup E(C_k)$. Since D is connected, any two of these cycles have a common vertex. Then the circuits $C_1, C_2, \ldots, C_k$ can be combined to obtain an Euler circuit in D. Hence D is an Eulergraph.

Hamiltonian Graphs

In 1857, Sir William Rowan Hamilton invented a game called Around the world. In this game, a solid regular dodecahedron (20 vertices, 30 edges and 12 faces) and a supply of string is given. Every vertex is given the name of an important city. The objective of the game is to find a route along the edges of dodecahedron that visits every city exactly once and terminates where it started.

D:

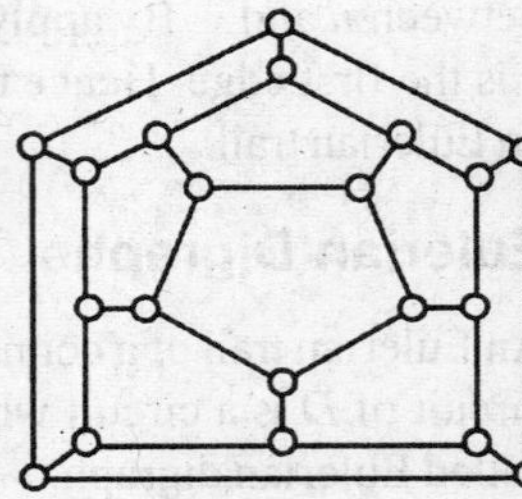

The graph D is a dodecahedron.

Another famous problem is 'The Knight's Puzzle'; Is it possible for a knight to tour the chess board, i.e. visit each square exactly once and return to its initial square?

It can be represented by a graph G, where the vertices u_i corresponds to squares S_i of the chess board and u_j is adjacent to u_i iff it is possible for a knight to proceed from S_i to S_j in a single move.

To solve 'around the world' & 'knight's puzzle', we must determine if the given graph is hamiltonian.

A path that contains every vertex of G is called a Hamilton path of G. Similarly a Hamilton cycle of G is a cycle that contains every vertex of G (in other words spanning cycle). A graph is Hamiltonian if it contains a Hamilton cycle (or spanning cycle)

Example 7.7 Prove that k_n has a Hamiltonian circuit, $\forall n \geq 3$.

Solution Let us construct the Hamiltonian circuit in k_n $(n \geq 3)$ as follows: Choose a vertex arbitrarily in k_n and beginning the Hamiltonian circuit at this vertex. Such a circuit can be built by traversing vertices in any order, as long as the path begins and terminates at the same vertex and visits each other vertex exactly once. This is possible in k_n, since every vertex is adjacent to all other vertices. [k_1, k_2 has only hamiltonian path, not circuit].

Graphical A sequence d = $(d_1, d_2, \ldots, d_n)$ is graphical if there exists a simple undirected graph on n vertices with the degrees of the vertices $d_1, d_2, \ldots, d_n$ respectively.

Example A graph with degree sequence (4, 4, 3, 2, 2, 1)

G :

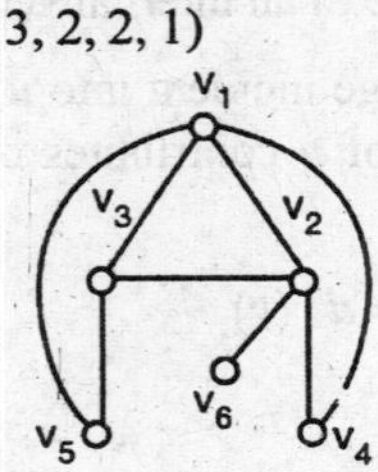

Closure A closure (CG) of a n-vertex graph G is a graph from G by recursively joining pairs of non-adjacent vertices whose degree sum is atleast n until no such pair remains.

Example

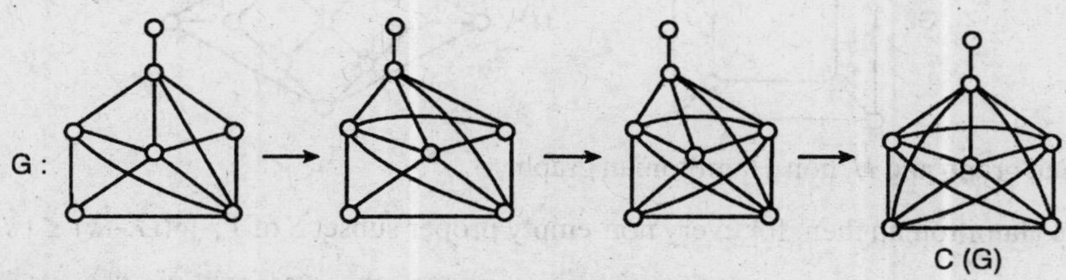

The above figures explain one way of constructing a closure of a graph.

Important Theorems

Theorem Let G be a n-vertex graph. Suppose G_1 and G_2 are two graphs obtained from G by recursively joining pairs of non-adjacent vertices whose degree sum is atleast n. Then $G_1 = G_2$. [In other words, $C(G)$, the closure of a graph G is unique]

Proof Let $e_1, e_2, \ldots, e_k$ and $f_1, f_2, f_3, \ldots, f_e$ be the edges added to G to obtain G_1 and G_2 respectively. We have to prove that every e_i $(1 \le i \le k)$ is an edge of G_2 and f_j $(1 \le j \le l)$ is an edge of G_1. Suppose that some edge in the sequence $e_1, e_2, \ldots, e_k$ does not belong to G_2. Let p be the smallest positive integer such that e_{p+1} is not an edge of G_2. Let $e_{p+1} = uv$. Let $H = G + \{e_1, e_2, \ldots, e_p\}$. Then H is a subgraph of G_1 and G_2. By the construction of G_1,

$$d_H(u) + d_H(v) \ge n$$

Therefore $d_{G_2}(u) + d_{G_2}(v) \ge d_H(u) + d_H(v) \ge n$

This is a contradiction since u and v are non-adjacent in G_2. Therefore each e_i is an edge of G_2. Similarly, each f_j belongs to G_1. Hence $G_1 = G_2$.

Theorem A graph G is hamiltonian iff its closure $C(G)$ is Hamiltonian.

Proof Let $e_1, e_2, \ldots, e_n$ be edges added to G to obtain its closure $C(G)$. Let G_i be the graph obtained from G by adding the edge e_i.

By repeated application of problem (5)

G is Hamiltonian $\Leftrightarrow C(G)$ is Hamiltonian.

Corollary 1 Let G be a graph with atleast 3 vertices. If $C(G) \cong k_n$, $(n \ge 3)$ then G is Hamiltonian.

Proof By result 1, k_n is Hamiltonian. Since $C(G) \cong k_n, C(G)$ is Hamiltonian and hence G is Hamiltonian.

Corollary 2 Let G be a graph with atleast 3 vertices if $d(u) + d(v) \ge n$ $(n \ge 3)$, for all pairs u and v of non-adjacent vertices of G, then G is Hamiltonian.

Proof Let G be a graph with atleast 3 vertices. Given that $d(u) + d(v) \ge n$ $(n \ge 3)$ for all pairs of non-adjacent vertices of G. Hence we can add edges between such pair of vertices to obtain $C(G)$. Since $C(G)$ is complete by Cor1, G is Hamiltonian.

Example

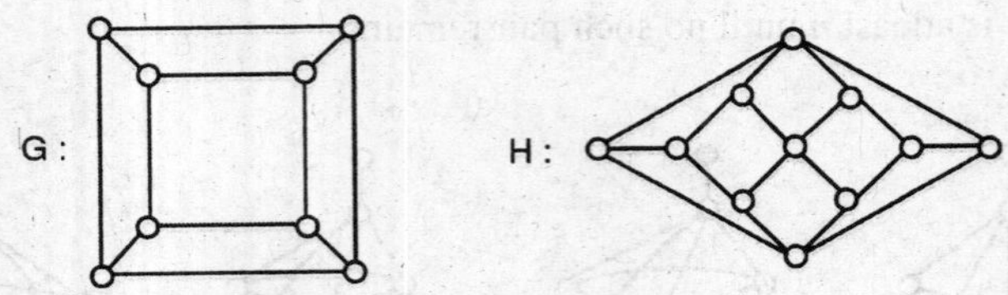

G, Hamiltonian graph and H, non-Hamiltonian graph.

Theorem If G is Hamiltonian then, for every non-empty proper subset S of V, $w(G-s) \le |s|$

Proof Let G be a Hamiltonian graph and S be a proper subset of V. Since G is Hamiltonian, G has a Hamiltonian cycle C. Suppose $\omega(G-S) = n$, where $G_1, G_2, \ldots, G_n$ are the components of G - S. Let u_i $(1 \le i \le v)$ be the last vertex of C that belongs to G_i and let v_i be the vertex that immediately follows u_i on C. Clearly $v_i \in S$. for each i and $v_j \ne v_k$ for $j \ne k$. Hence there are atleast as many vertices in S as components in $G-S$.

i.e. $\omega(G-S) \le |S|$.

Weight graph A graph G is called a weight graph if every edge of G is assigned with a real number.

Travelling Salesmen Problem (TSP)

Suppose that a salesman is expected to take a trip through a given collection of n cities $(n \ge 3)$. What route should he take to minimize the total distance travelled? This can be represented as a weight graph. Let G be a connected weight graph whose vertices represent the cities to be visited and let the weight of an edge $v_i\, v_j$ be the distance between the cities v_i and v_j. Now TSP is equivalent to finding a minimum Hamiltonian cycle in a connected weight graph.

EXERCISE III

1. Prove or disprove
   A connected digraph is eulerian iff each edge lies on odd number of cycles.
2. Prove that if G is a connected graph with $2n(n \ge 1)$ odd vertices, then edge set of G can be partitioned into n-edge disjoint trails.
3. Prove that no bipartite graph with odd number of vertices is Hamiltonian.
4. Let G be a n-vertex graph. $(n \ge 3)$. If $\delta(v) \ge n/2$. $\forall V \in V(G)$, then G is Hamiltonian.
5. Let G be a n-vertex graph $(n \ge 3)$. Suppose u and v are non-adjacent vertices of G such that $d(u) + d(v) \ge n$ Then G is Hamiltonian iff $G+uv$ is Hamiltonian.
   (Hint: if $d(v) \ge (n-1)/2$, $\forall V \in V(G)$ then G contains a Hamiltonian path).
6. Prove that the graphs $k_{n,n}$ are Hamiltonian for every $n \ge 2$.
7. Let G be a graph with atleast 3 vertices and has degree sequence $d_1, d_2, \ldots, d_n$ where $d_1 \le d_2 \le, \ldots, \le d_n$. If for every $k < n/2$, $d_k \le k \Rightarrow d_{n-k} \ge n-k$. Find if G is Hamiltonian.
8. Prove that if a graph with atleast 3 vertices has an isolated vertex or an end vertex, then its closure is not complete.

TREES

In this section we shall study the characteristics of a tree.

Acyclic Graph A graph G which has no cycles is called an acyclic graph.

Tree A connected acyclic graph G is called a tree.

Examples

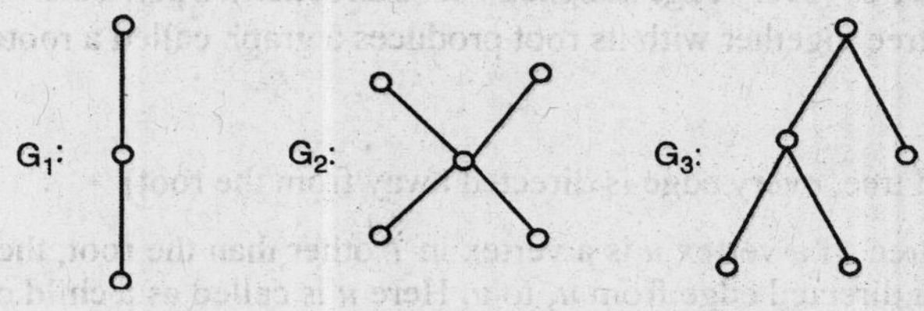

Note:

1 Trees are often known as open graphs.

2 Any organizational hierarchy is also an example of tree.

Theorem Every two vertices in a tree, are joined by a unique path.

Proof By contradiction: Let G be a tree and assume that there are two distinct (v,w) paths P_1 and P_2 in G. Since $P_1 \neq P_2$, there is an edge $e = V_1V_2$ of P_1 that is not in P_2. Clearly $(P_1 \cup P_2) - e$ is connected. Therefore it contains a $(V_1 - V_2)$ path P. Now $P + e$ is a cycle in the acyclic graph G, which is a contradiction to the fact that G is a tree.

Theorem If G is a tree on n vertices, then G has $(n-1)$ edges.

Proof By induction on the number of vertices.

When $n = 1$, $E(G) = 0 = n-1$ $(G \cong K_1)$

When $n = 2$, $E(G) = 1 = n-1$ $(G \cong K_2)$

Let us assume that this theorem is true for all trees of G with fewer than n vertices.

Now, let G be a tree on n vertices. Let $e = uv$ be an edge in G. Then $G - e$ is disconnected and G has two components say G_1 and G_2 of $G - e$. Since G is acyclic, G_1 & G_2 are also acyclic and hence G_1 & G_2 are also trees. Moreover G_1 and G_2 has fewer than n vertices say n_1 and n_2 respectively. Therefore, by induction hypothesis,

G_1 has $(n_1 - 1)$ edges and G_2 has $(n_2 - 1)$ edges.

$$\begin{aligned}\therefore E(G) &= E(G_1) + (G_2) + 1 \quad \text{(Here 1 in the sum corresponds to the edge } e\text{)}\\ &= (n_1 - 1) + (n_2 - 1) + 1\\ &= n_1 + n_2 - 1\\ &= n - 1\end{aligned}$$

Therefore an n vertex tree has $(n-1)$ edges.

Theorem Every tree has atleast two vertices of degree one (or) in a tree, there are atleast two pendant vertices.

Proof Let G be a tree on n vertices. Then

$$d(v) \geq 1, \ \forall v \in v(G) \qquad (1)$$

Already we have, $$\sum_{v \in v} d(v) = 2.E(G) = 2.e \qquad (2)$$

Since G is an n-vertex tree, it has $(n-1)$ edges.

$$\therefore \sum_{v \in v(G)} d(v) = (2n-2) \qquad (3)$$

From (1) and (3), it follows that $d(v) = 1$ for atleast two vertices.

Note: In a tree, every edge is a cut-edge.

Rooted tree In a directed tree (every edge assigned with a direction), a particular vertex is called a root if that vertex is of degree zero. A tree together with its root produces a graph called a rooted tree.

Example

(Note that in the rooted tree, every edge is directed away from the root)

Suppose T is a rooted tree. If a vertex u is a vertex in T other than the root, the parent of u is the unique vertex u_1 such that there is a directed edge from u_1 to u. Here u is called as a child of u_1. Vertices of the same parent are called as siblings. A vertex of a rooted tree is called as a leaf if it has no children and those vertices which have children, are called as internal vertices.

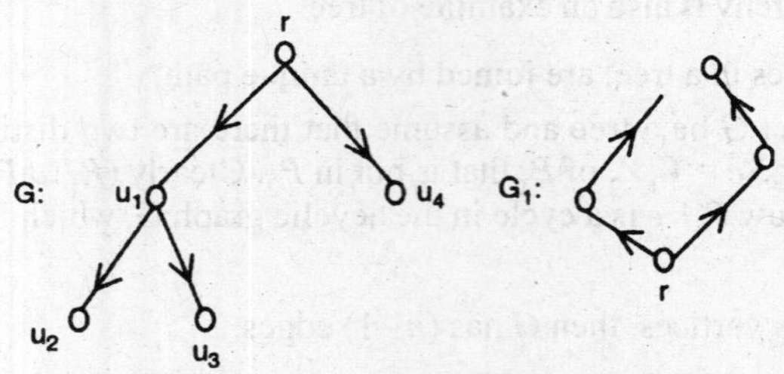

If v is a vertex in a tree, the subtree with v as its root is the subgraph of the tree consisting of v and its children and all edges incident to these children.

Example

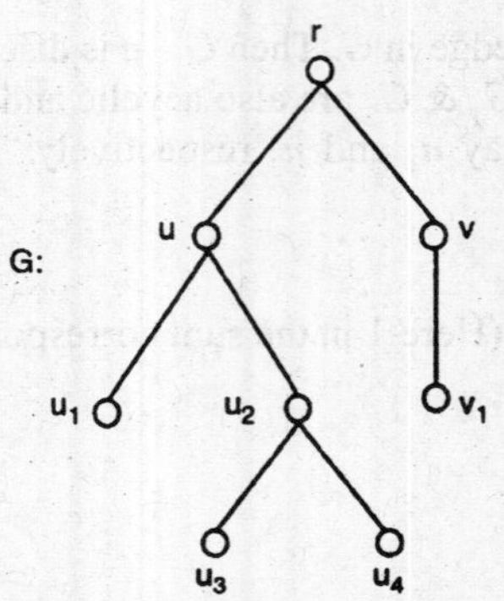

Rooted tree T

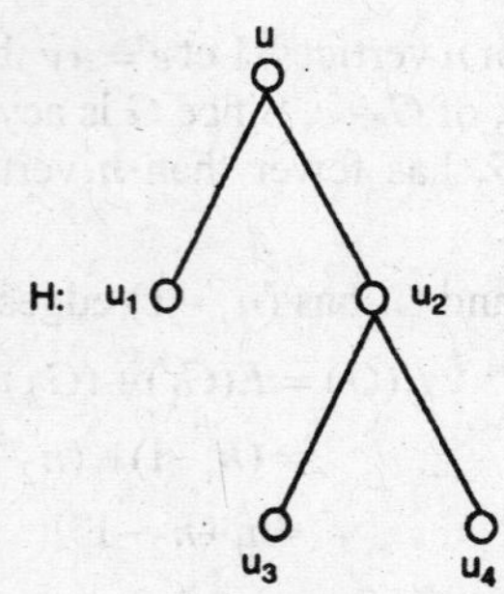

Subtree of T with its root u.

k-ary tree A rooted tree is called a k-ary tree if every internal vertex has, not more than k-children. The tree is called a full k-ary tree if every internal vertex has, exactly k-children. A k-ary tree with $k = 2$ is called a binary tree.

Example

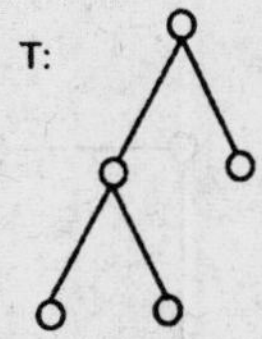

T, 2-ary tree

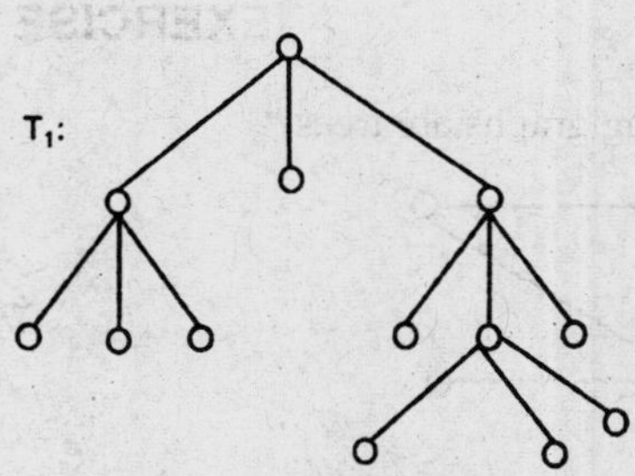

T_1, 3-ary tree

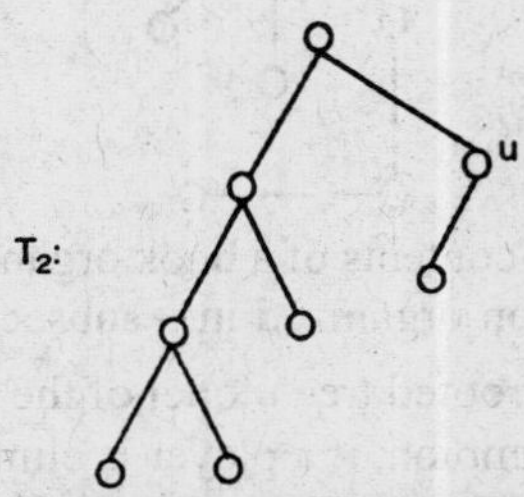

T_2, not a 2-ary tree. (vertex u has only one child, whereas all the other vertices have two children)

A tree T is called as a binary tree if there is only one vertex with degree 2 and the remaining vertices are of degree 1 or 3.

Example 7.8 Prove that a full k-ary tree with i-internal vertices contains k_{i+1} vertices.

Solution In a full k-ary tree, every internal vertex has k children and hence a full k-ary tree with i-internal vertices can have ki vertices. If we include the root, the tree has $ki + 1$ vertices. By looking at the fall k-ary tree, we can observe the following.

i. n vertices has $i = (n-1)/k$ internal vertices and $p = [(k-1)n + 1]/k$ leaves.

ii. i internal vertices has $n = ki + 1$ vertices and $p = (m-1)i + 1$ leaves.

iii. p leaves has $n = (kp-1)/(k-1)$ vertices and $i = (p-1)/(k-1)$ internal vertices.

Level and height in a rooted tree The level of a vertex v in a rooted tree is the length of the path from the root to this vertex. The height of a rooted tree is the length of the longest path from the root to any vertex.

Example

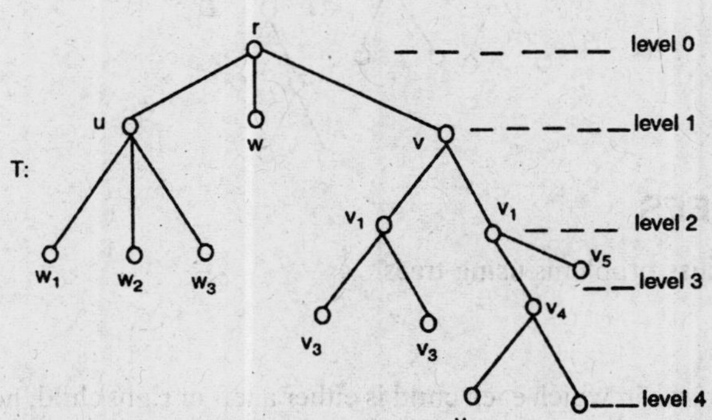

A rooted tree T with its levels. Height of T is 4.

Balanced tree A rooted k-ary tree of height h is balanced if all the leaves are at level h or $(h-1)$.

EXERCISE IV

1. Which of the following graphs are trees?

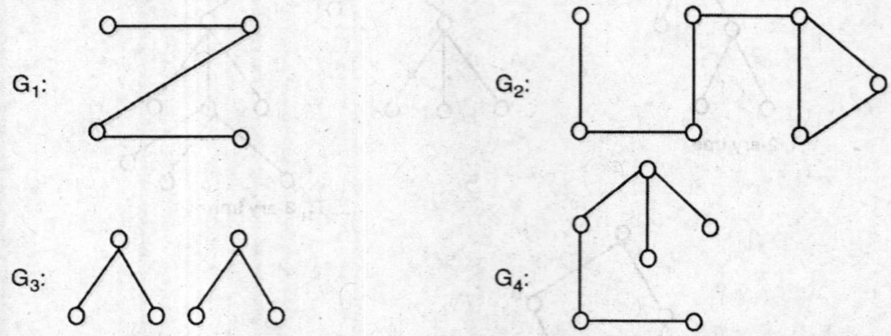

2. Explain how a tree can be used to represent the table of contents of a book organized into chapters, where each chapter is organized into sections and each section organized into subsections.
3. Suppose 1000 people enter a chess tournament. Use a rooted tree model of the tournament to determine how many games are to be played to determine the champion, if a player is eliminated after one loss and games are played until only one enterant has not lost. (Assume there are no ties).
4. Prove that a tree has either one or two centres. (centre of tree is the mini $\{e(v): v \in v(T)\}$.)
5. Which complete bipartite graphs $k_{m,n}$ are trees ? (*m* and *n* are positive integers)
6. How many vertices does a full 5-ary tree with 100 internal vertices have?
7. Label the tree, and answer the following.
   i. Identify the root.
   ii. Identify the children, siblings, of any parent
   iii. Draw a subtree.

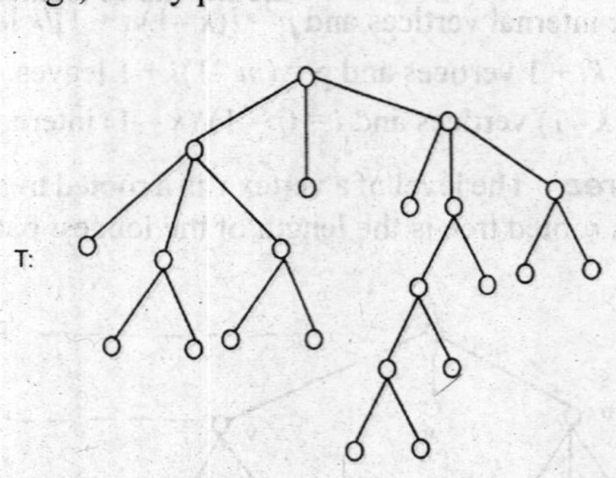

APPLICATION OF TREES

In this section, we shall discuss problems using trees.

Binary Search Trees

Binary search tree is a binary tree in which each child is either a left or right child; no vertex has more than one left child and one right child, and the data are associated with vertices.

Example 7.9 Build a binary search tree for the words banana, peach, apple, pear, coconut, mango and papaya using the alphabetical order.

Solution

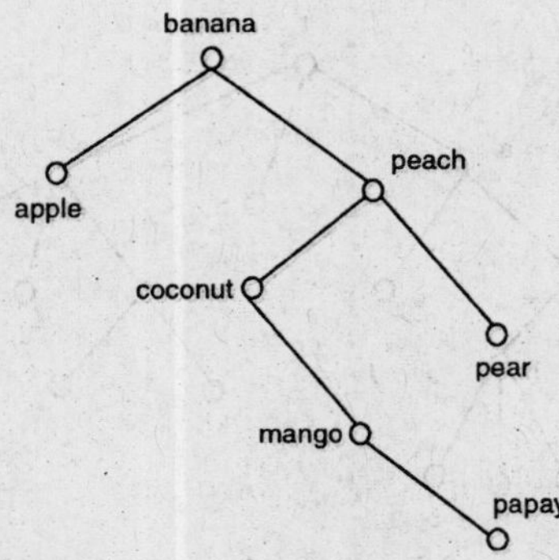

For if apple < peach, coconut < pear;

Further mango is the right child of coconut and papaya is the right child of mango.

Decision Trees

A rooted tree in which each internal vertex is assigned to a decision with a sub-tree at the vertices, then each possible outcome of the decision is called a decision tree.

Traversal of a Tree

A systematic method for visiting every vertex of an ordered rooted tree is called as a 'Traversal algorithm'.

Pre-order Let T be an ordered rooted tree with root r. Suppose T has one and only vertex say r, then r is the pre- order traversal of T. Suppose that T_1, T_2, ..., T_k are the subtrees at r from left to right in T, then pre-order traversal begins by visiting r. It continues by traversing T_1 in pre-order, then T_2 in pre-order and so on, until T_k is reached.

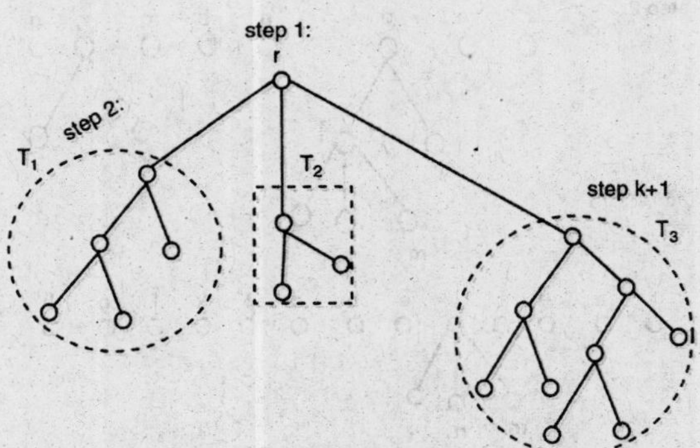

Pre-order traversal

Step 1 : Visit the root r

Step 2 : Visit T_1 in pre-order

Step 3 : Visit T_2 in pre-order

Step k+1 : Visit T_k in pre-order

Let us try to understand the above with an example.

T:

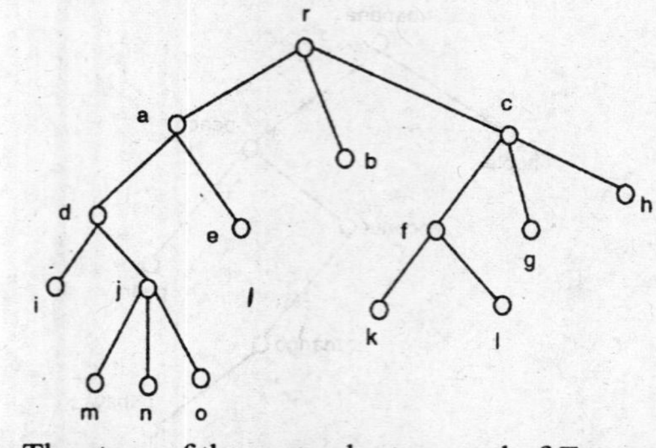

Let *T* be an ordered root tree. The steps of the pre-order traversal of *T* are as follows:

We traverse *T* in pre-order by listing the root *r*, followed by the pre-order list of subtree with root *a*, the pre-order list of subtree with root *b*, and the pre-order list of subtree with root *c*.

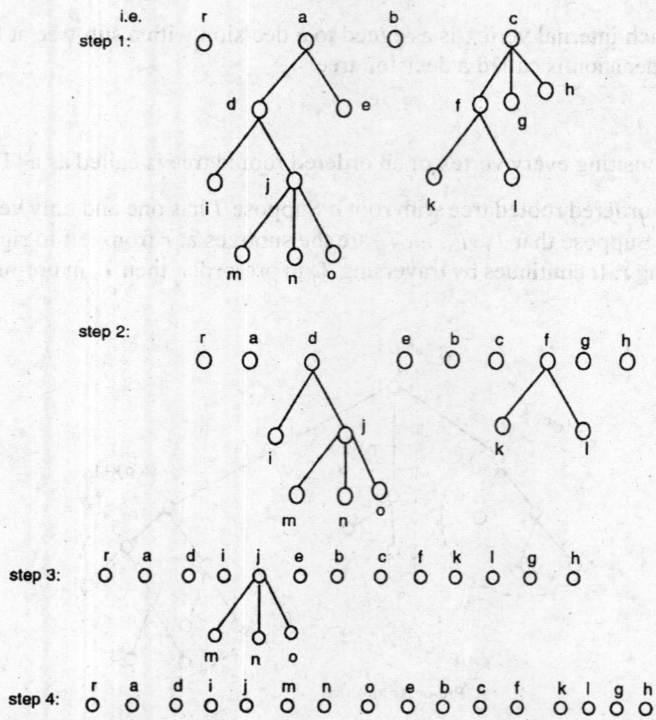

Algorithm: Pre-order Traversal

Step 1 : Visit root *r* and then list *r*.

Step 2 : For each child of *r* from left to right, list the root of first subtree then next sub-tree and so on until we complete listing the roots of subtrees at level 1.

Step 3 : Repeat step 2, until we arrive at the leaves of the given tree.

Step 4 : Stop.

In-order traversal Let T be an ordered rooted tree with its root at vertex r. Suppose T consists only root r, then r is the in-order traversal of T. If not, i.e., suppose T has subtrees $T_1, T_2, ..., T_k$ at r from left to right. The in-order traversal begins by traversing T_1 in-order, then visiting r. It continues by traversing T_2 in-order, then T_s inorder and so on and finally T_k in-order.

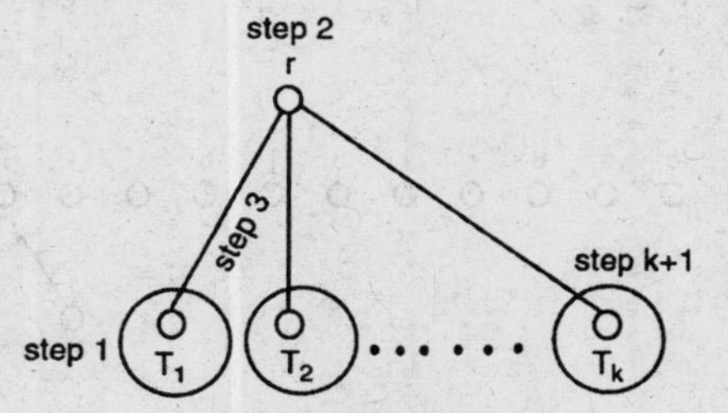

In-order traversal

Step 1 : Visit T_1 in-order

Step 2 : Visit root

Step 3 : Visit T_2 in-order

Step k+1 : Visit T_k in-order

Example 7.10 Determine the order in which the vertices of the following rooted tree is visited using an in-order traversal

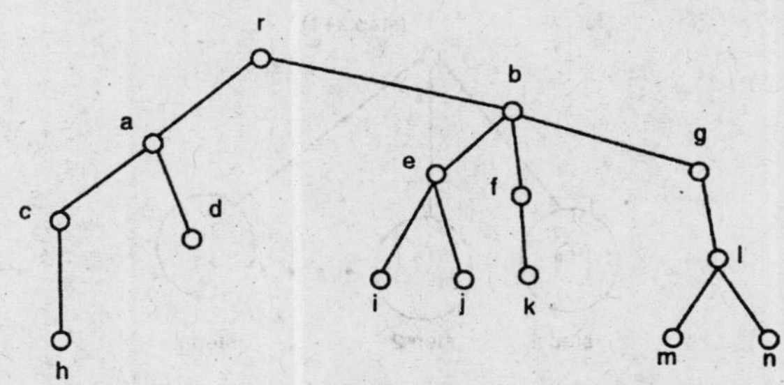

Solution The in-order traversal begins with an in-order traversal of the subtree with root at a, followed by the root r, and the in-order listing of the subtree with root b.

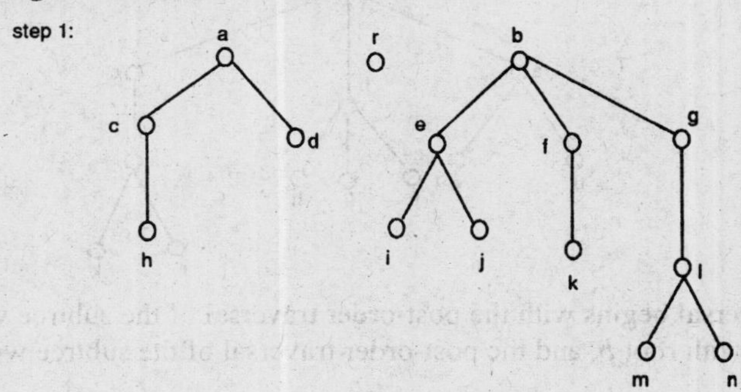

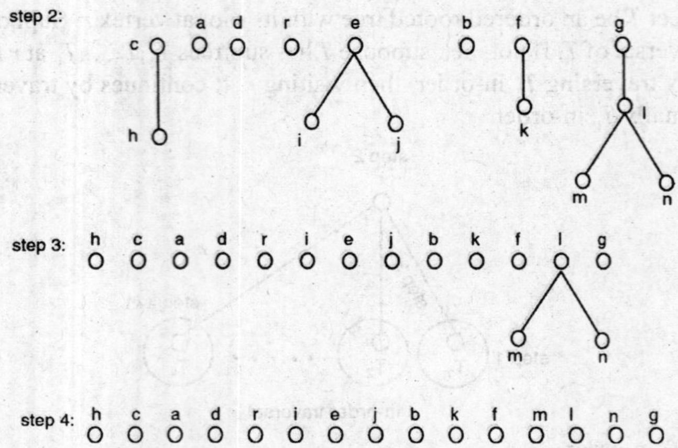

Definition (Post-order traversal)

Let T be an ordered rooted tree with root r. If T has only one vertex r, then r is the post-order traversal of T. But if T has subtrees $T_1, T_2, \ldots, T_k$ at r from left to right, the post-order traversal begins by traversing T_1 in post-order, then T_2 in post-order and so on until T_k and ends by visiting r.

Example

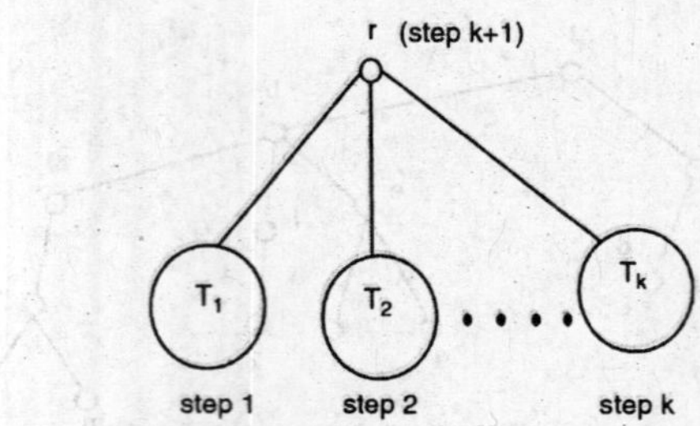

Example

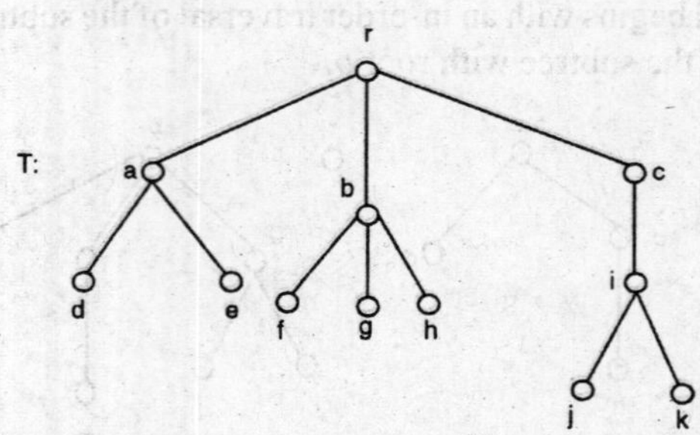

The post-order traversal begins with the post-order traversal of the subtree with root a, the post-order traversal of the subtree with root b, and the post-order traversal of the subtree with root c, followed by the root r.

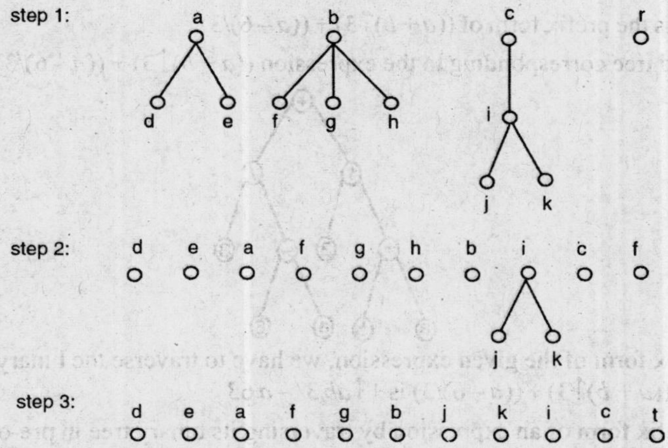

Infix, Prefix, and Postfix Notation

We can represent any expression (like arithmetic, compound proposition) using ordered rooted trees. An ordered rooted tree can be used to represent expressions, where the internal vertices represent operations, the leaves represent the variables or numerals.

Example 7.11 What is the ordered rooted tree that represents the expression $((a+b)\uparrow 3)+((a-6)/3)$

Solution First construct a subtree for $a + b$. Then this tree is included as a part of the next subtree of $((a + b) \uparrow 3)$. Similarly a subtree is constructed for $(a-6)$ then this tree is included as a part of the next subtree of $(a-6)/3$. Finally the subtrees $((a+b)\uparrow 3)$ and $(a-6)/3$ are combined to form the required tree corresponding to the given expression.

Ordered rooted tree corresponding to the expression $((a+b)\uparrow 3)$ $+((a-6)/3)$

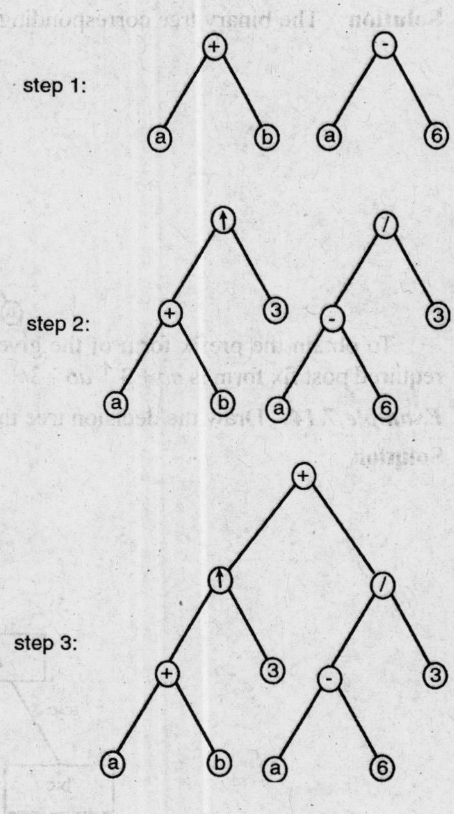

An in-order traversal of the binary tree representing an expression, produces the original expression with the elements and operations in the same order as they originally appeared (except unary operator).

If we use parenthesis, whenever we encounter an operation there will be no ambiguity. Such fully parenthesised expression is said to be infix form.

To get the *prefix form* of an expression, we traverse its rooted tree in pre-order.

Expressions written in prefix form are called Polish notation.

Example 7.12 What is the prefix form of $((a+b)\uparrow 3)+((a-6)/3)$

The ordered rooted tree corresponding to the expression $((a+b)\uparrow 3)+((a-6)/3)$ is

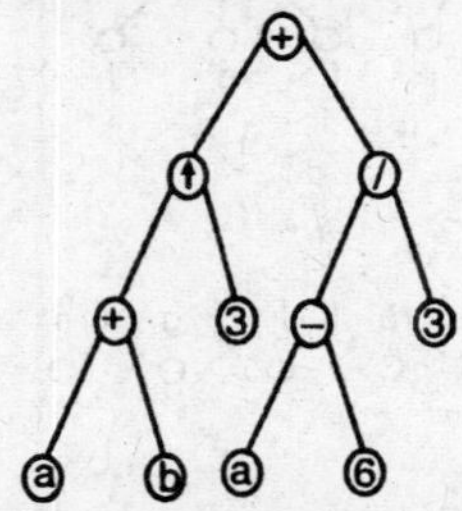

To obtain the prefix form of the given expression, we have to traverse the binary tree in pre-order. Prefix form of the expression $((a+b)\uparrow 3)+((a-6)/3)$ is $+\uparrow ab\,3/-a\,63$

We obtain the postfix form of an expression by traversing its binary tree in pre-order.

Example 7.13 What is the postfix form of $((a+b)\uparrow 3)+((a-6)/3)$?

Solution The binary tree corresponding to the expression is given as

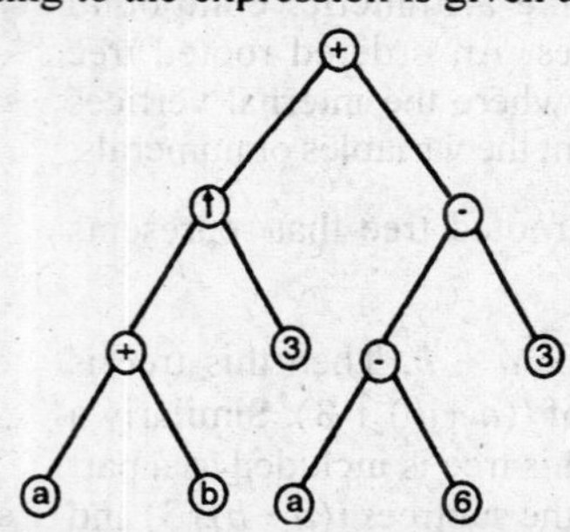

To obtain the prefix form of the given expression, we have to traverse its binary tree in pre-order. The required post fix form is $ab+3\uparrow a6-3/+$

Example 7.14 Draw the decision tree that orders the elements of the list a,b,c.

Solution

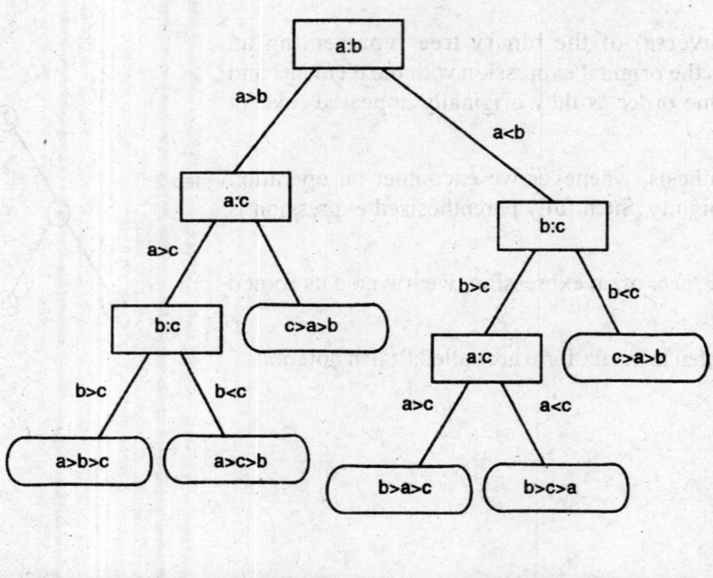

TREES AND SORTING

To sort a list of elements there are several methods. Here we shall see how trees are helpful in merge sort.

In general, a merge sort proceeds by iteratively splitting lists into two sublists of equal size (nearly) until each sublist consists of one element.

This succession of sublists can be represented by a balanced binary tree. The procedure continues by succesively merging pairs of lists (where both lists are in increasing order) into a larger list with elements in increasing order), until the original list is put into increasing order. The succession in a merged list can be represented by a balanced binary tree.

Example 7.15 Draw the recursive tree for merge sort of the list 9, 7, 11, 4, 5, 3, 6, 8, 12, 10.

Solution The list of elements can be represented as

$$a[0]=9;\ a[1]=7;\ a[2]=11;\ a[3]=4;\ a[4]=5;\ a[5]=3;\ a[6]=6;\ a[7]=8;\ a[8]=12;\ a[9]=10.$$

Let us denote 0-9 as the position of the elements. Given list is[9,7,11,4,5,3,6,8,12,10]

As a first step this list is splitted into two sublists of size 0-4 and 5-9 respectively. Then these two sublists are further splitted into two sublists until each sublist consist of one element. The required tree is given below.

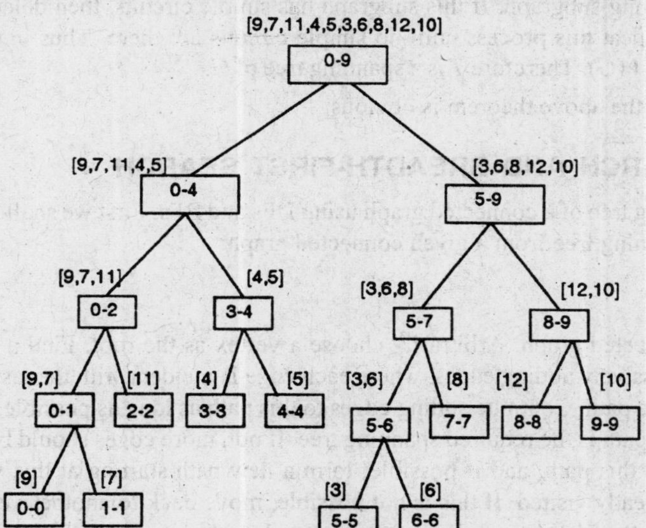

We shall study further applications of tree in the chapters on formal languages and automation theory such as parsing tree etc.

EXERCISE V

1. Using alphabetical order, build a binary search tree for the words in the sentence, "The quick brown fox jumps over the lazy dog"
2. Represent the expression $((a+5)\uparrow 4)+(b-(5+a))-4$ using a binary tree.
3. Write the expression in for the above problem using.
   (i) postfix notation (ii) prefix notation and (iii) infix notation.
4. Represent $(A\backslash B) \cup (B-A)$ using an ordered rooted tree.
5. Draw the ordered rooted tree corresponding to the expression + * + – 73213 and find its value.

SPANNING TREES

In this section, we shall study the spanning acyclic subgraph of a connected subgraph, and its optimality.

Let G be a simple connected graph. A spanning tree of G is a subgraph of G i.e. a tree containing every vertex of G.

Example

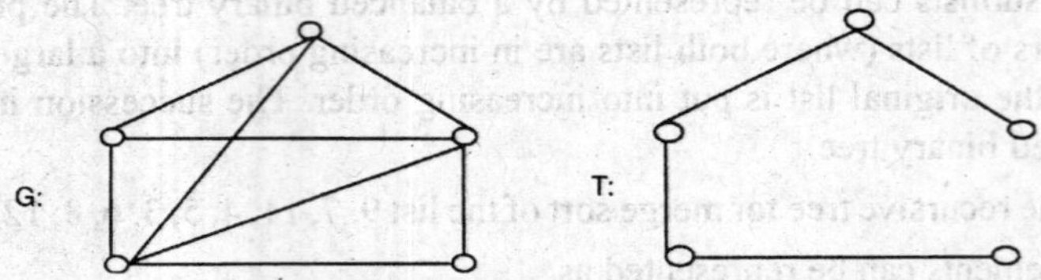

Simple graph G and its spanning tree T

Theorem A simple graph is connected if there exists atleast one spanning tree.

Proof Let G be a simple connected graph.Suppose G has no circuits then G itself is a spanning tree. Suppose G has a simple circuit. By deleting an edge from one of these simple circuits, the resulting subgraph is still connected if it is a spanning subgraph. If this subgraph has simple circuits, then delete an edge from one of these simple circuits. Repeat this process until no simple circuits are there. Thus in this manner a tree T is obtained in which $V(T) = V(G)$. Therefore T is a spanning tree of G.

Note: The converse of the above theorem is obvious.

DEPTH-FIRST SEARCH AND BREADTH-FIRST SEARCH

We can build the spanning tree of a connected graph using DFs and BFs. First we shall see how DFs are useful in construction of a spanning tree from a given connected graph.

Depth-First Search

Let G be the given connected graph. Arbitrarily choose a vertex as the root. Find a path starting from this choosen vertex by successively adding edges, where each edge is incident with the last vertex in the path and a vertex not already in the path. Continue adding edges to this path as long as possible. If this path consists of all the vertices of G, this path is the required spanning tree. If not, more edges should be added. Move back to the next to last vertex in this path, and if possible, form a new path starting at this vertex passing through vertices that were not already visited. If this is not possible, move back to another vertex in this path (i.e. 2 vertices back from the last) and try again. Repeat this procedure, beginning at the last vertex visited, moving back up the path one vertex at a time, forming new long paths until no more edges can be added. This process ends with a spanning tree.

When this procedure returns to vertices previously visited, it is also called as backtracking.

Example 7.16 Construct a spanning tree for the following graph G.

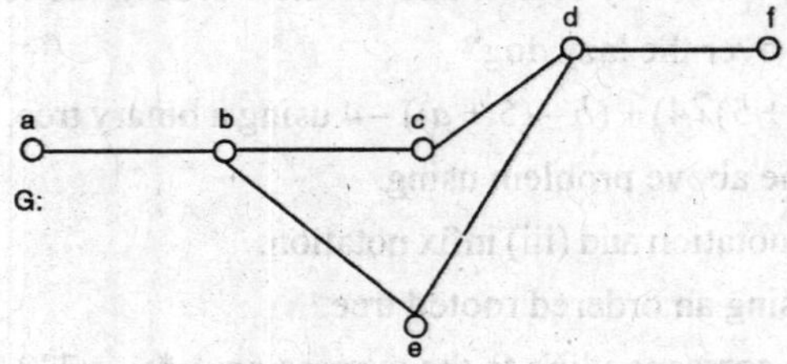

Solution First we choose arbitrarily a vertex say e as the root. Form a path at e. i.e. $c\ d\ f$ is the path. Backtrack to d. Form a path beginning at d in such a way that it has to visit the vertices which where not visited in the previous path., $d\ e\ b\ a$. Since all the vertices of G are visited, this procedure gives the spanning tree T.

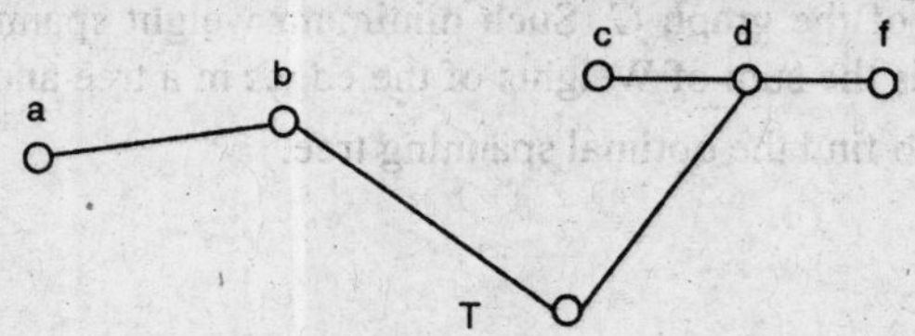

Breadth-first Search

First choose a vertex arbitrarily as the root. Add the edges of G which are incident with this vertex. The new vertices added at this stage becomes level 1 in the spanning tree. Order these vertices arbitrarily. Next, for each vertex at level 1 visited in order, add each edge incident to this vertex to the tree as long as it does not-create a simple circuit. Order the children of each vertex at level 1 arbitrarily. This produces the vertices at level 2 in the tree. Continue in this manner until all the vertices of G have been added. Ultimately we end with a spanning tree.

Example 7.17 Construct a spanning tree of the following graph G.

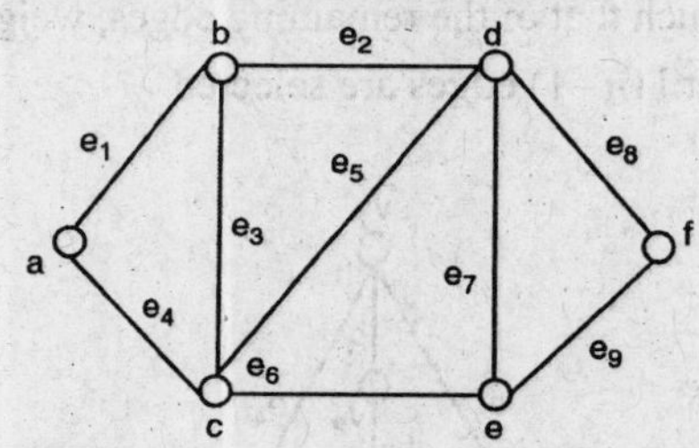

Solution First choose a vertex say d (arbitrarily) as the root. Add the edges incident to this vertex d. Hence the edges e_2, e_5, e_7, e_8 are incident with the vertex d. These vertices create level 1.

i.e.,

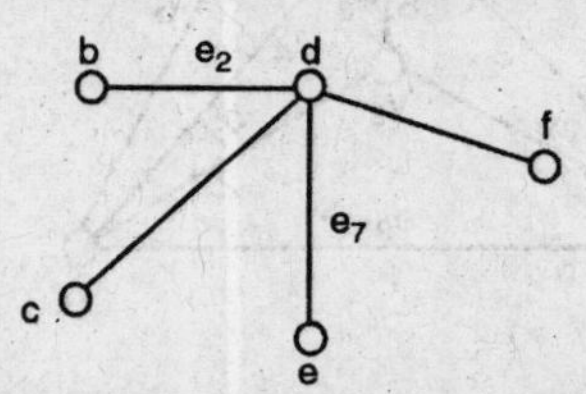

Now add the edges which are incident to the vertices b, c, e, f in such a way that the resulting graph does not contain any circuit.

T:

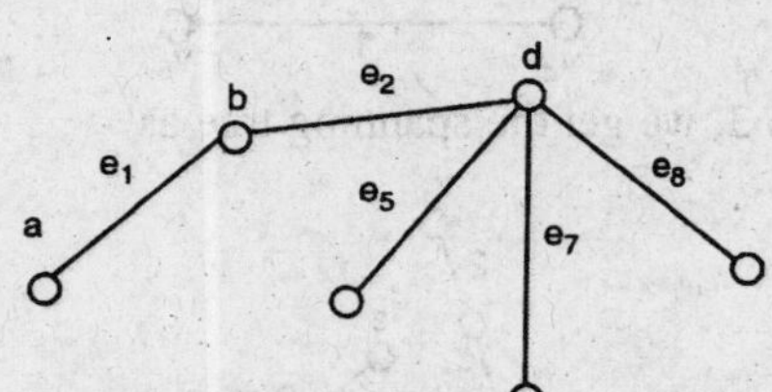

Thus at this level itself we have got the spanning tree T.

Note: If the given graph is directed graph then we construct the underlying undirected graph and apply DFs or BFs to obtain a spanning graph.

OPTIMAL SPANNING GRAPH

Let G be a weighted graph. (Every edge of the graph is associated with a real number). We have to find the minimum weight spanning tree of the graph G. Such minimum weight spanning tree is called an Optimal spanning tree. [weight of a tree is the sum of Weights of the edges in a tree and is denoted by $wt(T)$]

There are three algorithms to find the optimal spanning tree.

i. Kruskal's algorithm

ii. Prim's algorithm

iii. Boruvke's algorithm

Kruskal's Algorithm

Let G be a connected graph on n vertices.

Step 1 : Arrange the edges in ascending order according to their weights. Choose the minimum weight edge say e_1.

Step 2 : Having selected $e_1, e_2, \ldots, e_k$ in such a way that the subgraph formed by these edges $<e_1, e_2, \ldots, e_k>$ is acyclic, choose e_{k+1} such that of the remaining edges, weight of e_{k+1} is minimum.

Step 3 : Repeat steps 1 and 2 until $(n-1)$ edges are selected.

Example

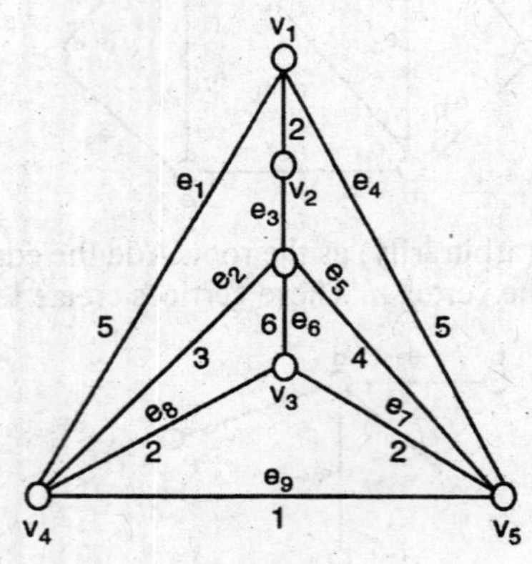

Step1 : $e_9, e_7, e_8, e_3, e_2, e_5, e_4, e_1, e_6$

Among these equations e_9 has the minimum weight 1.

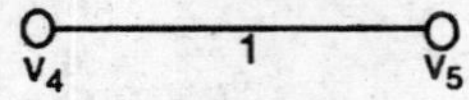

After applying step 2 and step 3, we get the spanning tree as

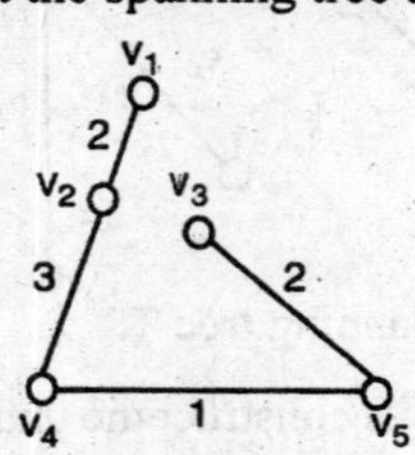

Weight of the optimal spanning tree is $2 + 3 + 1 + 2 = 8$

Prim's Algorithm

Let G be a connected graph.

Step 1 : Choose arbitrarily a vertex say v_1 and an edge e_1 with minimum weight among the edges incident with v_1.

Step 2 : Having selected the vertices $v_1, v_2, .., v_k$ and the edges $e_1, e_2, .., e_k$ choose the edge e_{k+1} as follows. e_{k+1} is incident with any one of the vertices $\{v_1, v_2, .., v_k\}$ and incident with $v(G) - \{v_1, v_2, .., v_k\}$. Moreover the subgraph formed with $v_1, v_2, ..., v_k, v_{k+1}$ and the edges $e_1, e_2, .. e_k, e_{k+1}$ is acyclic and of the remaining edges e_{k+1} has minimum weight.

Step 3 : Repeat steps 1 and 2 till $(n-1)$ edges are arrived at.

Example

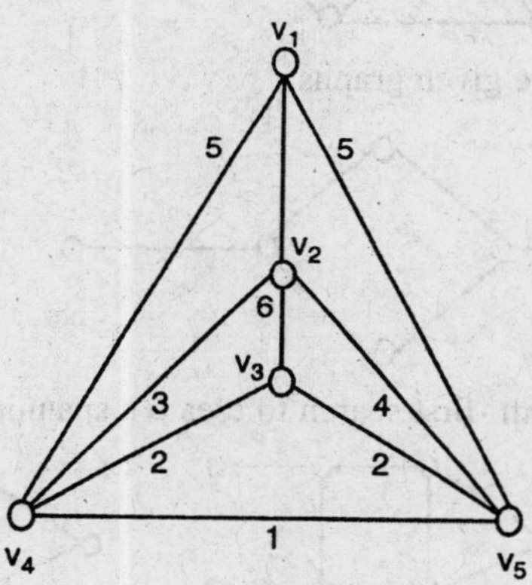

Step 1 : Choose arbitrarily vertex v_3 and apply step 2 and step 3. Now we get the spanning tree.

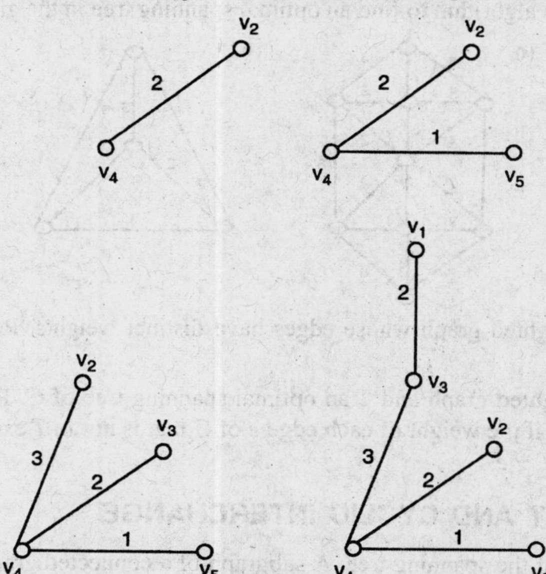

Weight of the spanning tree is 8.

EXERCISE VI

1. How many edges must be removed from a connected graph with n vertices and m edges to produce a spanning tree.
2. Find a spanning tree for the graphs shown by removing edges in a simple circuit.

(a)

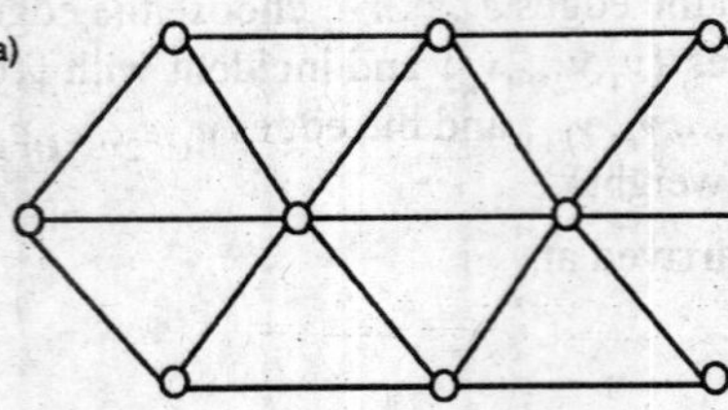

(b)

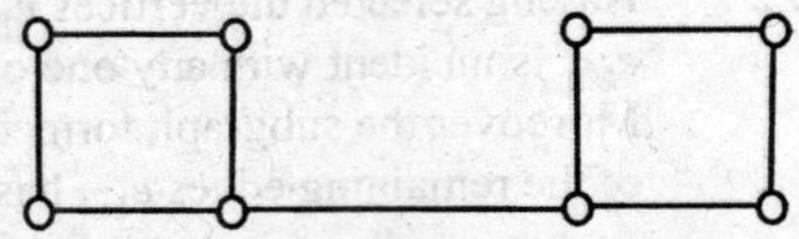

3. Draw all the spanning trees of the given graphs.

(a)

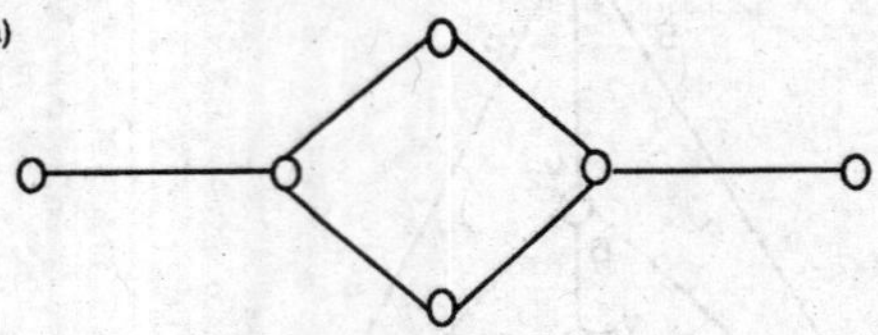

(b)

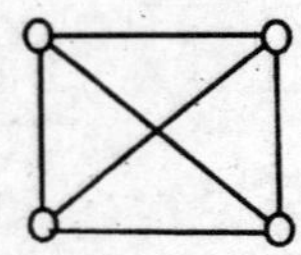

4. Use depth-first search and breadth -first search to create a spanning tree of the given graphs.

(a)

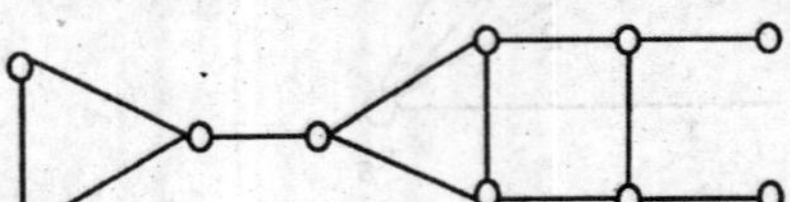

(b)

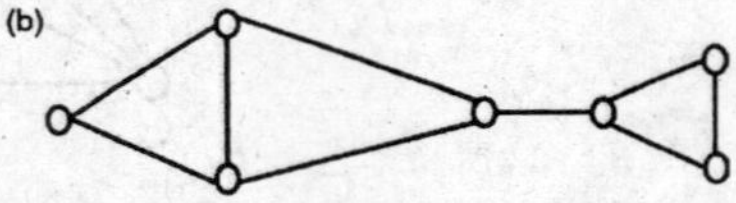

5. Apply Kruskal's and Prim's algorithm to find an optimal spanning tree in the given weighted graphs.

(a)

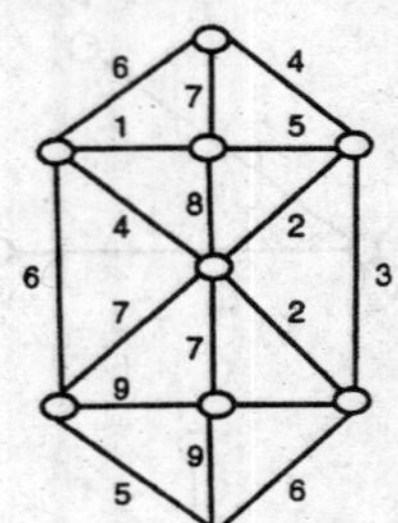

(b)

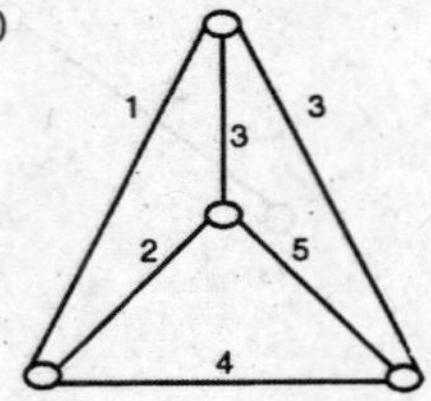

6. Let G be a connected weighted graph whose edges have distinct weights. Prove that G has a unique optimal spanning tree.
7. Let G be a connected weighted graph and T an optimal spanning tree of G. Prove that T is the unique optimal spanning tree of G if the weight of each edge e of G that is not in T exceeds the weight of every other edge on the cycle in $T + e$.

FUNDAMENTAL CIRCUIT AND CYCLIC INTERCHANGE

Let us recollect the definition of the spanning tree. A subgraph of a connected graph is called as a spanning tree if the subgraph is a tree containing all the vertices of the graph.

A connected graph might contain more than one spanning tree. Consider the following graphs.

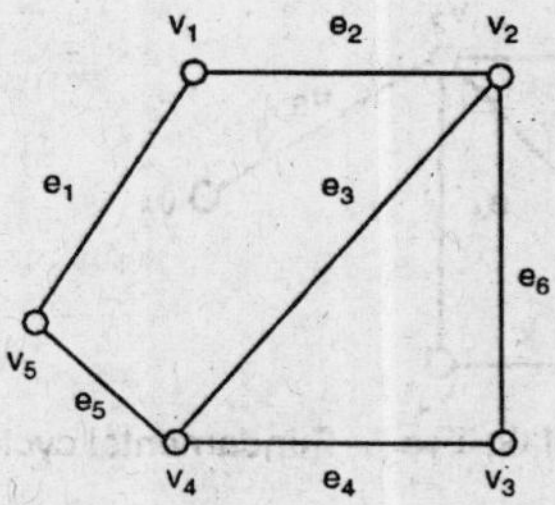

G:

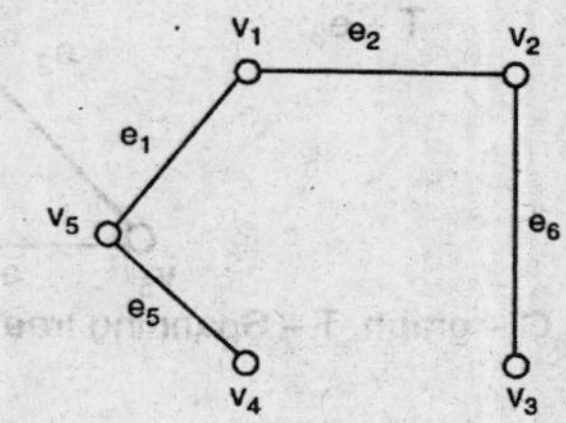

T1:

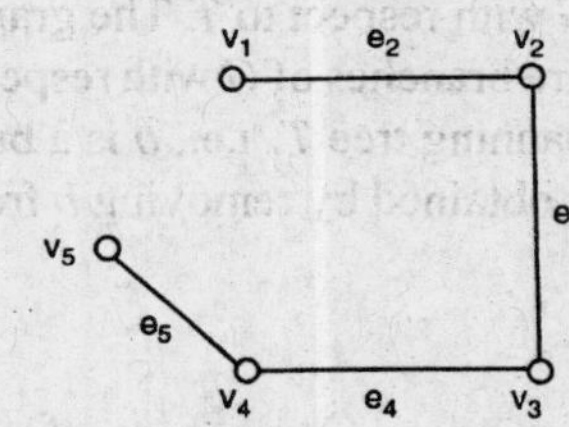

T2:

In T_1 the edges e_1, e_2, e_5, e_6 are present, whereas in T_2 the edges e_2, e_4, e_5, e_6 are present.

The edges of G which are present in a spanning tree T are called as the branches of G with respect to T. The edges of G which are not present in its spanning tree T are called the *chords* of G with respect to T.

In the above example, the branches of G are e_1, e_2, e_5, e_6 with respect to T_1: the branches of G are e_2, e_4, e_5, e_6, with respect to T_2.

Note: Let G be a connected graph on n vertices; e-edges and T be one of its spanning tree. Since T is a tree on n vertices, it has $(n-1)$ edges. i.e., the number of branches of G with respect to T is $(n-1)$; The number of chords of G with respect to T is $e-(n-1)$. Often the number of branches of G is called as rank of G and is denoted by $r(G)$; the number of chords of G is called as the nullity of G, denoted by $\mu(G)$. In general, for a connected graph on n-vertices and e-edges, $r(G)$, the rank of G is $(n-1)$; $\mu(G)$, the nullity of G is $e - n + 1$.

Fundamental Circuit

Let T be the spanning tree of a connected graph G, and e be a chord of G with respect to T. Since the spanning tree T is minimally acyclic, the graph $T+e$ contains a unique cycle. This cycle is called a fundamental cycle in G with respect to T.

Every chord of G gives rise to a fundamental cycle. Therefore, the number of fundamental cycles possible for a connected graph is atmost $\mu(G)$.

Example

G:

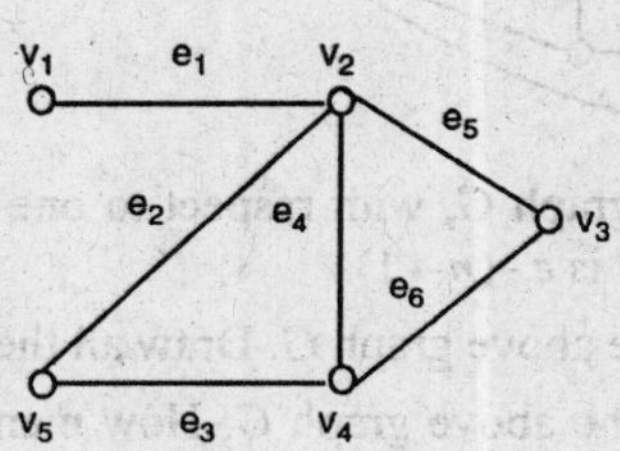

T:

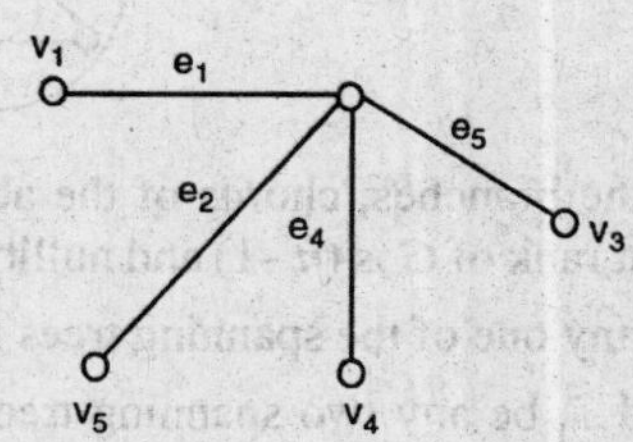

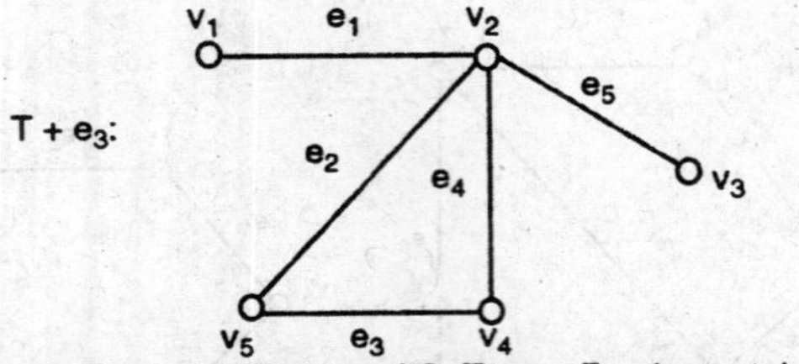

G – graph, T – Spanning tree of G, T+e_3 – Fundamental cycle

Cyclic Interchange

Let T be a spanning tree of G and e be a chord of G with respect to T. The graph $T+e$ is a fundamental circuit. In this circuit other than edge e, all the other edges are branches of G with respect to T. On removal of any of the branches from the fundamental circuit, we get a spanning tree T_1 i.e., b is a branch in the fundamental circuit (with respect to a chord e), then spanning tree T_1 is obtained by removing b from $T+e$ i.e, $T_1 = T + e - b$. This process is called cyclic interchange.

Example

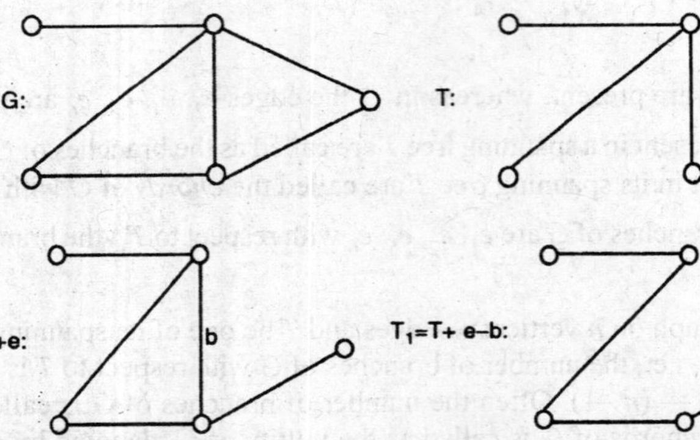

G - connected graph, T - spanning tree

T+e - fundamental circuit; T_1 - spanning tree obtained by cyclic interchange.

EXERCISE VII

1. a. Draw all possible spanning trees for the following graph G.

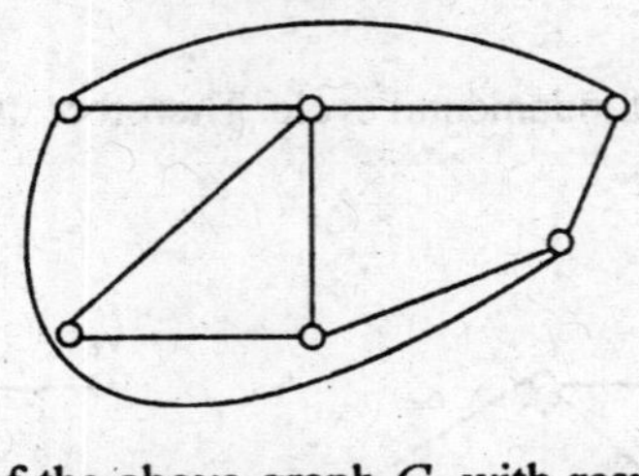

b. Identify the branches, chords of the above graph G, with respect to one of its spanning trees and verify that rank of G is $(n-1)$ and nullity of G is $e-(n-1)$.

c. Let T be any one of the spanning trees for the above graph G. Draw all the fundamental cycles.

d. Let T and T_1 be any two spanning trees of the above graph G. How many cyclic-interchanges are required to obtain the spanning tree T_1 from T.

CUT-SETS AND CONNECTIVITY OF GRAPHS

Let G be a connected graph. Let us recollect the definition of cut-edge (bridge) and cut-vertex. If G contains an edge e such that $G–e$ is disconnected, then e is a bridge of G. Further if G contains a vertex v such that $G–v$ is disconnected, then v is a cut-vertex of G.

Definition

Edge cut-set A subset S of the edge set of a connected graph G is called an edge cut-set or cut-set of G if

i. $G - S$ is disconnected

ii. $G - S_1$ is connected for every proper subset S_1 of S

Vertex cut-set A subset u of the vertex set of G is called a vertex cut-set if

i. $G - u$ is disconnected

ii. $G - u_1$ is connected for every proper subset u_1 of u

Example

G:

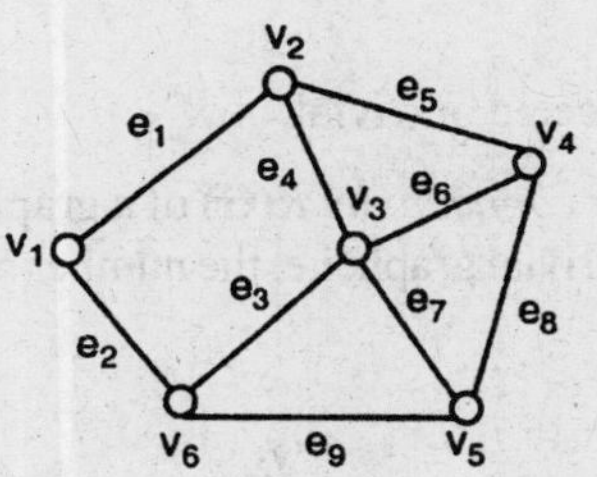

a) $S = \{e_1, e_4, e_6, e_8\}$ is a cut-set

G-S:

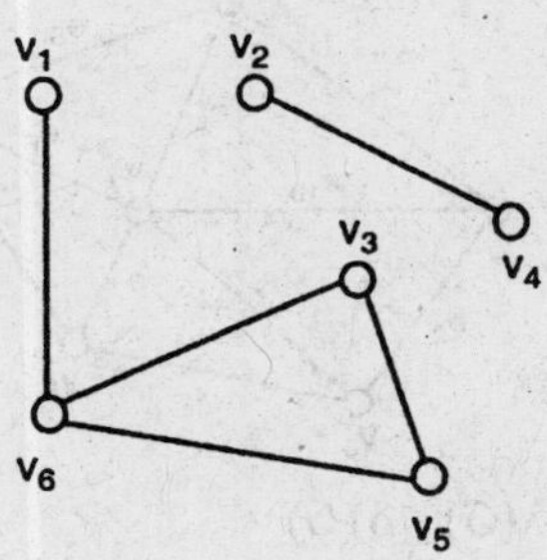

b) $u = \{v_1, v_3, v_5\}$ is a vertex-cut-set

G-U:

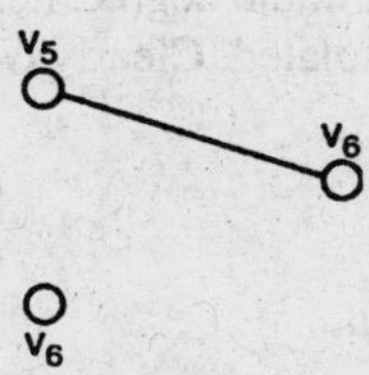

Note: For a connected graph, there may be more than one cut-set.

For example, consider the above graph G. Some cut-sets of G are

$S_1 = \{e_1, e_4, e_6, e_8\}$

$S_2 = \{e_1, e_2\}$, $S_3 = \{e_1, e_3, e_9\}$

From the above note, we are forced to introduce two more parameters for graphs viz edge-connectivity $\lambda(G)$ and vertex connectivity $k(G)$.

Edge connectivity The edge connectivity $\lambda(G)$ of a graph is the minimum cardinality of a set S of edges of G such that $G–S$ is disconnected i.e. the edge (line) connectivity of a connected graph is the number of edges in a minimum cut-set in the graph.

Example

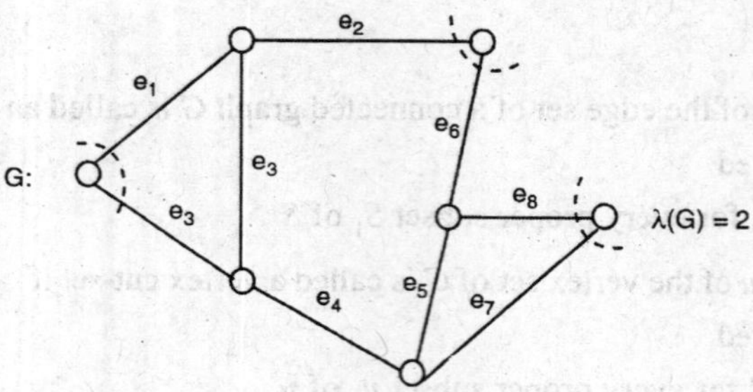

Note:

1 If G is a tree then $\lambda(G) = 1$.

2 G has $\lambda(G) = 0$ iff G is disconnected or trivial.

Vertex connectivity The *vertex connectivity* $K(G)$ of a graph G is the minimum number of vertices whose deletion makes G a disconnected or trivial graph. i.e. the number of vertices in a minimum *vertex* cut is called the connectivity of the graph.

Example

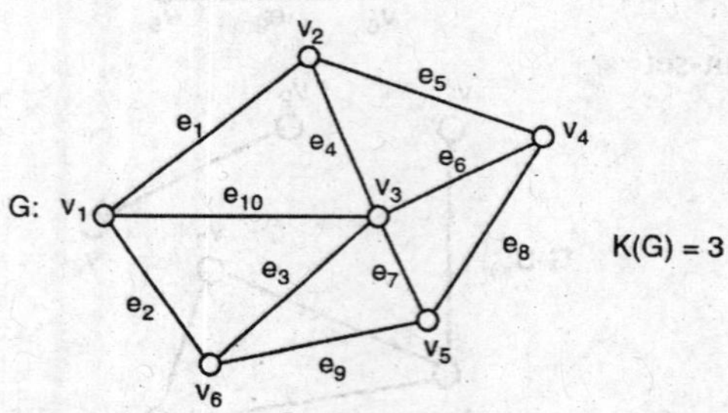

Result For every graph G, $K(G) \leq \lambda(G) \leq \delta(G)$

Proof Let v be a vertex of G with minimum degree. i.e., $d(v) = \delta(G)$. Removing $\delta(G)$ edges of G incident with v produces a graph G_1, in which v is isolated. Clearly G is disconnected or trivial.

$$\therefore \lambda(G) \leq \delta(G) \qquad (1)$$

Claim $K(G) \leq \delta(G)$

If $\delta(G) = 0$ then G is disconnected.

$\therefore K(G) = 0$.

If $\delta(G) = 1$ then G is a connected graph containing a cut-edge (bridge). Therefore either G is isomorphic to K_2 or G is a connected graph having atleast one cut-vertex.

$\therefore$ In both cases, $K(G) = 1$.

Now let us assume that $\lambda(G) \geq 2$. Let S be a cutset of G with $\lambda(G)$ edges and $e = xy$ be an edge in S. If the edges of $S–\{e\}$ are deleted from G, the resulting subgraph H_1 is connected and contains e as a cut-edge. Now select an incident vertex different from x and y for each and every edge in $S-\{e\}$.Remove these vertices from H_1, the resulting subgraph H_2 is disconnected, then

$K(G) \leq \lambda(G)-1 < \lambda(G)$.

Suppose the subgraph H_2 is connected, then H_2 is isomorphic to K_2 or the subgraph H_2 has a cut-vertex [since H_2 is an induced subgraph of H_1]. In any case, there exists a vertex of H_2 whose removal results in a disconnected graph. Therefore

$$K(G) \leq \lambda(G) \qquad (2)$$

From (1) and (2), $K(G) \leq \lambda(G) \leq \delta(G)$

G:

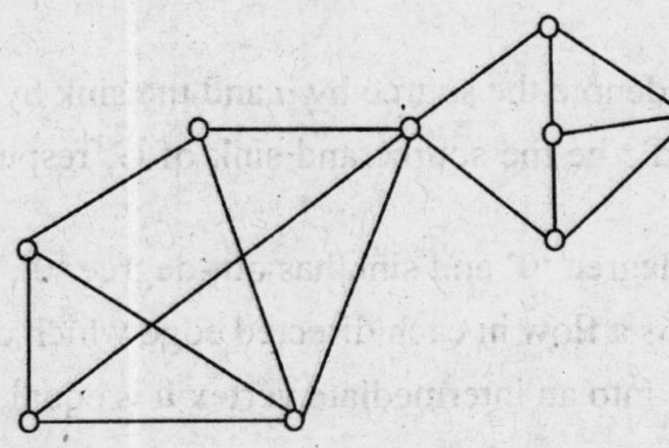

In the figure $K(G) = 1$; $\lambda(G) = 3$ and $\delta(G) = 3$.

n-edge connected A graph G is n-edge connected ($n \geq 1$) if $\lambda(G) \geq n$ and G is n-connected if $K(G) \geq n$.

EXERCISE VIII

1. Find $\lambda(K_{m,n})$ and $K(K_{m,n})$ where $1 \leq m \leq n$.
2. Prove that if G is a connected graph and S is a cut-set with $\lambda(G)$ edges, then $G-S$ contains exactly two components.
3. Prove or disprove
   a. If G is a graph having connectivity n, then $G-S$ is disconnected for every set S of n vertices of G.
   b. If G is a connected graph and u is a vertex cut-set with $K(G)$ vertices, then $G-u$ contains exactly two components.
   c. The connectivity of a complete graph on n vertices is $(n-1)$.
   d. $K(G) = 0 \Leftrightarrow G$ is disconnected or trival.
4. If G is a graph whose diameter is 2, then $\lambda(G) = \delta(G)$.
5. The connection number $c(G)$ of a graph on n vertices ($n \geq 2$) is the smallest integer k with $2 \leq k \leq n$ such that every induced with subgraph on K vertices in G is connected.

   Find the connection number of K_n, $K_{m,n}$ and a disconnected graph.

NETWORKS AND FLOWS

In this chapter we discuss networks model. Flow is a network and and is made of directed graphs. The network may be a transportation network through which commodities flow or a pipeline network through which oil flows or any number of other possibilities. In each case, the problem is to determine a maximal flow. Maximizing the flow in a network is a problem that belong is to graph theory and operations research.

Network Models

A transport network is a simple, weighted, directed graph G satisfying the following:

1. There are two distinguished vertices s and t of G, called the source and sink of G respectively, i.e., in G, only one vertex with no incoming edge is called as source, s and only one vertex with no outgoing edge is called as sink, t.
2. The weight C_{ij} of the directed edge (i,j) is a non-negative number. i.e, every edge is associated with non-negative real numbers called as capacity of that edge.
3. The underlying (undirected) graph is connected.

Note:

1. If G is a network, we will denote the source by a and the sink by z.
2. Let G be a network, a and z be the source and sink of G, respectively. The other vertices are called as intermediate vertices.
3. In general, source has in-degree '0' and sink has out-degree '0'.
4. A flow in a network assigns a flow in each directed edge which does not exceed the capacity of the edge.
5. It is assumed that the flow into an intermediate vertex u is equal to the flow out of u.

Flow

Let G be a transport network. Let C_{ij} denote the capacity of the directed edge (i,j). A flow F in G assigns each directed edge (i,j) a non-negative number F_{ij} such that

i. $F_{ij} \leq C_{ij}$

ii. For each intermediate vertex j,

$$\sum_i F_{ij} = \sum_i F_{ji}$$

(Here $\sum_i F_{ij}$ is the flow into j and $\sum_i F_{ji}$ is the flow out of j)

Value of the flow Let F be the flow in a network G. The value $\sum_i F_{ai} = \sum_i F_{iz}$ is called the value of the flow F.

Note: Whenever the flow along an edge is equal to the capacity of that edge, then that edge is said to be saturated. Otherwise it is unsaturated.

Example

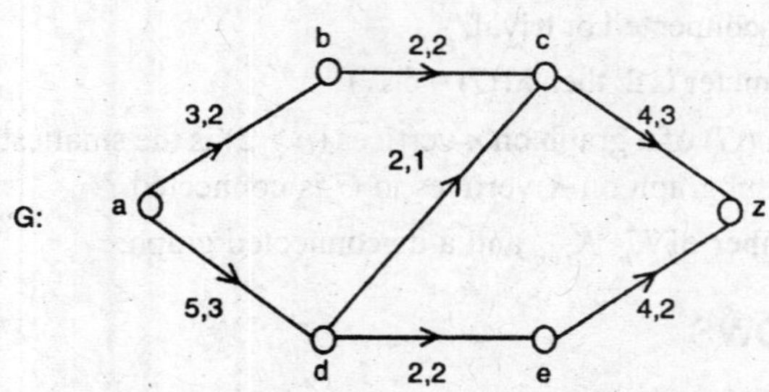

Here, the first quantity represents the capacity and the second quantity represents the flow.
The flow of the above network is 5.

Example 7.18 Find the flow for the following networks.

i.

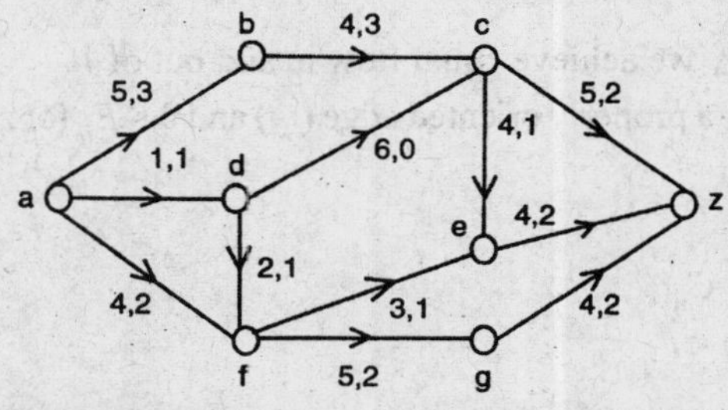

Solution The flow out of source a is $F_{ab} + F_{ad} + F_{af} = 3 + 1 + 2 = 6$.

The flow into the sink z is $F_{gz} + F_{ez} + F_{cz} = 2 + 2 + 2 = 6$.

All other intermediate vertices are given by, $\sum_i F_{ij} = \sum_i F_{ji}$

$\therefore$ Flow of the network is 6.

ii.

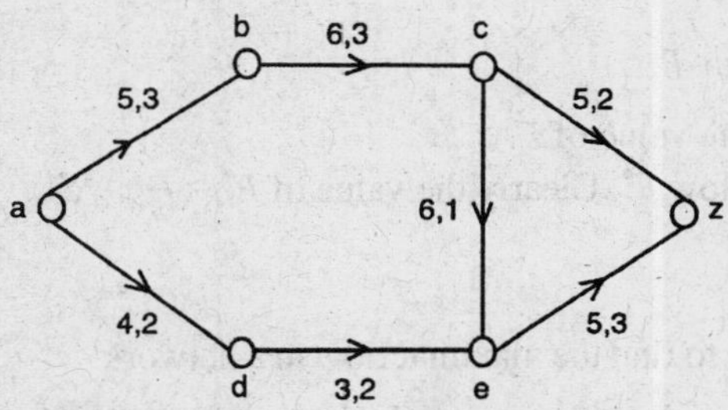

On similar grounds the flow is found to be 5.

Maximal flow A flow F in a network G is called a maximal flow if $|F| \geq |F'|$ for every flow F' in G.

Improve the value of flow F in a network Start with some initial flow and iteratively increase the value of flow until no improvement is possible in the given network.

We can take the initial flow to be the flow, when the flow in each edge is zero. To increase the value of a given flow, we have to determine a path from the source a to the sink z and increase the flow along this path. Let P be a path from a to z in the underlying graph of the network G. If an edge e in P is directed from v_i to v_{i+1}, we say e is properly oriented; otherwise e is improperly oriented.

If we can find a path P from a to z in which every edge in P is properly oriented and the flow in each edge is less than the capacity of the edge, then it is possible to increase the value of the flow.

Note: It is also possible to increase the flow in certain paths from a to z in which properly and improperly oriented edges are present.

Proof Let P be a path from a to z and let V be a vertex in P different from a and z. Since V is an intermediate vertex in the network N (also in P), two edges e_1 and e_2 are incident on this vertex V. There are now four cases for the orientations of the edges e_1 and e_2. They are as follows:

i. When flow is increased in each edge by Δ the flow into V will be equal to the flow out of V.

ii. When we increase the flow in e_2 by Δ, we must decrease the flow in e_1 by Δ so that the flow into V will still equal the flow out of V.

iii. When we increase the flow in e_1 by Δ, we must decrease the flow in e_2 by Δ to achieve equal flow in and out of V.

iv. When flow is decreased in both edges by Δ we achieve equal flow in and out of V.

All the above cases are possible since $F_{ij} < C_{ij}$ for a properly oriented edge (i,j) and $0 < F_{ij}$ for an improperly oriented edge (i,j)

Result Let P be a path from a to z in a network G.

i. For a properly oriented edge (i,j) in P, $F_{ij} < C_{ij}$

ii. For an improperly oriented edge (i,j) in P, $0 < F_{ij}$

Let $\Delta = \text{mini}\, x$

Where x consists of the numbers $C_{ij} - F_{ij}$ for properly oriented edges (i,j) in P and F_{ij} for improperly oriented edge (i,j) in P.

$$\text{Define } F^*_{ij} = \begin{cases} F_{ij} \text{ if } (i, j) \text{ is not in } P. \\ F_{ij}+\Delta \text{ if } (i, j) \text{ is properly oriented in } P. \\ F_{ij}-\Delta \text{ if } (i, j) \text{ is improperly oriented in } P. \end{cases}$$

Then F^* is a flow whose value is greater by Δ than the value of F.

In every case, the resulting edge assignments give a flow F^*. Clearly the value of $F^* = F + \Delta$

Labelling Procedure

Now we present an algorithm called as labelling procedure to find the maximal flow in a network.

We have a network G in which a is the source, z is the sink and capacity C. The capacity of each edge is a non-negative integer. The vertices of G are ordered $a = v_0, v_1, \ldots, v_n = z$.

i. Set $F_{ij} = 0$ for each edge (i,j) [initialization]

ii. Label vertex a by $(-, \alpha)$ [labelling the source]

iii. If the sink z is labelled, go to step (vi).

iv. Choose the not yet examined, labelled vertex V_i with smallest index i. Stop if there is no such vertex;otherwise set $v = v_i$.[Go to adjacent vertices of the source]

v. Let (α, Δ) be the label of v. Examine each edge of the form (v,w), (w,v) where w is unlabelled.

If $F_{vw} < C_{vw}$, label vertex w by (V, mini $(\Delta, C_{vw} - F_{vw})$

if $F_{vw} = C_{vw}$, do not label w.

For an edge of the form (w,v)

if $F_{wv} > 0$, label vertex w by (v, mini (Δ, F_{wv}))

if $F_{wv} = 0$, do not label w. Go to step (iii)

vi. Let (γ, Δ) be the label of z.

Let $w_0 = z$, $w_1 = \gamma$

If the label of $w_i = (\gamma^1, \Delta^1)$ put $w_i + 1 = \gamma^1$. Continue $w_k = a$. Now the path P: $a = w_k, w_{k-1}, \ldots, w_1, w_0 = z$. is the required path from a to z.

To change the flow of the edges in P find. If the edge e is properly oriented, then increase the flow in e by Δ. Otherwise decreases the flow in e by Δ. Remove all labels from vertices and go to step (ii).

Example 7.19 Find the maximal flow for the following network by labelling procedure.

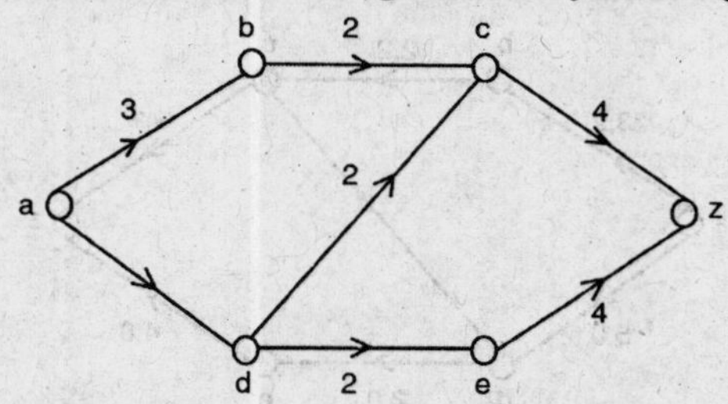

Solution Let us name the vertices as $V_0 = a$, $V_1 = b$, $V_2 = c$, $V_3 = d$, $V_4 = e$, $V_5 = z$.
Initially, set the flow as zero in every edge.

1st iteration

Label the vertex a as $(-, \infty)$.

Examine the edges (a,b) and (a,d).

For the edge (a,b), $F_{ab} = 0 < 3 = C_{ab}$

Label the vertex b as $(a, \text{mini}(\infty, C_{ab} - F_{ab}))$. i.e. b as $(a,3)$.

For the edge (a,d), $F_{ad} = 0 < 5 = C_{ad}$

Label the vertex d as $(a, 5)$.

Now, the next smallest unexamined vertex is b.

Examine the edge (b,c).

For the edge (b,c), $F_{bc} = 0 < 2 = C_{bc}$

Label the vertex c as $(b,2)$.

Now, the next smallest unexamined vertex is c.

Examine the edge (c, z).

For the edge (c,z), $F_{cz} = 0 < 4 = C_{cz}$

Label the vertex z as $(c, \text{mini}(2, 4-0)) = (c,2)$

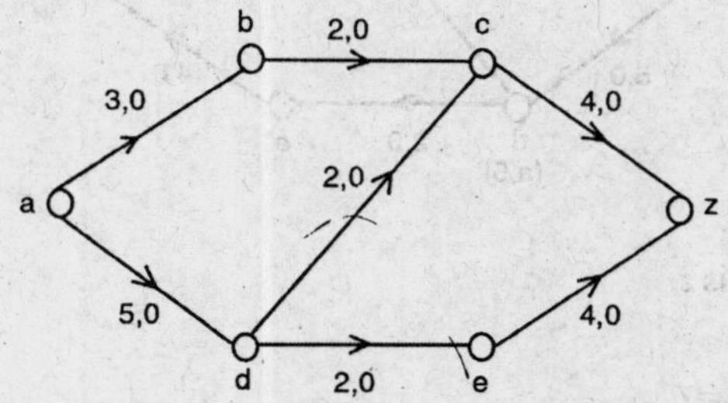

Since we reached the sink z, the required path is P: a,b,c,z. As all the edges in P are properly oriented, increase the flow in each edge of P by Δ,

where $\Delta = \text{mini}\,[C_{ab} - F_{ab}, C_{bc} - F_{bc}, C_{cz} - F_{cz}]$

$= \text{mini}\,[3-0, 2-0, 4-0] = 2.$

Now we have the following network.

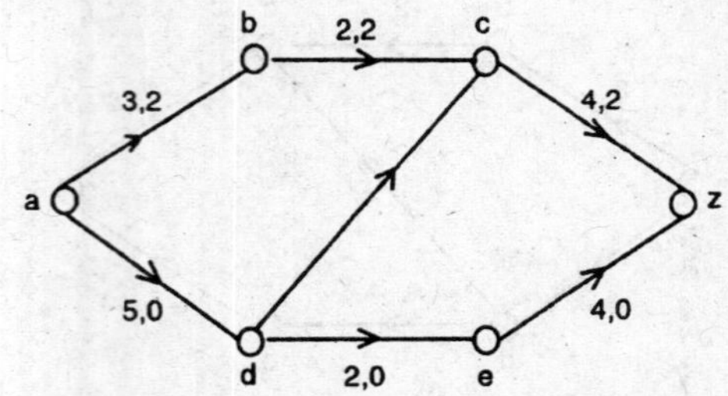

2nd iteration (Remove all the labels) **Start labelling the first vertex *a* as (–, ∞). Examine the egdes (*a*,*b*) and (*a*, *d*).**

For the edge (*a*,*b*), $F_{ab} = 2 < 3 = C_{ab}$

Label the vertex *b* as (*a*, 1).

For the edge (*a*, *d*), $F_{ab} = 0 < 5 = C_{ab}$

Label the vertex *d* as (5,0).

Now, the next smallest unexamined vertex is *b*. Examine the edge (*b*,*c*).

For the edge (*b*,*c*), $F_{bc} = 2 = C_{bc}$

Therefore the edge (*b*,*c*) is saturated and hence we cannot label the vertex *c* at this point. The next smallest unexamined vertex is *d*.

Examine the edges (*d*,*c*), and (*d*,*e*)

For the edge (*d*,*c*), $F_{dc} = 0 < 2 = C_{dc}$

Label the vertex *c* as (*d*,2).

For the edge (*d*,*e*), $F_{de} = 0 < 2 = c_{de}$

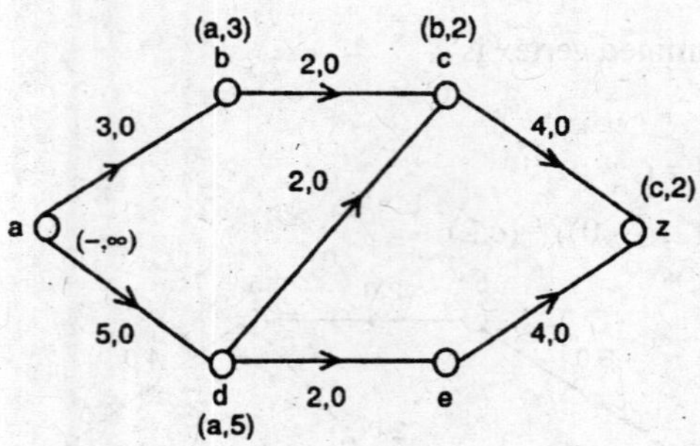

Label the vertex *e* as (*d*,2)

Now, the unexamined vertex is *z*.

Examine the edge (*c*,*z*)

For the edge (*c*,*z*), $F_{cz} = 2 < 4 = C_{cz}$

Label the vertex *z* as (*c*,2).

since we have reached the vertex *z*, labelling can be stopped. The following figure shows the labels obtained in this iteration.

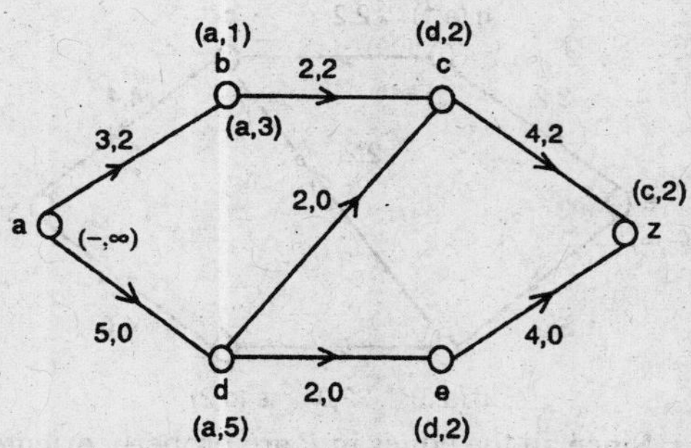

The path obtained is *P*: *adcz*. All the edges in *P* are properly oriented. Increase the flow in each edge of *P* by $\Delta = \text{mini } [C_{ad} - F_{ad}, C_{dc} - F_{dc}, C_{cz} - F_{cz}] = 2$

Now, we have the following network.

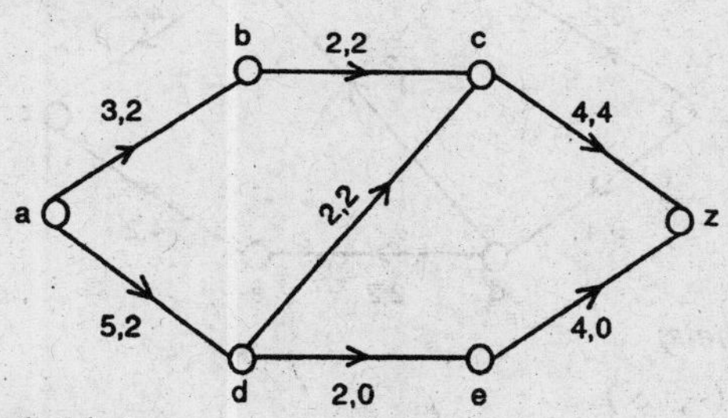

3rd iteration (Remove all the labels)

Start labeling the first vertex *a* as (–, ∞).

Examine the edge (*a*,*b*) and (*a*,*d*).

For the edge (*a*,*b*), $F_{ab} = 2 < 3 = C_{ab}$. Label this vertex *b* as (*a*,1).

For the edge (*a*,*d*), $F_{ad} = 2 < 5 = C_{ad}$. Label this vertex *d* as (*a*,3).

The next smallest unexamined vertex is *b*. Examine the edge (*b*,c).

$F_{bc} = 2 = C_{be}$. Therefore the edge is unsaturated and we cannot label *c* at this juncture.

The next smallest unexamined vertex is *d*.

Examine the edges (*d*,*c*) and (*d*,*e*)

For the edge (*d*,*c*), $F_{dc} = 0 < e_{dc} = 2$. Label the vertex *c* as (d,2)

For the edge (*d*, *e*), $F_{de} = 0 < 2 = C_{de}$.

Label this vertex *e* as (*d*,2).

The next smallest unexamined vertex is *z*.

Examine the edge (*e*,*z*).

For the edge (*e*,z), $F_{cz} = 0 < 4 = C_{cz}$.

Label this vertex *z* as (*c*,2).

Since we have reached the vertex *z*, labelling can be stopped. The following figure shows the labels obtained in this iteration.

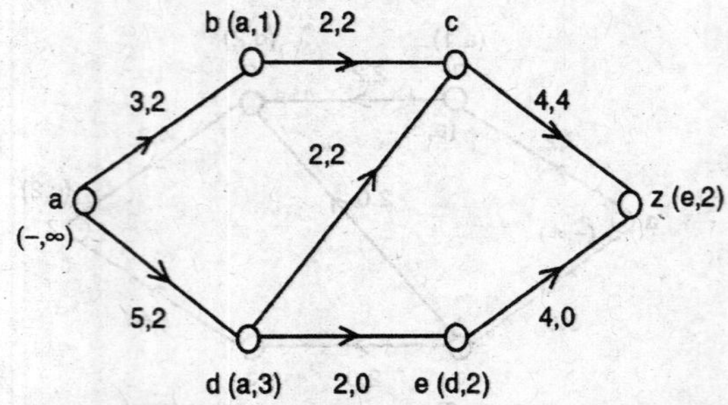

The path obtained in P : *adcz*. Since all the edges in P are properly oriented, we have to increase the flow in each edge of P by $\Delta = 2$. Now, we have the following network.

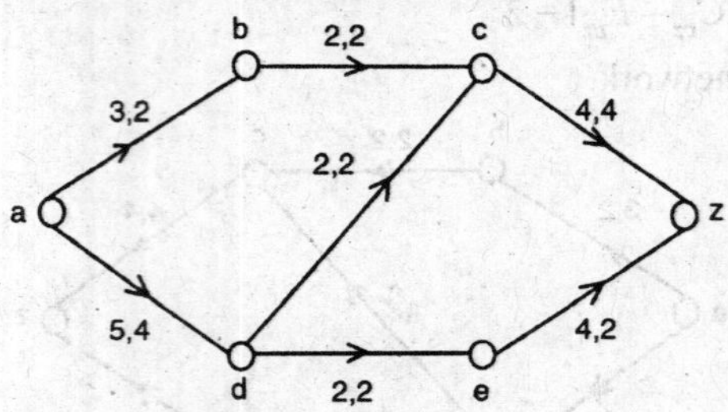

4th iteration (Remove all the labels)

Start labeling the vertex a as $(-, \infty)$.

Examine the edges (a,b) and (a,d). For the edge (a,b), $F_{ab} = 2 < 3 = c_{ab}$.

Label this vertex b as $(a,1)$.

for the edge (a,d), $F_{ad} = 4 < 5 = C_{ad}$.

Label this vertex d as $(a,1)$.

The next smallest unexamined vertex is b.

Examine the edge (b,c). For the edge (b,c),

$F_{bc} = 2 = c_{bc}$, we cannot label this vertex c.

The next smallest unexamined vertex is d.

Examine the edges (d,c) and (d,e)

For the edge (d,c), $F_{dc} = 2 = c_{dc}$, we cannot label c at this point.

For the edge (d,e), $F_{de} = 2 =$ cde, we can not label e at this moment.

Also no other vertices can be labelled. Hence the algorithm terminates and the flow obtained in the previous iteration is maximal.

Example 7.20 Find the maximal flow in the following network.

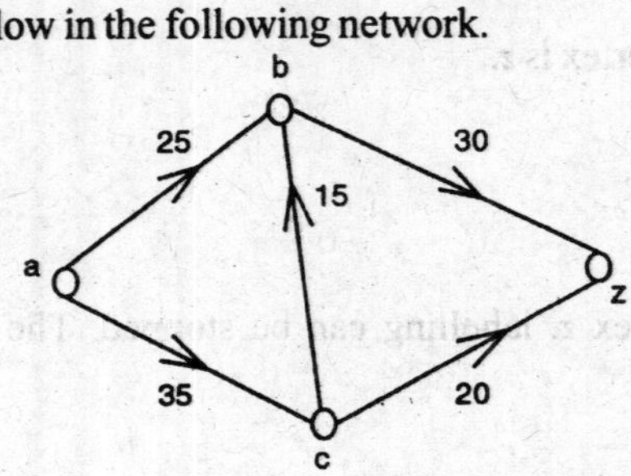

Solution Let us arrange the vertices in an order as $V_0 = a$; $V_1 = b$; $V_2 = c$; $V_3 = z$. Initially, set the flow in edge as zero.

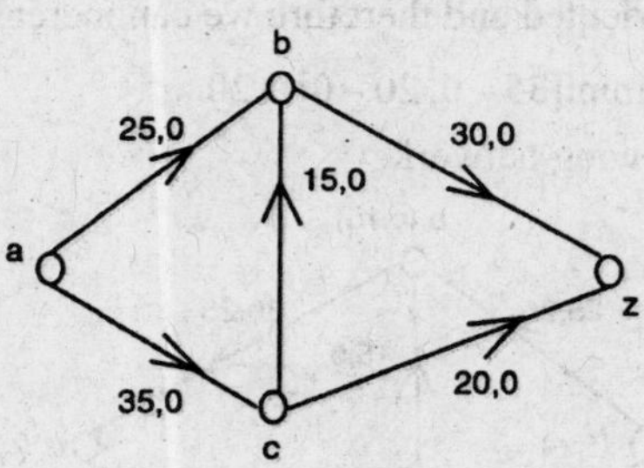

Fig. 1

1st iteration

Label the first vertex a as $(-, \infty)$. Examine the edges (a,b) and (a,c)

For the edge (a,b), $F_{ab} = 0 < 25 = c_{ab}$. Label the vertex b as $(a,25)$.

For the edge (a,c), $F_{ac} = 0 < 35 = c_{ac}$. Label the vertex c as $(a,35)$.

Now, the next smallest labelled unexamined vertex is b. Examine the edge (b,z).

For the edge (b,z), $F_{bz} = 0 < 30 = c_{az}$. Label this vertex z as $(c,25)$

Since we have reached the sink z, we can stop the labelling procedure and the required path is $P : abz$. Since all the edges in P are properly oriented, we can increase the flow by

$$\Delta = \text{mini}\,[c_{ab} - F_{ab}, c_{bz} - F_{bz}] = \text{mini}\,[25 - 0, 30 - 0] = 25$$

Thus we obtained the following network

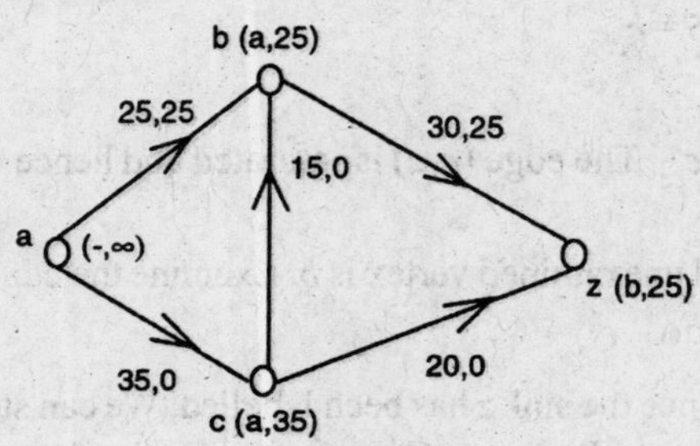

Fig. 2

2nd iteration (Remove all the labels in Fig. 2)

Label the vertex a as $(-, \infty)$. Examine the edges (a, b) and (a, c).

For the edge (a, b), $F_{ab} = 25 = c_{ab}$. We cannot label the vertex b at this moment.

For the edge (a, c), $F_{ac} = 0 < 35 = c_{ac}$. Label the vertex c as $(a, 35)$.

The next smallest labelled unexamined vertex is e.

Examine the edges (c, b) and (c, z)

For the edge (c, b), $F_{cb} = 0 < 15 = c_{cb}$. Label the vertex b as $(c, 15)$.

For the edge (c, z), $F_{cz} = 0 < 20 = c_{cz}$. Label the vertex z as $(c, 20)$.

Since the sink z has been reached, we can stop the labelling procedure and the required path is obtained $p: acz$. All edges in P are properly oriented and therefore we can increase the flow by Δ,

where $\Delta = \text{mini}[c_{ac} - F_{ac}, c_{cz} - F_{cz}] = \text{mini}[35-0, 20-0] = 20$.

Thus we have obtained the following network.

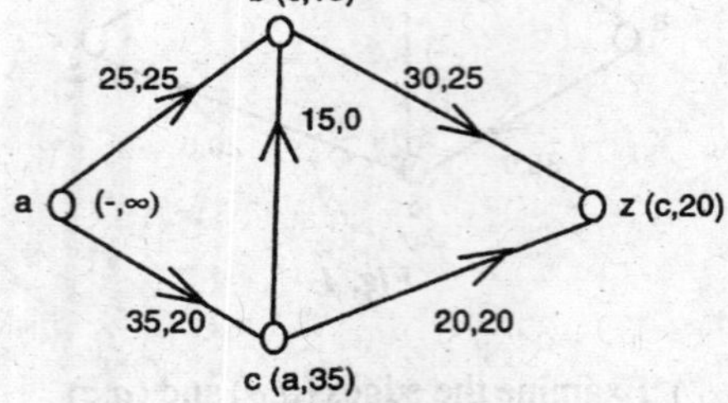

Fig. 3

3rd iteration (Remove all the labels in Fig. 3)

Label the vertex a as $(-, \infty)$. Examine the edge (a,b) and (a,c). For the edge (a,b),

$F_{ab} = 25 = c_{ab}$, the edge (a,b) is saturated and hence we cannot label the vertex b at this moment.

For the edge (a,c), $F_{ac} = 20 < 35 = c_{ac}$.

Label the vertex c as $(a, 15)$

Now the next smallest labelled unexamined vertex is c. Examine the edges (c, b) and (c, z).

For the edge (c,b), $F_{cb} = 0 < 15 = c_{cb}$.

Label this vertex b as $(c, 15)$

For the edge (c, z), $F_{cz} = 20 = c_{cz}$. The edge (c, z) is saturated and hence we cannot label the vertex z at this moment.

Now the next smallest labelled unexamined vertex is b. Examine the edge (b, z).

For the edge (b, z), $F_{bz} = 25 < 30 = C_{bz}$.

Label the vertex z as $(b,15)$. Since the sink z has been labelled. We can stop the labelling procedure and the required path is P: $acbz$. All the edges in P are properly oriented and hence we can increase the flow of all edges of P by $\Delta = 5$.

Thus we obtained the following network

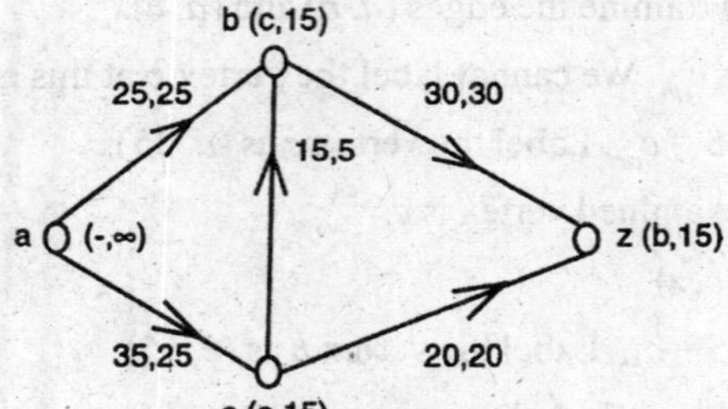

Fig. 4

4th iteration (Remove all the labels in Fig. 4)

Label the vertex a as $(-, \infty)$.

Examine the edges (a,b) and (a,c).

Already the edge (a,b) is saturated, we cannot increase the flow through the edge (a,b), and cannot label the vertex b at this moment.

For the edge (a,c),

$F_{ac} = 25 < 35 = c_{ac}$

Label the vertex c as $(a,10)$.

Now, the next smallest labelled unexamined vertex is c. Examine the edges (c,b) and (c,z).

For the edge (c,b),

$F_{cb} = 5 < 15 = c_{cb}$

Label the vertex b as $(c,10)$.

Already the edge (c,z) is saturated. We cannot the label the vertex z at this moment. No further vertices can be labelled. Hence the algorithm ends here. The flow obtained in the previous iteration is maximal. (i.e. $F = 50$)

EXERCISE IX

Find the Maximal flow for the following

1.

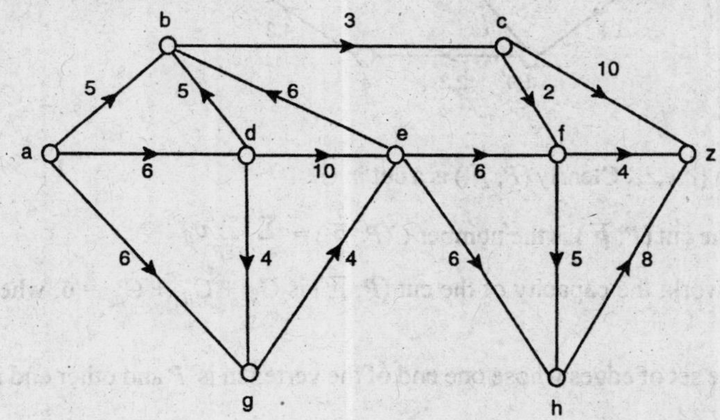

2.

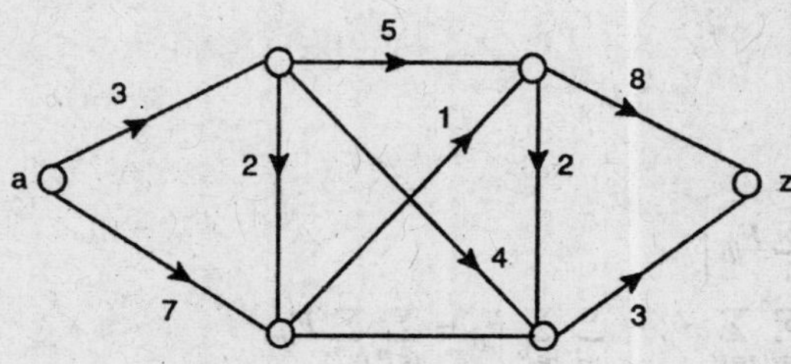

3.

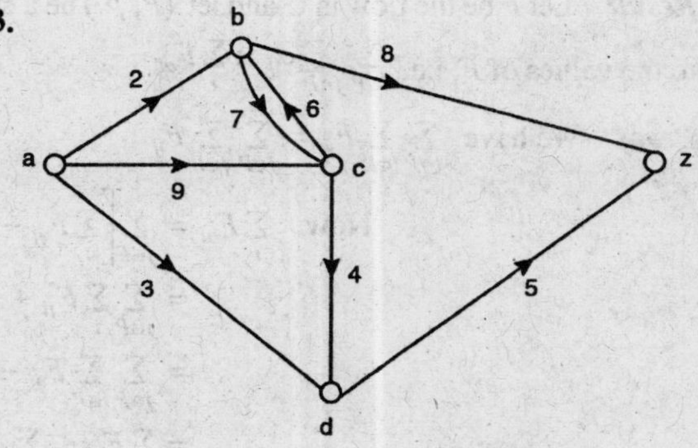

4.

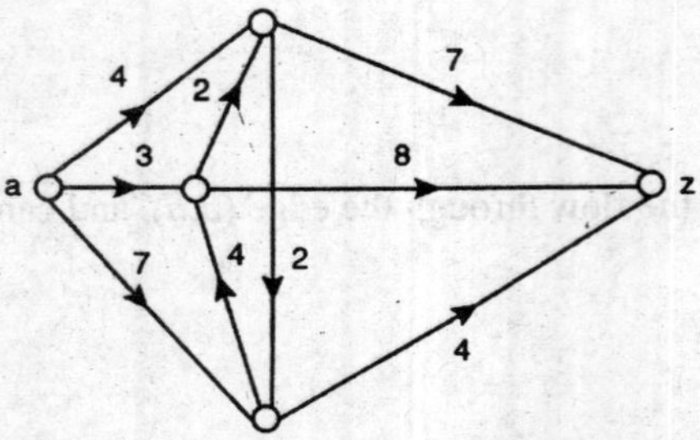

5.

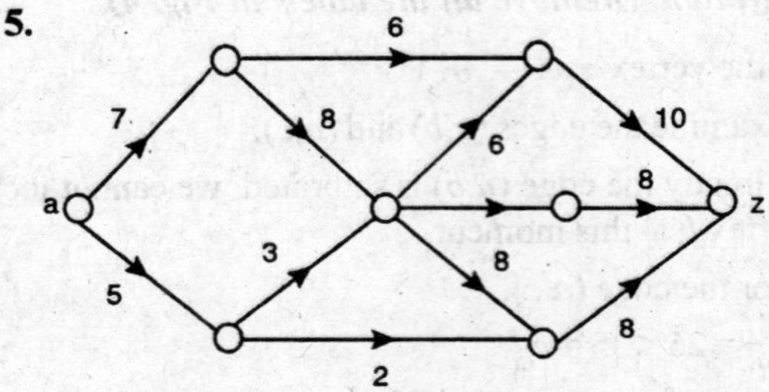

Let G be a network and its flow be F. When we apply the maximum flow algorithm for G, it terminates with some vertices labelled and some unlabelled. So we shall define a cut and the capacity of a cut.

Cut Let G be a network with source a and sink z. A cut $(P, \overline{P})$ in G consists of a set P of vertices and the complement $\overline{P}$ of P, $a \in P$ and $z \in \overline{P}$

Example

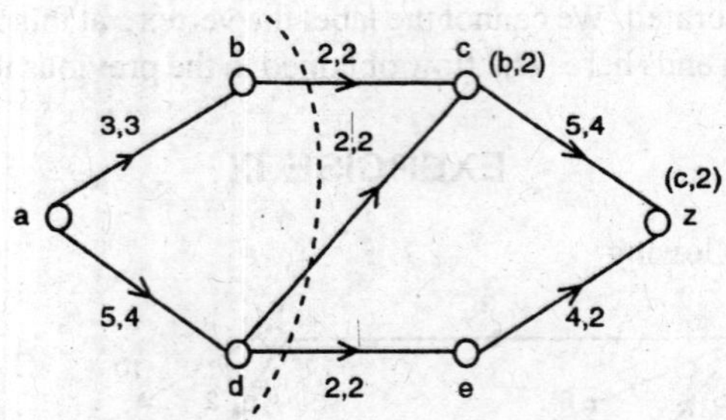

Consider the network G.

Here $P = \{a,b,d\}$ and $\overline{P} = \{c,e,z\}$. Clearly $(P, \overline{P})$ is a cut in G.

Capacity The capacity of the cut $(P, \overline{P})$ is the number $C(P, \overline{P}) = \sum_{i \in P} \sum_{j \in P} c_{ij}$.

Example For the above network, the capacity of the cut $(P, \overline{P})$ is $C_{bc} + C_{de} + C_{de} = 6$, where $P = \{a, b, d\}$, $\overline{P} = \{c, e, z\}$

Note: Cut can be defined as the set of edges whose one end of the vertex in is P and other end in $\overline{P}$ with $a \in P$ and $z \in \overline{P}$.

Result Let F be the flow in G and let $(P, \overline{P})$ be a cut in G. Then the capacity of $(P, \overline{P})$ is greater than or equal to the values of F. i.e. $\sum_{i \in P} \sum_{j \in \overline{P}} c_{ij} \geq \sum_i F_{ai}$

Proof We have $\sum_{j \in P} \sum_{i \in P} F_{ji} = \sum_{i \in P} \sum_{j \in P} F_{ij}$

$$\text{Now} \quad \sum_i F_{ai} = \sum_{j \in P} \left[\sum_i F_{ji} - \sum_i F_{ij} \right]$$

$$= \sum_{j \in P} \sum_i F_{ji} + \sum_{j \in P} \sum_{i \in P} F_{ji} - \sum_{j \in P} \sum_{i \in P} F_{ij} - \sum_{j \in P} \sum_{i \in P} F_{ij}$$

$$= \sum_{j \in P} \sum_{i \in P} F_{ji} - \sum_{j \in P} \sum_{i \in P} F_{ij} \sum \sum_{j \in P} \sum_{i \in P} F_{ji} \leq \sum_{j \in P} \sum_{i \in P} c_{ji}$$

$$= \sum_i F_{ai} \leq \sum_{i \in P} \sum_{j \in P} c_{ij}$$

MAX-FLOW MIN-CUT THEOREM

Let $G = (V, E)$ be a connected as well as a loop - free directed graph. Then G is called a network or transport network, if the following conditions are satisfied:

a. There exists an a unique vertex $a \in v$, with the in–degree of a which is equal to zero. Then this vertex a is called the source.

b. There is a unique vertex $Z \in V$, called the sink, where the out–degree of (Z) equals to zero.

c. The graph G is weighted, so that there is a function from E to the set of non-negative integers assigned to each edge.

$e = (v, w) \in E$, a capacity denoted by $c(e) = c(v,w)$

Flow

If $G = (V, E)$ is a transport network, a mapping $f: E \rightarrow I$ is called a flow for G if

a. $f(e) \leq c(e) \;\; \forall \, e \in E$

b. For any $v \in V$, other than the source a or the sink $\sum_{w \in v} f(w, v) = \sum_{w \in v} f(v, w)$
(If there is no edge (v, w), $f(v, w) = 0$)

Let 'f' be the flow for a transport network $G = (V,E)$

a. An edge e of the network is called saturated if $f(e) = c(e)$. When $f(e) < c(e)$, the edge is called unsaturated

b. If a is the source of G, then Val $(f) = \sum_{v \in V} (a,v)$ is called the value of the flow.

Example

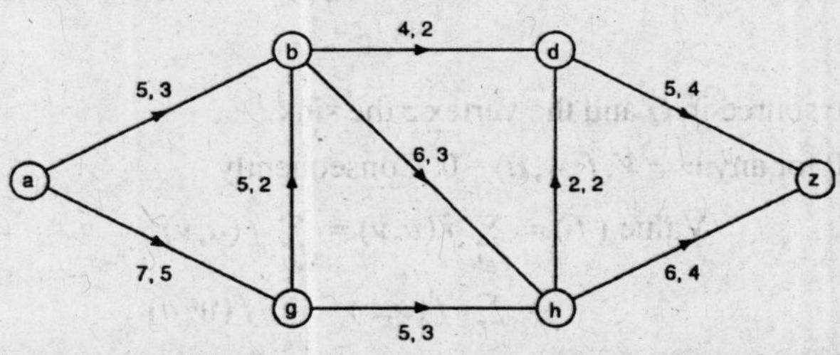

Only the edge (h,d) is saturated. All other edges are unsaturated. The value of the flow in this network is

$$\text{Val}(f) = \sum_{v \in V} \pm (a, v) = f(a, b) + f(a, g)$$

$$= 3 + 5 = 8$$

But is there another flow f_1 such that val$(f_1) > 8$? The determination of a flow that achieves the greatest possible value (maximal flow) from the above example is

$$\sum_{v \in V} f(a, v) = 3 + 5 = 8 = 4 + 4 = f(d, z) + f(h, z) = \sum_{v \in V} f(v, z)$$

$\therefore$ The total flow leaving the source a is equal to the total flow into the sink z.

Source - sink cut If $G = (V, E)$ is a network and C is a cut - set for the undirected graph associated with N, then C is called cut or a source - sink cut, if the removal of the edges in C from the network results in the separation of source and sink.

Example

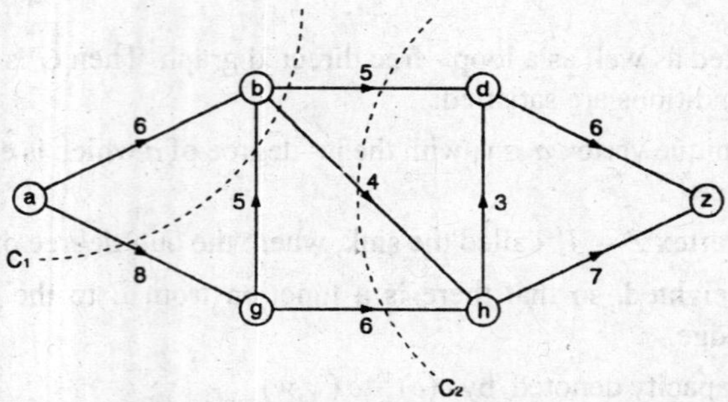

The cut C_1 consists of the undirected graph whose edges are $\{a, g\}$, $\{b, d\}$, $\{b, g\}$, and $\{b, h\}$. This cut, partitions the vertices of the network into the two sets $P = \{a, b\}$ and its complement $\overline{P} = \{d, g, h, z\}$, so that C, is denoted as $(P, \overline{P})$. The capacity of a cut, is then denoted as

$$C(P, \overline{P}) \ \sum_{\substack{V \in P \\ W \in \overline{P}}} C(v, W)$$

In this example, $c(P, \overline{P}) = c(a, g) + c(b, d) + c(b, h)$

$$= 8 + 5 + 7 = 20$$

The cut C_2 induces the vertex in the partition $\varphi = \{a, b, g\}$, $\overline{\varphi} = \{d, h, z\}$ and has the capacity

$$C(\varphi, \overline{\varphi}) = c(b, d) + c(b, h) + c(g, h) = 5 + 7 + 6 = 18$$

Theorem Let f be a flow in a network $G = (V, E)$. If $C = (P, \overline{P})$ is any cut in G, then the value (f) cannot exceed $C(P, \overline{P})$

Proof Let vertex a be the source in G and the vertex z the sink.

Since in-degree $(a) = 0$ for any $w \in V, f(w, a) = 0$, Consequently

$$\text{Value } (f) = \sum_{v \in V} f(a, v) = \sum_{v \in V} f(a, v)$$

$$= \sum_{v \in V} f(a, v) - \sum_{w \in V} f(w, a)$$

By condition (*b*) in definition of a flow, for any $x \in P, x \neq a$

$$\sum_{v \in V} f(x, v) - \sum_{w \in V} f(w, x) = 0$$

$$\therefore \text{Value } (f) = [\sum_{v \in V} f(a, v) - \sum_{w \in V} f(w, a)] + \sum_{\substack{x \in P \\ w \notin a}} f[\sum_{v \in V} f(x, v) - \sum_{w \in V} f(w, x)]$$

$$= \sum_{\substack{x \in P \\ v \in V}} f(x, v) - \sum_{\substack{x \in P \\ w \in V}} f(w, x)$$

$$= [\sum_{\substack{x \in P \\ v \in P}} f(x, v) + \sum_{\substack{x \in P \\ v \in \overline{P}}} f(x, v)] - [\sum_{\substack{x \in P \\ w \in P}} f(w, x) + \sum_{\substack{x \in P \\ w \in \overline{P}}} f(w, x)]$$

Since,

$$\sum_{\substack{x\in P\\ v\in \bar{P}}} f(x,v) \text{ and } \sum_{\substack{x\in P\\ w\in \bar{P}}} f(w,x)$$

are summed over the same set of all ordered pairs in $P \times P$, these summations are equal.

$$\text{Value }(f) = \sum_{\substack{x\in P\\ v\in \bar{P}}} f(x,v) - \sum_{\substack{x\in P\\ w\in \bar{P}}} f(w,x) \text{ for all } x, w \in v, f(w,x) \geq 0, \text{ so } \sum_{\substack{x\in P\\ w\in \bar{P}}} f(w,x) \geq 0 \text{ and}$$

$$\text{Value }(f) \leq \sum_{\substack{x\in P\\ v\in \bar{P}}} f(x,v)$$

$$\leq \sum_{\substack{x\in P\\ v\in \bar{P}}} c(x,v)$$

$$= C(P,\bar{P})$$

So we find that in a network G, the value for any flow is less than or equal to the capacity of any cut in that network. Hence the value of the maximum flow cannot exceed the minimum capacity over all cuts in a network.

$$\therefore \text{ Value}(f) = \sum_{\substack{x\in P\\ v\in \bar{P}}} f(x,v) - \sum_{\substack{x\in P\\ w\in \bar{P}}} f(w,x)$$

Where $(P, \bar{P})$ is any cut in N.

Therefore once a flow is constructed in a network, then for any cut $(P, \bar{P})$ in the network, the value of the flow equals the sum of the flows in the edge directed from the vertices in P to those in $\bar{P}$ minus the sum of the flows in the edges directed from the vertices in $\bar{P}$ to those in P.

Corollary If f is a flow in a network $G = \langle V, E\rangle$, then the value of the flow from the source a is equal to the value of the flow into the sink z

Proof

$$\sum_{\substack{x\in P\\ v\in \bar{P}}} f(x,v) - \sum_{\substack{x\in P\\ w\in \bar{P}}} f(w,x) = \text{Value }(f) = \sum_{\substack{y\in Q\\ v\in \bar{Q}}} f(y,v) - \sum_{\substack{y\in Q\\ w\in \bar{Q}}} f(w_1 y)$$

with $P = \{a\}$ and in-degree $(a) = 0$, we find that

similarly, for $\bar{Q} = \{z\}$ and out-degree $(z) = 0$, it follows that

$$\sum_{\substack{y\in Q\\ w\in \bar{Q}}} f(w,y) = \sum_{y\in Q} f(z,y) = 0$$

Consequently

$$\sum_{\substack{x\in P\\ v\in \bar{P}}} f(x,v) = \sum_{v\in \bar{P}} f(a,v) = \text{Value }(f) = \sum_{\substack{y\in Q\\ v\in \bar{Q}}} f(y,v) = \sum_{y\in Q} f(y,z)$$

The Max - Flow Min - cut Theorem

For a network $G = (v,E)$, the maximum flow that can be attained in G is equal to the minimum capacity over all the cuts in a network.

Proof Using the previous theorem if $(P, \overline{P})$ is a cut of minimum capactiy in G, then the value of any Flow '*F*' in *G* satisfies value (*F*) $\leq c(P, \overline{P})$. To verify this we use iterative algorithm

Step 1 Given a network *G*, define an initial flow *f* in *G* by $f(e) = 0$ for every *e* in *E*

Step 2 Label the source *a* with $(-, \alpha)$, (this indicates that from the source *a* as much materials is available as is needed to acheive a maximum flow)

Step 3 For any vertex *x* that is adjacent from a label *x*, we find.

a. if $c(a, x) - f(a, x) > 0$ define $\Delta(x) = c(a, x) - f(a, x)$ and label vertex *x* with $(a^*, \Delta(x))$

b. if $c(a, x) - f(a, x) = 0$, leaves vertex *x* unlabelled.

Note: [The label $(a^*, \Delta(x))$ indicates that the present flow from *a* to *x* can be increased by the amount $\Delta(x)$, with $\Delta(x)$ additional units supplied from the source *a*]

Step 4 As long as there exists $(x \neq a) \in v$ such that *x* is labelled, and there is an edge (*x*, *y*), where *y* is not labelled, label vertex *y* as follows.

a. If $c(x, y) - f(x, y) > 0$ define $\Delta(y) = \min\{\Delta(x), c(x, y) - f(x, y)\}$ and label vertex *y* as $(x^*, \Delta(y))$

b. If $c(x, y) - f(x, y) = 0$, leaves vertex *Y* unlabelled.

Note: The label $(x^*, \Delta(y))$ indicates that the present flow into vertex *y* can be increased by an amount $\Delta(y)$ taken from vertex *x*.

Step 5 Likewise, as long as there is a vertex $x \neq a$ where *x* is labelled, and there is an edge (*y*, *x*), where *y* is not labelled, label vertex *y* as follows:

a. If $f(y, x) > 0$, label vertex *y* as $(\overline{x}, \Delta(y))$, where $\Delta(y) = \min\{\Delta(x), f(y, x)\}$

b. If $f(y, x) = 0$, leave vertex *y* unlabelled

Note: The label $(x^-, \Delta(y))$ tells us that by decreasing the flow from *y* to *x*, the total flow out of *y* to the labelled vertices can be decreased by $\Delta(y)$. These $\Delta(y)$ units may then be used to increase the total flow from *y* to the unlabelled vertices.

Case (i) If the sink *z* is labelled as $(x^*, \Delta(z))$, then the flow in the edge (*x*, *z*) can be increased from $f(x, z)$ to $f(x, z) + \Delta(z)$, as indicated by the label.

Vertex *x* can be labelled either $(v^*, \Delta(x))$ or $(\overline{v}, \Delta(x))$, where $\Delta(x) \geq \Delta(z)$. For the label $(v^*, \Delta(x))$, we may regard vertex *V* as the source for increasing the flow in edge (*x*, *y*) by the amount of $\Delta(z)$. We increase the present flow in edge (*v*, *x*) from $f(v, x)$ to $f(v, x) + \Delta(z)$ (not $f(v, x) + \Delta(x)$). If *x* has the label $(\overline{v}, \Delta(x))$, then since a vertex *y* may be adjacent to or from more than one labelled vertex, the result of this procedure need not be unique.

Suppose if *x* is labelled, the network could include both of the edges (*x*, *y*) and (*y*, *x*), perhaps resulting in two possible labels for *y*.

Steps 4 and 5 are repeated as long as possible with regard to the current (and changing) set of labelled vertices. With each iteration, there are two possible cases to consider.

Flow in the edge (*x*, *v*) is changed from $f(x, v)$ to $f(x, v) - \Delta(z)$ in order to provide the additional flow of Δz units from *x* to *z*.

As this process continues back to source *a*, each directed edge along a path from *a* to *z* has its flow increased or decreased (for a vertex (on the path) with a negative label) by Δz units.

When this is accomplished, all vertex labels except for (–, ∞) and the source are removed, the process is repeated to see whether it is possible to increase the flow even more.

Case (ii) If the labelling procedure is carried out as far as possible and the sink z is unlabelled, then the maximum flow has been attained. Let P be the vertices in V that are labelled, and $\overline{P} = v - P$. Since the vertices in $\overline{P}$ are not labeled, the flows in the edges (x, y), where $x \in P$ are $y \in \overline{P}$, satisfy $f(x, y) = c(x, y)$. Also for any edge (w, v) with $w \in \overline{P}$ and $v \in P$, we have $f(w, v) = 0$. Consequently, there is a flow for the given network, where the value of the flow is equal to the capacity of the cut $(P, \overline{P})$. From the theorem it follows that the flow is maximum.

EXERCISE X

1. Find a maximum flow for the following network

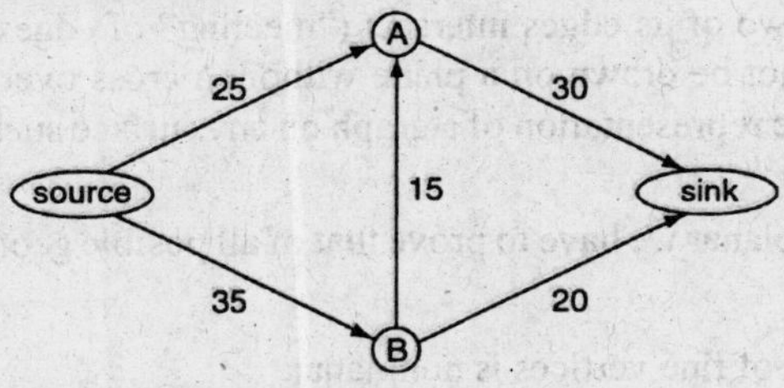

2. Find the maximum flow for the following networks

i.

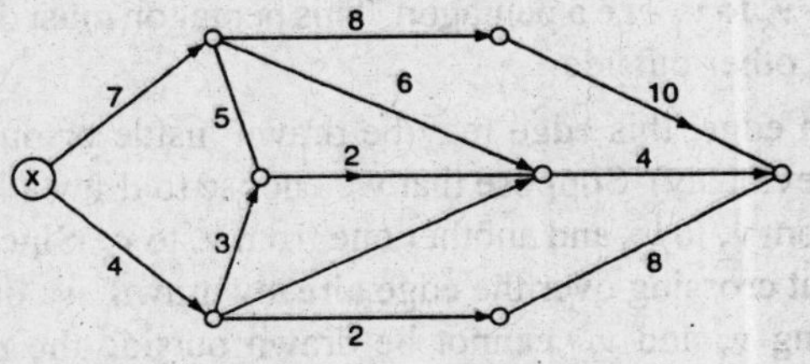

ii.

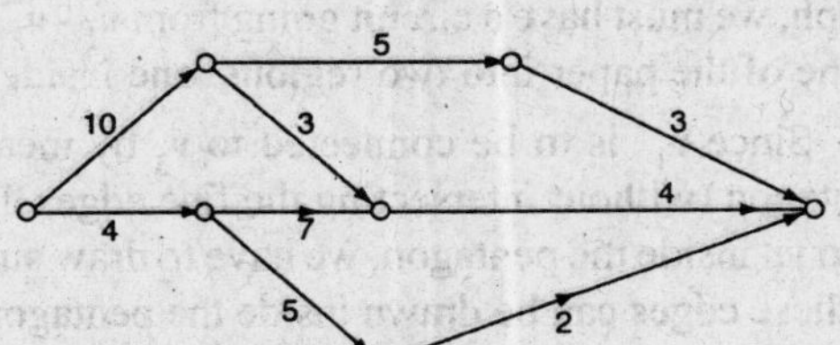

(3) Find a maximum flow for the network

i.

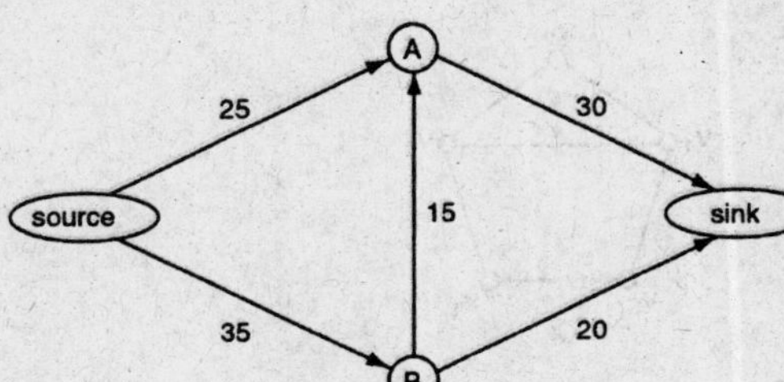

ii.

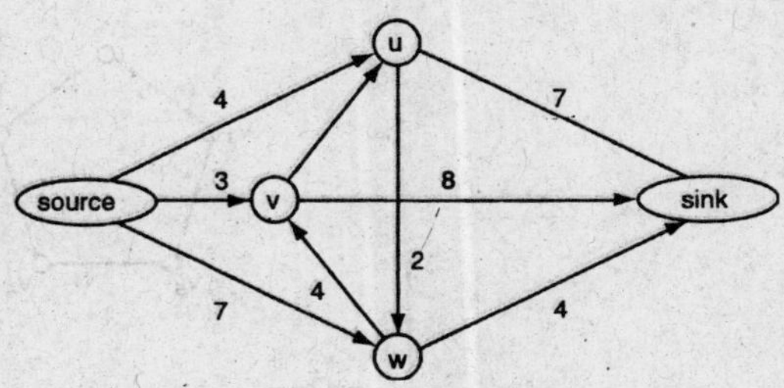

ANSWERS

(3)

i.

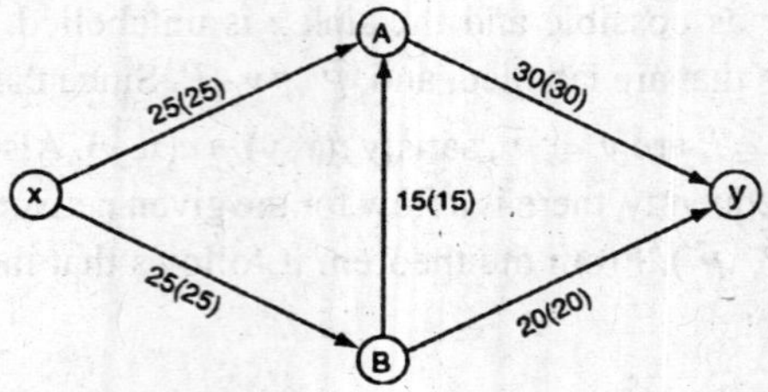

ii.

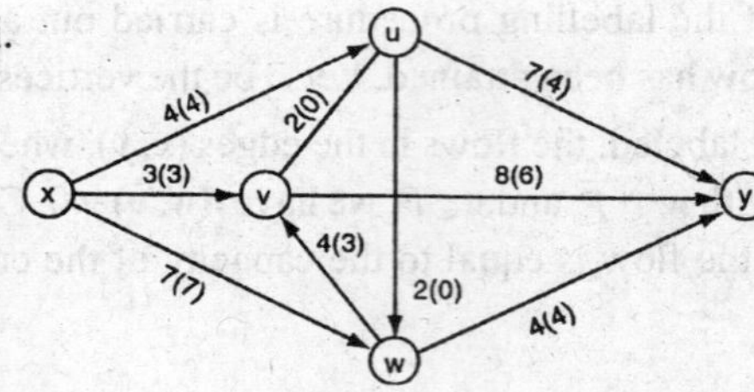

Planar Graphs

Introduction A graph G is said to be *planar* if there exists some geometric representation of G which can be drawn on a plane such that no two of its edges intersect ("meeting" of edges at a vertex is not considered an intersection). A graph that cannot be drawn on a plane withouc a cross over between its edges is called non planar. A drawing of a geometric representation of a graph on any surface such that no edges intersect is called embedding.

Note: To show a graph G is nonplanar we have to prove that of all posible geometric representations of G, none can be embedded in a plane.

Theorem The complete graph of fine vertices is nonplanar.

Proof Let the fine vertices in the complete graph be v_1, v_2, v_3, v_4 and v_5. Using the definition of complete graph, we must have a circuit going from $v_1 - v_2 - v_3 - v_4 - v_5$ to v_1 - i.e a pentagon. This pentagon must divide the plane of the paper into two regions, one inside and the other outside.

Since v_1 is to be connected to v_3 by means of an edge, this edge may be drawn inside or outside the pentagon (without intersecting the fine edges drawn previously). Suppose that we choose to draw a line from v_1 to v_3 inside the pentagon, we have to draw an edge from v_2 to v_H and another one from v_2 to v_5. Since neither of these edges can be drawn inside the pentagon without crossing over the edge already drawn, we draw both these edges outside the pentagon. The edge connecting v_3 and v_5 cannot be drawn outside the pentagon without crossing the edge between v_2 and v_4. Therefore v_3 and v_5 have to be connected with an edge inside the pentagon.

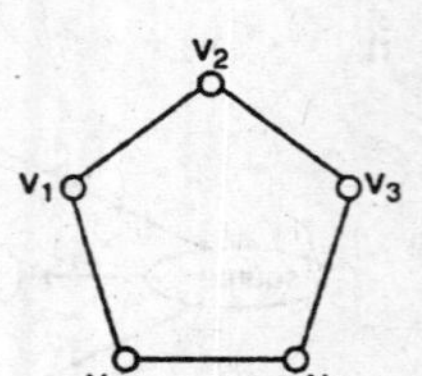

(i)

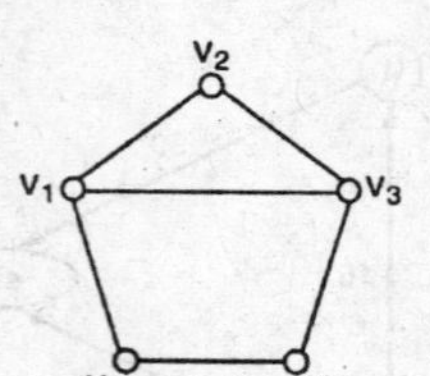

(ii)

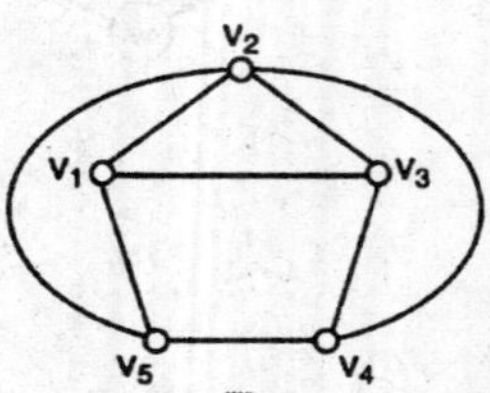

(iii)

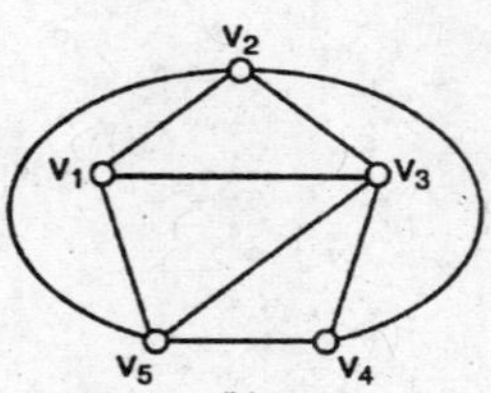

(iv)

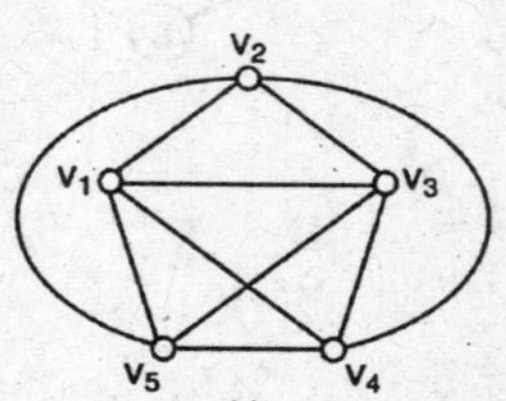

(v)

Note: Complete graph is nothing but a simple graph in which every vertex is joined to every other vertex by means of an edge.

Theorem Kurtowski's (Polish mathematician) second graph is also nonplanar. ($k_{3,3}$ is nonplanaı)

Note: In the plane, a continuous non - self intersecting curve whose origin and terminus coincide is said to be a jordan curve. If j is a jordan curve in the plane π , then $\pi - j$ is a union of two disjoint connected open sets called the interior and the exterior of j.

Example Prove that K_5 is non - planar.

Step 1: Draw a circuit c on 5 vertices. This circuit c devides the plane into two regions called nterior & exterior of c

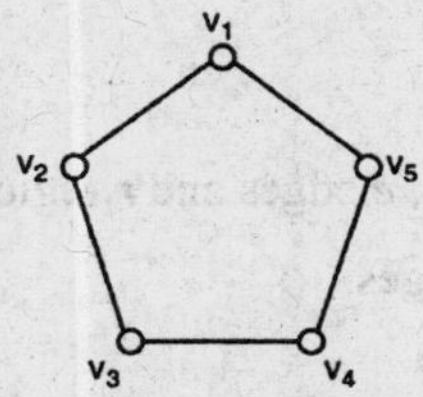

Step 2: Draw the edges v_1v_3, v_1v_4 in the interior. We cannot draw any more edge in the interior of c, without intersecting any edge.

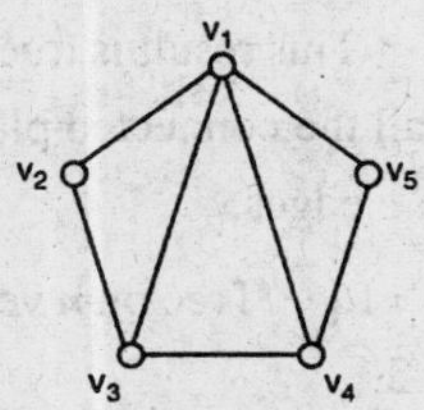

Now draw the edges v_2v_5, v_2v_4 in the exterior of c. But the edge v_3v_5 cannot be drawn in the interior or exterior of c, without intersecting the edge of c.

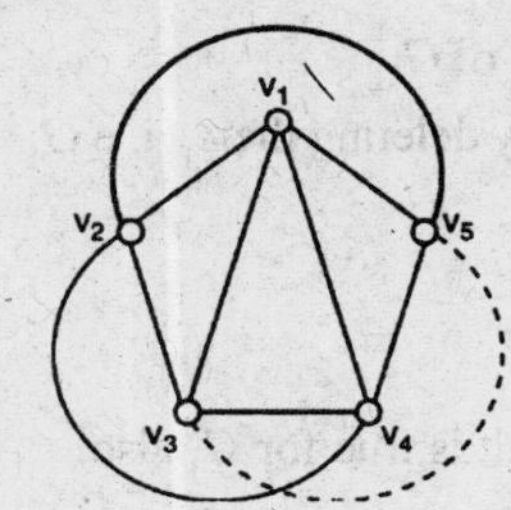

Thus k_5 is non - planar.

We can prove that $k_{3,3}$ is non-planar in the following manner.

Proof Assume that $k_{3,3}$ is planar. Let the vertex of $k_{3,3}$ is $\{v_1,...,v_6\}$. Let $P = \{v_1, v_3, v_5\}$ and $Q = \{v_2, v_4, v_6\}$.

Let C be the cycle $v_1\ v_2\ v_3\ v_4\ v_5\ v_6\ v_1$. It is a jordan curve. The other three edges v_1v_4, v_2v_5, v_3v_6 are chords of the cycle C. So either interior of C or Exterior of C contains two of there three chords., say there are two chords in Int c. These two chords must cross each other, which is a contradiction hence $k_{3,3}$ is non-plannar.

Contour Let G be a connected planar graph. A region of G is the domain of the plane surrounded by edges of the graph such that any two points in it can be joined by a line not crossing any edge. The edges "touching" a region contain a simple cycle called the contour of the region. Two regions are said to be adjacent if the contours of the two regions have atleast one edge in common.

Example

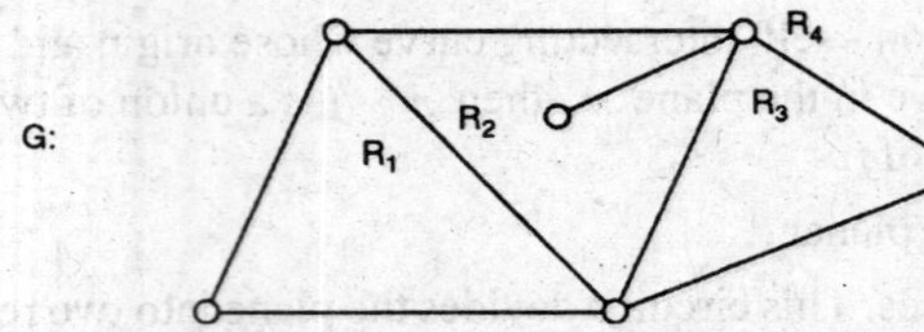

G, a planar graph: R_i, $i = 1, 2, 3, 4$ are the regions of G. Here R_4 is the infinite region.

Euler's Formula

If G, a connected planar graph has n vertices, e edges and r regions, then, $n - e + r = 2$

Proof By induction on e, the number of edges.

If $e = 0$, then $G = K_1$($\because$ G is connected)

$\therefore n = 1$; $r = 1$ (infinite face) $\quad \therefore n - e + r = 1 - 0 + 2 = 2$

If $e = 1$ then $n = 2$ ($\because$ G is connected) & $r = 1$ (infinite face)

$\therefore n - e + R = 2\text{-}1\text{+}1 = 2$ $\quad \therefore$ This result is true for $e = 0$ and $e = 1$.

Let us assume that this result is true for all the connected planer graphs on (e-1) edges.

Let G be a connected planar graphs with e edges.

Case (i) If G is a tree with e edges then $n = e + 1$ [$\because$ Tree on n vertices has (n–1) edges]. $r = 1$

$$\therefore n\text{-}e + r = e + 1 - e + 1 = 2.$$

Case (ii) If G is not a tree.

Since G is connected, it contains cycles.

Let e_1 be an edge in some simple circuit of G.

Let G_1 be the graph obtained from G by deleting the e_1 i. e $G_1 = G - e_1$

Now, number of vertices in $G_1 = n$

Number of edges in $G_1 = e–1$

Number of regions in $G_1 = r–1$

Since G_1 has less then e edges, the result is true for G_1 also.

$\therefore$ By induction hypothesis, $n_1 – e_1 + r_1 = 2$, where n_1 is the number of vertices, e_1 is the number of edges and r_1 is the number of regions of G_1 respectively.

$$\therefore n–(e–1) + r–1 = 2 \Rightarrow n–e + r = 2.$$

$\therefore$ In all the cases, the result in true.

Corollary If G is a connected simple planar graph without loops and has n vertices, $e \geq 2$ edges and r regions, then $3/2\ r \leq e \leq 3n–6$.

Proof If $r = 1$ then $3/2 \leq e \leq 3n–6$ is true, since $e \geq 2$.

If $r > 1$. Let K be the number of edges in the contours of the finite regions.

Since G is simple, each region (finite) is bounded by atleast 3 edges.

Therefore $k \geq 3(r-1)$ (1)

But, in a planar graph, an edge belongs to the contours of atmost two regions and atleast 3 edges touch the infinite region.

$\therefore k \leq 2e - 3$ (2)

From (1) & (2), $3r-3 \leq k \leq 2e-3$

$\Rightarrow 3r - 3 \leq 2e - 3$

$\Rightarrow 3r \leq 2e \Rightarrow 3/2\, r \leq e$ (3)

Since G is planar, $n-e+r = 2$, by Euler's Formula.

$\therefore n - e + 2/3\, e \geq 2$ [$\because$ From (3) $r \leq 2/3\, e$]

$\Rightarrow 3n - 3e + 2e \geq 6$

$\Rightarrow -e \geq -3n + 6$

$\Rightarrow e \leq 3n - 6$ (4)

From (3) & (4), $3/2r \leq e \leq 3n-6$

Example 7.21 (i) Prove that K_5 is non-planar.

Solution Suppose K_5 is planar, then by the above corollory, $e \leq 3n - 6$. In K_5 $n = 5$, $e = 10$;

$\therefore 10 \leq 3 \times 5 - 6 = 9$, which is absurd. $\therefore K_5$ is non-planar

Remark K_5, $K_{3,3}$ are called Kuratowski's first graph, second graph respectively.

Corollary If G is a simple connected planar graph on n vertices, e edges and r regions and does not contain any triangle, then $2r \leq e \leq (2n-4)$

Subdivision A subdivision of a graph G is obtained by inserting vertices (of degree 2) into the edges of G.

Example

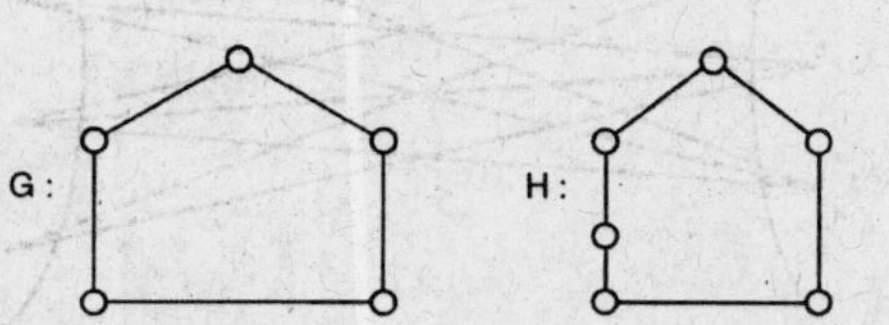

H is the subdivision of G.

Kuratowski theorem A graph is planar if it contains no subgraph that is isomorphic to or is a subdivision of K_5 or $K_{3,3}$. (Interested readers can see the proof in " Graph theory" - *W. T.* Tulte, Cambridge University Press, Pg. no: 320 – 325).

EXERCISE XI

1. Determine which of the following graphs are planar.

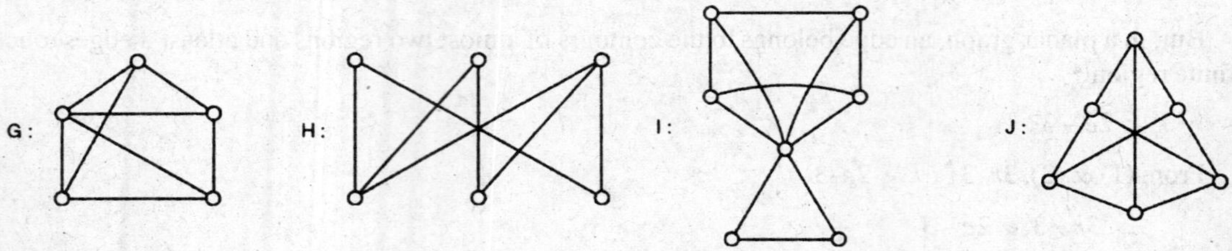

2. Prove that, for every positive integer n, the graph $k_{1,n}$ is planar. How many regions result, when the graph is embedded in the plane?
3. Show that any graph having four or fewer vertices is planar.
4. Show that any graph having five or fewer vertices of degree 2 is planar.
5. Show that if G is a simple graph on 11 or more vertices, then G or its complement $\overline{G}$ is non - planar.

Colouring

No two adjacent vertices having the same colour is called the proper colouring or colouring of a graph. A Graph G that requires K different colours (minimum number) for its proper colouring can be referred to as k - chromatic graph, and the number k is called the chromatic number of G.

A graph consisting of only isolated vertices is 1 - chromatic.

A graph with one or more edges (not a self-top) is atleast 2 - chromatic (also called bichromatic).

Bipartite graph A graph G is called bipartite if its vertex set V can be decomposed into two disjoint sub sets v_1 and v_2 such that every edge in G joins a vertex in v_1 with a vertex in v_2.

Note: Every tree is a bipartite graph.

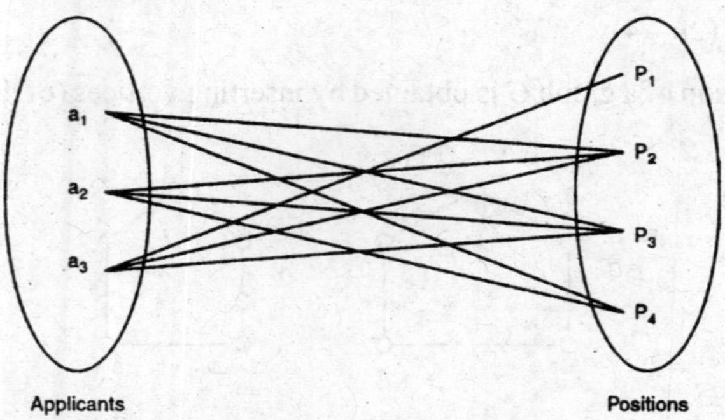

Theorem Every tree with two or more vertices is bichromatic.

Proof Choose any vertex r in the given tree T. Let T be a rooted tree at vertex v. Paint v with colour 1. Paint all vertices adjacent to v with colour 2. Next, paint the vertices adjacent to these using colour1. Repeat this process till every vertex in T has been painted. Hence all vertices at odd distances from v have colour 2 while v and vertices at even distances from v have colour 1.

Along any path in T the vertices are of alternating colours. Since there is one and only one path between any two vertices in a tree, no two adjacent vertices have the same colour.

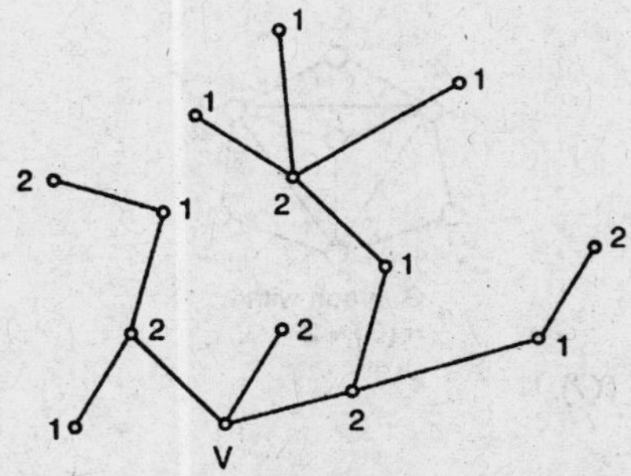

Proper colouring of a tree

Theorem A tree with atleast one edge is bichromatic.

Proof Let T be a tree & v be any vertex of T. For this vertex v assign colour 1. Assign colour 2 to be adjacent vertex of v. Continuing this process, we can see that all vertices at even distance from v are assigned colour 1 & the vertices at odd distance from v are assigned colour 2.

This gives a two. Colouring of T. Since T is a connected graph with atleast one edge, T is not 1 - coloured $\therefore$ T is bichromatic.

Theorem A graph is bipartite iff it is bichromatic.

Proof Let G a bipartite graph with atleast one edge. Let (v_1, v_2) be the partition of the vertex set of G. In v_1 as well as in v_2, no two vertices are adjacent. Now assign colour 1 to the vertices in v_1 & assign colour 2 to the vertices in v_2. Hence G is bichromatic.

Conversely, let us assume that G is a bichromatic, graph. Therefore $x(G) = 2$. Let v_1 be the set of all vertices for which colour 1 is assigned & v_2 be the set of all vertices for which colour 2 is assigned. Clearly (v_1, v_2) is the partition of the vertex set of G. Otherwise atleast two vertices in v_1 or v_2 have the same colour. Therefore G is a bipartite graph.

Note: A graph G is bipartite if it contains no odd cycles.

Independent Set A set U of vertices in a graph G is called an independent set if no two vertices in v are adjacent in G. An independent set U of vertices in a graph G is called a maximal independent set if U is not a proper subset of any other independent set of vertices of G. The cardinality of a maximal independent set is called as an independence number and is denoted by $\beta(G)$

Example Here $\{v_1, v_4, v_2, v_7\}$, $\{v_1, v_6\}$, $\{v_1, v_7\}$, $\{v_2, v_7\}$ are all independent set & the sets $\{v_1, v_4, v_2, v_7\}$ is a maximal independent set & $\beta(G) = 4$.

Dominating Set A set S of vertices in a graph G is a dominating *set* if every vertex not in S is adjacent to a vertex in S. A dominating set S is called as a minimal dominating set if no proper subset of S is a dominating set. The cordinality of a mimimal dominating set is called as domination number and is donoted by $\sigma(G)$.

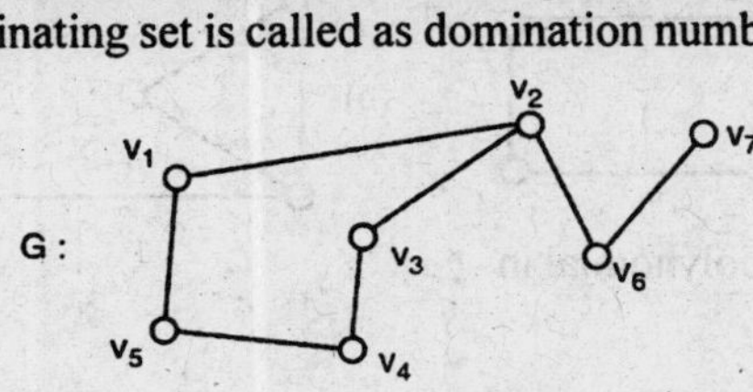

Example

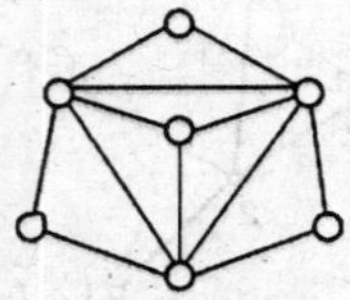

G. graph with
σ (G) = 2

Note: For every graph G, $\sigma(G) \leq \beta(G)$.

Chromatic Polynomial

A graph G of n vertices can be properly coloured in many different ways using a sufficiently large number of colours. This property of a graph is expressed by a polynomial. This polynomial can be referred to as chromatic polynomial.

The value of chromatic polynomial is given by

$$P_n(\lambda) = \sum_{i=1}^{n} c_i \begin{pmatrix} \lambda \\ \cdot \\ l \end{pmatrix}$$

Where C_i is the different ways of properly colouring G using exactly l colours out of λ colours in $\begin{pmatrix} \lambda \\ \cdot \\ i \end{pmatrix}$ different ways.

Theorem A graph of n vertices is a complete graph if and only if its chromatic polynomial is

$$P_n(\lambda) = \lambda(\lambda-1)(\lambda-2)\ldots(\lambda-n+1)$$

Proof Let us use λ colours, then there are λ different ways of colouring any selected vertex of a graph. A second vertex can be coloured properly in exactly (λ–1) ways, the third in (λ–2) ways, the fourth in (λ–3) ways etc., and n^{th} in (λ–n+1) ways iff every vertex is adjacent to every other. That is if and only if the graph is complete.

Note: An n - vertex graph is a tree if its chromatic polynomial $p_n(\lambda) = \lambda(\lambda-1)^{n-1}$.

EXERCISE XII

1. Find the chromatic polynomial of the following graphs

(a) (b)

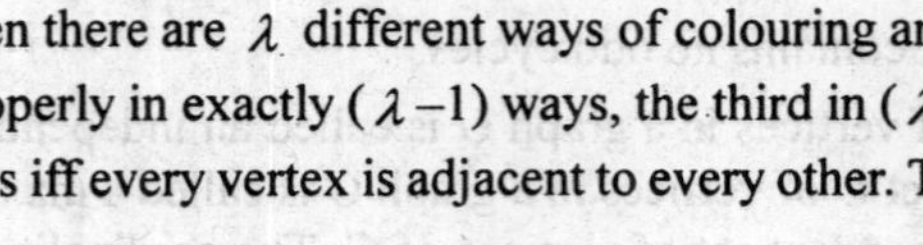

2. For every graph G, $P(G, \lambda)$ is a polynomial in λ.

3. Determine $\beta(G), \sigma(G), X(G)$ & $P(G, \lambda)$ for the following graph G.

G :

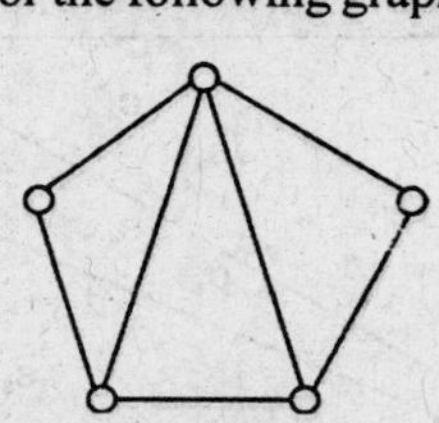

4. Determine the chromatic polynomial of a cycle of length 5.
5. Let W_n be a wheel on 'n' vertices and

$$W_n = C_n + K_1, \forall n \geq 3.$$
$$P(w_n, \lambda) = \lambda . P(c_n, \lambda - 1)$$

6. Find the chromatic number of the following graph.

a. $k_{4,4}$ b. $k_{n,n}$ c. peterson graph d.

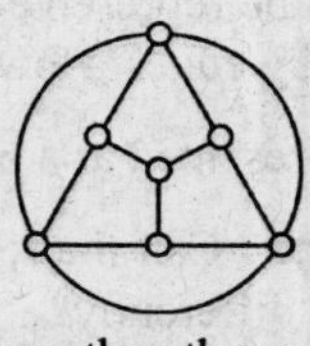

7. Prove that the chromatic number of a graph will not exceed by more than the maximum degree of the graph.
8. Give an example of graph G that has only one chromatic partition.
9. Find the following for the, given graph

i. β (G) ii. σ (G) iii. X(G)

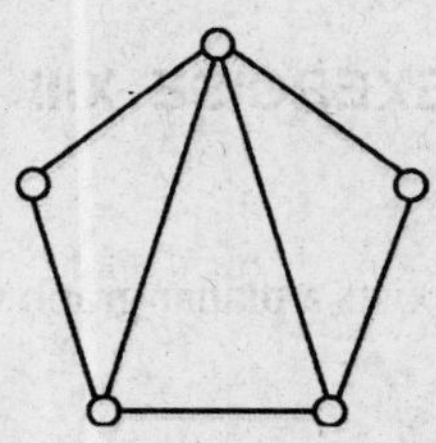

10. Prove that if G is a n-vertex graph then $n \leq X(G) . \beta(G)$
11. Prove or disprove : If G is a graph with $X(G) = n$, then $\delta(G) \leq n$-1.

Map Colourings

When we colour the map of a country we see to it that its neighbouring states are coloured differently. In such circumstances, one may ask what is the minimum number of colours needed to colour so that the adjacent states receive different colours?

From the following example, one can say that 3 colours are not sufficient.

Example

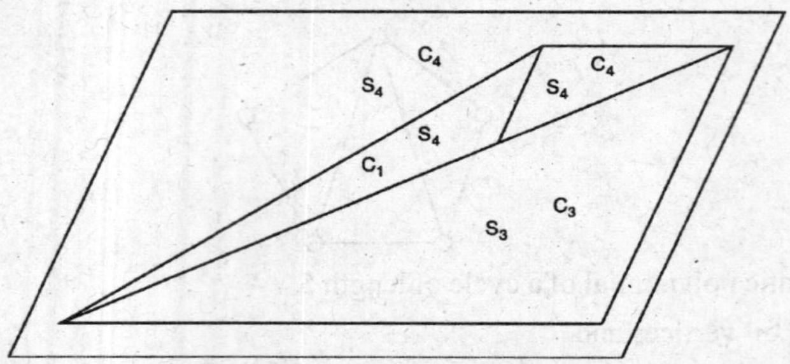

In the above map, S_i denotes the states & C_i denotes the colours received by S_i. clearly 3 colours are not sufficient for map colouring so that no two adjacent states receive the same colour.

i.e. The above map can be represented as a graph. The states are denoted by vertices and the adjacency between states are denoted by edges. Now the graph obtained in the above fashion is not 3 - colourable.

Note: Every map can be represented as a planar graph.

The four colour problem Can the regions of a planar graph be coloured with four colours so that the adjacent states are coloured differently?

Yes every planar graph is 4 - colourable.

Theorem Every simple planar graph is 5-colourable.

(Interested readers can refer "Graph theory with application" by Narsingh Deo for the proof (p. 188 – 189).

EXERCISE XIII

1. If $X(G) \leq 4$, then G is planar.
2. Prove that for every integer $n \leq 4$, there exists a planar graph with 'n' vertices such that $X(G) = 4$.

CHAPTER

8

Formal Languages and Automata

ALPHABET

An alphabet is a finite non-empty set. The letters I, Σ, V, N, T are used for alphabets.

A word or string over an alphabet I is a finite sequence $\{a_1 .... a_k\}$ of elements from I. The length of the word is denoted by $|w|$ and is equal to the number of symbols in w.

If $|w| = 0$, then it is referred to be an empty word and is denoted by $\in$. A non-empty word is of length $|w| \geq 1$.

Two words $a_1 \ldots a_k$ and $b_1 \ldots b_r$ are equal iff $k = r$ and $a_i = b_i$ for each i.

If $x = a_1 \ldots a_k$ and $y = b_1 \ldots b_r$ then xy, called the product or concatenation of x and y is the word.

The set of all words, including $\in$ over an alphabet I is denoted by I^* and I^+ where $I^+ = I^* - \{\in\}$.

GRAMMAR

A grammar is a non-empty finite set (say G) consisting (N, T, P, S) where $N \cap T = \phi$.

$N \rightarrow$ Finite set of symbols called non-terminals (or variables or non-terminal vocabulary or non-terminal alphabet).

$T \rightarrow$ Finite set of symbols called the terminals (or terminal vocabulary or terminal alphabet or terminal letters).

$S \rightarrow$ Start symbol (or sentence symbol or initial symbol).

$V = N \cup T$

$P \rightarrow$ Finite set of ordered pairs of the form (u, v) where $u \in v^+$ contains at least one non-terminal and $v \in v^*$. Each ordered pair (α, β) is generally written as $u \rightarrow v$ and is read u rewritten as v.

Notation

Terminals $\rightarrow a, b, c \ldots$

Non-terminals $\rightarrow A, B, C \ldots$

Strings of terminals $\rightarrow x, y, z, w \ldots$

Strings of Terminals and Non-Terminals

We write $\alpha u \beta \underset{G}{\Rightarrow} \alpha v \beta$ if $(u, v) \in P$ and $\alpha, \beta \in v^*$. We say that the rule $u \to v$ is applied to the string $\alpha u \beta$ to obtain the string $\alpha \vee \beta \overset{*}{\underset{G}{\Rightarrow}}$ which is the reflexive, transitive closure of $\Rightarrow$. In other words if $\alpha_1, \alpha_2, \ldots, \alpha_n$ are strings in V and $\alpha_1 \underset{G}{\Rightarrow} \alpha_2 \underset{G}{\Rightarrow} \alpha_3 \underset{G}{\Rightarrow} \ldots \underset{G}{\Rightarrow} \alpha_n$ we write $\alpha_1 \overset{*}{\underset{G}{\Rightarrow}} \alpha_n$. Now we can define the language generated by a phrase-structure grammar G.

Phrase-structure grammar (PSG)

Let

i. $G = (V, I, P, S)$ where $I \subset V$ or

ii. $G = (V, \Sigma, P, \sigma)$ where $\Sigma \subset V$ and σ is the start symbol or

iii. $G = (V_N, V_T, P, S)$ where $V_N \cap V_T = \varphi$ and $S \in V_N$.

Phase structure language If $G = (N, T, P, S)$ is a PSG, then $L(G) = \{w / s \overset{*}{\Rightarrow} w, w \in T^*\}$ is the phrase-structure language (PSL) generated by G.

Context-sensitive grammar A phrase-structure grammar $G = (N, T, P, S)$ is called a context-sensitive grammar (CSG) if each $(u, v) \in$, P such that $u = \alpha U \beta$, $v = \alpha y \beta$ where $U \in N$, $\alpha, \beta \in V^*$ and $y \in V^+$.

The phrase-structure language generated by a context-sensitive grammar is called a context-sensitive language.

The name context-sensitive refers to the fact that a rule $\alpha u \beta \to \alpha y \beta \in \rho$ means that U can be replaced by v in the context α, β.

We impose the restriction that $|u| \leq |v|$ (u must contain atleast one non-terminal)

Note: Context-sensitive language can also be generated when the rules in P are of the form (u, v) where u is in N^+ and v in V^+.

Production rule

$U \to V$, where U and V are strings of non-terminals only; or $A \to b$, where A is a non-terminal and b is a terminal.

Context free grammar Let $G = (N, T, P, S)$ be a PSG. If every rule in P is of the form $A \to \alpha$, where $A \in N$, and $\alpha \in v^*$ then G is called context-free grammar (CFG).

Note: Every rule in CFG is such that the left side is a single non-terminal and the right side any string of terminals and non-terminals. The language $L(G)$ generated by a context-free grammar G is called a context-free language.

Regular set If a phrase-structure grammar consists of only rules of the form $A \to a B$ or $A \to a$ then the grammar is called a regular grammar or right linear grammar. A language generated by a regular grammar is called a regular set.

Note: From the definition of the grammars and the corresponding languages, the family of PSL $\geq$ the family of CSL $\geq$ the family of CFL $\geq$ the family of regular sets.

Example

$L(G)=\{a^n b^n c^n \mid n \geq 1\}$

Let $G=(V_N, V_T, P, S)$

where $V_N=\{S, A, B\}, V_T=\{a, b, c\}$

P consists of the following productions:

(1) $S \rightarrow aS\,AB$ (2) $S \rightarrow aAB$ (3) $BA \rightarrow AB$ (4) $aA \rightarrow aa$ (5) $bA \rightarrow ba$ (6) $bB \rightarrow bb$ (7) $cB \rightarrow cb$

If we use the production 1,(n–1) times, we get

$S \overset{*}{\Rightarrow} a^{n-1} S (AB)^{n-1}$ $\{P_1\,(n-1)$ times$\}$

$S \overset{*}{\Rightarrow} a^n\ (AB)^n$ $\{P_2\ n$ times$\}$

$S \overset{*}{\Rightarrow} a^n\ A^n B^n$ $\{P_3\}$

$S \overset{*}{\Rightarrow} a^n b^n B^n$ $\{P_4$ once and $P_5\,(n-1)$ times$\}$

$S \overset{*}{\Rightarrow} a^n b^n c^n$ $\{P_6$ and $P_7\}$

Therefore $L(G)=\{a^n b^n c^n \mid n \geq 1\}$. It is a phrase-structure language or type 1 language.

Example

$L(G)=\{a^n b^n \mid n \geq 1\}$

Let $G=\{V_N, V_T, P, S\}$, $V_N=\{S\}$, $V_T=\{a, b\}$

and $P=\{S \rightarrow aSb,\ S \rightarrow ab\}$

$S \rightarrow aSb$

$\rightarrow aabb \rightarrow a^2 b^2$

Hence $n=2>1$

Example The grammar generating

$L=\{wcw^R \mid w \text{ in } (a, b)^*)$

Let $G=\{V_N, V_T, P, S\}$, $V_N=\{S\}$

$V_T=\{a, b\}$ and $P=\{S \rightarrow aS\,a,\ S \rightarrow bSb,\ S \rightarrow c\}$

$S \rightarrow aSa$

$\rightarrow abSba$

$\rightarrow abcba$

Example $L=\{a^i b^j \mid j>i\}$

L is generated by G where

$G=\{V_N, V_T, P, S\}$, $V_N=\{S\}$

$V_T=\{a, b\}$ and $P=\{S \rightarrow aS\,b,\ S \rightarrow Sb,\ S \rightarrow b\}$

Example 8.1 Obtain grammar for Language

$L=\{a^m b^n : n \neq m,\ n>0\}$

Solution Define a Grammar 'G'

$$G = \{V_N, V_T, P, S\} \quad \text{where } V_N = \{S, A\}, \quad V_T = \{a, b\},$$

P for

Case (i) $m > n$

$$S \to a\ S_1$$
$$S_1 \to a\ A$$
$$A \to aAb$$
$$A \to ab$$

e.g.

$$S \to a S_1$$
$$\to aaA$$
$$\to aaaAb$$
$$\to aaaabb \to a^4 b^2$$

Case (ii) $m < n$

$$P \Rightarrow \quad S \to S_2$$
$$S_2 \to S_2 b$$
$$S_2 \to Ab$$
$$A \to aAb$$
$$A \to ab$$

$$a^m \left(\frac{b^m}{b^m}\right) b^n \Rightarrow a^m b^m b^{n-m}$$

$$\Rightarrow b^{n-m} a^m b^m$$

$$S \to S_2$$

$$\overset{*}{\Rightarrow} S_2\, b^{n-m-1}\ [S_2 \to S_2 b\ (n-m-1) \text{ times}]$$

$$\Rightarrow bAb^{n-m-1}\ [S_2 \to Ab \text{ once}]$$

$$\to b^{n-m}\ A$$

$$\overset{*}{\Rightarrow} b^{n-m} a^{m-1} Ab^{m-1} [A \to aAb\ (m-1) \text{ times}]$$

$$\Rightarrow b^{n-m}\ a^{m-1}\ ab^{m-1} b [A \to ab \text{ once}]$$

$$\Rightarrow b^{n-m} a^m b^m$$

Hence $a^m b^n$ belongs to the Language given hence $L(G)_2\ L_1 \cup L_2$

Example 8.2 Construct a grammar which generates all even integers upto 998.

Solution

Let $G = \{V_N, V_T, P, S\}$

$V_N \to \{S, S_1 A, B\}$

$V_T \to \Sigma$ where $\Sigma = \{0, 1, 2, ..., 9\}$

and P consists of $S \to 0 / 2 / 4 / 6 / 8$

$$S \to A\ S_1$$
$$A \to 1/2/3/.../9$$
$$S_1 \to 0/2/4/6/8$$
$$S \to AB\ S_1$$
$$B \to 1/2/.../9$$

$S \to 0/2/4/6/8$ generate even integers with one digit

$S \to A\ S_1$ generate all even integers with two digit remaining producetions can be used to generate all even integers with three digits.

$$S \to \underline{A}\ S_1$$
$$\to (1/2/3/.../9)S_1$$
$$\to (1/2/3/.../9)\ (0/2/4/6/8)$$

i.e 10, 12, 14, 16, 18 or 20, 22, 24, 26, 28, etc., 90, 92, 94, 96, 98

$$S \to A\ B\ S_1$$
$$\to (1/2/3/...9)\ BS_1$$
$$\to (1/2/3/...9)\ (1/2/3/...9)\ S_1$$
$$\to (1/2/3/...9)(1/2/3/.../9)(0/2/4/.../8)$$

e.g. 110, 122, etc.,

Example 8.3 Canstruct a grammar for the language $L = \{a^x b^y,\ x > y > 0\}$

Solution

Let $G = \{V_N, V_T S, P\}$

where $V_N = \{S, A\}$ and $V_T = \{a, b\}$

P consists of $S \to a\ S$

$$S \to aA$$
$$A \to aAb$$
$$A \to ab$$
$$a^x b^y = a^{x-y} a^y b^y$$

$A \to aAb$ and $A \to ab$ can be used to generate $a^x b^y$. For getting, a^{x-y} in $a^x b^y$ we can use the production $S \to aS$, $S \to aA$. Now

$$S \overset{*}{\Rightarrow} a^{x-y-1}S \text{ (by applying) } S \to aS\ x - y - 1 \text{ times)}$$
$$\Rightarrow a^{x-y-1}aA \text{ (by applying} A \to aAb(j-1) \text{ times)}$$
$$\overset{*}{\Rightarrow} a^{x-y}a^{y-1}Ab^{y-1} \text{ (by applying } A \to aAb\ (j-1) \text{ times)}$$
$$\Rightarrow a^{x-y}a^y b^y \text{ (by applying } A \to ab \text{ once)}$$

Hence $a^x b^y \in L(G)$ where $x > y > o$.

FINITE AUTOMATON

A finite automaton (FA) consists of a finite set of states and a set of transitions from one state to the other. It occurs on input symbols chosen from alphabet. For each input symbol there is exactly one transition from of each state (possibly back to the state itself). q_0 is the initial state, in which the automaton starts. Some states are designated as final or accepting states.

A directed graph, called a transition diagram, is associated with a finite automaton. Therefore we now define finite automaton by a 5-tuple $(Q, \Sigma, \delta, q_o, F)$ where Q is a finite set of states, Σ is a finite input alphabet, $q_o \in Q$ is the initial state, $F \leq Q$ is a set of final states and $\delta : Q \times \Sigma \rightarrow Q$. That is $\delta(q, a)$ is a state for each state q and input symbol a.

A string x is said to be accepted by a finite automaton $M = (Q, \Sigma, \delta, q_o, F)$ if $\delta(q_0, x) = p$ for some p in the final state. The language accepted by M, designated by $L(M)$, is the set $\{x / \delta(q_0, x) \text{ is in final state}\}$. A language is a regular set (or just regular) if it is the set accepted by same finite automaton.

The transition function which maps $Q \times \Sigma^*$ into Q i.e., (maps a state and a string of input symbols including the empty string into a state) is called the indirect transition function. We shall use the same symbol δ to represent both types of transition functions and the difference can be easily identified by the nature of mapping (symbol or a string) i.e., by argument.

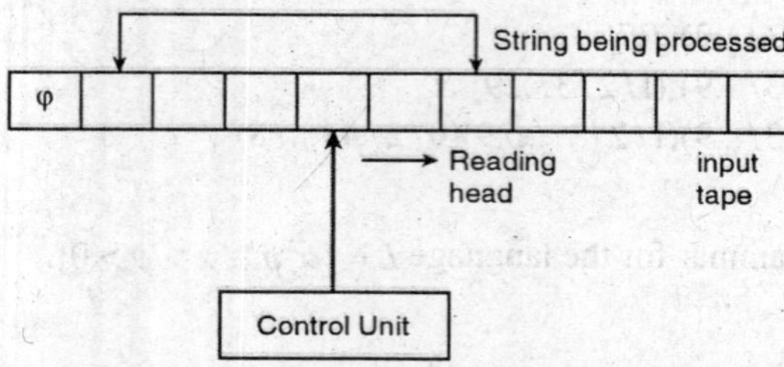

Block diagram of a finite automaton

Properties of Transition Function

1. $\delta(q, \in) = q$ in a finite automaton. That is the state of the system can be changed only by an input symbol.
2. For all strings (w) and input symbols (a),

$$\delta(q, aw) = \delta(\delta(q, a), w)$$
$$\delta(q, wa) = \delta(\delta(q, w), a)$$

That is the state after the automaton consumes or reads the first symbol of a string aw and the state after the automaton consumes a prefix of the string wa.

Example Consider the diagram

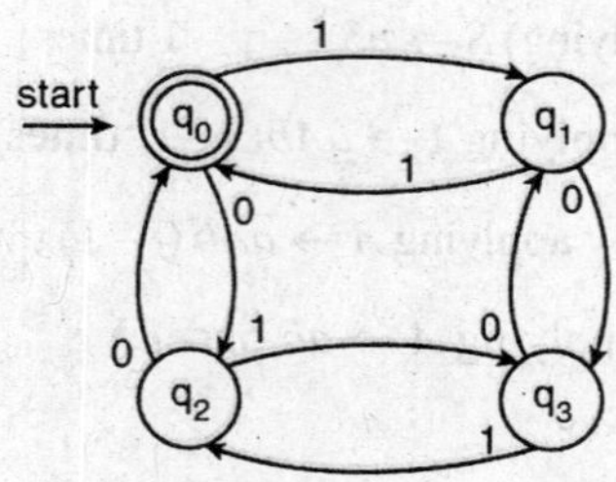

Finite automaton:

$$M = (Q, \Sigma, \delta, q_o, F)$$

where $Q = \{q_0, q_1, q_2, q_3\}$, $\Sigma = \{0,1\}$, $F = \{q_0\}$

where $\delta(q_0, 0) = q_2$

$\delta(q_0, 1) = q_1$

$\delta(q_1, 0) = q_3$

$\delta(q_1, 1) = q_0$

$\delta(q_2, 0) = q_0$

$\delta(q_2, 1) = q_3$

$\delta(q_3, 0) = q_1$

$\delta(q_3, 1) = q_2$

Consider the input 110101 to M

$$\begin{aligned}\delta(q_0, \underline{1}10101) &= \delta(q_1, \underline{1}0101)\\ &= \delta(q_0, \underline{0}101)\\ &= \delta(q_2, \underline{1}01)\\ &= \delta(q_3, \underline{0}1)\\ &= \delta(q_1, 1)\\ &= q_0\end{aligned}$$

or we can use it the other way

$$\delta(q_0, 11) = \delta(\delta(q_0, 1), 1) = \delta(q_1, 1) = q_0$$

We might remove this 11 in $L(M)$ by continuing as follows

$$\begin{aligned}\delta(q_0, 110) &= \delta(\delta(q_0, 11), 0)\\ &= \delta(q_0, 0)\\ &= q_2\end{aligned}$$

again

$$\delta(q_0, 1101) = q_3,\ \delta(q_0, 11010) = q_1 \text{ and finally } \delta(q_0, 110101) = q_0$$

∴ The entire sequence of state is $q_0\ q_1\ q_0\ q_2\ q_3\ q_1\ q_0$.

∴ $L(M)$ is the set of strings with an even number of 0's and an even number of 1's.

Example 8.4 Prove that for any transition function δ and for any two input strings x and y.

$$\delta(q_1\, xy) = \delta(\delta(q_1 x), y)$$

Solution By the method of induction on $|y|$

$$|y| = \text{length of } y$$

$$\text{when } |y| = 1, \quad y = b \in \Sigma$$

$$\text{LHS} = \delta(q_1, xb) = \delta(\delta(q_1 x), b) \text{ (using property 2)}$$

$$= \textbf{RHS}$$

Assume the result, is $\delta(q_1\, xy) = \delta(\delta(q_1\, x), y)$ for all strings x and y with $|y| = n$. Let y be the string of length $n + 1$

$$\therefore\ y = y_1\, a \text{ where } |y_1| = n$$

$$\delta(q_1\, x\, y) = \delta(q_1\, x\, y_1\, a)$$
$$= \delta(q_1, x_1 a) \text{ where } x_1 = xy_1$$
$$= \delta(\delta(\delta(q_1 x), y_1)a) \text{ by induction hypothesis}$$

Similarly

$$\delta(\delta(q_1\, x), y) = \delta(\delta(q, x), y_1 a)$$
$$= \delta(\delta(\delta(q_1 x), y_1), a) \text{ (using property 2)}$$

$\therefore$ *LHS* = *RHS*. This proves that

$\delta(q_1, xy) = \delta(\delta(q_1 x), y)$ for string y of length $n + 1$. By principle of induction the above result is true for all strings.

Non-Deterministic Finite Automata

Let us modify the finite automaton model to allow zero, one or more transitions from a state in the same input symbol. This new model is called a non deterministic finite automaton (NFA).

A non deterministic finite automaton (NFA) is a 5-tuple $(Q, \Sigma, \delta, q_0, F)$, where

i. $Q \rightarrow$ is a finite non-empty internal state
ii. $\Sigma \rightarrow$ is a finite non-empty set of inputs
iii. $\delta \rightarrow$ next state function. Here $\delta : Q \times \Sigma \rightarrow 2^Q$
iv. $q_0 \in Q$ is the initial state and $F \leq Q$ is the set of final states.

Note: Difference between DFA and NFA is only in δ. For deterministic automaton (DFA) the outcome is a state i.e a state is an element of Q. For non deterministic automaton (NFA) the outcome is a subset of Q.

Example Consider

$$M = (Q, \Sigma, q_0, \delta, F) \quad \text{where } Q = \{q_0, q_1, q_2, q_3, q_4\}$$

$F_2\{q_4\}$ $\delta(q_0, 0) = \{q_0, q_3\}$

$\delta(q_0, 1) = \{q_0, q_1\}$

$\delta(q_1, 0) = q_1$

$\delta(q_1, 1) = q_2$

$\delta(q_2, 0) = q_2$

$\delta(q_2, 1) = q_2$

$\delta(q_3, 0) = q_3$

$\delta(q_4, 0) = q_4$

$\delta(q_4, 1) = q_4$

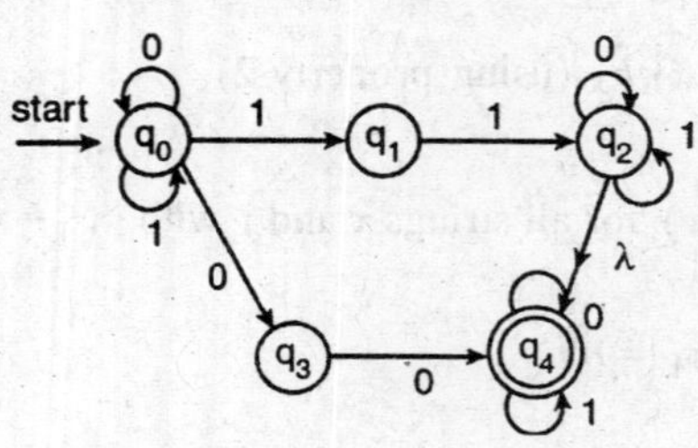

Let us consider an input string 0100

$$\begin{aligned}\delta(q_0,\underline{0}100) &\Rightarrow \delta(q_0,\underline{1}00)\\ &\Rightarrow \delta(q_0,\underline{0}0)\\ &\Rightarrow \delta(q_3,0)\\ &\Rightarrow q_4, \in F\end{aligned}$$

Note:

1. A final state is also called an accepting state.

   Since q_4 is an accepting state for the input string, 0100 will be accepted by the non deterministic automaton.

2. A string $w \in \Sigma^*$ is accepted by NFA M is $\delta(q_0, w)$ containing some final state.

   Since M is non deterministic, $\delta(q_0, w)$ may have more than one state. So w is accepted by M if a final state is one among the possible states. M can reach an appreciation of w.

3. The set accepted by an automaton M (deterministic or non deterministic) is the set of all input strings accepted by M. It is denoted by $T(M)$.

Transition System Containing ∈ Moves

∈-transitions or ∈-moves are associated with a null symbol ∈. These transitions can occur when no input is applied. But it is possible to convert a transition system with ∈ moves into an equivalent transition system without ∈ moves.

Suppose we want to replace a ∈-move from vertex v_1 to vertex v_2, then we proceed as follows.

Step 1 Find all the edges starting from v_2.

Step 2 Duplicate all these edges starting from v_1, without changing the edge labels.

Step 3 If v_1 is an initial state, make v_2 also as an initial state.

Step 4 If v_2 is a final state, make v_1 also as the final state.

Example 8.5

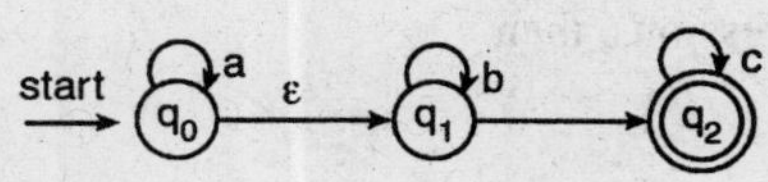

Solution Eliminate ∈-move from q_0 to q_1

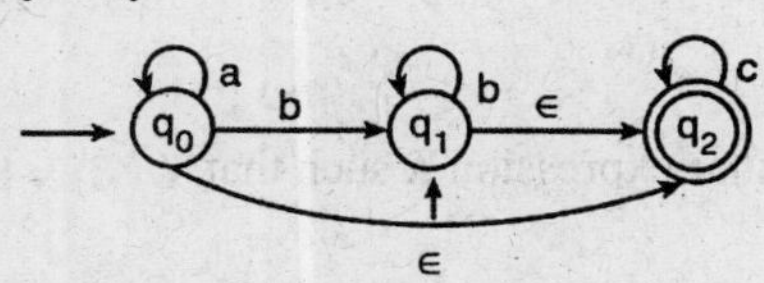

then q_1 is made the initial state. Now we eliminate the ∈-move from q_0 to q_2.

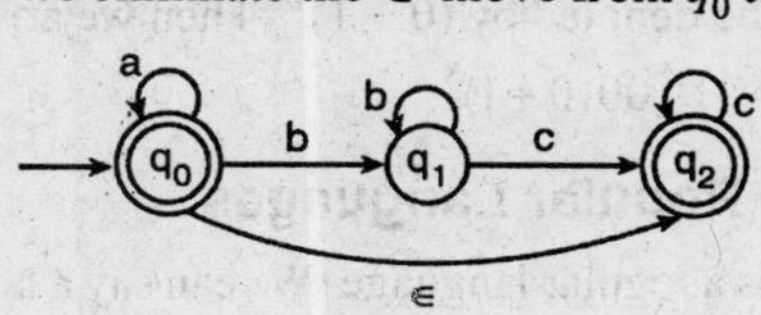

As q_2 is the final state, q_0 is also made a final state. Finally the $\in$-move from q_1 to q_2 is eliminated.

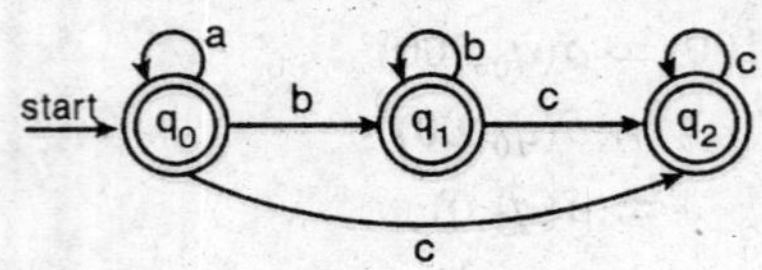

REGULAR EXPRESSIONS

It is easy to describe regular languages using regular expressions. This notation involves a combinations of strings and symbols from some alphabet Σ, parenthesis and the operators +, . , and *.

Example The expressions $(a + b.c)^*$ stands for the star closure of $\{a\}\ U\ \{bc\}$

$L = \{\lambda, a, bc, aa, abc, bca, bcbc, aaa, aabc, ...\}$

Let Σ be a given alphabet. Then

1. φ, λ, and $a \in \Sigma$ are all regular expressions. More ever it may be referred to as primitive regular expressions.
2. If R_1 and R_2 are regular expressions so are $R_1 + R_2$, $R_1.R_2$, R_1^* and (R_1)
3. A string is a regular expression if it can be derived from the primitive regular expressions within a finite number of applications. (using rule 2).

Example For $\Sigma = \{a, b, c\}$, the string $(a + b.c)^* \cdot (C + \varphi)$ is a regular expression, since it is constructed by application of the above rules.

Let $R_1 = c$ and $R_2 = \varphi$ we find that $c + \varphi$ and $(c + \varphi)$ are also regular expressions. Repeating this, we eventually generate the whole string.

The language $L(R)$ denoted by any regular expression R is defined by the following rules.

1. φ is a regular expression denoting the empty set.
2. λ is a regular expression denoting $\{\lambda\}$.
3. For every $a \in \Sigma$, a is a regular expression denoting $\{a\}$.

   If R_1 and R_2 are regular expressions, then
4. $L(R_1 + R_2) = L(R_1)U\ L(R_2)$
5. $L(R_1.R_2) = L(R_1)L(R_2)$
6. $L((R_1)) = L(R_1)$
7. $L(R_1^*) = (L(R_1))^*$

Example For $\Sigma = \{0, 1\}$, give a regular expression R such that $L(R) = \{w \in \Sigma^* : W$ has atleast one pair of consecutive zeros$\}$.

Every string in $L(R)$ must contain zero some where, but what comes before and after it is completely arbitrary. An arbitrary string on $\{0, 1\}$ can be denoted by $(0 + 1)^*$. Then we arrive at the solution

$$R = (0+1)^* 00 (0+1)^*.$$

Regular Expressions Denote Regular Languages

If R is a regular expression, then $L(R)$ is a regular language. We can say a language is regular if it is accepted by some non deterministic finite automaton. If we have any regular expression R, we can find a non deterministic finite automaton that accepts $L(R)$. Let us consider an automaton that accept the languages for the simple regular expressions φ, λ, and $a \in \Sigma$.

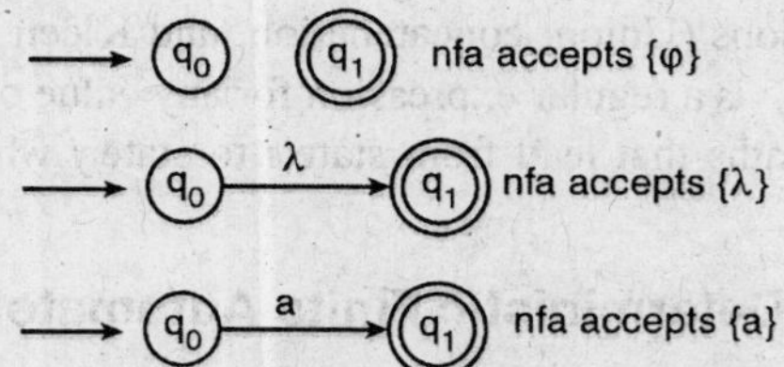

Assume now that we have automata $M(R_1)$ and $M(R_2)$ that accept languages denoted by regular expressions R_1 and R_2 respectively.

The graph vertex at the left represents the initial state, and the diagram encircled by double circle is referred to as final state. We construct automata for the regular expressions $R_1 + R_2$, R_1R_2 and R_1^*.

Schematic representation of an nfa accepting for L (R)

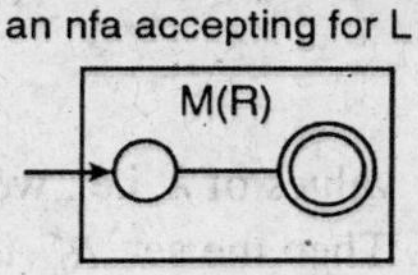

Automaton for L (R_1 + R_2)

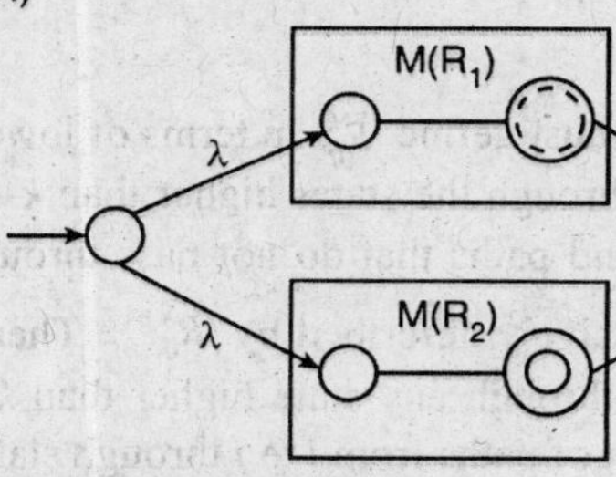

Automaton for L (R_1R_2)

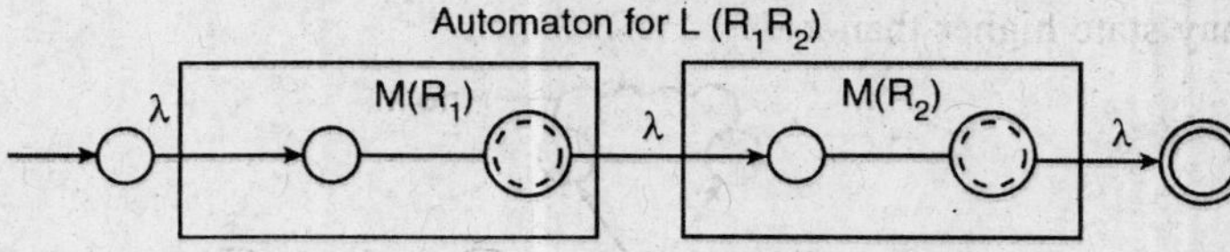

As indicated in the above drawings, the initial and final states of the constituent machines lose their status and are replaced by new initial and final states. By stringing together several such steps, we can build automata for arbitrarily complex regular expressions.

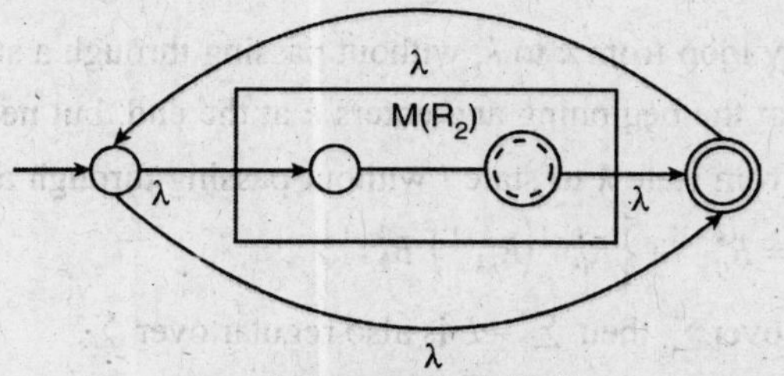

Automaton for L (R_1*)

With the help of above graphs the result may be obviously true. Let R be a regular expression. Then there exists some non deterministic finite accepter that accepts $L(R)$. Consequently $L(R)$ is a regular language. To argue more rigorously, we can give a formal method for constructing the states and transitions of the parts, then an automaton that accepts the language denoted by any particular regular expression. Where * denotes the particular expression to any number of iterations of this loop. Therefore it must be $\left(R_{KK}{}^{k-1}\right)^*$.

With this inductive step, we can proceed to build all possible paths in the machine from the expressions for the base cases. Since R^{KS} are built from the regular expressions for the various R^{KS-1} *s* using only operations that are closed for regular expressions (Union, concatenation, and Kleen closure), the R^{KS} are also regular expressions. ∴ We can state that R_{ij}^{k} is a regular expression for any value of *i*, *j* and *k*, with $1 \le i,\ j, k \le n$ and that these expressions denote all paths that lead from state *i* to state *j* while not passing through any state numbered higher than *k*.

Regular Expressions from Deterministic Finite Automaton

Base Let R_{ij}^{0} be the set of paths for all pairs of nodes *i* and *j* in the deterministic finite automaton. For a specific pair of nodes *i* and *j*, these are the paths that go directly from node *i* to node *j* without passing through any intermediate states.

$$\Sigma = \{\text{if } i = j\};\ \text{and} \mid \text{or}$$
$$a = \{\text{if } \delta(q_L, a) = q_j\}$$

including the case *i* = *j* with a self loop

Note: {is the path of length zero.

Inductive The inductive step must define R_{ij}^{k} in terms of lower values of *k*. i.e., we want to know how to move from *i* to *j* without going through the states higher than *k*–1. Then the set R_{ij}^{k} is nothing but union of paths that pass through state *k* and paths that do not pass through state *k*.

The second set can be recursively described by R_{ij}^{k-1}. There will be a difficulty about paths that pass through state *k* without passing through any state higher than *k*–1, even thought *k* is higher than *k*–1. To resolve this difficulty, split the set of paths from *i* to *j* through state *k* into three separate components, none of which passes through any state higher than *k*–1.

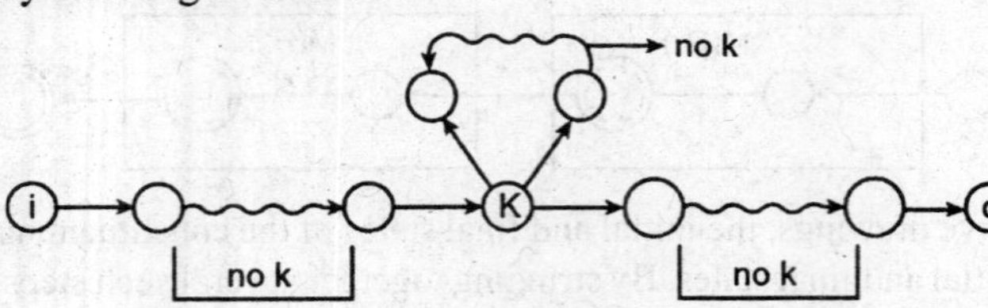

R_{ik}^{k-1} → Paths that go from i to *k* without passing through a state higher than *k*–1

R_{kk}^{k-1} → One iteration of any loop from *k* to *k*, without passing through a state higher than *k*–1 (The paths exits *k* at the beginning and enters *k* at the end, but never passes through *k*)

R_{kj}^{k-1} → The paths that go from state *k* to state *j* without passing through a state higher than *k*–1.

$$\therefore R_{ij}^{k} = R_{ij}^{k-1} + \left\{ R_{ik}^{k-1} \left(R_{kk}^{k-1} \right)^{*} R_{kj}^{k-1} \right\}$$

Theorem If *L* is a regular set over Σ, then $\Sigma^{*} - L$ is also regular over Σ.

Proof Let us consider a *DFA M* where $M = (Q, \Sigma, \delta, q_0, F)$ accepting *L* i.e., *L* = *T*(*M*).

We construct another DFA say $M' = (Q, \Sigma, \delta, q_0, F')$ by defining $F' = Q - F$, i.e., *M* and M' differ only in their final states. A final state of M' is a non final state of *M* and vice versa. The state diagrams of *M* and M' are the same except for the final states.

$$w \in T(M') \text{ iff } \delta(q_o, w) \in F' = Q - F$$

i.e. if $w \notin L$. This proves that $T(M') = \Sigma^{*} - X$.

Theorem If X and Y are regular sets over Σ, then $X \cap Y$ is also regular over Σ

Proof Using De Morgan's law for sets $X \cap Y = \Sigma^* - \{(\Sigma^* - X) \cup (\Sigma^* - Y)\}$. Then using the theorem $\Sigma^* - x$ and $\Sigma^* - y$ are regular. So $(\Sigma^* - X) \cup (\Sigma^* - Y)$ is also regular. By applying the theorem once again we have $\Sigma^* - \{(\Sigma^* - X) \cup (\Sigma^* - Y)\}$ is regular i.e. $X \subset Y$ is regular.

Pumping Lemma for Regular Sets

If a Language is regular, it is accepted by a DFA say $M = (Q, \Sigma, d, q_o, F)$ with some particular number of states say n.

Let us consider the input symbol $a_1, a_2, ..., a_m$ where $m >, n$ and for $i = 1, 2, ...m$

Let $\delta(q_0, a_1, a_2, ...a_i) = q_i$.

It is not possible for each of the $(n + 1)$ states $q_0, q_1, ..., q_n$ to be distinct, since there are only 'n' different states. Let us consider j and k be two integers where $0 \le j < k \le n$ such that $q_j = q_k$

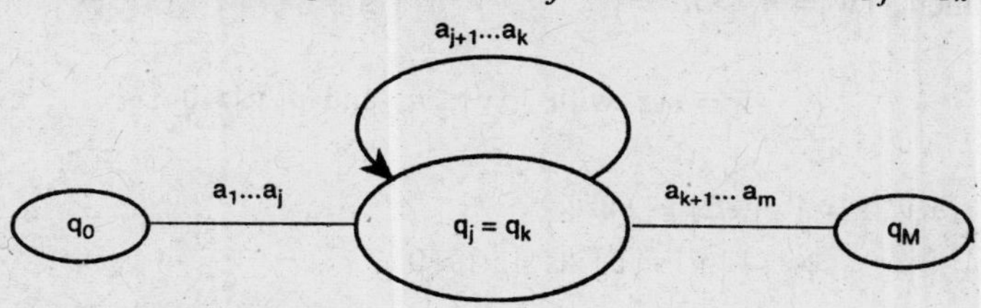

$a_{j+1}...a_k$ is of length atleast 1 since $j < k$.

$a_{j+1}...a_k$ is of length not more than n since $k < n$.

$$\delta(q_0, a_1, ..., a_j \ \ a_{k+1}, ...a_m) = \delta\left(\partial(q_0, a_1, ..., a_j), \ a_{k+1}, ..., a_m\right)$$
$$= \delta(q_j, a_{k+1} ..., a_m)$$
$$= \delta(q_k, a_{k+1}, ..., a_m) = q_m$$
$$q_m \in F$$

Similarly we could go around the loop more than once-infact as many times as we like.

Thus $a_1, ..., a_j (a_{j+1}, ...a_k)^i \ a_{k+1}, ..., a_m$ is in L(M) for any $i \ge 0$.

Note: Given any long string accepted by an finite authomaton we can find a substring near the begining of the string that may be "Pumped" ie repeated as many times as we like.

Lemma

Let L be a regular set. Then there is a constant n such that if Z is any word in L and $|Z| \ge n$, we may write $Z = uvw$ in such a way that $|uv| \le n$, $|V| \ge 1$ and for all i $i \ge 0$, $uv^i w$ is in L. Further more, n is no greater that the number of states of the smallest finite automaton accepting L.

Example 8.6 Show that $L = \{a^p \,/\, P \text{ is a prime}\}$ is not regular

Solution Let p be a prime number greater than n. Let $w = a^p$, using pumping lemma, we can write $w = xyz$ with $|xy| \le n$ and $|y| > 0. x, y$ and z are simply strings of 'a' hence $y = a^m$ for sence $m \ge 1$ and $m \le n$. Let $i = p + 1$ and $xy^i z \in L$ then

$$|xy^i z| = |xyz| + |y^{i-1}| = p + (i-1)m$$
$$= p + pm = p(1+m)$$

where $P(1+m)$ is not a prime. So $xy^i z \notin L$. This is a cantradiction. Thus L is not regular.

Applications of Pumping Lemma

This theorem can be used to prove that certain sets are not regular.

Step 1 Assume L is regular. Let n be the number of states in the corresponding finite automaton.

Step 2 Choose a string w such that $|w| \geq n$. Use pumping lemma to write $w = xyz$, with $|xy| \leq n$ and $|y| > 0$.

Step 3 Find a suitable integer i such that $XY^iZ \notin L$. This contradicts our assumption. Therefore L is not regular.

Example 8.7 Show that the set $L = \{a^{L^2} / i \geq 1\}$ is not regular.

Solution

Step 1 Assume L is regular. Let n be the number of states in FA accepting L.

Step 2

Let $w = a^{n^2}$ then $|w| = n^2 > n$

Using pumping lemma

$$w = xyz \text{ with } |xy| \leq n \text{ and } |y| > 0$$

Step 3 Consider xy^2z

$$|xy^2z| = |x| + 2|y| + |z|$$
$$> |x| + |y| + |z| \text{ as } |y| > 0$$

This means

$$n^2 = |xyz|$$
$$= |x| + |y| + |z|$$
$$< |xy^2z|$$
$$\text{as } |xy| \leq n, |y| \leq n$$

Therefore,

$$|xy^2z| = |x| + 2|y| + |z| \leq n^2 + n$$
$$\text{i.e., } n^2 < |xy^2z| \leq n^2 + n$$
$$< n^2 + n + n + 1$$

Hence $|xy^2z|$ strictly lies between n^2 and $(n+1)^2$, but it is not equal to any of them.

Thus $|xy^2z|$ is not a perfect square and so $xy^2z \notin L$. But by pumping lemma $xy^2z \notin L$. This is a contradiction.

Push Down Automata

We have seen that the regular languages are precisely those accepted by finite automata. Finite automata have strictly finite memories, whereas the recognition of a context-free language may require storing unbounded amount of information. For example let $L = \{a^nb^n / n \geq 0\}$. When scanning a string from the above language, we must not only check that all a's precede the first b,we must also count the number of a's. Since n is unbounded, this counting cannot be done with a finite memory. We want a machine that can count without limits. Let us consider another example (ww^R), we need more than unlimited counting ability. We need the ability to store and match a sequence of symbols in reverse order. This suggests that we might try a stack as a storage structure, allowing unbounded storage that is restricted to operating like a stack. This type of arrangement where a finite automaton has a stack, leads to the generation of a push down automaton.

Let us consider the components of a push down automaton and the way it operates.

Components

1. Read-only input tape
2. Input alphabet
3. Finite state control
4. A set of final states
5. Initial state
6. A stack called pushdown store (abbreviated as PDS)

Operation A finite automaton is in some state and on reading, an input symbol moves to a new state. The push down automaton is also in some state and on reading an input symbol and the topmost symbol in PDS, it moves to a new state and writes (adds) a string of symbols in PDS.

Note: The storage structure stack is a read-write pushdown store as we add elements to PDS or remove elements from PDS.

A pushdown automaton M is a system $(Q, \Sigma, \rho, \delta, q_0, S_0, F)$

where

1. $Q \rightarrow$ finite set of states
2. $\Sigma \rightarrow$ input alphabet
3. $\rho \rightarrow$ stack alphabet
4. $q_0 \in Q \rightarrow$ initial state
5. $S_0 \in \rho \rightarrow$ particular stack symbol called the start symbol
6. $F \leq Q \rightarrow$ the set of final states.
7. $\delta : Qx(\Sigma \cup \{\lambda\}) \times \rho \rightarrow$ finite subsets of $Q \times \rho^*$ is the transition function.

Note: The arguments of δ are the current state of the control unit, the current input symbol, and the current symbol on top of the stack. The result is (q, x), where q is the next state of the control unit and x is a string which is put on top of the stack in place of the single symbol available there before. Note that the second argument of δ may be λ, indicating that a move that does not consume an input symbol is possible. We can say that such a move can be represented as λ-transition. Note also that δ is defined so that it needs a stack symbol; no move is possible if the stack is empty. The range of δ which is a finite subset and necessary because $Q \times \rho^*$ is an infinite set and therefore has infinite subsets.

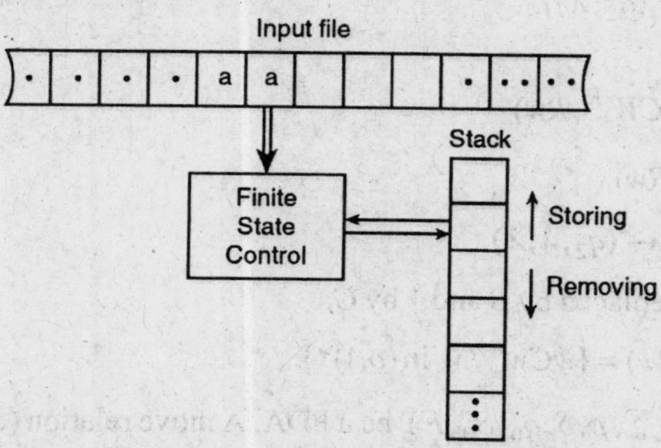

Model of Pushdown automaton

Example

Let $M = (q_1, q_2, q_3), \Sigma, \rho, \delta, So, P_1, \{P_3\}$

where $\Sigma = \{a, b\}, \rho = \{a, So, S_1\}$

$$\delta(q_1, a, S_0) = \{(q_1, S_1)\}$$
$$\delta(q_1, a, S_1) = \{(q_1, S_1 a)\}$$
$$\delta(q_1, a, a) = \{(q_1, aa)\}$$
$$\delta(q_1, b, a) = \{(q_2, \lambda)\}$$
$$\delta(q_1, b, S_1) = \{(q_3, S_1)\}$$
$$\delta(q_2, b, a) = \{(q_2, \lambda)\}$$

and $\delta(q_2, b, S_1) = \{(q_3, S_1)\}$

Then $(q_1, a^n b^n, S_0) \vdash (q_1, a^{n-1} b^n, S_1)$

$$\overset{*}{\vdash} (q_1, b^n, S_1, a^{n-1})$$

$$\overset{*}{\vdash} (q_2, b, S_1) \overset{*}{\vdash} (q_3, \lambda, S_1)$$

so that a^n b^n is in $T(M)$

$$T(M) = \{a^n b^n / n \geq 1\}$$

Example Let $M = \{(q_1, q_2)\}, \{0, 1, c\}, \{R, B, G\}, \delta, R, q_1, \varphi)$

where δ is defined as follows

$$\delta(q_1, 0, R) = \{q_1, RB)\}, \delta(q_1, 0, B) = \{(q_1, BB)\},$$
$$\delta(q_1, 0, G) = \{q_1, GB)\}, \delta(q_1, C, R) = \{(q_2, R)\},$$
$$\delta(q_1, C, B) = \{q_2, B)\}, \delta(q_2, 0, B) = \{(q_2, \lambda)\},$$
$$\delta(q_2, \lambda, R) = \{q_2, \lambda)\},$$
$$\delta(q_1, 1, R) = \{q_1, RG)\},$$
$$\delta(q_1, 1, B) = \{q_1, BG)\},$$
$$\delta(q_1, 1, G) = \{q_1, GG)\},$$
$$\delta(q_2, 1, G) = \{q_2, \lambda)\},$$

if w is (0, 1)*

$$(q_1, WCW^R, R)x \overset{*}{\vdash} (q_1, CW^R, Rw)$$
$$\vdash (q_2, W^R, Rw)$$
$$\vdash (q_2, \lambda, R) \leftarrow (q_2, \lambda, \lambda)$$

where W is w, where O is replaced by B and 1 by G.

$$N(M) = \{wCw^R / w \text{ in } (o, 1)^*\}$$

Move relation i. Let $A = (Q, \Sigma, \rho, \delta, q_0, S_0, F)$ be a PDA. A move relation (denoted by $\vdash$) between IDs is defined as

$$(q, a_1 a_2 ... a_n, S_1 S_2 ... S_n) \leftarrow (q_1, a_2 ... a_n, \beta S_2 S_m)$$

If $\delta(q_1, a_1, S_1)$ contains $(q_1 \beta)$

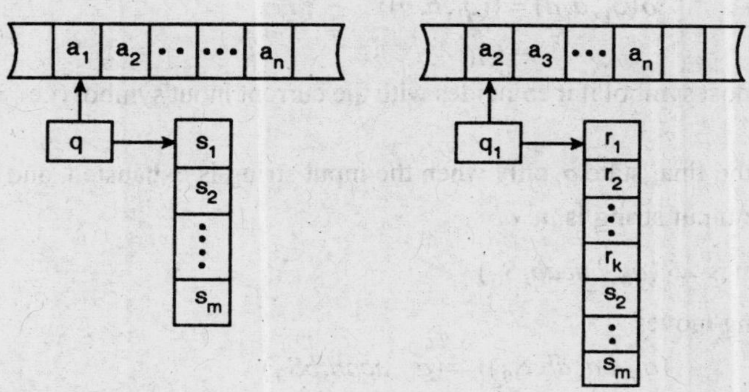

Properties of the Relation $\vdash^*$

Property 1 If $\delta(q_1, x, \alpha) \vdash (q_2, \lambda, \beta)$ then for every $y \in \Sigma^*$,

$(q_1, xy, a) \vdash^* (q_2, y, B)$ conversely, if $(q_1, xy, a) \vdash^* (q_2, y, b) \vdash^* (q_2, y, b)$ for some $y \in \Sigma^*$ then

$$(q_1, x, a) \vdash^* (q_2, \lambda_1, \beta)$$

Property 2 If $(q_1 x, \alpha) \vdash^* (q_1, \lambda, r)$ then for every $\beta \in \rho^*$,

$$(q, x, \alpha\beta) \vdash^* (q_1, \lambda, \gamma\beta)$$

Example Let $A = (\{q_0, q_1, q_f\}), \{a, b, c\}, \{a, b, \delta_0\}, \delta, q_0, S_0, \{q_f\})$ is a *Pda*, where δ is defined as

$$\delta(q_0, a, S_0) = \{(q_{0,} a, S_0)\}$$
$$\delta(q_0, S, S_0) = \{(q_{0,} b, S_0)\}$$
$$\delta(q_0, a, a) = \{(q_{0,} aa)\}$$
$$\delta(q_0, b, a) = \{(q_{0,} ba)\}$$
$$\delta(q_0, a, b) = \{(q_{0,} ab)\}$$
$$\delta(q_0, b, b) = \{(q_{0,} bb)\}$$
$$\delta(q_0, c, a) = \{(q_{1,} a)\}$$
$$\delta(q_0, c, b) = \{(q_{1,} b), \delta(q_0, c, b)\}$$
$$= \{(q_1, S_0)\}$$
$$\delta(q_1, a, a) = \delta(q_1, b, b) = \{(q_1, \lambda)\}$$
$$\delta(q_1, \lambda, S_0) = \{(q_f, S_0)\}$$

Explanation of δ

If A is in initial ID then $\delta(q_0, a, S_0) = \Sigma(q_0, aS_0)$ using $\delta(q_0, b, S_0) = \{(q_0, bS_0)\}$, A pushes the first symbol of the input string on PDS if it is a or b. By $\delta(q_0, b, a) = \{(q_0, ba)\}$ and $\delta(q_0, a, a) = \{(q_0, aa)\}$, Further $\delta(q_0, a, b) = \{(q_0, ab)\}$ and $\delta(q_0, b, b) = \{(q_0, bb)\}$, the symbols of the input string are pushed on PDS until it sees the centre marker C. By using $\delta(q_0, c, a) = \{(q_1, a)\}$ and $\delta(q_0, c, b) = \{(q_1, b)\}, \delta(q_0, c, S_0) = \{(q_1, S_0)$ On seeing C, the PDA moves to state q_1 without marking any changes in PDS.

By using

$$\delta(q_1, a, a) = \{(q_1, b, b)\}$$
$$= (q_1, \lambda)$$

the PDA erases the topmost symbol if it coincides with the current input symbol (i.e., if they do not match, then the PDA halts).

The PDA reaches the final state q_f only when the input string is exhausted, and the PDS has only S_o.

Let us consider the input string is *ac ab*

Initial configuration is $\Rightarrow (q_0, bacab, S_0)$

We get the following moves

$$(q_0, bacab, S_0) \vdash (q_0, acab, bS_0)$$
$$\vdash (q_0, cab, abS_0)$$
$$\vdash (q_1, ab, abS_0)$$
$$\vdash (q_1, b, bS_0)$$
$$\vdash (q_1, \lambda, S_0)$$
$$\vdash (q_f, \lambda, S_0)$$

i.e. $(q_0, bacab, S_0) \overset{*}{\vdash} (q_f, S_0)$

Proceeding in a similar way

We can show that $(q_0, wcm^T, S_0) \overset{*}{\leftarrow} (q_f, S_0)$ for all $w \in \{a, b\}^*$

Suppose an initial configuration is $(q_0, abcbb, S_0)$. Then we have

$$(q_0, abcbb, S_0) \vdash (q_0, bcbb, aS_0)$$
$$\vdash (q_0, cbb, baS_0)$$
$$\vdash (q_1, bb, baS_0)$$
$$\vdash (q_1, b, aS_0)$$

Once the PDA is in ID (q_1, b, aS_0), it has to halt as $\delta(q_1, b, a) = \varphi$.

Hence we have $(q_0, abcbb, S_0) \overset{*}{\leftarrow} (q_1, b, aS_0)$

As $\delta(q_0, c, S_0) = \varphi$ the PDA cannot make any transition if its starts with an ID of the form (q_0, cw, S_0)

Acceptance by PDA

A PDA has final states like an NFA and has also the additional structure viz. PDS. So we can define acceptance of input strings by PDA in terms of final states or in terms of PDS.

i. Let $A = (Q, \Sigma, \rho, \delta, q_0, S_0, F)$ be a PDA. The set accepted by PDA by final state is defined by $T(A) = \{w \in \Sigma^* \mid (q_0, w, S_0) \overset{*}{\vdash} (q_f, \lambda, \alpha)$ for some $q_f \in F$ and $\alpha \in \rho^*\}$

ii. Let $A = (Q, \Sigma, \rho, \delta, q_0, S_0, F)$ be a PDA. The set $N(A)$ accepted by null store (or empty store) is defined by

$$N(A) = \{w \in \Sigma^* (q_0, w, S_0) \overset{*}{\vdash} (q_f, \lambda_1, \lambda) \text{ for some } q \in Q$$

In other words, w is in $N(A)$ if A is in the initial ID (q_0, w, S_0) and empties the PDS after processing all the symbols of w. So in defining $N(A)$, we consider the change brought about on PDS by application of w, and not the transition of states.

TURING MACHINE

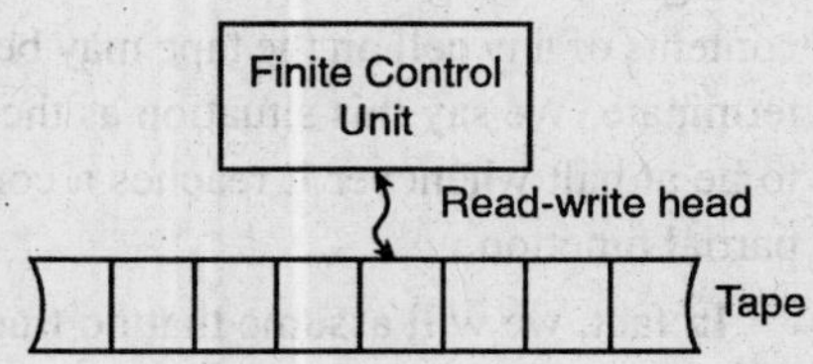

It can be visualized as a single, one dimensional array of cells, each of which can hold a single symbol. This array extends indefinitely in both directions and is therefore capable of holding an unlimited amount of information. This information can be read and changed in any order. It has a storage device called Tape which is quite analogous to the magnetic tapes used in actual computers. Read-Write head is associated with the tape that can travel right or left on the tape and that can read and write a single symbol on each move. The automaton that we use as a Turing machine will have neither an input file nor any output file. Whatever input and output is necessary will be done on the machine's tape.

A turing machine M is defined by $M = \{Q, \Sigma, \rho, \delta, q_0, B, F\}$

where Q is the set of internal states,

Σ is the input alphabet,

ρ is a finite set of symbols called the tape alphabet,

δ is the transition function,

$B \in \rho$ is a special symbol called blank, $q_0 \in Q$ is the initial state, and

$F \leq Q$ is the set of final states.

Assumption We assume $\Sigma \subseteq \rho - \{B\}$; that is, the input alphabet is a subset of the tape alphabet, not including the blank.

Note: The transition function δ is defined as $\delta : Q \times \rho \to Q \times \rho \times \{L, R\}$

Where δ is a partial function on $Q \times \rho$. Its interpretation gives the principle by which a Turing machine operates. The arguments of δ are the current state of the control unit and the current tape symbol being read. Then the result is a new state of the control unit, a new tape symbol, which replaces the old one, and a move symbol L or R. The move symbol indicates whether the read-write head moves left or right, one cell after the new symbol has been written on the tape.

Example

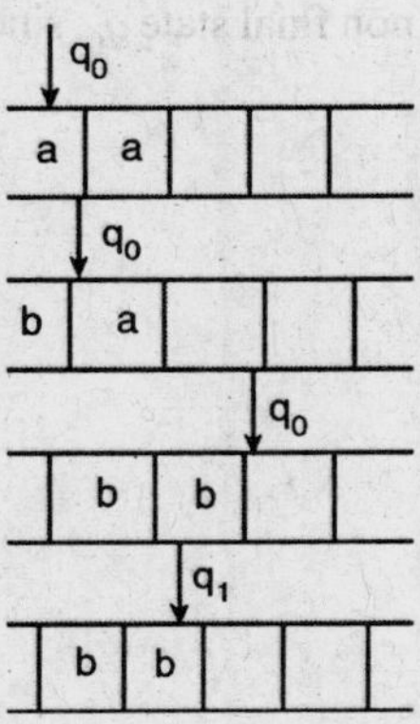

$$M = \{Q, \Sigma, \rho, q_0, F, B\}$$

where $Q = \{q_0, q_1\}, \Sigma = \{a, b\}, \rho = \{a, b, B\}$ and $F = \{q_1\}$

and
$$\delta(q_0, a) = (q_0, b, R)$$
$$\delta(q_0, b) = (q_0, b, R)$$
$$\delta(q_0, B) = (q_0, B, L)$$

We can think of a Turing machine as a simple computer. It has a processing unit, which has a finite memory, and the tape, is a secondary storage of unlimited capacity. Here δ defines how this computer acts and is often called the "Program" of the machine.

We know that the automaton always start in the given initial state with some information on the tape. It then goes through a sequence of steps controlled by the transition function δ. During this process, the contents of any cell on the tape may be examined and changed many times. Eventually, the whole process may terminate. We say this situation as the turing machine is brought into halting state. A Turing machine is said to be at halt whenever it reaches a configuration for which δ is not defined; this is possible because δ is a partial function.

In fact, we will assume that no transitions are defined for any final state. So the Turing machine will halt whenever it enters a final state.

Note: If the read-write head moves alternatively between right and left, but makes no modifications to the tape, this is an instance of the Turing machine not at halt. As an analogy with programming terminology, we say that the Turing machine is in an infinite loop.

Instantaneous Description of a Turing Machine

An instantaneous description (ID) is defined as $\alpha_1 q \alpha_2$ where α_1 is the string processed already, α_2 is the string to be processed when the turing machine is in the state q.

Language accepted by a Turing machine Let M be a Turing machine. Then the language accepted by M denoted by $T(M) = \{w \in \Sigma^+ : q_0 w \overset{*}{\vdash} \alpha_1 q_f \alpha_2$ for some $q_f \in F$ and $\alpha_1, \alpha_2 \in \rho^*\}$

Example 8.8 Let $\Sigma = \{0,1\}$. Design a Turing machine that accepts the language denoted by the regular expression 00^*

Solution

$$\delta(q_0, 0) = (q_0, 0, R)$$

$$\delta(q_0, B) = (q_0, B, R)$$

Starting at the left end of the input, we read each symbol and check that it is a 0. If it is, we continue by moving right. If we reach a blank without encountering it, we continue by moving right. If we reach blank without encountering any thing but 0, we terminate and accept that string. If the input contains 1 anywhere, the string is not in $L(00^*)$ and we halt in a non final state. To keep track of the computation, two internal states $Q = (q_0, q_1)$ and one final state $F = (q_1)$ are sufficient. As long as *0* appears under the read-write head, the head will move to the right. If at any time 1 is read, the machine will halt in the non final state q_0, since $\delta(q_0, 1)$ is undefined.

Example 8.9 Design a turing machine that accepts $L = \{a^n b^n; n \geq 1\}$

Solution

Let $\Sigma\{a,b\}$, $Q\{q_0, q_1, q_2, q_3, q_4\}$, $F = \{q_4\}$ and $P = \{a, b, x, y, B\}$

The transitors can be broken into several parts. The set

$$\partial(q_0, a) = (q, x, R)$$

$$\partial(q_1, a) = (q_1, a, R)$$

$$\partial(q_1, y) = (q_1, y, R)$$

$$\partial(q_1, b) = (q_2, y, L)$$

replaces the leftmost a with an x, then canses the read - write head to travel right to the first b, replacing it with a y when y is written, the machine enters a state q_2 indicating that a has been successfully poined with ab.

The next set of transition reverses the direction until an x is encountered, repositions the read-write head over the leftmost a, and returns to the initial state.

$$\partial(q_2, y) = (q_2, y, L)$$
$$\partial(q_2, a) = (q_2, a, L)$$
$$\partial(q_2, x) = (q_0, x, R)$$

This gives the initial state q_0, ready to deal with next a and b.

Computation after first pass

$$q_0 aa...abb...b \overset{*}{\vdash} xq_0 a...ayb...b$$

After second pass

$$q_0 aa...abb...b \overset{*}{\vdash} xxq_0 ...ayy...b$$

To detect input where an a follows ab

$$\partial(q_0, y) = (q_3, y, R)$$
$$\partial(q_3, y) = (q_3, y, R)$$
$$\partial(q_3, B) = (q_4, B, R)$$

the particular input $aabb$ gives the following successive instantaneous descriptions.

$$q_0\, aabb \vdash x\, q_1 aabb \vdash xaq_1 bb$$
$$\vdash x\, q_2\, ayb$$
$$\vdash q_2\, x\, ayb$$
$$\vdash x\, q_0\, ayb$$
$$\vdash xx\, q_1\, yb$$
$$\vdash xx\, yq_1 b \vdash xxq_2 yy$$
$$\vdash x\, q_2\, xyy$$
$$\vdash xx\, q_0\, yy$$
$$\vdash xxy\, q_3\, y$$
$$\vdash xxyy\, q_3\, B$$
$$\vdash xxyy\, Bq_4\, B$$

At this point the turing machine halts in a final state, so the string $aabb$ is accepted.

Example 8.10 Let x and y be two positive integers represented in Mary notation. Construct a Turing machine that halts in a final state q_y if $x \geq y$ and that will halt in non final state q_n if $x < y$.

i.e. $q_0 w(x) O w(y) \overset{*}{\vdash} q_y w(x) O w(y)$ if $x \geq y$

$q_0 w(x) O w(y) \overset{*}{\vdash} q_n w(x) O w(y)$ if $x < y$

Solution To solve this problem in the previous example instead of matching *a*'s, we match each 1 on the left of the dividing 0 with 1 on the right. At the end of matching, we will have on the tape either

$$xx...110xxx...xB \text{ or } xx...xx0xx...x11B,$$

depending an whether $x > y$ or $y > x$. In the first case, when we attempt to match another 1, we encounter the blank at the right of the working space. This can be used as a signal to enter the state q_y.

We still find 1 on the right when all 1's on the left have been replaced. We use this to get into the other state q_n (case two).

Example 8.11 Desingn a turing machine that computes the function

$$f(x, y) = x + y \text{ if } x \geq y$$
$$= 0 \text{ if } x < y$$

Solution

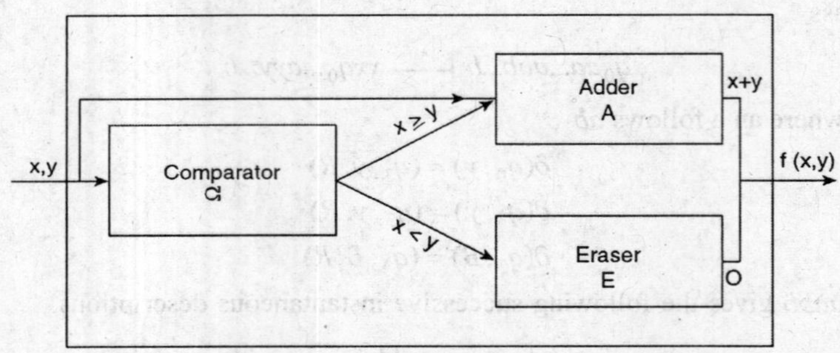

The diagram shows that first we use a comparing machine to determine whether or not $x \geq y$. If so, the comparator sends a start signal to the adder, which then computes $x + y$. If not, an erasing program is started that changes every 1 to blank. For the eraser *E*, we construct a Turing machine having states indexed with *E*. The computations to be done by *C* are

$$q_{c,o}\, w(x) O\, w(y) \overset{*}{\vdash} q_{A,o}\, w(x) O\, w(y) \text{if } x \geq y$$

and

$$q_{c,o}\, w(x) O\, w(y) \overset{*}{\vdash} q_{E,o}\, w(x) O\, w(y) \text{if } x < y$$

If we take $q_{A,o}$ and $q_{E,o}$ as the initial states of *A* and *E* respectively, we see that *C* starts either *A* or *E*.

Computations performed by the adder A

$$q_{A,o}\, w(x) O w(y) \overset{*}{\vdash} q_{A,f} w(x+y) O,$$

Computations performed by Eraser E

$$q_{E,o}\, w(x) O w(y) \overset{*}{\vdash} q_{E,f} O.$$

Objective and Short Answer Questions

1. Enumerate the elements of the set $\{x \in R \mid x^2 - 3x + 2 = 0\}$

   Ans: $x^2 - 3x + 2 = 0 \Rightarrow (x-1)(x-2) = 0 \Rightarrow x = 1$ or 2

   Hence $\{x \in R \mid x^2 - 3x + 2 = 0\} = \{1, 2\}$.

2. Enumerate the elements of the sets $\{x \in R \mid x^2 + 1 = 0\}$ and $\{x \in C \mid x^2 + 1 = 0\}$.

   Ans: $x^2 + 1 = 0 \Rightarrow x^2 = -1 \Rightarrow x = i$ or $-i$

   Hence $\{x \in R \mid x^2 + 1 = 0\} = \phi$, and $\{x \in R \mid x^2 + 1 = 0\} = \{i, -i\}$.

3. Enumerate the set $A = \{x \in \cup \mid 81x^4 - 16 = 0\}$ if (i) $\cup = R$, and (ii) $\cup = C$.

   Ans: $81x^4 - 16 = 0 \Rightarrow 81x^4 = 16 \Rightarrow x^4 = \dfrac{16}{81}$

   $$\Rightarrow x^2 = \frac{4}{9} \text{ or } -\frac{4}{9} \Rightarrow x = \frac{2}{3} \text{ or } -\frac{2}{3} \text{ or } \frac{2i}{3} \text{ or } -\frac{2i}{3}.$$

   Hence,

   i. if $\cup = R$, then $A = \left\{\dfrac{2}{3}, \dfrac{-2}{3}\right\}$ and

   ii. if $\cup = C$, then $A = \left\{\dfrac{2}{3}, \dfrac{-2}{3}, \dfrac{2i}{3}, \dfrac{-2i}{3}\right\}$.

4. If $A = \{(x, y) \mid x + y = 7; x, y \in R\}$ and $B = \{(x, y) \mid x - y = 3; x, y \in R\}$, find $A \cap B$.

   Ans: Solving the equations $x + y = 7$, $x - y = 3$ we obtain $x = 5, y = 2$. Hence $A \cap B = \{(5, 2)\}$

5. Find the intersection of the sets $\{(x, y) \mid x^2 + y^2 = 25; x \in R, y \in R\}$ and $\{(x, y) \mid x - y = 0; x \in R, y \in R\}$.

   Ans: Solving the equation $x^2 + y^2 = 25$ and $x - y = 0$, we obtain $x = y = \pm\dfrac{5}{\sqrt{2}}$. Hence the intersection of the sets is $\left\{\left(\dfrac{5}{\sqrt{2}}, \dfrac{5}{\sqrt{2}}\right), \left(\dfrac{-5}{\sqrt{2}}, \dfrac{-5}{\sqrt{2}}\right)\right\}$.

6. If $\cup = \{0, 1, 2, 3, 4, 5, 6, 7, 8, 9\}$, $A = \{x \in \cup : x$ is a multiple of $3\}$, and $B = \{x \in \cup : x^2 - 5 \geq 0\}$, determine $A \cup B, A \cap B$, and B^c.

   Ans: Since $\{A = 0, 3, 6, 9\}$ and $B = 3, 4, 5, 6, 7, 8, 9$ we obtain

   $$A \cup B = \{0, 3, 4, 5, 6, 7, 8, 9\}, \quad A \cap B = \{3, 6, 7\} \text{ and } B^c = \{0, 1, 2\}$$

7. List all the nonempty proper subsets of $\{1, 2, 3\}$.

   Ans: $\{1\}, \{2\}, \{3\}, \{1, 2\}, \{1, 3\}, \{2, 3\}$

8. How many nonempty proper subsets are there for $\{1, 2, 3, 4\}$.
Ans: Total number of subsets is $2^4 = 16$. But this includes the full set and the empty set. So number of nonempty proper subsets in $16 - 2 = 14$.

9. If $A = \{+, -\}$ and $B = \{00, 01, 10, 11\}$ find A^2 and $A \times B$.
Ans: $A^2 = A \times A = \{(+,+), (+,-), (-,+), (-,-)\}$, and
$A \times B = \{(+,00), (+,01), (+,10), (+,11), (-,00), (-,01), (-,10), (-,11),\}$

10. If $A = \{\alpha, \beta\}$ and $B = \{1, 2, 3\}$ find $A \times B$ and $B \times A$.
Ans: $A \times B = \{(\alpha,1), (\alpha,2), (\alpha,3), (\beta,1), (\beta,2), (\beta,3)\}$
$B \times A = \{(1,\alpha), (1,\beta), (2,\alpha), (2,\beta), (3,\alpha), (3,\beta)\}$

11. If A, B, C are sets such that $A \cup B = A \cup C$ and $A \cap B = A \cap C$, show that $B = C$.
Ans:
$$\begin{aligned} B &= B \cap (A \cup B) \\ &= B \cap (A \cup C) && (\because A \cup B = A \cup C) \\ &= (B \cap A) \cup (B \cap C) \\ &= (A \cap C) \cup (B \cap C) && (\because A \cap B = A \cap C) \\ &= (A \cup B) \cap C \\ &= (A \cup C) \cap C && (\because A \cup B = A \cup C) \\ &= C \end{aligned}$$

12. Show that for any two sets A and B, $A - (A \cap B) = A - B$.
Ans: $x \in A - (A \cap B) \Leftrightarrow x \in A$ and $x \notin A \cap B$
$\Leftrightarrow x \in A$ and $x \notin B \Leftrightarrow x \in A - B$. Hence, $A - (A \cap B) = A - B$

13. State the rule of products.
Ans: If two operations must be performed, and if the first operations can be performed in P_1 different ways and the second operation in P_2 different ways, then the two operations can be performed in $P_1 P_2$ different ways.

14. If the cardinality of a set A in n then the cardinality of its power set $P(A)$ is
Ans: 2^n

15. If the cardinality of the power set $P(A)$ is m then find the cardinality of A.
Ans: Let n be the cardinality of A. Then cardinality of $P(A)$ is 2^n. hence $2^n = m$.
Therefore, $n = \log_2 m$. i.e., cardinality of A is $\log_2 m$.

16. Define the power set $P(A)$ of a set A.
Ans: The power set is the collection of all subsets of A. Hence $P(A) = \{B : B \subseteq A\}$.

17. If $A = \{a, b, c\}$, determine $P(A)$.
Ans: $P(A) = \{\phi, \{a\}, \{b\}, \{c\}, \{a,b\}, \{a,c\}, \{b,c\}, \{a,b,c\}\}$.

18. How many subsets of $\{1, 2, 3, ..., 10\}$ contain at least 7 elements.
Ans: ${}^{10}C_7 + {}^{10}C_8 + {}^{10}C_9 + {}^{10}C_{10} = 176$.

19. How many proper subset of $\{1, 2, 3, 4, 5\}$ contain the numbers 1 and 5. How many of them also do not contain the number 2.
Ans: (i) Number of proper subsets contaning 1 and 5 is $2^3 - 1 = 8 - 1 = 7$.
(ii) Number of such subsets not containing 2 is $2^2 = 4$.

20. Suppose that you are about to flip a coin and then roll a dice.

Let $A = \{HEAD, TAIL\}$ and $B = \{1, 2, 3, 4, 5, 6\}$. Find $\#(A \times B)$. How do you interpret the set $A \times B$.

Ans: $\#(A \times B) = 2 \times 6 = 12$ and $A \times B$ is the set of all possible outcomes of the experiment.

21. Suppose that you walk into a theatre that has a single row of 6 seats. If 2 seats are to be occupied, how many seating arrangements are possible.

Ans: $6 \times 5 = 30$ possible arrangements.

22. A questionnaire contains 4 questions that have 2 possible answers and 3 questions with 5 possible answers. In how many ways the questionnaire can be answered.

Ans: $2^4 \times 5^3 = 2000$ different ways.

23. if ${}^n c_2 = {}^n c_5$ then $n =$ ________.

Ans: $n = 2 + 5 = 7$.

24. How many ways can a student do a 10 -question true-false exam if he or she can choose to not answer any number of questions.

Ans: For each question the student has 3 choices: true, false, or do not answer.

Hence the number of ways $= 3^{10} = 59049$.

25. If a raffle has 3 different prizes and there are 1000 raffle tickets sold, how many different ways the prizes can be distributed.

Ans: ${}^{1000}P_3 = 1000 \times 999 \times 998$ different ways.

26. In how many ways can you arrange the letters in the words (i) COMBINE (ii) SUBSET (iii) NOON.

Ans: i. $7! = 5040$ ways. ii. $\frac{6!}{2!} = 360$ ways. iii. $\frac{4!}{2!\,2!} = 6$ ways.

27. A questionnaire contains 6 questions each having yes – no answers. For each yes response there is a follow-up question with 4 possible responses. In how many ways can the questionnaire be answered.

Ans: Since each question can be answered in 5 different ways, the questionnaire can be answered in $5^6 = 15625$ ways.

28. If $T = \{1, 2, 3, 4, 5\}$, how many subsets of T have less than 4 elements.

Ans: ${}^5c_0 + {}^5c_1 + {}^5c_2 + {}^5c_3 = 1 + 5 + 10 + 10 = 26$ subsets. $[OR: 2^5 - ({}^5C_4 + {}^5C_5) = 32 - (5+1) = 26]$

29. The judiciary committee at a college is mode up of 3 faculty members and 4 students. If 10 faculty members and 25 students have been dominated for the committee, how many different committees are possible.

Ans: ${}^{10}C_3 \times {}^{25}C_4 = 120 \times 12650 = 1518000$

30. State the law of addition.

Ans: If a finite set A can be written as $A = B_1 \cup B_2 \cup ... \cup B_n$ such that $B_i \cap B_j = \phi$ for all $i \neq j$, then the cardinality of A is the sum of the cardinalities of $B_1, ..., B_n$, i.e., $\#A = \#B_1 + \#B_2 + ... + \#B_n$.

31. Define counter example.

Ans: An example which disproves a statement is called a counter-example.

32. Define partition of a set.

Ans: Let A be a nonempty set. A partition of A is a collection of nonempty subsets $A_1, A_2, ...$ of A such that (i) $A = A_1 \cup A_2 \cup ...$; and (ii) $A_i \cap A_j = \phi$ for all $i \neq j$.

33. Find all partitions of the set $\{a, b, c\}$.

Ans : $\{\{a\},\{b\},\{c\}\}$, $\{\{a\},\{b,c\}\}$, $\{\{b\},\{a,c\}\}$, $\{\{c\},\{a,b\}\}$

34. If $A = \{1,2,3,4,5,6\}$, $B_1 = \{1,3,5\}$ and $B_2 = \{1,2,3\}$, find all the minsets of A generated by B_1 and B_2.

Ans: The minsets are i. $B_1 \cap B_2 = \{1,3\}$ ii. $B_1^c \cap B_2 = \{2\}$ iii. $B_1 \cap B_2^c = \{5\}$ iv. $B_1^c \cap B_2^c = \{4,6\}$

35. Partition the set of strings of 0s and 1s of length two or less, using the minsets generated by $B_1 = \{s: s$ has length 2$\}$, and $B_2 = \{s: s$ starts with a 0$\}$.

Ans: $B = \{\lambda, 0, 1, 00, 01, 10, 11\}$, $B_1 = \{00, 01, 10, 11\}$, $B_2 = \{0, 00, 01\}$

The minsets are: $B_1 \cap B_2 = \{00, 01\}$; $B_1^c \cap B_2 = \{0\}$; $B_1 \cap B_2^c = \{10, 11\}$; $B_1^c \cap B_2^c = \{\lambda, \phi\}$

so the partition is $\{\{\lambda, \phi\}, \{0\}, \{00, 01\}, \{10, 11\}\}$.

36. Explain the duality principle for sets.

Ans: Let S be any identity involving sets and the operations $C, \cap$, and $\cup$. If S^* is obtained from S by making the substitutions $\cap \to \cup$, $\cup \to \cap$, $\Phi \to U$, $U \to \Phi$, then the statement S^* in also an identity.

37. State the dual of $A \cup ((B^c \cup A) \cap B)^c = \cup$.

Ans: $A \cap ((B^c \cap A) \cup B)^c = \phi$.

40. Let r and s be relations defined on the set $\{1, 3, 5\}$ defined by xry iff $y = x + 2$, and xsy iff $x \leq y$. Find rs and sr.

Ans: $r = \{(1, 3), (3, 5)\}$ and $s = \{(1, 1), (1, 3), (1, 5), (3, 3), (3, 5), (5, 5)\}$. Hence,

$rs = \{(1, 3), (1, 5), (3, 5)\}$

$sr = \{(1, 3), (1, 5), (3, 5)\}$.

41. Let $A = \{1, 2, 3, 4, 5\}$ and define r on A by xry iff $x + 1 = y$. Find (i) r, (ii), r^2, and (iii) r^3.

Ans: $r = \{(1, 2), (2, 3), (3, 4), (4, 5)\}$

$r^2 = \{(1, 3), (2, 4), (3, 5)\}$

$r^3 = \{(1, 4), (2, 5)\}$

42. Given s and t, relation on Z, $s = \{(1, n) : n \in Z\}$ and $t = \{(n, 1) : n \in Z\}$, what are st and ts.

Ans: $st = \{(1, 1)\}$, $ts = \{(n, m) : n, m \in Z\} = Z \times Z$.

43. Let r_1, r_2, and r_3 be relations on a set A Prove: if $r_1 \subseteq r_2$ then $r_1 r_3 \subseteq r_2 r_3$:

Ans: Let $(x, y) \in r_1 r_3$. The $\exists Z$ such that $(x, z) \in r_1$ and $(z, y) \in r_3$. Since $r_1 \subseteq r_2$, we obtain $(x, z) \in r_2$. Hence $(x, y) \in r_2 r_3$ $(\because (x, z) \in r_2$ and $(z, y) \in r_3)$.

44. Let $A = \{1, 2, 3, 4\}$, and let r be the relation $<$ on A. Draw the digraph and the Hasse diagram of r.

Ans:

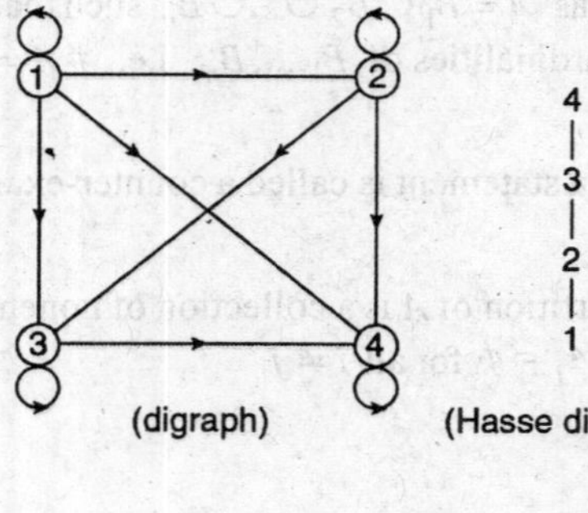

(digraph) (Hasse diagram)

45. Let $B = \{2, 3, 4, 5, 6, 12, 36, 48\}$, and s be the relation 1, "divides", on B. Draw the Hasse diagram of S.

Ans:

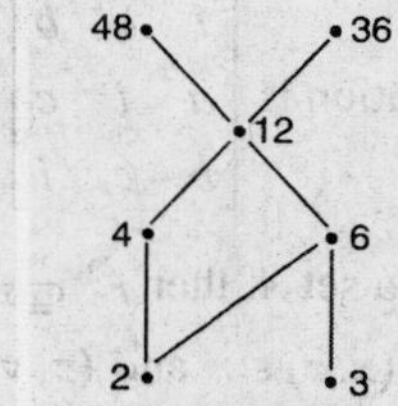

46. Draw the Hasse diagram of the relation $\subseteq$ on $P(A)$, where $A = \{a, b, c\}$.

Ans:

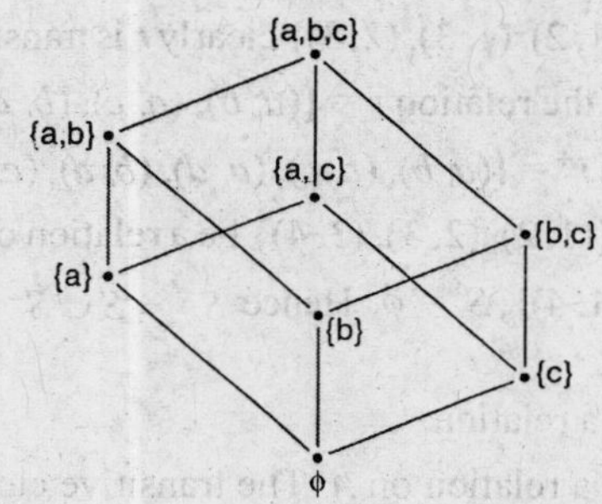

47. Define partial ordering.

Ans: A relation on a set A which is reflexive, antisymmetric, and transitive is called a partial ordering on A. More precisely, a relation r on a set A is a partial ordering if
(i) $ara\ \forall a \in A$, (ii) arb and $bra \Rightarrow arb$ for all $a, b \in A$; and (iii) are and $brc \Rightarrow arc$ for all $a, b, c \in A$.

48. Define an equivalence relation.

Ans: A relation defined on a set which is reflexive, symmetric, and transitive is called an equivalence relation.
More precisely, a relation r on a set A is an equivalence relation if (i) $ara\ \forall a \in A$; (ii) $arb \Rightarrow bra\ \forall a, b \in A$; and (iii) $arb, brc \Rightarrow arc,\ \forall a, b, c \in A$.

49. Let $X = \{1, 2, 3, 4, 5, 6, 7\}$ and $r = \{(x, y)\}\ x - y$ is divisible by 3}. Show that r is an equivalence relation on X.

Ans: (i) Since for each $x \in X$, $x - x = 0$ is divisible by 3, $(x, x) \in r$. so r is reflexive.

(ii) if $(x, y) \in r$ then $x - y$ is divisible by $3 \Rightarrow y - x$ is divisible by $3 \Rightarrow (y, x) \notin r$. Hence r is symmetric.

(iii) $(x, y) \in r$ and $(y, z) \in r \Rightarrow x - y$ and $y - z$ are divisible by 3 $\Rightarrow$ their sum $x - z$ is also divisible by $3 \Rightarrow (x, z) \in r$. So r is transitive. Hence r is an equivalence relation.

50. Give an example of a relation which is reflexive and symmetric but not transitive.

Ans: Let $A = \{1, 2, 3\}$ and define a relation r on A by $r = \{(1, 1), (2, 2), (3, 3), (1, 2), (2, 1), (2, 3), (3, 2)\}$. Then r is reflexive and symmetric, but not transitive. ($\because (1, 2) \in r, (2, 3) \in r$ but $(1, 3) \notin r$)

51. Find the adjacency matrix of the relation $r = \{(2, 2), (2, 5), (5, 6), (6, 6)\}$ on the set $A = \{2, 5, 6\}$.

Ans: $R = \begin{bmatrix} 1 & 1 & 0 \\ 0 & 0 & 1 \\ 0 & 0 & 1 \end{bmatrix}$

52. How many different reflexive, symmetric relations are there on a set with three elements.

Ans: The akjacency matrix of such a relation is $\begin{bmatrix} i & a & b \\ a & i & c \\ b & c & i \end{bmatrix}$, where a,b,c are 0's and 1's. Hence number of such matrices is $2 \times 2 \times 2 = 8$.

53. Prove that if r is a transitive relation on a set A, then $r^2 \subseteq r$.

Ans: Let $(x, y) \in r^2$. Then $\exists z \in A$ with $(x, z) \in r$ and $(z, y) \in r$. Since r is transitive, $(x, z) \in r$, $(z, y) \in r$ imply $(x, y) \in r$. Hence $r^2 \subseteq r$.

54. Find an example of a transitive relation for which $r^2 \neq r$.

Ans: Let $A = \{1, 2, 3\}$ and $r = \{(1, 2), (1, 3), (2, 3)\}$ clearly r is transitive. But $r^2 = \{(1, 3)\} \neq r$.

55. Find the transitive closure r^+ of the relation $r = \{(a, b), (a, c), (b, d), (c, d)\}$ on the set $\{a, b, c, d\}$.

Ans: $r^2 = \{a, d\}$, $r^3 = \phi$. Hence $r^+ = \{(a, b), (a, c), (a, d), (b, d), (c, d)\}$.

56. Let $A = \{1, 2, 3, 4\}$ and Let $S = \{(1, 2), (2, 3), (3, 4)\}$ be a relation on A. Find the transitive closure S^+.

Ans: $S^2 = \{(1, 3), (2, 4)\}$; $S^3 = \{(1, 4)\}$; $S^4 = \phi$. Hence $S^+ = S \cup S^2 \cup S^3 = \{(1, 2), (1, 3), (1, 4) (2, 3), (2, 4), (3, 4)\}$.

57. Define the transitive closure of a relation.

Ans: Let A be a set and let r be a relation on A. The transitive closure of r, denoted by r^+, is the smallest relation which contains r as a subset and which is transitive.

58. S is a relation on Z defined by asb iff $a + 1 = b$. Find the transitive closure of S.

Ans: The transitive closure is the relation < (less than) on Z.

59. What is the transitive closure of the relation R defined on Z by $a \, R \, b \Leftrightarrow |a - b| = 2$.

Ans: The transitive closure R^+ is $a R^+ b \Leftrightarrow a - b$ is even.

60. Explain Warshall's Algorithm

Ans: Let R be a known relation matrix. The following algorithm which computer the transitive closure R^+ is called Warshalls alyorithm:

1.0 $T := R$

2.0 Repeat

   2.1 $S := T$

   2.2 $T := S(I + S)$ * Using Booleam Arithmetic*

   Until $T = S$

3.0 Terminate with $R^+ = T$

61. Let $f : R \to R$ and of $g : R \to R$ be defined by $f(x) = x^2 - 2$ and $g(x) = x + 4$. Find $f \circ g$ and $g \circ f$.

Ans: $(f \circ g)(x) = f(g(x)) = g(x)^2 - 2 = (x + 4)^2 - 2 = x^2 + 8x + 14$.

$(g \circ f)(x) = g(f(x)) = f(x) + 4 = x^2 - 2 + 4 = x^2 + 2$.

62. Let $f : R \to R$ be given by $f(x) = x^3 - 2$. Find f^{-1}.

Ans: Let $y = f(x) = x^3 - 2$. Then $x^3 = y + 2 \Rightarrow x = (y + 2)^{1/3}$

$\therefore \; f^{-1}(y) = (y + 2)^{1/3}$ or $f^{-1}(x) = \sqrt[3]{x + 2}$.

63. Define an inverse function.

Ans: Let $f: X \to Y$ be a function. If there exists a function $g: Y \to X$ such that $(f \circ g)(y) = y\ \forall y \in y$ and $(g \circ f)(x) = x\ \forall x \in x$, then g is called the inverse of f, are we write $g = f^{-1}$.

64. Let $g = R \to R$ be defined by $g(x) = x^2$. What are $g^{-1}(4)$ and $g^{-1}(-1)$.

Ans : $g^{-1}(4) = \{x : g(x) = 4\} = \{x : x^2 = 4\} = \{2,-2\}$

$g^{-1}(-1) = \{x : g(x) = -1\} = \{x : x^2 = -1\} = \{\phi\ (\because x \in R)$.

65. A one -to-one function is also known as ______.

Ans: Injective function.

66 An onto function is also known as ______.

Ans: Surjective function.

67. Define a one-to-one function.

Ans: A function $f: X \to Y$ is said to be one-to-one if $f(a) = f(b) \Rightarrow a = b\ \ \forall a, b \in X$.

68. Define an onto function.

Ans: A function $f: X \to Y$ is said to be onto if for each $b \in y$ there corresponts $a \in x$ with $f(a) = b$.

69. Define permutation.

Ans: A bijection from a set *A* onto itself is called a permutation.

70. How many permutations are there on set with *n!* elements.

Ans: n!

71. Let *f* and *g* be functions from *Z* into *Z* defined by $f(n) = n+5,\ \ g(n) = n-2$. Find (i) *fg* (ii) f^3.

Ans : $(fg)(n) = f(g(n)) = g(n)+5 = n-2+5 = n+3$.

$f^3(n) = f(f^2(n)) = f^2(n)+5 = f(f(n))+5$

$= f(n)+5+5 = n+5+10 = n+15$

[or $f^2(n) = f(f(n)) = f(n+5) = n+5+5 = n+10$,

so $f^3(n) = f(f^2(n)) = f(n+10) = n+10+5 = n+15$]

72. If $f: A \to B$ and $g: B \to C$ are injections, show that $gof: A \to C$ is also an injection.

Ans: Let $x, y \in A$ be such that $(g \circ f)(x) = (g \circ f)(y)$. Then $g(f(x)) = g(f(y))$. But since g is an injection, this implies $f(x) = f(y)$. This inture implies $x = y$, since *f* is an injection. Hence gof is an injection.

73. If $f: A \to B$ and $g: B \to C$ are surjective functions prove that $g \circ f: A \to C$ is also surjective.

Ans: Let $z \in C$. Since $g: B \to C$ is surjective, there exists $y \in B$ such that $g(y) = z$. Again since $y \in B$ and $f: A \to B$ is surjective, there exists $x \in A$ with $f(x) = y$. Now $(g \circ f)(x) = g(f(x)) = g(y) = z$. Hence $g \circ f$ in surjective.

74. Let *f* and *g* be functions whose inverses exist. Prove that $(g \circ f)^{-1} = f^{-1} \circ g^{-1}$.

Ans: Let $h = f^{-1} \circ g^{-1}$. Then

$$ho(g \circ f) = (f^{-1} \circ g^{-1}) \circ (g \circ f) = f^{-1} o(g^{-1} o(gof))$$

$$= f^{-1} \circ \left((g^{-1} \circ g) \circ f\right) = f^{-1} \circ (i \circ f)$$

$$= f^{-1} \circ f = i$$

$$= (g \circ f) \circ h = (g \circ f) \circ (f^{-1} \circ g^{-1}) = g \circ \left(f \circ (f^{-1} \circ g^{-1})\right)$$
$$= g \circ \left((f \circ f^{-1}) \circ g^{-1}\right) = g \circ (i \circ g^{-1})$$
$$= g \circ g^{-1} = i$$

Hence $(g \circ f) = h$

75. Of $f : A \to B$ has an inverse, then prove that the inverse is unique.

Ans: Suppose g and h are both inverse of f.

So $g : B \to A$ and $h : B \to A$, are $g \circ f = i_A, f \circ g = i_B;\ \ h \circ f = i_A, f \circ h = i_B$.

Now $g = g \circ i_B = g \circ (f \circ h) = (g \circ f) \circ h = i_A \circ h = h$. So $g = h$.

76. Define Proposition.

Ans: A proposition is a sentence to which one and only one of the terms 'true' or 'false' can be meaningfully applied.

77. Write the truth tables for $\wedge$ and $\vee$ operators.

Ans:

p	q	p∧q
0	0	0
0	1	0
1	0	0
1	1	1

p	q	p∨q
0	0	0
0	1	1
1	0	1
1	1	1

78. Write the truth table of the conditional operator.

Ans:

p	q	p→q
0	0	1
0	1	1
1	0	0
1	1	1

79. Write the truth table for the biconditional.

Ans:

p	q	p→q
0	0	1
0	1	1
1	0	0
1	1	1

80. Define proposition generated by a set.

Ans: Let S be a set of propositions A proposition generated by S is any valid combination of propositions in S with conjunction, disjunction, and negation.

81. Write the truth table of c: $q \wedge (p \to q) \to p$.

Ans:

p	q	p→q	q∧(p→q)	c
0	0	1	0	1
0	1	1	1	0
1	0	0	0	1
1	1	1	1	1

82. construct the truth table for $c:(p \wedge q) \vee (\sim q \wedge r)$.

Ans:

			a		b	
p	q	r	p ∧ q	~q	~q ∧ r	c = a ∨ b
0	0	0	0	1	0	0
0	0	1	0	1	1	1
0	1	0	0	0	0	0
0	1	1	0	0	0	0
1	0	0	0	1	0	0
1	0	1	0	1	1	1
1	1	0	1	0	0	1
1	1	1	1	0	0	1

83. If *P* is necessary for *q* the which of the following is time : (i) $p \to q$ (ii) $q \to p$

Ans: $q \to p$.

84. If *P* is sufficient for *q* then which of the following is true : (i) $p \to q$ (ii) $q \to p$.

Ans: $p \to q$.

85. Rewrite the following using "if ... then" Language: A neccessary and sufficient condition for a finite monoid to be a group is that its identity element is its only idempotent.

Ans: A finite monoid is a group if and only if the identify el ment is its only idempotent.

86. Rewrite the following using if ...then: Continuity of $f(x)$ is a necessary condition for the differentiability of $f(x)$.

Ans: Of $f(x)$ is differentiable then $f(x)$ is continuous.

87. Define a tautology.

Ans: A logical expression that is true for all cases of its truth table is called a tautology.

88. Define a contradiction.

Ans: A logical expression that is false for all cses of its truth table is called a contradiction.

89. Define equivalence of propositions genereted by a set.

Ans: Let S be a set of propositions and let *r* and *s* be propositions generated by *S*.

We say that *r* and *s* are equivalent, $r \Leftrightarrow s$, if $r \leftrightarrow s$, is a tautology.

90. Define implication.

Ans: Let *r* and *S* be propositions generated by a set *S*.

We say that *r* implies *S*, $r \Rightarrow s$, if $r \to s$ is a tautology.

91. Prove using truth tables : $(p \to q) \wedge \sim q \Rightarrow \sim p$.

Ans:

p	q	p→q	~q	a (p→q)∧~q	~p	a→~p
0	0	1	1	1	1	1
0	1	1	0	0	1	1
1	0	0	1	0	0	1
1	1	1	0	0	0	1

Since $a \to \sim p$ has always value 1, $a \Rightarrow \sim p$; i.e., $(p \to q) \wedge \sim q \Rightarrow \sim p$.

92. Define theorem.

Ans: A true proposition derived from the axioms of a mathematical system is called a theorem.

93. Define proof.

Ans: A proof of a theorem is a finite sequence of logically valid steps that demonstrate that the premises of a theorem imply the conclusion.

94. Give a direct proof of $\sim p \vee q,\ s \vee p,\ \sim q \Rightarrow s$.

Ans:

Step	*Proposition*	*Justification*
1.	$\sim p \vee q$	Premise
2.	$\sim q$	Premise
3.	$\sim p$	(1), (2), Disjunctive simplification.
4.	$s \vee p$	Premise
5.	s	(3), (4), Disjunctive simplification.

95. Give an indirect proof of $a \to b,\ \sim (b \vee c) \Rightarrow \sim a$.

Ans:

Step	*Proposition*	*Justification*
1.	a	Negated conclusion
2.	$a \to b$	Premise
3.	b	(1), (2), detachment
4.	$b \vee c$	(3), disjunctive addition
5.	$\sim (b \vee c)$	Premise
6.	o	(4), (5), #

96. Find the truth set of $4x^2 - 3x = 0$ in the universe of (i) Z, (ii) Q.

Ans: $4x^2 - 3x = 0 \Rightarrow x(4x-3) = 0 \Rightarrow x = 0$ or $\frac{3}{4}$

i. In Z, Truth set = $\{0\}$; ii. In Q, Truth set = $\left\{0, \frac{3}{4}\right\}$.

97. If p and q are proposition over a universe U, express the truth sets $T_{p\to q}, T_{p\leftrightarrow q}$ in terms of T_p and T_q.

Ans: $T_{p\to q} = T_p^c \cup T_q; T_{p\leftrightarrow q} = (T_p \cap T_q) \cup (T_p^c \cap T_q^c)$

98. Suppose s is a proposition over $\{1, 2, ..., 8\}$. If $T_s = \{1, 2, 5, 7\}$, give two examples of propositions that are equivalent to s.

Ans: (i) $p(n) : (n-1)(n-2)(n-5)(n-7) = 0$

(ii) $q(n) : n^2 \in \{1,4,25,49\}$ clearly Tp = Tq = $\{1, 2, 5, 7\}$ over $\{1, 2, ..., 8\}$

99. Let the universe be Z, the set of integers. Which of the following propositions are equivalent over Z

(*a*) $0 < n^2 \le 4$ (*b*) $0 < n^3 \le 8$ (*c*) $0 < n \le 2$.

Ans: Truth set of : $0 < n^2 \le 4$ in $\{-1, 1, -2, 2\}$

Truth set of : $0 < n^3 \le 8$ in $\{1, 2\}$

Truth set of : $0 < n \le 2$ in $\{1, 2\}$.

Hence $0 < n^3 \le 8$ are $0 < n \le 2$ are equivalent.

100. Determine the truth sets of the following propositions over the positive integers:

$p(n)$: n is perfect square and $n < 100$

$q(n)$: $n = \# P(A)$ for some set A. Also determine $T_{p \wedge q}$.

Ans: $T_p = \{1, 4, 9, 16, 25, 36, 49, 64, 81\}$

$T_q = \{1, 2, 4, 8, 16, 32, 64, 128, ...\} = \{2^n \mid n \ge o\}$

$T_{p \wedge q} = T_p \cap T_q = \{1, 4, 16, 64\}$.

101. State the principle of mathematical induction.

Ans: Let p(n) be a proposition over the positive integers, then $p(n)$ is a tautology if (i) $p(1)$ in true, and (ii) $n \ge 1$, and $p(n) \Rightarrow p(n+1)$.

102. What are the two types of quantifiers.

Ans: The two quantifiers are (1) the existential quantifier, $\exists$; and (2) the universal quantifier.

103. Use quantifiers to say that $\sqrt{3}$ is not a rational number.

Ans: $(\forall x)_Q (x^2 \ne 3)$.

104. Over the universe of real numbers, use quantifiers to say that the equation $a + x = b$ has a solution for all values of a and b.

Ans: $(\forall a)_R ((\forall b)_R ((\exists x)_R (a + x = b)))$

105. Translate into your own words and indicate whether it is true or false that $(\exists u)_z (4u^2 - 9 = 0)$.

Ans: The equation $4u^2 - 9 = 0$ has an integer solution. Since $4u^2 - 9 = 0 \Leftrightarrow 4u^2 = 9 \Leftrightarrow u^2 = \frac{9}{4} \Leftrightarrow u = \pm\frac{3}{2}$, we conclude that the given statement is false.

106. Use universal quantifiers to state that the sum of any two rational numbers in rational.

Ans: $(\forall x)_Q (\forall y)_Q (x + y \in Q)$.

107. Let $M(x)$ be "x is a mammal", let $A(x)$ be "x is an animal", and let w(x) be "x is warm - blooded".

i. Transtlate into English: $(\exists x)(A(x) \wedge \sim M(x))$

ii. Translate into a formula : Every mammal is warm blooded.

Ans: (i) There exists an animal which is not a mammal and (ii) $(\forall x)(M(x) \rightarrow W(x))$.

108. True or false : If $A = \{1, 2, 3, 4, 5\}$ then $(\exists x \in A)(x + 3 < 5)$.

Ans: True, because when $x = 1$, $x + 3 = 1 + 3 = 4 < 5$.

109. True of False: If $A = \{1, 2, 3, 4, 5\}$ then $(\exists x \in A)(x + 3 = 10)$.

Ans: False, because $x + 3 = 10 \Rightarrow x = 7 \notin A$.

110. True or False : If $A = \{1, 3, 5, 7, 9\}$ then $(\forall x \in A)(x+2 \text{ is prime})$.

Ans: False, because when $x = 7, x+2 = 7+2 = 9$ which is not prime.

111. Use quantifiers to state that the system of equations $2x - y = 2$ and $4x - 2y = 5$ has no solution.

Ans: Let $q(x, y) = (2x - y = 2) \wedge (4x - 2y = 5)$.

Then $(\forall x)_R((\forall y)_R(\sim q(x, y)))$;

or, equivalently,

$\sim((\exists x)_R((\exists y)_R(q(x, y))))$.

112. Two systems of linear equations are called equivalent if ________.

Ans: they have the same set of solutions.

113. What are the elementary row operations on a matrix.

Ans: 1. Interchange any two rows.
2. Multiply any row of the matrix by a nonzero constant.
3. Multiply any row by a nonzero constant and add the result to a second row, with the sum replacing the latter row.

114. Define row equivalence of matrices.

Ans: Two matrices A and B are said to be row equivalent if one can be obtained from the other by a sequence of elementary row operations.

115. Define eigen values and eigen vectors of a $n \times n$ matrix A.

Ans: A scalar λ is called an ligen value if it satisfies $Ax = \lambda x$ for some nonzero $x \in R^n$. The nonzero vector x in R^n is called the eigen vector corresponding to the eigen value λ

116. Define diagonalizable matrix.

Ans: An $n \times n$ matrix A is called digonalizable if there exists an $n \times n$ invertible matrix P such that $P^{-1}AP$ is a diagonal matrix.

117. Determine whether the matrix $A = \begin{pmatrix} 1 & 1 \\ 0 & 1 \end{pmatrix}$ is diagonalizable or not.

Ans: Characteristic equation $\det(A - \lambda I) = 0$ is $\begin{vmatrix} 1-\lambda & 1 \\ 0 & 1-\lambda \end{vmatrix} = 0 \Rightarrow (1-\lambda)^2 = 0 \Rightarrow \lambda = 1{,}1$.

Now $Ax = \lambda x \Rightarrow \begin{bmatrix} 1 & 1 \\ 0 & 1 \end{bmatrix}\begin{bmatrix} x \\ y \end{bmatrix} = \begin{bmatrix} x \\ y \end{bmatrix}$

$\therefore \begin{bmatrix} x \\ y \end{bmatrix} = \begin{bmatrix} x \\ 0 \end{bmatrix} = x\begin{bmatrix} 1 \\ 0 \end{bmatrix}$ so there is only one independent sign vector for $\lambda = 1$. Hence the matrix is not diagonalizable.

118 State the recursive definition of a polynomial (or) Define a polynomial recursively.

Ans: Let S be a non-empty set, called the set of coefficients.

i. A zeroth degree polynomial is an element of S.

ii. For $n \geq 1$, an nth degree polynomial is an expression of the form $P(x)\, x + a$, where $P(x)$ is an $(n-1)$ the degree polynomial and $a \in S$.

119. Express $5x^4 + 12x^3 - 6x^2 + x + 6$ in telescoping form.

Ans: The telescoping form is $((((5)x+12)x-6)x+1)x+6$.

120. What is the recurrence relation for the nth Fibonacci number F_n.

Ans: $F_n = F_{n-1} + F_{n-2}, n \geq 2$, with $F_0 = 0, F_1 = 1$.

121. What is the recurrence relation for the binomical coefficients $c(n, k)$.

Ans: $(n, k) = V(n-1, k) + C(n-1, k-1)$ if $n > k > 0$,

$C(n, n) = 1$ and $C(n, o) = 1$.

122. What is wrong with the following definition of $f : R \rightarrow R$? $f(0) = 1$ and $f(x) = \frac{1}{2} f\left(\frac{x}{2}\right)$.

Ans: Since f is not a discrete function, computation of $f(x)$ requinses infinite number of steps. Inded since $f(x) = \frac{1}{2} f\left(\frac{x}{2}\right) = \frac{1}{4} f\left(\frac{x}{4}\right) = \ldots \frac{1}{2^n} f\left(\frac{x}{2^n}\right)$ value of $f(x)$ cannot be computed in n steps for any $n \geq 1$.

123. What do you mean by the closed form of an algebraic expression.

Ans: Let $E = E(x_1, x_2, \ldots, x_n)$ be an algebraic expression involving variables $x_1, \ldots, x_n$. We say that E is a closedform expression if there exists a number $B > 0$ such that the evalution of E with any allowed value of $x_1, \ldots x_n$ will take no more than B operations.

124. Find a closed form expression for the sequence $B(k) = B(k-1) + 3$ for $k \geq 1$; $B(0) = 2$.

Ans: since $B(k) - B(k-1) = 3$, the auxiliary equation is $\lambda - 1 = 0$ or $\lambda = 1$.

$\therefore B_h(k) = b$ (a constant)

A particular solution is $B_p(k) = dk$.

$B_p(k) - B_p(k-1) = 3 \Rightarrow dk - d(k-1) = 3 \Rightarrow d = 3$

$\therefore B_p(k) = 3k$.

Hence $B(k) = B_h(k) + B_p(k) = b + 3k$.

Now $B(o) = 2 \Rightarrow b = 2 \quad \therefore B(k) = 2 + 3k$.

125. What is the characteristic equation of $Q(k) + 2Q(k-1) - 3Q(k-2) - 6Q(k-4) = 0$.

Ans: $\lambda^4 + 2\lambda^3 - 3\lambda^2 - 6 = 0$.

126. The recurrence relation for the closed form expression $u_n = 2(5^n)$ is ________.

Ans: $u_n + 1 = 2(5^{n+1}) = 2(5^n . 5) = 5 . 2(5^n) = 5u_n$.

$u_0 = 2(5^0) = 2$. Hence : $u_{n+1} = 5u_n, \quad u_0 = 2$

127. The recurrence relation for the closed form expression $A(k) = k^2 - k$ is ________.

Ans :

$$A(k+1) = (k+1)^2 - (k+1)$$
$$= k^2 + k$$
$$\therefore A(k+1) - A(k) = (k^2 + k) - (k^2 - k) = 2k$$
$$\Rightarrow A(k+2) - A(k+1) = 2(k+1) = 2k + 2.$$
$$\therefore (A(k+2) - A(k+1)) - (A(k+1) - A(k)) = (2k+2) - 2k = 2$$
$$\text{or } A(k+2) - 2A(k+1) + A(k) = 2;\ A(0) = 0,\ A(1) = 0.$$

128. Solve : $S(n)-ns(n-1)=0,\ n\geq 1;\ S(0)=1$.

Ans : $S(n)=n.S(n-1)=n.(n-1)S(n-2)$

$$= n(n-1)(n-2)...3\cdot 2\cdot 1\,S(o)$$

$$= n(n-1)(n-2)...3.2.1 = n!$$

129. Explain the term 'Derangement'.

Ans: A bijection $f: A\to A$ such that $f(a)\neq a\forall a\in A$, is called a derangement.

130. Define generating function of a sequence.

Ans: The generating function of a sequence S with terms $S_0, S_1, S_2,...$ is the infinite sum

$$G(S;z)=\sum_{n=0}^{\infty} S_n z^n = S_0+S_1 z+S_2 z^2+...$$

131. Find the generating function of $S_n=3^n, n\geq 0$.

Ans: $G(s;z)=\sum_{n=0}^{\infty}3^n z^n=\sum_{n=0}^{\infty}(3z)^n=\dfrac{1}{1-3z}$

132. What is the generating function of the sequence $s(n)=\dfrac{1}{n!}$.

Ans: $G(s;z)=\sum_{n=0}^{\infty}\dfrac{1}{n!}z^n=e^z$

133. What is the generating function of the sequence $S(k) = ba^k$.

Ans : $G(s;z)=\sum_{k=0}^{\infty} ba^k z^k = b\sum_{k=0}^{\infty}(az)^k = b\dfrac{1}{1-az}=\dfrac{b}{1-az}$

134. What is the generating function of the sequence of binomial coefficients $C(n,0), C(n,1),..., C(n,n), n\geq 1$.

Ans: Generating function $=\sum_{k=0}^{n} C(n,k)z^k = (1+z)^n$

135. Translate the recurrence relation $S(n)-25(n-1)-3S(n-2)=0, n\geq 2$ into an equation about generating functions.

Ans: Replacing n by $n+2$:

$$s(n+2)-2s(n+1)-3s(n)=0,\ n\geq 0$$

$$\Rightarrow (s\uparrow 2)(n)-2(s\uparrow 1)(n)-3s(n)=0$$

$$\Rightarrow G(s\uparrow 2;z)-2G(s\uparrow 1;z)-3G(s;z)=0$$

$$\Rightarrow \frac{G(s;z)-s(0)-s(1)z}{z^2}-2.\frac{G(s;z)-s(0)}{z}-3G(s;z)=0$$

$$\Rightarrow \{G(s;z)-s(0)-s(1)z\}-2z\{G(s;z)-s(0)\}-3z^2G(s;z)=0$$

$$\Rightarrow (1-2z-3z^2)G(s;z)-s(0)-s(1)z+2s(0)z=0$$

$$\therefore G(s;z)=\frac{s(0)+(s(1)-2s(0))z}{1-2z-3z^2}.$$

136. Find the generating function of the Fibonacci sequence.

Ans: The Fibonacci sequence is

$$F(k)=F(k-1)+F(k-2);\ F(0)=0, F(1)=1$$

$$\Rightarrow F(k+2)=F(k+1)+F(k)$$

$$\Rightarrow (F \uparrow 2)(k) = (F \uparrow 1)(k) + F(k)$$
$$\Rightarrow G(F \uparrow 2; z) = G(F \uparrow 1; z) + G(F; z)$$
$$\Rightarrow \frac{G(F;z) - F(0) - F(1)z}{z^2} = \frac{G(F;z) - F(0)}{z} + G(F;z)$$
$$\Rightarrow \frac{G(F;z) - z}{z^2} = \frac{G(F;z)}{z} + G(F;z)$$
$$\text{or } G(F;z) - z = zG(F;z) + z^2G(F;z)$$
$$\text{or } G(F;z)(1 - z - z^2) = z \Rightarrow G(F;z) = \frac{z}{1 - z - z^2}$$

137. Let $D(n)$ be the number of ways that the set $\{1, 2, ..., n\}$, $n \geq 1$ can be partitioned into two nonempty subsets. Find a recurrence relation for D.

Ans: For each partition $\{A, B\}$ of $\{1, 2, ..., n-1\}$ we can construct two partition of $\{1, 2, ..., n\}$:

$$(A \cup \{n\}, B) \text{ or } (A, B \cup \{n\})$$

Further $(\{1, 2, ..., n-1\}, \{n\})$ is a partition. Hence $D(n) = 2D(n-1) + 1$.

[Alternate method: $D(n) = \frac{1}{2}(2^n - 2) = 2^{n-1} - 1$

$$\therefore D(n+1) = 2^n - 1 = 2.2^{n-1} - 1 = 2(D(n) + 1) - 1$$
$$\text{or } D(n+1) = 2D(n) + 1.]$$

138. Define a Group.

Ans: A group $(G, *)$ consists of a nonempty set G an a linary operation $* : G \times G \to G$ satisfying the properties:

i. $*$ is associative on G; i.e., $a * (b * c) = (a * b) * c \forall a, b, c \in G$.

ii. There exists an identify element $e \in G$ such that $a * e = e * a = a \forall a \in G$.

iii. For each $a \in G$ there exists an inverse $a^{-1} \in G$ such that $a * a^{-1} = a^{-1} * a = e$.

139. Define an abelian group.

Ans: A group $[G; *]$ is said to be abelian if the linary operation $*$ is commutative; i.e., $a * b = b * a \forall a, b \in G$.

140. Prove that the identify element of a group is unique.

Ans: Let $[G; *]$ be a group suppose $e_1, e_2 \in G$ are the identifies. Then

$$a * e_1 = e_1 * a = a \forall a \in G; \quad \text{and} \quad a * e_2 = e_2 * a = a \forall a \in G.$$

Putting $a = e_2$ in the first equation; and $a = e_1$ in the second.

$e_2 * e_1 = e_1 * e_2 = e_2$

$e_1 * e_2 = e_2 * e_1 = e_1$

Thus $e_1 = e_2 = e_2 * e_1$

Hence the identify element is unique.

141. Prove that the inverse of any element of a group is unique.

Ans: Let $[G; *]$ be a group. Let $a \in G$. Suppose $b, c \in G$ are the inverses of a then

$$a * b = b * a = e, \text{ and } \quad a * c = c * a = e$$

Now $b = b * e = b * (a * c) = (b * a) * c = e * c = c$. Hence $b = c$. Therefore the inverse is unique.

142. If a is an element of a group G then $(a^{-1})^{-1} = a$.

Ans: Let $b = a^{-1}$ and $c = b^{-1}$.

Then, $a = a * e = a * (b * c)$ $[\because b*c = b*b^{-1} = e]$

$= (a * b) * c = e * c$ $[\because a*b = a*a^{-1} = e]$

$= c$

$\Rightarrow a = c = b^{-1} = (a^{-1})^{-1}$

143. If $a, b \in G$ then $(a*b)^{-1} = b^{-1} * a^{-1}$.

Ans: Let $x = (a*b)^{-1}$. Then $(a*b)*x = c$.

$\Rightarrow a*(b*x) = e \Rightarrow b*x = a^{-1}$

$\Rightarrow x = b^{-1} * a^{-1}$. Hence $(a*b)^{-1} = b^{-1} * a^{-1}$.

144. Prove that {1, 2, 3, 4} is a group under the operation X_5.

Ans: The multiplication table for X_5 is

X_5	1	2	3	4
1	1	2	3	4
2	2	4	1	3
3	3	1	4	2
4	4	3	2	1

Hence X_5 is closed. Identity is 1. Inverse of 2 is 3; Inverse of 3 is 2; Inverse of 4 is 4.

145. Define a cyclic group.

Ans: A group G is cyclic if there exists $a \in G$. Such that $G = \{na \mid n \in z\}$.

Here

$$na = a*a*...*a \text{ (} n \text{ times) if } n \geq 0$$

$$\text{and } (-n)a = (-a)*(-a)*...*(-a), \text{ (} n \text{ times) if } n \geq 0.$$

146. If G is a group and $a \in G$ then prove that $(a) = (-a)$.

Ans : $(a) = \{na \mid n \in z\} = \{-na \mid n \in z\} = \{n(-a) \mid n \in z\} = (-a)$

147. Describe in simpler terms the following subgoups of Z : (1) $5Z \cap 4Z$, (2) $4Z \cap 6Z$, (3) the only finite subgroup $7Z$.

Ans: (1) $5Z \cap 4Z = 20Z$ (2) $4Z \cap 6Z = 12Z$ (3) {0}

148. List all cyclic subgroups of Z_{12}.

Ans: $Z_{12} = \{0, 1, 2, 3, 4, 5, 6, 7, 8, 9, 10, 11\}$

Cyclic subgroups are:

i. $(0) = \{0\}$

ii. $(1) = Z_{12}$

iii. $(2) = \{0, 2, 4, 6, 8, 10\}$

iv. $(3) = \{0, 3, 6, 9\}$

v. $(4) = \{0, 4, 8\}$

vi. $(6) = \{0, 6\}$

149. If H and K are subgroups of a group G, then show that $H \cap K$ is also a subgroup of G.

Ans: Let $*$ be the binary operation on G.

i. $H \cap K$ is closed under $*$:

Let $x, y \in H \cap K$. Then $x, y \in H$ and $x, y \in K$

$\Rightarrow x * y \in H$ and $x * y \in K$ ($\because$ H, K are subgroups)

$\Rightarrow x * y \in H \cap K$

ii. The identify $e \in H$ are also $e \in K$.

$\therefore e \in H \cap K$

iii. If $x \in H \cap K$ then $x \in H$ and $x \in K$

$\Rightarrow x^{-1} \in H$ and $x^{-1} \in K$ ($\because$ H, K are subgroups)

$\Rightarrow x^{-1} \in H \cap K$

Thus $H \cap K$ is a subgroup.

150. Say true or false: If *H* and *K* are subgroups of a group *G* then $H \cup K$ is also a subgroup of *G*.

Ans: False, in general.

Let $G = Z_6 = \{0, 1, 2, 3, 4, 5\}$, $H = \{0, 2, 4\}$, $K = \{0, 3\}$

Then *H* and *K* are subgroups, but $H \cup K = \{0, 2, 3, 4\}$ is not a subgroup.

($\because 2, 3 \in H \cup K$ but $2 + 3 = 5 \notin H \cup K$).

151. Show that $Z_2 \times Z_2 \times Z_2$ is not isomorphic to Z_8.

Ans: If $x = (x_1, x_2, x_3) \in Z_2 \times Z_2 \times Z_2$, then $x + x = (x_1 + x_1, x_2 + x_2, x_3 + x_3) = (0,0,0) = 0$. So $x + x = 0$ has 8 solutions in $Z_2 \times Z_2 \times Z_2$. But in Z_8 the equation $x + x = 0$ has only 2 solution, namely $x = 0$ or $x = 4$. So Z_8 is not isomorphic to $Z_2 \times Z_2 \times Z_2$.

152. Show that *Z* and $Z \times Z$ are not isomorphic.

Ans: *Z* is cyclic, because $Z = (1)$; but $Z \times Z$ is not cyclic. So *Z* is not isomorphic to $Z \times Z$.

153. Prove that every cyclic group is abelian.

Ans: Let $G = (a)$ be a cyclic group Let $x, y \in G$. The $x = na$, $y = ma$ for some $n, m \in Z$.

$$\therefore x * y = na * ma = (n+m)a = (m+n)a = ma * na = y * x.$$

Hence *G* is abelian.

154. State Lagrange's theorem.

Ans: The order of a subgroup of a finite group must divide the order of the group.

155. If $G = Z_4 \times Z_2$ and $H = ((2,0))$ then find *G/H*.

Ans: Since $G = \{(0,0), (1,0), (2,0), (3,0), (1,1), (2,1), (3,1), (0,1)\}$, $H = \{(0,0), (2,0)\}$,

$\# G/H = \frac{8}{2} = 4$. So *G/H* contains 4 elements:

$$H = \{(0,0), (2,0)\}$$
$$(1,0) + H = \{(1,0), (3,0)\}$$
$$(1,1) + H = \{(1,1), (3,1)\}$$
$$(0,1) + H = \{(0,1), (2,1)\}.$$

i.e., $G/H = \{H, (1,0) + H, (0,1) + H, (1,1) + H\}$.

156. Define a cozet.

Ans: Let G be a group are let H be a subgroup G. The set $a * H = \{a * x \mid x \in H\}$ is called a left cozet of H for each $a \in G$.

157. Define direct product of two groups G and H.

Ans: Let $[G;*_1]$ and $[H;*_2]$ be two groups. Let $G \times H$ denote the carterian product of G and H. oin $G \times H$ we define a binary operation $*$ as follows:

$$(x_1, y_1) * (x_2, y_2) = (x_1 *_1 x_2,\ y_1 *_2 y_2) \text{ for } x_1, x_2 \in G \text{ and } y_1, y_2 \in H.$$

Then $G \times H$ becomes a group called the direct product of G and H.

158. Define isomorphic groups.

Ans: $[G_1, *_1]$ and $[G_2, *_2]$ be two groups. Suppose $\exists\ f: G_1 \to G_2$, a bijection which satisfies $f(x *_1 y) = f(x) *_2 f(y)\ \forall x \in G_1, y \in G_2$. Then the groups G_1 and G_2 are said to be isomorphic.

159. Define Factor group.

Ans: Suppose G is a group and H is a subgroup of G. The collection of all left cosets of H in G will be denoted by G/H. i.e., $G/H = \{a * H \mid a \in G\}$.

If G/H form a group under the binary operation $(a * H)(b * H) = (a * b) * H$, then G/H is called the factor group.

160. Define the symmetric group S_n.

Ans: Let $A = \{1, 2, ..., n\}$. The set of all bijections on A form a group with the binary operation "composition of functions". This group is denoted by S_n, and is called the symmetric group or permutation group.

161. What is the order of S_n.

Ans: $n!$

162. Prove that S_n is nonabelian for $n \geq 3$.

Ans: Let $f = (1, 2, 3)$ and $g = (2, 3)$

Then $fg = (1, 2, 3)(2, 3) = (1, 2)$ and $gf = (2, 5)(1, 2, 3) = (1, 3)$

$\therefore fg \neq gf$. Since $f, g \in S_n\ (\forall n \geq 3)$, S_n is nonabelian.

163. What is the smallest non abelian group and what is its order.

Ans: S_3. Its order is $3! = 6$.

164. Write $\begin{pmatrix} 1 & 2 & 3 & 4 & 5 & 6 & 7 & 8 \\ 8 & 2 & 7 & 6 & 5 & 4 & 1 & 3 \end{pmatrix}$ as a product of disjoint cycles.

Ans: $(1, 8, 3, 7)(4, 6)$.

165. Define even and odd permutations.

Ans: A permutation can always be written as a product of either even or odd number of transpositions. If it is a product of even number of transposition then the permutation is said to be even, otherwise odd.

166. Define the alternating group A_n, and what is its order.

Ans: The set of all even permutations in S_n forms a group, called the alternating group, denoted by A_n. So $A_n = \{\sigma \in S_n \mid \sigma \text{ is even}\}$.

order of $A_n = \dfrac{n!}{2}$.

167. How many left-coset does $A_n (n \geq 3)$ have?

Ans: Two (of left cosets $= \dfrac{\# S_n}{\# A_n} = \dfrac{n!}{(n!/2)} = 2$).

168. If $f = \begin{pmatrix} 1 & 2 & 3 & 4 \\ 2 & 1 & 3 & 4 \end{pmatrix}$, $g = \begin{pmatrix} 1 & 2 & 3 & 4 \\ 2 & 3 & 4 & 1 \end{pmatrix}$, and $h = \begin{pmatrix} 1 & 2 & 3 & 4 \\ 3 & 2 & 4 & 1 \end{pmatrix}$, find (i) fg (ii) h^{-1}.

Ans: i. $fg = \begin{pmatrix} 1 & 2 & 3 & 4 \\ 3 & 2 & 1 & 4 \end{pmatrix}$. ii. $h^{-1} = \begin{pmatrix} 1 & 2 & 3 & 4 \\ 4 & 2 & 1 & 3 \end{pmatrix}$.

169. Define a normal subgroup.

Ans: Let G be a group and let $H \leq G$. We say that H is normal in G, and write $H \Delta G$, if

$axa^{-1} \in H \ \forall x \in H \forall a \in G.$

i.e., $aHa^{-1} \subseteq H \ \forall a \in G.$

170. Give three equivalent condition for a subgroup H to be normal in the group G.

Ans:

i. $a * h * a^{-1} \in H \ \forall h \in H \forall a \in G.$

ii. If $h \in H$ and $a \in G, \exists h^{-1} \in H$ such that $h * a = a * h^{1}$.

iii. Every left coset of H in G is also a right coset. *i.e.,* $a * H = H * a \ \forall a \in G$

171. Define a group homomorphism.

Ans: Let $(G_1, *_1)$ and $(G_2, *_2)$ be groups. A mapping $f: G_1 \to G_2$ is called a homomorphism if $f(x *_1 y) = f(x) *_2 f(y) \ \forall x, y \in G_1$.

172. If $f: G_1 \to G_2$ is a group homomorphism, show that kerf ΔG_1.

Ans: kerf $= \{x \in G_1 \mid f(x) = e_2\}$ where e_2 is the identity in G_2.

i. If $x, y \in$ kerf, then $f(x*, y) = f(n) *_2 f(y) = e_2 *_2 e_2 = e_2 \Rightarrow x *_1 y \in$ kerf

So kerf is closed under $*_1$.

ii. If e_1 = identity of G_1, then $f(e_1) = e_2$, $e_1 \in$ kerf

iii. If $x \in$ kerf, then $f(x) = e_2$

$$\Rightarrow f(x^{-1}) = f(x^{-1}) = e_2^{-1} = e_2 \Rightarrow x^{-1} \in \text{kerf}$$

Hence kerf $\leq G_1$.

Finally, if $a \in G_1$, $x \in$ kerf, then

$$\begin{aligned} f(a *_1 x *_1 a^{-1}) &= f(a) *_2 f(x) *_2 f(a^{-1}) \\ &= f(a) *_2 e_2 *_2 f(a^{-1}) \qquad (\because x \in \text{kerf}) \\ &= f(a) *_2 f(a^{-1}) = f(a) *_2 f(a)^{-1} = e^2 \\ &\Rightarrow a *_1 x *_1 a^{-1} \in \text{kerf} \end{aligned}$$

So kerf ΔG_1.

173. How many homomorphisms are there from Z_{12} into Z_{12}.

Ans: Since $Z_{12} = \{0, 1, 2, 3, 4, 5, 6, 7, 8, 9, 10, 11\}$ has generator 1, and a homomorphism can be defined by specifying its value on any one generator, there are 12 homomorphisms.

174. Let [G, *] be a group and a be an element of G. Define $f: G \to G$ by $f(x) = a * x$. Prove that f is a bijection.

Ans: f is one – one: $f(x) = f(y) \Rightarrow a * x = a * y$

$\Rightarrow x = y$ (left - cancellation)

f is onto : Let $y \in G$. Let $x = a^{-1} * y$

Then $f(x) = (a * x) = a * (a^{-1} * y) = (a * a^{-1}) * y = e * y = y$

Hence f is a bijection.

175. Define an enconding function.

Ans: Let $e: B^m \to B^n\ (m < n)$ be a one-to-one map. The function e is called (*m, n*) - encoding function.

176. Define the Hamming distance.

Ans: If x any y belong to B^n, then the Hamming distance $H(x, y) := |x \oplus y|$,

where $|W|$ mean the weight of w, and $x \oplus y$ is defined as follows:

$x = x_1 x_2 \ldots x_n,\ y = y_1 y_2 \ldots y_n,$

$\Rightarrow x \oplus y = x_1 +_2 y_1\ x_2 +_2 y \ldots x_n +_2 y_n.$

177. Find the Hamming distance between $x = 110110$ and $y = 000101$.

Ans: $H(x, y) = |x \oplus y| = |110011| = 4.$

178. Define the minimum distance of an encoding function.

Ans: Let e: $B^m \to B^n$ be an encoding function.

The minimum distance of e is min $\{H(e(x), e(y) \mid x, y \in B^m\}$

179. Define a group code.

Ans: An encoding function $e: B^m \to B^n$ is called a group code if $e(B^m)$ is a subgroup of B^n.

180. Define a simple graph and a multi graph.

Ans: A graph that has neither self loops nor parallel edges is called a simple graph. A graph that has either self loops or parallel edges or both is called a multi graph.

181. Define a regular graph and a complete graph.

Ans: A graph in which every vertex has the some degree is called a regular graph. A simple graph G with n vertices is said to be a complete graph if the degree of every vertex is $n - 1$.

182. The sum of the degrees of all vertices of a graph is equal to _______.

Ans: Twice the number of edges.

183. The sum of the indegrees of all vertices of a digraph is _______.

Ans: equal to the sum of the oct degrees of half the number of edges.

184. Define a tree.

Ans: An undirected graph is a tree if it is connected and extains no cycles or self-loops.

185. Define spanning tree of a graph.

Ans: If E^1 is a spanning set for a graph (*GE*) and $T = (V, E')$ is a free, then T is called a spanning tree for G.

186. Using kruskal's algorithm, One cau find _______ Spanning tree in a graph G.

Ans: minimal.

187. Define a rooted tree.

Ans: A single vertex with no children is a rooted tree.

188. Define a binary tree.

Ans: A rooted tree inwhich every vertex has either a left child, a right child, a left and a right child or no children is called as a binary tree.

189. The number of internal vertices in a binary tree with *n*-vertices is ________.

Ans: $\frac{n-1}{2}$

190. The chromatic number of any complete graph with *n* vertices is ________.

Ans: $n-1$

191. If $G(V, E)$ is a connected planar graph with n vertices and *q* edges and with '*r*' regions ther $n + r - q =$ ________.

Ans: 2

192. A binary tree with 2*K* vertices of level *K* has atleast ________ vertices.

Ans: $2^{k+1}-1$

193. The maximum number of vertex at level k of a binary tree is ________.

Ans: 2^k.

194. How many different directed trees are there with three nodes?

Ans: unlabelled – 3 and labelled – 18.

195. The maximum number of edges in a simple graph with *n*-vertices is ________.

Ans: n^2

196. The maximum number of degree of a vertex in a simple graph of *n*-vertices is ________.

Ans: *n*!

197. Draw the free representation of *a* (*b* + *c*).

Ans:

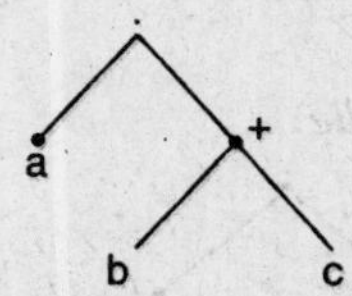

198. Draw the tree representation of $p \to (p \cup q)$.

Ans:

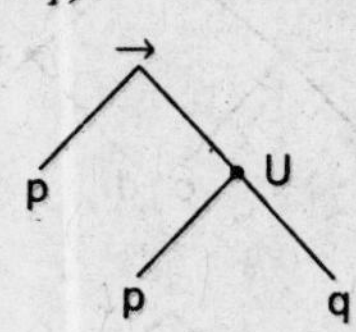

199. Define isomorphic graphs.

Ans: Let G(V,E) and $G'(V', E')$ be two graphs. They are ismorphic if there exists a bijection *f* from *V* into V' such that $(V_i, V_j) \in E$ if and only if $(f(V_i), f(V_j)) \in E'$.

200. A directed tree is ________ digraph in which one node has indegree zero and all other has indegree one.

Ans: an acyclic.

201. Define a bridge.

Ans: Let *G* (*V*, *E*) be an undirected graph and let {*L*, *R*} be a partition of *V*. A bridge between *L* and *R* is an edge in *E* that connects a vertex in *L* of a vertex in *R*.

202. **Draw the expression tree for $a^2 - b^2$ and $(a+b)*(a-b)$.**

Ans:

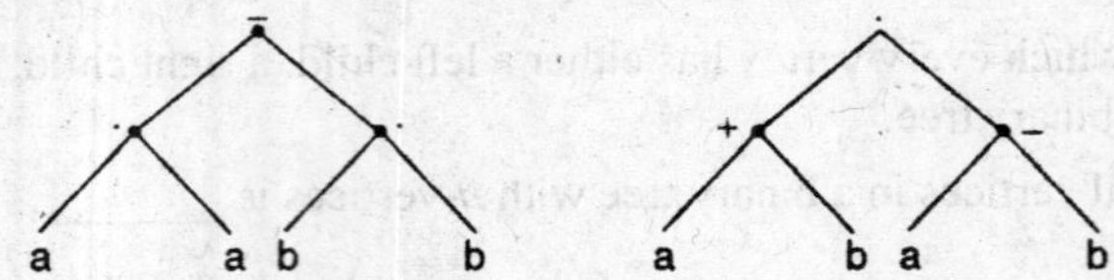

($a^2 - b^2$)

203. **In the travelling sales man problem with 6 cities, the number of complete circuits is ________.**

Ans: $(6-1)!$

204. **A connected graph with no circuit is known as ________**

Ans: Tree

205. **The number of Hamiltonian circuits to be enumerated to solve a travelling salesman problem with *n* cities is ________.**

Ans: $(n-1)!$

206. **Define a Tournament graph.**

Ans: A Tournament graph is a directed graph with the property that no edge connects a vertex to itself, and between any two vertices there is at most one edge.

207. **Can a simple undirected graph of eight vertices have forty edges excluding self-loops?**

Ans: No, the maximum number of edges would be $\frac{7\times 8}{2} = 28$.

208. **How many edges does a complete tournament graph with *n* vertices have?**

Ans: $\frac{n(n-1)}{2}$

209. **Define a Eulerian path.**

Ans: A Eulerian path through a graph is a path whose edge list contains each edge of the graph exactly once.

210. **Draw the expression tree for a $A(b-c)/(dte)$.**

Ans:

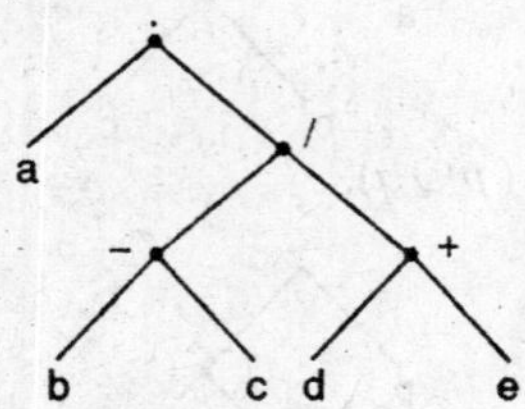

211. **Draw the expression tree for *ab* + *ac*.**

Ans:

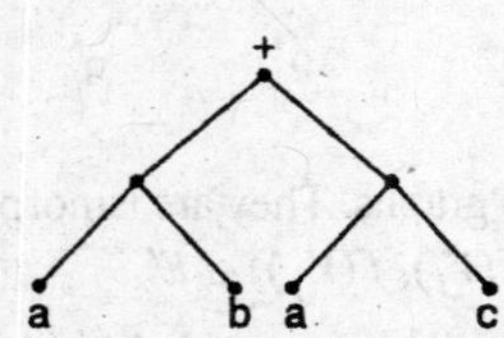

212. **Draw a binary tree with seven vertices and as many leaves as possible.**

Ans:

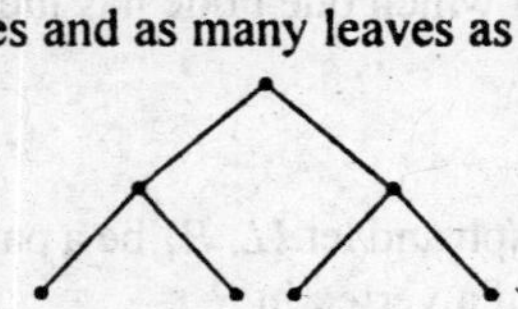

213. A vertex of degree zero is called _______ vertex.

Ans: an isolated

214. Define a complete graph.

Ans: A graph G is said to be complete if every vertex in G is connected to every other vertex in G.

215. Find the number of trees with seven vertices.

Ans: Ten.

216. Find the number of spanning trees in 

Ans: 15.

217. Find the number of connected graphs with four vertices.

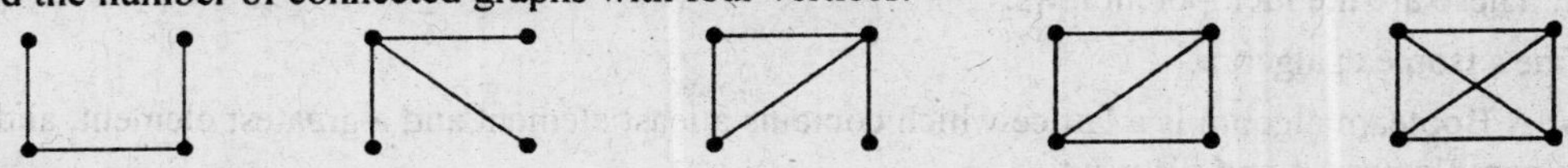

Ans: 5.

218. A vertex V with zero indegree is called a _______ and a vertex V with zero outdegree is called a _______.

Ans: source, sink.

219. Suppose a finite digraph G is cycle-free, thus G contains a _______ and a _______.

Ans: source, sink.

220. If G is an undirected graph, thus the adjacency matrix A of G will be a _______ matrix.

Ans: symmetric

221. Find the indegree and out degree of each vertex of

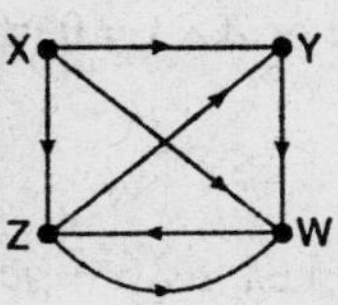

Ans:

Vertex:	X	Y	Z	W
Indegree:	0	2	2	3
Outdegree:	3	1	2	1

222. Define a poset

Ans: Let L be a nonempty set. Let $\leq$ be a relation defined on L that satisfies:

i. $a \leq a \ \forall a \in L$ (reflexive),

ii. $a \leq b$ and $b \leq a \Rightarrow a = b$ for $a, b \in L$ (antisymmetric),

iii. $a \leq b$ and $b \leq c \Rightarrow a \leq c$ for $a, b, c \in L$ (transitive).

The set L with such a relation $\leq$ is called a partially ordered set or a poset. The relation $\leq$ is called a partial ordering on L.

223. Define a Lattice.

Ans: Let $[L, \leq]$ be a poset. We say that L is a Lattice if for every pair of elements $a, b \in L,$ the glb $(a \wedge b)$ and the lub $(a \vee b)$ are in L. i.e., $\forall a, b \in L,$ $a \wedge b$ and $a \vee b$ exist and belowy to L.

224. Define a distributive lattice.

Ans: Let $[L; \vee, \wedge]$ be a lattice (under $\leq$). We say that the lattice is distributive if for all $a, b, c \in L,$

$$a \vee (b \wedge c) = (a \vee b) \wedge (a \vee c) \quad \text{and} \quad a \wedge (b \vee c) = (a \wedge b) \vee (a \wedge c)$$

225. Give examples of lattices where the distributive laws do not hold.

Ans: Two non distributive lattices are

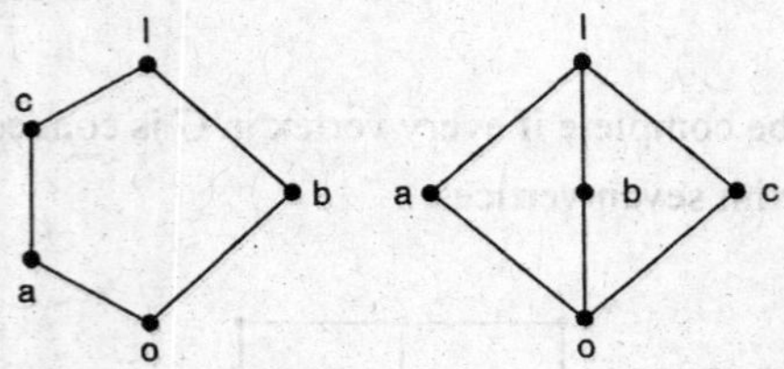

In both examples $a \vee (b \wedge c) \neq (a \vee b) \wedge (a \vee c)$

226. In a lattice $a \vee a = a$, $a \wedge a = a$ is called ________ property.

Ans: These are the idempotent laws.

227. Define a Boolean algebra.

Ans: A Booleam algebra is a lattice which contains a least element and a greatest element, and which is both complemented and distributive.

228. State the principle of duality for Boolean Algebras.

Ans: Let $[B;-,\vee,\wedge]$ be a Boolean algebra (under $\leq$) and let S be a trul statement in B. If $S*$ is obtained from S by replacing $\leq$ by $\geq$, $\vee$ by $\wedge$, $\wedge$ by $\vee$, 0 by 1, 1 by 0, then $S*$ is also a true statement. This principle is called duality principle.

229. State the dual of $a \vee \overline{((\bar{b} \vee a \wedge b)} = 1$.

Ans: $a \wedge \overline{((\bar{b} \wedge a \vee b)} = 0$.

230. Define the atoms in a Boolean algebra.

Ans: A nonzero element a in a Boolean algebra $[B;-,\vee,\wedge]$ is called an atom if $a \wedge x = a$ or $a \wedge x = 0\ \forall x \in B$.

231. Find all atoms of the Boolean algebra $[D_{30};-,\vee,\wedge]$.

Ans: $D_{30} = \{1, 2, 3, 5, 6, 10, 15, 30\}$ with partial ordering (divides).

The atoms are 2, 3 and 5, because in this Boolean algebra the operator $\wedge$ is *gcd*, i.e., $a \wedge x = gcd$ of a and x.

232. The number of elements of any finite Boolean algebra is ________.

Ans: 2^n for some positive integer n.

233. What is the graph of the Boolean algebra of order 2^n.

Ans: The n-cube.

234. List all atoms of B_2^4.

Ans: (1, 0, 0, 0), (0, 1, 0, 0), (0, 0, 1, 0), and (0, 0, 0, 1).

235. Determine the minterm normal form (principal disjunctive normal form) of the Boolean expression $f(x_1, x_2) = x_1 \vee x_2$.

Ans: $(x_1 \wedge \bar{x}_2) \vee (x_1 \wedge x_2) \vee (x_1 \wedge x_2)$.

236. The disjunctive normal form of the Boolean function $f(A, B) = A + B$ is ________.

Ans : $A + B = A.1 + 1.B = A.(B + \overline{B}) + (A + \overline{A})B = AB + A\overline{B} + AB + \overline{A}B = AB + A\overline{B} + \overline{A}B$.

(**Note:** Problem 234 & 235 are the same, only notations are different).

237. The simplified form of the Boolean function $a + a\bar{b}$ is ________.

Ans: $a + a\bar{b} = a.1 + a.\bar{b} = a(1+5) = a.1 = a.$

238. Define a monoid.

Ans: A monoid is a set M together with a binary operation $*$ with the properties:

a. $a*(b*c) = (a*b)*c,\ a,b,c \in M$ (associative)

b. $\exists e \in M$ such that $a*e = e*a = a\ \forall a \in M$. (identify element)

239. Define a string.

Ans: Let be a nonempty set elements of which are called alphabets. A string of lengh n, $n \geq 1$, over A is a finite requence of n letters from $A : a_1 a_2 ... a_n$.

The null string, λ, is defined as the string of length zero containing no letter.

240. Define a formal language.

Ans: Let A be an alphabet and let $A*$ be the set of all strings of finite length over A. Any subset L of $A*$ is called a formal language over A.

241. What is the recognition problem.

Ans: Let L be a formal language over A. We want to design an algorithm that determines the truth of $a \in L$ in a finite number of steps for all $a \in A^*$. Any such algorithm is called a Recognition algorithm. The problem of finding such an algorithm is called the recognition problem.

242. Define a recursive language.

A language is said to be recursive if there exists a recognition algorithm for it.

243. If a and b are strings of length m and n respectively, then the concatenation of a and b is a string of length ________.

Ans: $m + n$

244. What is the generating problem.

Ans: Let L be a formal language over A. We want to design an algorithm that generates all the strings in L. The problem of finding such an algorithm is called the generating problem.

245. Define a phase structure grammar.

Ans: A phase structure grammar consists of four components (V_T, V_N, S, Φ), where:

i. V_T is a nonempty finite set of terminal characters;

ii. V_N is a finite set of nonterminal characters;

iii. $S \in V_N$ is called the starting symbol;

iv. Φ is a set of production rules, each of the form $X \to Y$ where X and Y are strings over $V = V_T \cup V_N$ and X contains atleast one nonterminal symbol.

246. Define the language "the strings with an odd number of characters over B" with a phase structure grammar.

Ans: $V_T = \cup\{0,1\}$: $V_N = \{S, E\}$;

Starting Symbol : S;

Production rules:

$S \to 00S,\ S \to 01S,\ S \to 10S,\ S \to 11S, S \to E,\ E \to 0,\ E \to 1.$

247. What are the different types of grammar.

Ans: **Let (V_T, V_N, S, Φ) be a phase structure grammar.**

i. **If every production $\alpha \rightarrow \beta$ satisier $|\alpha| \leq |\beta|$, then the grammar is said to be context-sensitive or Type - 1 grammar.**

ii. **If every production is of the form $A \rightarrow \beta$ where $A \in V_T$ is a single nonterminal character, then the grammar is context-free, or of Type -2.**

iii. **If every production is either of the form $A \rightarrow a$ or $A \rightarrow aB$ where $a \in V_T$ and $A, B \in V_N$ then the grammar is regular or of Type - 3.**

iv. **If there are no conditions on the production rules, then the grammar is of Type - 0.**

248. Say true of false: Type - 3 grammar is also called regular grammars.

Ans: **True.**

249. Grammar of Type - 1 is often called _______ grammar.

Ans: **Context - sensitive grammar.**

250. Say true of false: Type - II grammar is sometimes called context-free grammar.

Ans: **True.**

251. Define the set of "Strings in $B*$ for which all 0s precade all 1s" with a regular grammar.

Ans: $V_T = \{0, 1, \lambda\}$, $\quad V_N\{S, E\}$,

Starting symbol = S.

Production rules: $S \rightarrow 0s$, $S \rightarrow 0$, $S \rightarrow E$, $E \rightarrow 1E$, $E \rightarrow 1$, $S \rightarrow \lambda$.

252. Define the set of "Strings over B with an even number of 1s" with a regular grammar.

Ans: $V_T = \{0, 1, \lambda\}$, $\quad V_N\{S, E\}$,

Starting Symbol = S,

Production rules: $S \rightarrow \lambda$, $S \rightarrow 0$, $S \rightarrow 0S$, $S \rightarrow 1A$, $A \rightarrow 0A$, $A \rightarrow 1S$

253. Define a finite state machine.

Ans: **A finite state machine is defined by a quintet (S, X, Z, W, T) where**

i. **$S = \{s_1, s_2, ..., s_r\}$ is the state set, a finite set that corresponds to the set of memory configurations that a machine can have at any time;**

ii. **$X = \{x_1, x_2, ..., x_m\}$ is the input alphabet;**

iii. **$Z = \{z_1, z_2, ..., z_n\}$ is the output alphabet;**

iv. **$W : X \times S \rightarrow Z$ is the output function, which specificies which output symbol $W(x, s) \in Z$ is written onto the output tape when the machine is in state s and the input symbol x is read; (s) $t : X \times S \rightarrow S$ is the next-state (or transition) function, which specifies which state $t(x, s) \in S$ the machine should enter when it is in state s and it reads the symbol x.**

254. Finite-state machines can recognize strings of a language only if it is a _______ type of language.

Ans: **Regular or Type-3 Language.**

255. Draw a transition diagram for a machine that recognizes whether or not a string in B^* contains an even number of 1s.

Ans:

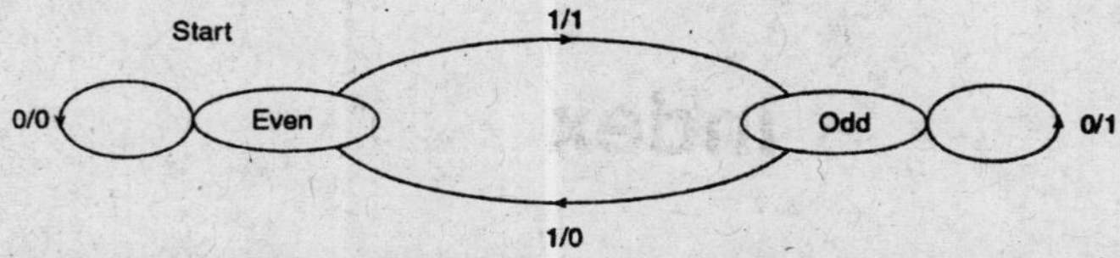

If the output is 1 then the string contains an odd number of 1s; and if the output is 0, then the string contains an even number of 1's.

Index